OSCILLATION THEORY FOR FUNCTIONAL DIFFERENTIAL EQUATIONS

PURE AND APPLIED MATHEMATICS

A Program of Monographs, Textbooks, and Lecture Notes

MONOGRAPHS AND TEXTBOOKS IN PURE AND APPLIED MATHEMATICS

1. *K. Yano,* Integral Formulas in Riemannian Geometry (1970)
2. *S. Kobayashi,* Hyperbolic Manifolds and Holomorphic Mappings (1970)
3. *V. S. Vladimirov,* Equations of Mathematical Physics (A. Jeffrey, ed.; A. Littlewood, trans.) (1970)
4. *B. N. Pshenichnyi,* Necessary Conditions for an Extremum (L. Neustadt, translation ed.; K. Makowski, trans.) (1971)
5. *L. Narici et al.,* Functional Analysis and Valuation Theory (1971)
6. *S. S. Passman,* Infinite Group Rings (1971)
7. *L. Dornhoff,* Group Representation Theory. Part A: Ordinary Representation Theory. Part B: Modular Representation Theory (1971, 1972)
8. *W. Boothby and G. L. Weiss, eds.,* Symmetric Spaces (1972)
9. *Y. Matsushima,* Differentiable Manifolds (E. T. Kobayashi, trans.) (1972)
10. *L. E. Ward, Jr.,* Topology (1972)
11. *A. Babakhanian,* Cohomological Methods in Group Theory (1972)
12. *R. Gilmer,* Multiplicative Ideal Theory (1972)
13. *J. Yeh,* Stochastic Processes and the Wiener Integral (1973)
14. *J. Barros-Neto,* Introduction to the Theory of Distributions (1973)
15. *R. Larsen,* Functional Analysis (1973)
16. *K. Yano and S. Ishihara,* Tangent and Cotangent Bundles (1973)
17. *C. Procesi,* Rings with Polynomial Identities (1973)
18. *R. Hermann,* Geometry, Physics, and Systems (1973)
19. *N. R. Wallach,* Harmonic Analysis on Homogeneous Spaces (1973)
20. *J. Dieudonné,* Introduction to the Theory of Formal Groups (1973)
21. *I. Vaisman,* Cohomology and Differential Forms (1973)
22. *B.-Y. Chen,* Geometry of Submanifolds (1973)
23. *M. Marcus,* Finite Dimensional Multilinear Algebra (in two parts) (1973, 1975)
24. *R. Larsen,* Banach Algebras (1973)
25. *R. O. Kujala and A. L. Vitter, eds.,* Value Distribution Theory: Part A; Part B: Deficit and Bezout Estimates by Wilhelm Stoll (1973)
26. *K. B. Stolarsky,* Algebraic Numbers and Diophantine Approximation (1974)
27. *A. R. Magid,* The Separable Galois Theory of Commutative Rings (1974)
28. *B. R. McDonald,* Finite Rings with Identity (1974)
29. *J. Satake,* Linear Algebra (S. Koh et al., trans.) (1975)
30. *J. S. Golan,* Localization of Noncommutative Rings (1975)
31. *G. Klambauer,* Mathematical Analysis (1975)
32. *M. K. Agoston,* Algebraic Topology (1976)
33. *K. R. Goodearl,* Ring Theory (1976)
34. *L. E. Mansfield,* Linear Algebra with Geometric Applications (1976)
35. *N. J. Pullman,* Matrix Theory and Its Applications (1976)
36. *B. R. McDonald,* Geometric Algebra Over Local Rings (1976)
37. *C. W. Groetsch,* Generalized Inverses of Linear Operators (1977)
38. *J. E. Kuczkowski and J. L. Gersting,* Abstract Algebra (1977)
39. *C. O. Christenson and W. L. Voxman,* Aspects of Topology (1977)
40. *M. Nagata,* Field Theory (1977)
41. *R. L. Long,* Algebraic Number Theory (1977)
42. *W. F. Pfeffer,* Integrals and Measures (1977)
43. *R. L. Wheeden and A. Zygmund,* Measure and Integral (1977)
44. *J. H. Curtiss,* Introduction to Functions of a Complex Variable (1978)
45. *K. Hrbacek and T. Jech,* Introduction to Set Theory (1978)
46. *W. S. Massey,* Homology and Cohomology Theory (1978)
47. *M. Marcus,* Introduction to Modern Algebra (1978)
48. *E. C. Young,* Vector and Tensor Analysis (1978)
49. *S. B. Nadler, Jr.,* Hyperspaces of Sets (1978)
50. *S. K. Segal,* Topics in Group Kings (1978)
51. *A. C. M. van Rooij,* Non-Archimedean Functional Analysis (1978)
52. *L. Corwin and R. Szczarba,* Calculus in Vector Spaces (1979)

53. *C. Sadosky,* Interpolation of Operators and Singular Integrals (1979)
54. *J. Cronin,* Differential Equations (1980)
55. *C. W. Groetsch,* Elements of Applicable Functional Analysis (1980)
56. *I. Vaisman,* Foundations of Three-Dimensional Euclidean Geometry (1980)
57. *H. I. Freedan,* Deterministic Mathematical Models in Population Ecology (1980)
58. *S. B. Chae,* Lebesgue Integration (1980)
59. *C. S. Rees et al.,* Theory and Applications of Fourier Analysis (1981)
60. *L. Nachbin,* Introduction to Functional Analysis (R. M. Aron, trans.) (1981)
61. *G. Orzech and M. Orzech,* Plane Algebraic Curves (1981)
62. *R. Johnsonbaugh and W. E. Pfaffenberger,* Foundations of Mathematical Analysis (1981)
63. *W. L. Voxman and R. H. Goetschel,* Advanced Calculus (1981)
64. *L. J. Corwin and R. H. Szczarba,* Multivariable Calculus (1982)
65. *V. I. Istrăţescu,* Introduction to Linear Operator Theory (1981)
66. *R. D. Järvinen,* Finite and Infinite Dimensional Linear Spaces (1981)
67. *J. K. Beem and P. E. Ehrlich,* Global Lorentzian Geometry (1981)
68. *D. L. Armacost,* The Structure of Locally Compact Abelian Groups (1981)
69. *J. W. Brewer and M. K. Smith, eds.,* Emily Noether: A Tribute (1981)
70. *K. H. Kim,* Boolean Matrix Theory and Applications (1982)
71. *T. W. Wieting,* The Mathematical Theory of Chromatic Plane Ornaments (1982)
72. *D. B.Gauld,* Differential Topology (1982)
73. *R. L. Faber,* Foundations of Euclidean and Non-Euclidean Geometry (1983)
74. *M. Carmeli,* Statistical Theory and Random Matrices (1983)
75. *J. H. Carruth et al.,* The Theory of Topological Semigroups (1983)
76. *R. L. Faber,* Differential Geometry and Relativity Theory (1983)
77. *S. Barnett,* Polynomials and Linear Control Systems (1983)
78. *G. Karpilovsky,* Commutative Group Algebras (1983)
79. *F. Van Oystaeyen and A. Verschoren,* Relative Invariants of Rings (1983)
80. *I. Vaisman,* A First Course in Differential Geometry (1984)
81. *G. W. Swan,* Applications of Optimal Control Theory in Biomedicine (1984)
82. *T. Petrie and J. D. Randall,* Transformation Groups on Manifolds (1984)
83. *K. Goebel and S. Reich,* Uniform Convexity, Hyperbolic Geometry, and Nonexpansive Mappings (1984)
84. *T. Albu and C. Năstăsescu,* Relative Finiteness in Module Theory (1984)
85. *K. Hrbacek and T. Jech,* Introduction to Set Theory: Second Edition (1984)
86. *F. Van Oystaeyen and A. Verschoren,* Relative Invariants of Rings (1984)
87. *B. R. McDonald,* Linear Algebra Over Commutative Rings (1984)
88. *M. Namba,* Geometry of Projective Algebraic Curves (1984)
89. *G. F. Webb,* Theory of Nonlinear Age-Dependent Population Dynamics (1985)
90. *M. R. Bremner et al.,* Tables of Dominant Weight Multiplicities for Representations of Simple Lie Algebras (1985)
91. *A. E. Fekete,* Real Linear Algebra (1985)
92. *S. B. Chae,* Holomorphy and Calculus in Normed Spaces (1985)
93. *A. J. Jerri,* Introduction to Integral Equations with Applications (1985)
94. *G. Karpilovsky,* Projective Representations of Finite Groups (1985)
95. *L. Narici and E. Beckenstein,* Topological Vector Spaces (1985)
96. *J. Weeks,* The Shape of Space (1985)
97. *P. R. Gribik and K. O. Kortanek,* Extremal Methods of Operations Research (1985)
98. *J.-A. Chao and W. A. Woyczynski, eds.,* Probability Theory and Harmonic Analysis (1986)
99. *G. D. Crown et al.,* Abstract Algebra (1986)
100. *J. H. Carruth et al.,* The Theory of Topological Semigroups, Volume 2 (1986)
101. *R. S. Doran and V. A. Belfi,* Characterizations of C*-Algebras (1986)
102. *M. W. Jeter,* Mathematical Programming (1986)
103. *M. Altman,* A Unified Theory of Nonlinear Operator and Evolution Equations with Applications (1986)
104. *A. Verschoren,* Relative Invariants of Sheaves (1987)
105. *R. A. Usmani,* Applied Linear Algebra (1987)
106. *P. Blass and J. Lang,* Zariski Surfaces and Differential Equations in Characteristic $p > 0$ (1987)
107. *J. A. Reneke et al.,* Structured Hereditary Systems (1987)

108. *H. Busemann and B. B. Phadke,* Spaces with Distinguished Geodesics (1987)
109. *R. Harte,* Invertibility and Singularity for Bounded Linear Operators (1988)
110. *G. S. Ladde et al.,* Oscillation Theory of Differential Equations with Deviating Arguments (1987)
111. *L. Dudkin et al.,* Iterative Aggregation Theory (1987)
112. *T. Okubo,* Differential Geometry (1987)
113. *D. L. Stancl and M. L. Stancl,* Real Analysis with Point-Set Topology (1987)
114. *T. C. Gard,* Introduction to Stochastic Differential Equations (1988)
115. *S. S. Abhyankar,* Enumerative Combinatorics of Young Tableaux (1988)
116. *H. Strade and R. Farnsteiner,* Modular Lie Algebras and Their Representations (1988)
117. *J. A. Huckaba,* Commutative Rings with Zero Divisors (1988)
118. *W. D. Wallis,* Combinatorial Designs (1988)
119. *W. Więsław,* Topological Fields (1988)
120. *G. Karpilovsky,* Field Theory (1988)
121. *S. Caenepeel and F. Van Oystaeyen,* Brauer Groups and the Cohomology of Graded Rings (1989)
122. *W. Kozlowski,* Modular Function Spaces (1988)
123. *E. Lowen-Colebunders,* Function Classes of Cauchy Continuous Maps (1989)
124. *M. Pavel,* Fundamentals of Pattern Recognition (1989)
125. *V. Lakshmikantham et al.,* Stability Analysis of Nonlinear Systems (1989)
126. *R. Sivaramakrishnan,* The Classical Theory of Arithmetic Functions (1989)
127. *N. A. Watson,* Parabolic Equations on an Infinite Strip (1989)
128. *K. J. Hastings,* Introduction to the Mathematics of Operations Research (1989)
129. *B. Fine,* Algebraic Theory of the Bianchi Groups (1989)
130. *D. N. Dikranjan et al.,* Topological Groups (1989)
131. *J. C. Morgan II,* Point Set Theory (1990)
132. *P. Biler and A. Witkowski,* Problems in Mathematical Analysis (1990)
133. *H. J. Sussmann,* Nonlinear Controllability and Optimal Control (1990)
134. *J.-P. Florens et al.,* Elements of Bayesian Statistics (1990)
135. *N. Shell,* Topological Fields and Near Valuations (1990)
136. *B. F. Doolin and C. F. Martin,* Introduction to Differential Geometry for Engineers (1990)
137. *S. S. Holland, Jr.,* Applied Analysis by the Hilbert Space Method (1990)
138. *J. Okniński,* Semigroup Algebras (1990)
139. *K. Zhu,* Operator Theory in Function Spaces (1990)
140. *G. B. Price,* An Introduction to Multicomplex Spaces and Functions (1991)
141. *R. B. Darst,* Introduction to Linear Programming (1991)
142. *P. L. Sachdev,* Nonlinear Ordinary Differential Equations and Their Applications (1991)
143. *T. Husain,* Orthogonal Schauder Bases (1991)
144. *J. Foran,* Fundamentals of Real Analysis (1991)
145. *W. C. Brown,* Matrices and Vector Spaces (1991)
146. *M. M. Rao and Z. D. Ren,* Theory of Orlicz Spaces (1991)
147. *J. S. Golan and T. Head,* Modules and the Structures of Rings (1991)
148. *C. Small,* Arithmetic of Finite Fields (1991)
149. *K. Yang,* Complex Algebraic Geometry (1991)
150. *D. G. Hoffman et al.,* Coding Theory (1991)
151. *M. O. González,* Classical Complex Analysis (1992)
152. *M. O. González,* Complex Analysis (1992)
153. *L. W. Baggett,* Functional Analysis (1992)
154. *M. Sniedovich,* Dynamic Programming (1992)
155. *R. P. Agarwal,* Difference Equations and Inequalities (1992)
156. *C. Brezinski,* Biorthogonality and Its Applications to Numerical Analysis (1992)
157. *C. Swartz,* An Introduction to Functional Analysis (1992)
158. *S. B. Nadler, Jr.,* Continuum Theory (1992)
159. *M. A. Al-Gwaiz,* Theory of Distributions (1992)
160. *E. Perry,* Geometry: Axiomatic Developments with Problem Solving (1992)
161. *E. Castillo and M. R. Ruiz-Cobo,* Functional Equations and Modelling in Science and Engineering (1992)
162. *A. J. Jerri,* Integral and Discrete Transforms with Applications and Error Analysis (1992)
163. *A. Charlier et al.,* Tensors and the Clifford Algebra (1992)

164. *P. Biler and T. Nadzieja*, Problems and Examples in Differential Equations (1992)
165. *E. Hansen*, Global Optimization Using Interval Analysis (1992)
166. *S. Guerre-Delabrière*, Classical Sequences in Banach Spaces (1992)
167. *Y. C. Wong*, Introductory Theory of Topological Vector Spaces (1992)
168. *S. H. Kulkarni and B. V. Limaye*, Real Function Algebras (1992)
169. *W. C. Brown*, Matrices Over Commutative Rings (1993)
170. *J. Loustau and M. Dillon*, Linear Geometry with Computer Graphics (1993)
171. *W. V. Petryshyn*, Approximation-Solvability of Nonlinear Functional and Differential Equations (1993)
172. *E. C. Young*, Vector and Tensor Analysis: Second Edition (1993)
173. *T. A. Bick*, Elementary Boundary Value Problems (1993)
174. *M. Pavel*, Fundamentals of Pattern Recognition: Second Edition (1993)
175. *S. A. Albeverio et al.*, Noncommutative Distributions (1993)
176. *W. Fulks*, Complex Variables (1993)
177. *M. M. Rao*, Conditional Measures and Applications (1993)
178. *A. Janicki and A. Weron*, Simulation and Chaotic Behavior of α-Stable Stochastic Processes (1994)
179. *P. Neittaanmäki and D. Tiba*, Optimal Control of Nonlinear Parabolic Systems (1994)
180. *J. Cronin*, Differential Equations: Introduction and Qualitative Theory, Second Edition (1994)
181. *S. Heikkilä and V. Lakshmikantham*, Monotone Iterative Techniques for Discontinuous Nonlinear Differential Equations (1994)
182. *X. Mao*, Exponential Stability of Stochastic Differential Equations (1994)
183. *B. S. Thomson*, Symmetric Properties of Real Functions (1994)
184. *J. E. Rubio*, Optimization and Nonstandard Analysis (1994)
185. *J. L. Bueso, P. Jara, and A. Verschoren*, Compatibility, Stability, and Sheaves (1995)
186. *A. N. Michel and K. Wang*, Qualitative Theory of Dynamical Systems (1995)
187. *M. R. Darnel*, Theory of Lattice-Ordered Groups (1995)
188. *Z. Naniewicz and P. D. Panagiotopoulos*, Mathematical Theory of Hemivariational Inequalities and Applications (1995)
189. *L. J. Corwin and R. H. Szczarba*, Calculus in Vector Spaces: Second Edition (1995)
190. *L. H. Erbe, Q. Kong, B. G. Zhang*, Oscillation Theory for Functional Differential Equations (1995)

Additional Volumes in Preparation

OSCILLATION THEORY FOR FUNCTIONAL DIFFERENTIAL EQUATIONS

L. H. Erbe
University of Alberta
Edmonton, Alberta, Canada

Qingkai Kong
Northeastern Illinois University
DeKalb, Illinois

B. G. Zhang
Ocean University of Qingdao
Qingdao, Shandong, People's Republic of China

Marcel Dekker, Inc. New York • Basel • Hong Kong

Library of Congress Cataloging-in-Publication Data

Erbe, L. H.
Oscillation theory for functional differential equations / L. H. Erbe, Qingkai Kong, B. G. Zhang.
p. cm. — (Monographs and textbooks in pure and applied mathematics ; 190)
Includes bibliographical references and index.
ISBN 0-8247-9598-9 (acid-free)
1. Functional differential equations—Numerical solutions. 2. Oscillations. I. Kong, Qingkai. II. Zhang, B. G. III. Title. IV. Series.
QA372.E63 1994
515'.352—dc20 94-38757
CIP

The publisher offers discounts on this book when ordered in bulk quantities. For more information, write to Special Sales/Professional Marketing at the address below.

This book is printed on acid-free paper.

MARCEL DEKKER, INC.
270 Madison Avenue, New York, New York 10016

Current printing (last digit):
10 9 8 7 6 5 4 3 2 1

PRINTED IN THE UNITED STATES OF AMERICA

Preface

Various models of functional differential equations (FDEs) have been posed in different sciences, strongly motivating research in the qualitative theory of FDEs. As a part of this approach, oscillation theory of FDEs has developed very rapidly during the last decade. It has concerned itself largely with the oscillatory and nonoscillatory properties of solutions. Specifically, the topics dealt with include the following:

1. Finding conditions for oscillation of all solutions;
2. Finding conditions for the existence of one nonoscillatory solution;
3. Finding conditions for the existence of one oscillatory solution;
4. Finding conditions for the nonexistence of oscillatory solutions;
5. Estimating the distance between zeros of oscillatory solutions;
6. Finding an asymptotic classification of nonoscillatory solutions and finding conditions for existence of solutions with designated asymptotic properties.

The first monograph using this approach was that of G. S. Ladde, V. Lakshmikantham, and B. G. Zhang, and it has received much attention. This book covers the work up to 1984 and concerns mainly the effect of the deviating argument on the oscillation of solutions. However, it does not touch upon FDEs of neutral type.

In the past decade research on the oscillation theory of FDEs of neutral type has been very active and fruitful, and has attracted the attention of many mathematicians worldwide. The recent interesting book by I. Györi and G. Ladas summarizes some important work in this area, especially the contributions of the group centered at the University of Rhode Island. For instance, the relation between the distribution of the roots of characteristic equations and the oscillation of all solutions, and linearized oscillation theory have been developed largely by this group. However, their book does not deal with some other important problems, such as topics 3–6 mentioned above, although much progress has been made in these areas. Also, their book does not consider equations with nonlinear neutral terms of the form

$$\big(x(t) + cx^{\alpha}(t-\tau)\big)' + p(t)x^{\beta}(t-\tau) = 0, \quad 0 < \alpha \neq 1, \quad \beta \neq 0$$

in which the nonlinear neutral term has a definite effect on the oscillation of solutions.

The purpose of this book is to reflect the overall new developments in this area, including some contributions of the authors and their colleagues. Some of the topics in this book are not emphasized in the book of Györi and Ladas, such as the existence of oscillatory solutions, estimates of the distance between zeros, asymptotic classification of nonoscillatory solutions and the criteria for certain types of nonoscillatory solutions, oscillation of equations with nonlinear neutral terms, and oscillation of systems of equations.

The results on oscillation criteria in this book are also different from those in the above books and some new techniques are introduced that also yield substantial improvements of previous work in this area. To avoid unnecessary repetition, in our presentation some fundamental results will be stated without proof.

The last part of this book concerns boundary value problems (BVPs) of FDEs. As is well known, oscillation theory of ordinary differential equations (ODEs) is closely related to BVPs of ODEs. But for FDEs this relation has not been explored in detail. One of our purposes in discussing these topics is to draw the attention of readers to this problem. The BVPs of FDEs have recently been developed further; however, we did not attempt to include all related work. Instead, we present results that reflect, to some extent, our personal interests. We discuss BVPs of equations of second order, both nonsingular and singular, illustrating how to extend the techniques for BVPs of ODEs to BVPs of FDEs. Finally, we discuss the BVPs of equations of nth order. We hope this book will stimulate further research.

We wish to express our thanks to the staff of Marcel Dekker, Inc. for their cooperation during the preparation of this book for publication. We also acknowledge with gratitude the support of NSERC-Canada. Thanks are due to Vivian Spak for her assistance in typing portions of the book and a very special thank you to Elizabeth Leonard for her help in typing, proof-reading, and coordinating the final version.

L. H. Erbe
Qingkai Kong
B. G. Zhang

Contents

1

Preliminaries

1.0. Introduction

In this chapter we introduce some basic concepts from the theory of functional differential equations and discuss briefly the problems in oscillation theory and boundary value problems of functional differential equations which will be studied in detail in this book.

Section 1.1 is concerned with the statement of the basic initial value problems and classification of equations with deviating arguments.

In Section 1.2 we will show the main problems in the oscillation theory of differential equations with deviating arguments.

In Section 1.3 we will state the formulation of the boundary value problem of functional differential equations.

Finally, we will introduce some fixed point theorems which are important tools when we study the existence of solutions for initial value problems and boundary value problems.

1.1. Initial Value Problems

We consider the scalar differential equation of nth order with ℓ deviating arguments of the form

$$f(t, x(t), \ldots, x^{(m_0-1)}(t), x^{(m_0)}(t), x(t-\tau_1(t)), \ldots, x^{(m_1)}(t-\tau_1(t)),$$
$$\ldots, x(t-\tau_\ell(t)), \ldots, x^{(m_\ell)}(t-\tau_\ell(t))) = 0 \qquad (1.1.1)$$

where the deviations $\tau_i(t) > 0$, and $\max\limits_{0 \leq i \leq \ell} m_i = n$.

Let t_0 be the given initial point. Each *deviation* $\tau_i(t)$ defines the inital set $E_{t_0}^{(i)}$ given by

$$E_{t_0}^{(i)} = \{t_0\} \cup \{t - \tau_i(t) : \ t - \tau_i(t) < t, \ t \geq t_0\}. \qquad (1.1.2)$$

We denote $E_{t_0} = \bigcup\limits_{i=1}^{\ell} E_{t_0}^{(i)}$, and on t_0 we must be given continuous functions $\varphi_k(t)$, $k = 0, 1, \ldots, \mu$, with $\mu = \max\limits_{1 \leq i \leq \ell} m_i$. In applications, it is most natural to consider the case where on E_{t_0}

$$\varphi_k(t) = \varphi_0^{(k)}(t), \quad 0, 1, 2, \ldots, \mu \qquad (1.1.3)$$

but it is not generally necessary.

The nth order differential equation should be given initial values $x_0^{(k)}$, $k = 0, 1, 2, \ldots, n-1$. Now let $x_0^{(k)} = \varphi_k(t_0)$, $\quad k = 0, 1, \ldots, \mu$. If $\mu < n-1$, then, in addition, the numbers $x_0^{(\mu+1)}, \ldots, x_0^{(n-1)}$ are given. If the point t_0 is an isolated point of E_{t_0}, then $x_0^{(0)}, \ldots, x_0^{(n-1)}$ are also given.

For equation (1.1.1), the basic initial value problem consists of the determination of an $(n-1)$ times continuously differentiable function x that satisfies Eq. (1.1.1) for $t > t_0$ and the conditions

$$\begin{aligned} x^{(k)}(t_0+0) &= x_0^{(k)}, & k &= 0, 1, 2, \ldots, n-1 \\ x^{(k)}(t-\tau_i(t)) &= \varphi_k(t-\tau_i(t)), & \text{if} \quad & t - \tau_i(t) < t_0 \\ k &= 0, 1, \ldots, \mu, & i &= 1, 2, \ldots, \ell. \end{aligned} \qquad (1.1.4)$$

At a point $t_0 + (k-1)\tau$ the derivative $x^{(k)}(t)$, generally speaking, is discontinuous, but the derivatives of lower order are continuous.

Let $\lambda = m_0 - \mu$. If $\lambda > 0$, (1.1.1) is called a delay equation. If $\lambda = 0$, it is called a neutral equation. If $\lambda < 0$, it is called an advanced equation.

Example 1.1.1. The equations

$$x'(t) + ax(t - \tau) = 0 \tag{1.1.5}$$

$$x'(t) + ax(t + \tau) = 0 \tag{1.1.6}$$

and

$$x'(t) + ax(t) + bx'(t - \tau) = 0 \tag{1.1.7}$$

are delay equations, an advanced equation and a neutral equation respectively, where $\tau > 0$, a and b are constants.

The above classification is incomplete. For example, the equation

$$x'(t) + ax(t - \tau) + bx(t + \sigma) = 0 \tag{1.1.8}$$

does not belong to the above three classes where $\tau > 0$, $\sigma > 0$ are constants. Sometimes (1.1.8) is called an equation with mixed type.

In oscillation theory we study the solutions defined on an infinite interval $[t_0, \infty)$. Therefore we are most interested in those equations for which the global existence and uniqueness theorems to the initial value problem can be established. In particular, we will study the neutral delay differential system

$$\begin{aligned} &\frac{d}{dt}\left(x(t) + g(t, x(t - \tau_i(t))), \ldots, x(t - \tau_\ell(t))\right) \\ &\qquad + f(t, x(t - \sigma_1(t))), \ldots, x(t - \sigma_n(t)) = 0 \end{aligned} \tag{1.1.9}$$

where for some $\bar{t}_0 \in R$ and some positive integer m,

$$g,\ f \in C([\bar{t}_0, \infty) \times R^m \times \cdots \times R^m, R^m) \tag{1.1.10}$$

$$\tau_i,\ \sigma_j \in C\big([\bar{t}_0, \infty), (0, \infty)\big) \tag{1.1.11}$$

$$\lim_{t \to \infty} \big(t - \tau_i(t)\big) = \infty, \quad \lim_{t \to \infty} \big(t - \sigma_j(t)\big) = \infty, \quad \text{for} \quad i \in I_\ell,\ j \in I_n$$

where $I_\ell = [1, 2, \ldots, \ell]$.

For a given initial point $t_0 \geq \bar{t}_0$ set

$$t_{-1} = \min \inf_{t \geq t_0} \big(t - \tau_i(t)\big).$$

We associate (1.1.9) with the initial condition

$$x(t) = \varphi(t) \quad \text{for} \quad t_{-1} \le t \le t_0 \tag{1.1.13}$$

where $\varphi : [t_{-1}, t_0] \to R^m$ is a given initial function.

A function x is said to be a solution of Eq. (1.1.9) on the interval $[t_0, \infty)$, if $x : [t_{-1}, \infty) \to R^m$ is continuous, $x(t) + g\big(t, t - \tau_1(t)), \ldots, x(t - \tau_\ell(t)\big)$ is continuously differentiable for $t \in [t_0, \infty)$ and x satisfies (1.1.9) for all $t \in (t_0, \infty)$.

A function x is said to be a solution of the initial value problem (1.1.9) and (1.1.13) on $[t_0, \infty)$, if x is a solution of Eq. (1.1.9) on $[t_0, \infty)$ and x satisfies (1.1.13).

Using the method of steps we can prove the following global existence and uniqueness theorem for the neutral delay differential system (1.1.9).

Theorem 1.1.1. *Assume that (1.1.10) and (1.1.11) hold. Let $t_0 \ge \bar{t}_0$ and $\varphi \in C(t_{-1}, t_0], R^m)$ be given. Then the initial value problem (1.1.9) and (1.1.13) has exactly one solution in the interval $[t_0, \infty)$.*

In oscillation theory we only discuss those solutions y of (1.1.9) which exist on some ray $[T_y, \infty)$ and satisfy $\sup\{|y(t)| : t \ge T\} > 0$ for every $T \ge T_y$.

1.2. Oscillation and Nonoscillation

We first give the definition of the oscillation of solutions for the scalar differential equation, for example

$$\big(x(t) + p(t)x(t - \tau)\big)' + Q(t)x(t - \sigma) = 0, \quad t \ge t_0. \tag{1.2.1}$$

A nontrivial solution x of Eq. (1.2.1) is said to be oscillatory if x has arbitrarily large zeros. Otherwise, x is said to be nonoscillatory. That is, x is nonoscillatory if there exists a $t_1 > t_0$ such that $x(t) \ne 0$ for $t \ge t_1$. In other words, a nonoscillatory solution must be eventually positive or eventually negative. When we say every solution of Eq. (1.2.1) is oscillatory, we mean that for every initial point $t_1 \ge t_0$ and every initial function $\varphi \in C$ on the inital set E_{t_1} the unique solution x of the initial value problem of Eq. (1.2.1) oscillates. Therefore when we prove that every solution of Eq. (1.2.1) is oscillatory we do not care about the initial condition. For the sake of convenience, we say Eq. (1.2.1) is oscillatory if every solution of Eq. (1.2.1) is oscillatory. When we want to prove that Eq. (1.2.1) has a nonoscillatory solution, it is sufficient to prove that there exists a point $t_2 \geq t_1 \geq t_0$ such that x is a solution on $[t_2, \infty)$ and $x(t) > 0$ for all $t \geq t_2$ or $x(t) < 0$ for all $t \geq t_2$.

For equations with complex deviating arguments we do not have existence and uniqueness theorems for global solutions. In fact, the formulation of the initial value problem for such equations is not always clear. In these cases we say all solutions oscillate in case every continuous function which satisfies the equation on some infinite interval $[\tau, \infty)$ has arbitrarily large zeros and we do not care about the formulation of the initial value problems.

For the system of first order equations with deviating arguments of the form (1.1.9) the solution $\left(x_1(t), \ldots, x_m(t)\right)^T$ is said to be oscillatory (strongly oscillatory) if at least one (each one) of its components is oscillatory.

We shall now present some examples to describe the oscillation and nonoscillation problems in this book.

Example 1.2.1. The delay equation

$$y'(t) + y(t - \frac{\pi}{2}) = 0 \tag{1.2.1}$$

has an oscillatory solution $y(t) = \sin t$ which is caused by the delay.

Example 1.2.2. Consider the system

$$\begin{aligned} x'(t) &= 2x(t) - y(t) \\ y'(t) &= x(t) + y(t). \end{aligned} \tag{1.2.2}$$

From the analysis of the characteristic equation we know that every solution $\left(x(t), y(t)\right)$ of (1.2.2) oscillates. But the delay system

$$\begin{aligned} x'(t) &= 2x(t) - y(t - \frac{1}{3}\,\ell\text{n}\,4) \\ y'(t) &= x(t - \frac{1}{3}\,\ell\text{n}\,4) + y(t) \end{aligned} \tag{1.2.3}$$

has a nonoscillatory solution $x(t) = \exp((3/2)t)$, $y(t) = \exp((3/2)t)$. This nonoscillation is caused by delay. Therefore we must study the effect of deviating arguments on the oscillation of systems.

Example 1.2.3. In Chapter 2, we shall show that the equation

$$x'(t) + px(t-1) = 0, \quad t \geq t_0, \quad p > 0 \tag{1.2.4}$$

has a nonoscillatory solution if and only if

$$pe \leq 1. \tag{1.2.5}$$

In Chapter 3, we shall prove that every solution of

$$\big(x(t) - cx(t-1)\big)' + px(t-1) = 0, \quad t \geq t_0 \tag{1.2.6}$$

is oscillatory if

$$pe > 1 - c, \quad c \in (0,1). \tag{1.2.7}$$

Assume that $1 - c < pe \leq 1$, $c \in (0,1)$, then the oscillations of (1.2.6) are caused by the neutral term.

If $pe > 1$, then every solution of (1.2.4) oscillates. In Chapter 3, we shall show that if $c < 0$, $pe > 1$, then Eq. (1.2.6) has a nonoscillatory solution. This nonoscillation is caused by the neutral term. Therefore the presence of the neutral term can cause or destroy the oscillation of the solutions.

In this book we will investigate systematically the effect of deviating arguments and neutral terms on the oscillation of systems in the sense that their presence causes or destroys the oscillation of the solutions. We shall present many of the recent developments in the following problems

(1) Criteria for all solutions to be oscillatory.
(2) The classification and existence of nonoscillatory solutions.
(3) The existence of oscillatory solutions.
(4) The distance between adjacent zeros of the oscillatory solutions.
(5) Some applications to ecology models.

1.3. Formulation of Boundary Value Problems for Functional Differential Equations

The main sources of boundary value problems for ordinary differential equations without deviating arguments are boundary value problems for partial differential equations.

For functional differential equations the situation is different. The usual sources of boundary value problems for functional differential equations arise from variational problems for these equations.

For example, we consider the problem of an extremum of the functional

$$V(x(t)) = \int_{t_0}^{t_1} F(t, x(t), x(t-\tau), \dot{x}(t), \dot{x}(t-\tau))dt \tag{1.3.1}$$

with fixed or movable boundary values and initial functions. These variational problems lead to the following simple boundary value problem statements.

1. Find the continuous and smooth solutions of the equation

$$\begin{aligned} x''(t) = f\big(t, x(t), x(t-\tau_1(t)), \ldots, x(t-\tau_m(t))\big), \\ x'(t), x'(t-\tau_1(t)), \ldots, x'(t-\tau_m(t)), \\ x''(t-\tau_1(t)), \ldots, x''(t-\tau_m(t)) \end{aligned} \tag{1.3.2}$$

for $t \in (t_0, t_1)$ having a given initial function $\varphi(t)$ on E_{t_0} and a given boundary value $x(t_1) = x_1$. It is assumed that $\varphi(t_0 - 0) = x(t_0 + 0)$, but smoothness at the point t_0 is not required.

2. Find the solution $x(t)$ of equation (1.3.2) satisfying

$$x(t) = \varphi(t), \quad t \in E_{t_0} \quad \text{and} \quad \psi(t_1, x(t_1), x'(t_1)) = 0 \tag{1.3.3}$$

with $\psi = \psi(t, x, y)$ a given function. Smoothness is not required at the point t_0.

From this example it is natural to pose the following problems.

(P_1) Does a solution exist?

(P_2) If so, is that solution unique?

In Chapter 7, we consider mainly the following boundary value problem

$$x''(t) = f(t, x_t, x'(t)) \quad 0 \le t \le T \tag{1.3.4}$$

$$\begin{aligned} x_0 &= \varphi \\ x(T) &= A \end{aligned} \tag{1.3.5}$$

where $x_t(\theta) = x(t+\theta), \quad \theta \in [-r, 0]$.

By a solution of the BVP (1.3.4) and (1.3.5) we mean a function $x \in C([-r,T],R^n) \cap C^2([0,T],R^n)$ which satisfies the equation (1.3.4) and the boundary conditions (1.3.5).

As in ordinary differential equations the BVP (1.3.4) and (1.3.5) can be transformed into the integral equation

$$x(t) = \begin{cases} \displaystyle\int_0^T G(t,s) f(s, x_s, x'(s))ds + \frac{A-\varphi(0)}{T}t + \varphi(0), & 0 \leq t < T \\ \varphi(t) & -r \leq t \leq 0 \end{cases} \tag{1.3.6}$$

where $G(t,s)$ is the Green's function for the BVP

$$x''(t) = 0, \quad x(0) = 0, \quad x(T) = 0$$

and is given by the formula

$$G(t,s) = \frac{1}{T} \begin{cases} (t-T)s, & 0 \leq s \leq t \leq T \\ t(s-T), & 0 \leq t \leq s \leq T. \end{cases} \tag{1.3.7}$$

Then the BVP (1.3.4) and (1.3.5) reduces to finding a fixed point for the mapping $\mathcal{S}$ defined by

$$(\mathcal{S}x)(t) = \begin{cases} \displaystyle\int_0^T G(t,s) f(s, x_s, x'(s))ds + \frac{A-\varphi(0)}{T}t + \varphi(0), & 0 \leq t \leq T \\ \varphi(t), & -r \leq t \leq 0. \end{cases}$$

The method which is used to prove that the equation $x = \mathcal{S}x$ has a solution, is usually the application of a suitable fixed point theorem.

1.4. Fixed Point Theorems

We will introduce some fixed point theorems which are employed throughout the book.

A vector space X on which we have defined a norm $\|\cdot\|$ is called a normed vector space. A subset S of a normed vector space X is said to be bounded if there is a number M such that $\|x\| \leq M$ for all $x \in S$. A subset S of a normed vector space X is called convex if, for any $x, y \in S$, $as + (1-a)y \in S$ for all $a \in [0,1]$.

A sequence $\{x_n\}$ in a normed vector space X is said to converge to the vector $x \in X$ if and only if the sequence $\|x_n - x\|$ converges to zero as $n \to \infty$.

A sequence $\{x_n\}$ in a normed vector space X is a Cauchy sequence in X if for every $\varepsilon > 0$ there exists an $N = N(\varepsilon)$ such that $\|x_n - x_m\| < \varepsilon$ for all $n, m \geq N(\varepsilon)$.

Clearly a convergent sequence is a Cauchy sequence but the converse may not be true. A space X where every Cauchy sequence of elements of X converges to an element of X is called a complete space. A complete normed vector space is said to be a Banach space.

Let M be a subset of a Banach space X. A point $x \in X$ is said to be a limit point of M if there exists a sequence of vectors in M which converges to x. We say a subset M is closed if M contains all of its limit points. The union of M and its limit points is called the closure of M and will be denoted by $\overline{M}$.

Let N, M be normed spaces, and X be a subset of N. An operator $T : X \to M$ is continuous at a point $x \in X$ if and only if for any $\varepsilon > 0$ there is a $\delta > 0$ such that $\|Tx - Ty\| < \varepsilon$ for all $y \in X$ such that $\|x - y\| < \delta$. T is continuous on X, or simply continuous, if it is continuous at all points of X.

The following result is well-known.

Theorem 1.4.1. *Every continuous mapping of a closed bounded convex set in R^n into itself has a fixed point.*

A subset S of a Banach space X is compact, if and only if every infinite sequence of elements of S has a subsequence which converges to an element of S. We say M is relatively compact if every infinite sequence in S contains a subsequence which converges to an element in X. That is, M is relatively compact, if $\overline{M}$ is compact.

A subset S in $C([a, b], R)$ with norm

$$\|f\| = \sup_{x \in [a,b]} |f(x)|$$

is relatively compact if and only if it is uniformly bounded and equicontinuous on $[a, b]$ (Arzela-Ascoli Theorem). A family S in $C([a, b], R)$ is called uniformly bounded if there exists a positive number M such that

$$|f(t)| \leq M \quad \text{for all} \quad t \in [a, b], \quad \text{all} \quad f \in S.$$

S is called equicontinuous if for every $\varepsilon > 0$ there exists a $\delta = \delta(\varepsilon) > 0$ such that $|f(t_1) - f(t_2)| < \varepsilon$ for all $t_1, t_2 \in [a, b]$ with $|t_1 - t_2| < \delta$ and for all $f \in S$.

Theorem 1.4.2. (Schauder's Fixed Point Theorem). *Let S be a closed, convex and nonempty subset of a Banach space X. Let $T : S \to S$ be a continuous mapping such that TS is a relatively compact subset of X. Then T has at least one fixed point in S. That is, there exists an $x \in S$ such that $Tx = x$.*

Remark 1.4.1. In oscillation theory we usually want to prove that the family of functions is uniformly bounded and equicontinuous on $[t_0, +\infty)$. According to Levitan's result, the family S is equicontinuous on $[t_0, \infty)$ if for any given $\varepsilon > 0$, the interval $[t_0, \infty)$ can be decomposed into a finite number of subintervals in such a way that on each subinterval all functions of the family S have oscillations less than ε.

A topology $\mathcal{T}$ on a linear space E is called locally convex if and only if every neighborhood of the element 0 includes a convex neighborhood of 0.

A real valued function $\rho(x)$ defined on a linear space X is called a seminorm on X iff the following conditions are satisfied:

$$\rho(x + y) \leq \rho(x) + \rho(y)$$
$$\rho(\alpha x) = |\alpha|\rho(x), \quad \alpha \quad \text{is scalar.}$$

From this definition, we can prove that a seminorm $\rho(x)$ satisfies $\rho(0) = 0$, $\rho(x_1 - x_2) \geq |\rho(x_1) - \rho(x_2)|$. In particular, $\rho(x) \geq 0$. However, it may happen that $\rho(x) = 0$ for $x \neq 0$.

A family P of seminorms on X is said to be separating iff to each $x \neq 0$ there corresponds at least one $\rho \in P$ with $\rho(x) \neq 0$. For a separating seminorm family P, if $\rho(x) = 0$ for every $\rho \in P$, then $x = 0$.

A locally convex topology $\mathcal{T}$ on a linear space is determined by a family of seminorms $\{\rho_\alpha : \alpha \in I\}$, I being the index set.

Let E be a locally convex space, $x \in E$, $\{x_n\} \in E$. Then $x_n \to x$ in E if and only if $\rho_\alpha(x_n - x) \to 0$ as $n \to \infty$, for every $\alpha \in I$.

A set $S \subset E$ is bounded if and only if the set of numbers $\{\rho_\alpha(x), x \in S\}$ is bounded for every $\alpha \in I$.

A complete metrizable locally convex space is called a Fréchet space.

Theorem 1.4.3. (Schauder-Tychonov Theorem). *Let X be a locally convex linear space, let S be a compact convex subset of X, and let $T : S \to S$ be a continuous mapping with $T(S)$ compact. Then T has a fixed point in S.*

For example $C([t_0, \infty), R)$ is a locally convex space consisting of the set of all continuous functions. The topology of C is the topology of uniform convergence on every compact interval of $[t_0, \infty)$. The seminorm of the space $C([t_0, \infty), R)$ is defined by

$$\rho_\alpha(x) = \max_{x \in [t_0, \alpha]} |x(t)|, \quad x \in C, \quad \alpha \in (t_0, \infty).$$

Let X be any set. A metric in X is a function $d : X \times X \to R$ having the following properties for all $x, y, z \in X$:

1. $d(x, y) \geq 0$ and $d(x, y) = 0$ if and only if $x = y$
2. $d(y, x) = d(x, y)$
3. $d(x, z) \leq d(x, y) + d(y, z)$

A metric space is a set X together with a given metric in X. A complete metric space is a metric space X in which every Cauchy sequence converges to a point in X.

Let (X, d) be a metric space and let $T : X \to X$. If there exists a number $r \in [0, 1)$ such that

$$d(Tx, Ty) \leq r d(x, y) \quad \text{for every} \quad x, y \in X,$$

then we say T is a contraction mapping on X.

Theorem 1.4.4. (The Banach Contraction Mapping Principle). *A contraction mapping on a complete metric space has exactly one fixed point.*

Theorem 1.4.5. (Krasnoselskii's Fixed Point Theorem). *Let X be a Banach space, let Ω be a bounded closed convex subset of X and let A, B be maps of Ω into X such that $Ax + By \in \Omega$ for every pair $x, y \in \Omega$. If A is a contraction and B is completely continuous, then the equation*

$$Ax + Bx = x$$

has a solution in Ω.

A nonempty and closed subset K of a Banach space X is called a cone if it possesses the following properties:

(i) if $\alpha \in R^+$ and $x \in K$, then $\alpha x \in K$
(ii) if $x, y \in K$ then $x + y \in K$
(iii) if $x \in K - \{0\}$, then $-x \notin K$.

We say that a Banach space X is partially ordered if X contains a cone K with a nonempty interior. The ordering $\leq$ in X is then defined as follows:

$$x \leq y \quad \text{if and only if} \quad y - x \in K.$$

Let M be a subset of a partially ordered Banach space X. Set

$$\overline{M} = \{x \in X : \; y \leq x \quad \text{for every} \quad y \in M\}.$$

We say that the point $x_0 \in X$ is the infimum of $\overline{M}$ if $x_0 \in \widetilde{M}$ and for every $x \in \overline{M}$, $x_0 \leq x$. The supremum of $\overline{M}$ is defined in a similar way.

Theorem 1.4.6. (Knaster's Fixed Point Theorem). *Let X be a partially ordered Banach space with ordering $\leq$. Let M be a subset of X with the following properties: The infimum of M belongs to M and every nonempty subset of M has a supremum which belongs to M. Let $T : M \to M$ be an increasing mapping, i.e., $x \leq y$ implies $Tx \leq Ty$. Then T has a fixed point in M.*

Let X be a Banach space let K be a cone in X, and let $\leq$ be the order in X induced by K, i.e, $x \leq y$ if and only if $y - x$ is an element of K. Let D be a subset of K, and $T : \; D \to K$ a mapping.

We denote by $\langle x, y \rangle$ the closed order interval between x and y, i.e.,

$$\langle x, y \rangle = \{x \in Z : \; x \leq z \leq y\}.$$

We assume that the cone K is normal in X, which implies that order intervals are norm bounded. The cones of nonnegative functions are normal in the space of continuous functions with supremum norm and in the space L_p.

Theorem 1.4.7. *Let X be a Banach space, K a normal cone in X, D a subset of K such that if x, y are elements of D, $x \leq y$, then $\langle x, y \rangle$ is contained in D, and let $T : \; D \to K$ be a continuous decreasing mapping which is compact on any closed order interval contained in D. Suppose that there exists $x_0 \in D$ such* *that T^2x_0 is defined (where $T^2x_0 = T(Tx_0))$ and further more Tx_0, T^2x_0 are (order) comparable to x_0. Then T has a fixed point in D provided that either:*

(I) $Tx_0 \leq x_0$ and $T^2x_0 \leq x_0$ or $Tx_0 \geq x_0$ and $T^2x_0 \geq x_0$, or
(II) The complete sequence of iterates $\{T^n x_0\}_{n=0}^{\infty}$ is defined and there exists $y_0 \in D$ such that $Ty_0 \in D$ and $y_0 \leq T^n x_0$ for all n.

Theorem 1.4.8. *Let X be a Banach space, $A : X \to X$ be a completely continuous mapping such that $I - A$ is one to one, and let Ω be a bounded set such that $0 \in (I - A)(\Omega)$. Then the completely continuous mapping $S : \Omega \to X$ has a fixed point in $\overline{\Omega}$ if for any $\lambda \in (0, 1)$, the equation*

$$x = \lambda Sx + (1 - \lambda)Ax$$

has no solution x on the boundary $\partial\Omega$ of Ω.

The following theorem is a very useful version of the Topological Transversality Theorem of Granas.

Theorem 1.4.9. *Suppose that X is a normed space (possibly incomplete). Let S be a bounded, closed, convex subset of X, containing the origin 0 in its interior D. Let $H : [0, 1] \times S \to X$ be a homotopy of compact transformations such that $H(0, \partial S) \subset S$ and $H(t, x) \neq x \quad (0 \leq t \leq 1,\ x \in \partial S)$. Then $\varphi = H(1, \cdot)$ has a fixed point in S.*

Theorem 1.4.10. (Leray–Schauder Alternative). *Let C be a convex subset of a normed linear space E and assume $0 \in C$. Let $F : C \to C$ be a completely continuous operator and let*

$$\mathcal{E}(F) = \{x \in C : x = \lambda Fx \quad \text{for some} \quad 0 < \lambda < 1\}.$$

Then either $\mathcal{E}(F)$ is unbounded or F has a fixed point.

2

Oscillations of First Order Delay Differential Equations

2.0. Introduction

In this chapter, we will describe some of the recent developments in the oscillation theory of first order delay differential equations. This theory is interesting from the theoretical as well as the practical point of view. It is well known that homogeneous ordinary differential equations (ODEs) of first order do not possess oscillatory solutions. But the presence of deviating arguments can cause the oscillation of solutions. In this chapter, we will see these phenomena and we will show various techniques used in the oscillation and nonoscillation theory of differential equations with deviating arguments. We will present some criteria for oscillation, for the existence of positive solutions and results in the distribution of zeros of oscillatory solutions of DDEs of first order. In the last section we will discuss the oscillation and nonoscillation behavior of some basic ecological delay equations.

2.1. Stable Type Equations with a Single Delay

2.1.1. Oscillation results

We consider the linear delay differential inequalities and equations of the form

$$x'(t) + p(t)x\big(\tau(t)\big) \le 0 \tag{2.1.1}$$

$$x'(t) + p(t)x\big(\tau(t)\big) \ge 0 \tag{2.1.2}$$

$$x'(t) + p(t)x\big(\tau(t)\big) = 0 \tag{2.1.3}$$

where $p,\ \tau \in C([t_0, \infty), R_+)$, $\quad R_+ = [0, \infty)$, $\quad \tau(t) \le t$ and $\lim_{t\to\infty} \tau(t) = \infty$.

Set

$$m = \liminf_{t\to\infty} \int_{\tau(t)}^{t} p(s)ds \tag{2.1.4}$$

and

$$M = \limsup_{t\to\infty} \int_{\tau(t)}^{t} p(s)ds. \tag{2.1.5}$$

The following lemmas will be used to prove the main results in this section. All the inequalities in this section and in the latter parts hold eventually if it is not mentioned specifically.

Lemma 2.1.1. *Set*

$$\delta(t) = \max\ \big\{T(s);\ s \in [t_0, t]\big\} \tag{2.1.6}$$

and $m > 0$. *Then we have*

$$\liminf_{t\to\infty} \int_{\delta(t)}^{t} p(s)ds = \liminf_{t\to\infty} \int_{\tau(t)}^{t} p(s)ds = m. \tag{2.1.7}$$

Proof: Clearly, $\delta(t) \ge \tau(t)$ and so

$$\int_{\delta(t)}^{t} p(s)ds \le \int_{\tau(t)}^{t} p(s)ds.$$

Hence

$$\liminf_{t\to\infty} \int_{\delta(t)}^{t} p(s)ds \le \liminf_{t\to\infty} \int_{\tau(t)}^{t} p(s)ds.$$

If (2.1.7) does not hold, then there exist $m' > 0$ and a sequence $\{t_n\}$ such that $t_n \to \infty$ as $n \to \infty$ and

$$\lim_{n\to\infty} \int_{\delta(t_n)}^{t_n} p(s)ds \le m' < m.$$

By definition, $\delta(t_n) = \max\,(\tau(s) : s \in [t_0, t_n])$, and hence there exists $t'_n \in [t_0, t_n]$ such that $\delta(t_n) = \tau(t'_n)$. Hence

$$\int_{\delta(t_n)}^{t_n} p(s)ds = \int_{\tau(t'_n)}^{t_n} p(s)ds > \int_{\tau(t'_n)}^{t'_n} p(s)ds.$$

It follows that $\{\int_{\tau(t'_n)}^{t'_n} p(s)ds\}_{n=1}^{\infty}$ is a bounded sequence having a convergent subsequence, say

$$\int_{\tau(t'_{n_k})}^{t'_{n_k}} p(s)ds \to c \le m', \quad \text{as} \quad k \to \infty,$$

which implies that

$$\liminf_{t\to\infty} \int_{\tau(t)}^{t} p(s)ds \le m',$$

contradicting (2.1.4). □

Lemma 2.1.2. *Let $x(t)$ be an eventually positive solution of (2.1.1). If $m > \frac{1}{e}$, then*

$$\lim_{t\to\infty} \frac{x(\tau(t))}{x(t)} = \infty. \tag{2.1.8}$$

If $m \le \frac{1}{e}$, then

$$\liminf_{t\to\infty} \frac{x(\tau(t))}{x(t)} \ge \lambda \tag{2.1.9}$$

where λ is the smaller positive root of the equation

$$\lambda = e^{m\lambda}. \tag{2.1.10}$$

Proof: Let t_1 be a sufficiently large number so that $x(\tau(t)) > 0$ for $t \geq t_1$. Hence

$$\frac{x'(t)}{x(t)} \leq -p(t)\,\frac{x(\tau(t))}{x(t)} \leq -p(t). \tag{2.1.11}$$

Integrating it from $\tau(t)$ to t we have that eventually

$$\frac{x(\tau(t))}{x(t)} \geq \exp\left(\int_{\tau(t)}^{t} p(s)ds\right).$$

Then for any $\varepsilon > 0$, there exists T_ε such that

$$\frac{x(\tau(t))}{x(t)} \geq e^m - \varepsilon, \quad t \geq T_\varepsilon. \tag{2.1.12}$$

Substituting (2.1.12) into (2.1.11) we have $\frac{x'(t)}{x(t)} \leq -(e^m - \varepsilon)p(t), \quad t \geq T_\varepsilon$, and hence

$$\liminf_{t\to\infty} \frac{x(\tau(t))}{x(t)} \geq \exp(m \exp m).$$

Set $\lambda_0 = 1, \quad \lambda_1 = \exp(m\lambda_0), \ldots, \lambda_n = \exp(m\lambda_{n-1}), \ldots$. For a sequence $\{\varepsilon_n\}$ with $\varepsilon_n > 0$ and $\varepsilon_n \to 0$ as $n \to \infty$, there exists a sequence $\{t_n\}$ such that $t_n \to \infty$ as $n \to \infty$ and

$$\frac{x(\tau(t))}{x(t)} \geq \lambda_n - \varepsilon_n \quad \text{for} \quad t \geq t_n. \tag{2.1.13}$$

If $m > \frac{1}{e}$, then $\lim_{n\to\infty} \lambda_n = \infty$, and (2.1.8) holds. If $m = \frac{1}{e}$, then $\lim_{n\to\infty} \lambda_n = e$; and if $m < \frac{1}{e}$, then λ_n tends to the smaller root of Eq. (2.1.10). □

Remark 2.1.1. From Theorem 2.1.1 we will see that (2.1.1) has no eventually positive solutions if $m > \frac{1}{e}$.

Lemma 2.1.3. *In addition to the hypotheses of Lemma 2.1.1, assume τ is nondecreasing, $0 \le m \le \frac{1}{e}$, and $x(t)$ is an eventually positive solution of (2.1.1). Set*

$$r = \liminf_{t\to\infty} \frac{x(t)}{x(\tau(t))}\,. \tag{2.1.14}$$

Then

$$A(m) := \frac{1 - m - \sqrt{1 - 2m - m^2}}{2} \le r \le 1. \tag{2.1.15}$$

Proof: Assume that $x(t) > 0$ for $t > T_1 \ge t_0$, and there exists a sequence $\{T_n\}$ such that $T_1 < T_2 < T_3 < \dots$ and $\tau(t) > T_n$ for $t > T_{n+1}$, $\quad n = 1, 2, \dots$. Hence $x(\tau(t)) > 0$ for $t > T_2$. In view of (2.1.1), $x'(t) \le 0$ on $(T_2, +\infty)$. Clearly, (2.1.15) holds for $m = 0$. If $0 < m \le \frac{1}{e}$, for any $\varepsilon \in (0, m)$, there exists N_ε such that

$$\int_{\tau(t)}^{t} p(s)ds > m - \varepsilon, \quad \text{for} \quad t > N_\varepsilon. \tag{2.1.16}$$

For a fixed ε we will show that for each $t > N_\varepsilon$ there exists a λ_t such that $\tau(\lambda_t) < t < \lambda_t$ and

$$\int_{t}^{\lambda_t} p(s)ds = m - \varepsilon. \tag{2.1.17}$$

In fact, for a given t, $\quad f(\lambda) := \int_t^\lambda p(s)ds$ is continuous and $\lim\limits_{\lambda\to\infty} f(\lambda) > m - \varepsilon > 0 = f(t)$. Hence there exists $\lambda_t > t$ such that $f(\lambda_t) = m - \varepsilon$, i.e., (2.1.17) holds. From (2.1.16) we have

$$\int_{\tau(\lambda_t)}^{\lambda_t} p(s)ds > m - \varepsilon = \int_{t}^{\lambda_t} p(s)ds,$$

therefore $\tau(\lambda_t) < t$.

Integrating (2.1.1) from t ($> \max\{T_4, N_\varepsilon\}$) to λ_t we have

$$x(t) - x(\lambda_t) \ge \int_{t}^{\lambda_t} p(y)x(\tau(y))dy. \tag{2.1.18}$$

We see that $\tau(t) \le \tau(y) \le \tau(\lambda_t) < t$ for $t \le y \le \lambda_t$.

Integrating (2.1.1) from $\tau(y)$ to t we have that for $t \leq y \leq \lambda_t$

$$\begin{aligned} x\big(\tau(y)\big) - x(t) &\geq \int_{\tau(y)}^{t} p(u)x\big(\tau(u)\big)du \\ &\geq x\big(\tau(t)\big)\int_{\tau(y)}^{t} p(u)du \\ &= x\big(\tau(t)\big)\left(\int_{\tau(y)}^{t} p(u)du - \int_{t}^{y} p(u)du\right) \\ &> x\big(\tau(t)\big)\left[(m-\varepsilon) - \int_{t}^{y} p(u)du\right]. \end{aligned} \tag{2.1.19}$$

From (2.1.18) and (2.1.19) we have

$$\begin{aligned} x(t) &\geq x(\lambda_t) + \int_{t}^{\lambda_t} p(y)x\big(\tau(y)\big)dy \\ &> x(\lambda_t) + \int_{t}^{\lambda_t} py\left\{x(t) + x\big(\tau(t)\big)\left[(m-\varepsilon) - \int_{t}^{y} p(u)du\right]\right\}dy \\ &= x(\lambda_t) + x(t)(m-\varepsilon) \\ &\qquad + x\big(\tau(t)\big)\left[(m-\varepsilon)^2 - \int_{t}^{\lambda_t} p(y)\int_{t}^{y} p(u)\,du\,dy\right]. \end{aligned} \tag{2.1.20}$$

Noting the known formula

$$\int_{t}^{\lambda_t} dy \int_{t}^{y} p(y)p(u)du = \int_{t}^{\lambda_t} du \int_{u}^{\lambda_t} p(y)p(u)dy,$$

or

$$\int_{t}^{\lambda_t} dy \int_{t}^{y} p(y)p(u)du = \int_{t}^{\lambda_t} dy \int_{y}^{\lambda_t} p(y)p(u)du,$$

we have

$$\int_t^{\lambda_t} dy \int_t^y p(y)p(u)du = \frac{1}{2}\left[\int_t^{\lambda_t} dy \int_t^y p(y)p(u)du + \int_t^{\lambda_t} dy \int_y^{\lambda_t} p(u)p(y)du\right]$$
$$= \frac{1}{2}\int_t^{\lambda_t} dy \int_t^{\lambda_t} p(y)p(u)du$$
$$= \frac{1}{2}\left[\int_t^{\lambda_t} p(s)ds\right]^2 = \frac{1}{2}(m-\varepsilon)^2.$$

Substituting this into (2.1.20) we have

$$x(t) > x(\lambda_t) + x(t)(m-\varepsilon) + \tfrac{1}{2}\,(m-\varepsilon)^2 x(\tau(t)). \tag{2.1.21}$$

Hence

$$\frac{x(t)}{x(\tau(t))} > \frac{(m-\varepsilon)^2}{2(1-m+\varepsilon)} := d_1, \tag{2.1.22}$$

and then

$$x(\lambda_t) > \frac{(m-\varepsilon)^2}{2(1-m+\varepsilon)}\, x(\tau(\lambda_t)) = d_1 x(\tau(\lambda_t)) \ge d_1 x(t).$$

Substituting this into (2.1.21) we obtain

$$x(t) > x(t)(m+d_1-\varepsilon) + \tfrac{1}{2}\,(m-\varepsilon)^2 x(\tau(t)),$$

and hence

$$\frac{x(t)}{x(\tau(t))} > \frac{(m-\varepsilon)^2}{2(1-m-d_1+\varepsilon)} := d_2.$$

In general we have

$$\frac{x(t)}{x(\tau(t))} > \frac{(m-\varepsilon)^2}{2(1-m-d_n+\varepsilon)} := d_{n+1}, \quad n = 1, 2, \ldots .$$

It is not difficult to see that if ε is small enough, $1 \ge d_n > d_{n-1}$, $n = 2, 3, \ldots$. Hence $\lim_{n\to\infty} d_n = d$ exists and satisfies

$$-2d^2 + 2d(1-m+\varepsilon) = (m-\varepsilon)^2,$$

i.e.,

$$d = \frac{1-m+\varepsilon \pm \sqrt{1-2(m-\varepsilon)-(m-\varepsilon)^2}}{2}.$$

Therefore, for all large t

$$\frac{x(t)}{x(\tau(t))} > \frac{1-m+\varepsilon - \sqrt{1-2(m-\varepsilon)-(m-\varepsilon)^2}}{2}.$$

Letting $\varepsilon \to 0$, we obtain that

$$\frac{x(t)}{x(\tau(t))} \geq \frac{1-m-\sqrt{1-2m-m^2}}{2} = A(m).$$

This shows that (2.1.15) holds. □

Lemma 2.1.4. *Assume that* $0 < M \leq 1$, τ *is nondecreasing. Let* $x(t)$ *be an eventually positive solution of (2.1.1). Set*

$$\liminf_{t\to\infty} \frac{x(\tau(t))}{x(t)} = \ell.$$

Then

$$\ell \leq B(M) := \left(\frac{1+\sqrt{1-M}}{M}\right)^2. \tag{2.1.23}$$

Proof: For a given $\varepsilon \in (0, M)$, there exists a sequence $\{t_n\}$ such that $t_n \to \infty$ as $n \to \infty$ and

$$\int_{\tau(t_n)}^{t_n} p(s)ds > M - \varepsilon, \quad t_n > T, \quad n = 1, 2, \dots .$$

Set $\theta_\varepsilon = 1 - \sqrt{1-(M-\varepsilon)}$. It is easy to see that $0 < \theta_\varepsilon < M - \varepsilon$ for small ε. Hence there exists $\{\lambda_n\}$ such that $\tau(t_n) < \lambda_n < t_n$ and

$$\int_{\lambda_n}^{t_n} p(s)ds = \theta_\varepsilon, \quad n = 1, 2, \dots .$$

Integrating (2.1.11) from λ_n to t_n we obtain

$$\begin{aligned} x(\lambda_n) - x(t_n) &\geq \int_{\lambda_n}^{t_n} p(s)x(\tau(s))ds \\ &\geq x(\tau(t_n)) \int_{\lambda_n}^{t_n} p(s)ds \\ &= \theta_\varepsilon x(\tau(t_n)). \end{aligned}$$

Similarly, we have

$$\begin{aligned} x(\tau(t_n)) - x(\lambda_n) &\geq \int_{\tau(t_n)}^{\lambda_n} p(s)x(\tau(s))ds \\ &\geq x(\tau(\lambda_n)) \int_{\tau(t_n)}^{\lambda_n} p(s)ds \\ &= x(\tau(\lambda_n)) \left[\int_{\tau(t_n)}^{t_n} p(s)ds - \int_{\lambda_n}^{t_n} p(s)ds \right] \\ &> x(\tau(\lambda_n))[(M-\varepsilon) - \theta_\varepsilon]. \end{aligned}$$

From the above inequalities we get

$$\begin{aligned} x(\lambda_n) &> \theta_\varepsilon x(\tau(t_n)) \\ &> \theta_\varepsilon[x(\lambda_n) + x(\tau(\lambda_n))(M - \varepsilon - \theta_\varepsilon)] \end{aligned}$$

and then

$$\frac{x(\tau(\lambda_n))}{x(\lambda_n)} < \frac{1-\theta_\varepsilon}{\theta_\varepsilon(M-\varepsilon-\theta_\varepsilon)}, \quad n = 1, 2, \ldots$$

which implies that

$$\ell \leq \frac{1-\theta_\varepsilon}{\theta_\varepsilon(M-\varepsilon-\theta_\varepsilon)}, \quad \varepsilon \in (0, M).$$

When $\varepsilon \to 0$, $\theta_\varepsilon \to 1 - \sqrt{1-M}$, then we obtain

$$\ell \leq \frac{\sqrt{1-M}}{(1-\sqrt{1-M})(M-1+\sqrt{1-M})} = \left(\frac{1+\sqrt{1-M}}{M}\right)^2. \qquad \square$$

We are now in a position to state the oscillation criteria for Eq. (2.1.3).

Theorem 2.1.1. *Assume $m > \frac{1}{e}$. Then*
(i) *(2.1.1) has no eventually positive solutions,*
(ii) *(2.1.2) has no eventually negative solutions,*
(iii) *every solution of Eq. (2.1.3) is oscillatory.*

Proof: It is sufficient to prove (i), (ii) and (iii) follow from (i). Suppose the contrary is true, and let $x(t)$ be an eventually positive solution of (2.1.1).

By Lemma 2.1.1, we may assume that τ is nondecreasing. In view of Lemma 2.1.2,

$$\liminf_{t\to\infty} \frac{x\big(\tau(t)\big)}{x(t)} = \infty.$$

On the other hand, from (2.1.22), $\frac{x(\tau(t))}{x(t)}$ is bounded above. This contradiction proves (i). □

Remark 2.1.2. If in addition to the conditions of Theorem 2.1.1, τ is nondecreasing, $M \in (0,1]$. Then the condition $m > \frac{1}{e}$ can be replaced by

$$m > \frac{\ln\, b}{b}, \quad b = \min\,\big(e, B(M)\big), \tag{2.1.24}$$

where $B(M)$ is defined by (2.1.23).

In fact, under the above assumptions Lemma 2.1.4 holds. Let $x(t)$ be a positive solution of (2.1.1) and $w(t) = \frac{x(\tau(t))}{x(t)}$. By Lemma 2.1.4, $\liminf_{t\to\infty} w(t) = \ell \le B(M)$. From (2.1.1), we obtain

$$-\frac{x'(t)}{x(t)} \ge p(t)w(t), \quad t \ge T.$$

Integrating it from $\tau(t)$ to t we obtain

$$\ln w(t) \ge \int_{\tau(t)}^{t} p(s)w(s)ds = w(\xi_t)\int_{\tau(t)}^{t} p(s)ds,$$

and hence

$$\liminf_{t\to\infty}\, \ln w(t) \ge \ell m$$

and

$$m \leq \frac{\ln \ell}{\ell} \leq \frac{\ln b}{b}$$

which contradicts (2.1.24). □

Theorem 2.1.2. *Assume $0 \leq m \leq \frac{1}{e}$, and τ is nondecreasing. Furthermore*

$$M > 1 - A(m) \tag{2.1.25}$$

where $A(m)$ is defined by (2.1.15), or

$$M > \frac{\ln \lambda + 1}{\lambda} \tag{2.1.26}$$

where λ is the smaller positive root of the equation

$$\lambda = e^{m\lambda}. \tag{2.1.27}$$

Then the conclusion of Theorem 2.1.1 is true.

Proof: The same as in Theorem 2.1.1 it is sufficient to show that under our assumptions (2.1.1) has no eventually positive solutions.

Integrating (2.1.1) from $\tau(t)$ to t we obtain

$$x(\tau(t)) - x(t) \geq \int_{\tau(t)}^{t} p(s)x(\tau(s))ds \geq x(\tau(t)) \int_{\tau(t)}^{t} p(s)ds.$$

Then if (2.1.25) holds, by Lemma 2.1.3, we have

$$\begin{aligned} M &= \limsup_{t\to\infty} \int_{\tau(t)}^{t} p(s)ds \leq \limsup_{t\to\infty} \left[1 - \frac{x(t)}{x(\tau(t))}\right] \\ &= 1 - \liminf_{t\to\infty} \frac{x(t)}{x(\tau(t))} = 1 - r \leq 1 - A(m), \end{aligned} \tag{2.1.28}$$

which contradicts with (2.1.15).

If (2.1.26) holds, choose $m' < m$ sufficiently close to m such that

$$M = \limsup_{t\to\infty} \int_{\tau(t)}^{t} p(s)ds > \frac{\ln \lambda' + 1}{\lambda'} \tag{2.1.29}$$

where λ' is a smaller root of the equation $\lambda = e^{m'\lambda}$.

Clearly $\lambda' < \lambda$ and hence $\frac{\ln \lambda'+1}{\lambda'} > \frac{\ln \lambda+1}{\lambda}$. By Lemma 2.1.2, we have

$$\frac{x(\tau(t))}{x(t)} > \lambda' \quad \text{for all large} \quad t. \tag{2.1.30}$$

From (2.1.29), there exists a t_1 so large that (2.1.30) holds for all $t > \tau(\tau(t_1))$, and

$$\int_{\tau(t_1)}^{t_1} p(s)ds > \frac{\ln \lambda' + 1}{\lambda'}. \tag{2.1.31}$$

Without loss of generality denote $t_0 = \tau(t_1)$. We shall show that $x(t) > 0$ on $[t_0, t_1]$ will lead to a contradiction. In fact, let $t_2 \in [t_0, t_1]$ be a point at which $x(t_0)/x(t_2) = \lambda'$. If such a point does not exist, take $t_2 = t_1$. Integrating (2.1.1) over $[t_2, t_1]$ and noting that $x(\tau(t)) \geq x(t_0)$, we have

$$\int_{t_2}^{t_1} p(s)ds \leq \frac{1}{\lambda'}. \tag{2.1.32}$$

On the other hand, dividing (2.1.1) by $x(t)$ and integrating it over $[t_1, t_2]$ we have

$$\int_{t_0}^{t_2} p(s)ds \leq -\frac{1}{\lambda'} \int_{t_0}^{t_2} \frac{x'(s)}{x(s)}\, ds = \frac{\ln \lambda'}{\lambda'}. \tag{2.1.33}$$

Combining (2.1.32) and (2.1.33) we get

$$\int_{t_0}^{t_1} p(s)ds \leq \frac{\ln \lambda' + 1}{\lambda'} \tag{2.1.34}$$

which contradicts (2.1.31). □

Example 2.1.1. Consider the equation

$$x'(t) + \frac{0.85}{a\pi + \sqrt{2}}(2a + \cos t)x(t - \tfrac{\pi}{2}) = 0 \tag{2.1.35}$$

where $a = 1.137$. We have

$$p(t) = \frac{0.85}{a\pi + \sqrt{2}}(2a + \cos t)$$

and

$$\int_{t-\frac{\pi}{2}}^{t} p(s)ds = \frac{0.85}{a\pi + \sqrt{2}} \left(a\pi + \sqrt{2}\ \cos(t - \tfrac{\pi}{4})\right).$$

Hence

$$m = \liminf_{t\to\infty} \int_{t-\frac{\pi}{2}}^{t} p(s)ds = 0.85\,\frac{a\pi - \sqrt{2}}{a\pi + \sqrt{2}} = 0.367837 < \frac{1}{e}$$

and

$$M = \limsup_{t\to\infty} \int_{t-\frac{\pi}{2}}^{t} p(s)ds = 0.85.$$

It is easy to see that (2.1.25) holds. Therefore every solution of Eq. (2.1.35) is oscillatory.

2.1.2. The existence of positive solution

First we consider a linear delay differential equation of the form

$$x'(t) + x\big(t - \tau(t)\big) = 0 \tag{2.1.36}$$

where $\tau \in C([t_0,\infty), R_+)$ and $\lim_{t\to\infty}\big(t - \tau(t)\big) = \infty$. Set $T_0 = \inf_{t\geq t_0}\{t - \tau(t)\}$.

Definition 2.1.1. A solution x is called a positive solution with respect to the initial point t_0, if $x(t)$ is a solution of (2.1.36) on $[t_0,\infty)$ and $x(t) > 0$, on $[T_0,\infty)$.

Theorem 2.1.3. *Eq. (2.1.3) has a positive solution with respect to t_0 if and only if there exists a real continuous function $\lambda_0(t)$ on $[T_0,+\infty)$ such that $\lambda_0(t) > 0$ for $t \geq t_0$ and*

$$\tau(t) \leq t - \Lambda_0^{-1}\big(\Lambda_0(t) - \ln\ \lambda_0(t)\big), \quad t \geq t_0, \tag{2.1.37}$$

where $\Lambda_0(t) = \int_{T_0}^{t} \lambda_0(s)ds$, $\Lambda_0^{-1}(u)$ denotes the converse function of Λ_0.

Proof: i) *Necessity.* Let $x_0(t)$ be a positive solution of (2.1.36) with respect to

t_0. That is, $x(t) > 0$ on $[T_0, +\infty)$. Set

$$x_0(t) = x_0(T_0)\exp\Big(-\int_{T_0}^{t} \lambda_0(s)\,ds\Big), \quad t \geq T_0. \tag{2.1.38}$$

Then $\lambda_0(t)$ satisfies the equation

$$\lambda_0(t) = \exp\Big(\int_{t-\tau(t)}^{t} \lambda_0(s)\,ds\Big), \quad t \geq t_0. \tag{2.1.39}$$

Clearly $\lambda_0(t) > 0$ for $t \geq t_0$. From (2.1.39) we have

$$\Lambda_0(t) - \Lambda_0\big(t - \tau(t)\big) = \ln \lambda_0(t).$$

Then

$$t - \tau(t) = \Lambda_0^{-1}\big(\Lambda_0(t) - \ln \lambda_0(t)\big)$$

and

$$\tau(t) = t - \Lambda_0^{-1}\big(\Lambda_0(t) - \ln \lambda_0(t)\big).$$

ii) *Sufficiency.* If there exists a function $\lambda_0(t)$ such that (2.1.37) holds. Then

$$\Lambda_0\big(t - \tau(t)\big) \geq \Lambda_0(t) - \ln \lambda_0(t),$$

and

$$\lambda_0(t) \geq \exp\Big(\int_{t-\tau(t)}^{t} \lambda_0(s)\,ds\Big).$$

Define

$$\lambda_1(t) = \begin{cases} \exp\Big(\int_{t-\tau(t)}^{t} \lambda_0(s)ds\Big), & t \geq t_0 \\ \lambda_1(t_0) + \lambda_0(t) - \lambda_0(t_0), & t \in [T_0, t_0). \end{cases}$$

Clearly, $\lambda_1(t) \leq \lambda_0(t), \quad t \geq T_0$ and $0 \leq \lambda_1(t) \leq \lambda_0(t), \quad t \geq t_0$.

In general, we define

$$\lambda_n(t) = \begin{cases} \exp\left(\int_{t-\tau(t)}^{t} \lambda_{n-1}(s)ds\right), & t \geq t_0 \\ \lambda_n(t_0) + \lambda_0(t) - \lambda_0(t_0), & t \in [T_0, t_0). \end{cases} \tag{2.1.40}$$

Thus

$$\lambda_0(t) - \lambda_0(t_0) \leq \lambda_n(t) \leq \lambda_{n-1}(t) \leq \cdots \leq \lambda_0(t), \quad t \geq T_0,$$

and $\lambda_n(t) \geq 0, \quad t \geq t_0$. Then $\lim_{n\to\infty} \lambda_n(t) = \lambda(t), \quad t \geq T_0$ exists and

$$\lim_{n\to\infty} \int_{t-\tau(t)}^{t} \lambda_n(s)ds = \int_{t-\tau(t)}^{t} \lambda(s)ds, \quad t \geq t_0.$$

Hence

$$\lambda(t) = \exp \int_{t-\tau(t)}^{t} \lambda(s)\,ds, \quad t \geq t_0.$$

Set

$$x(t) = \exp\left(- \int_{T_0}^{t} \lambda(s)\,ds \right), \quad t \geq T_0. \tag{2.1.41}$$

Then x is a positive solution of (2.1.36) with respect to t_0. □

Remark 2.1.3. We take $\lambda(t) = \lambda > 0$ in Theorem 2.1.3. Then condition (2.1.37) becomes

$$\tau(t) \leq \frac{\ln \lambda}{\lambda}, \quad t \geq t_0. \tag{2.1.42}$$

In particular, set $\lambda = e$. (2.1.42) becomes

$$\tau(t) \leq \frac{1}{e}, \qquad t \geq t_0, \tag{2.1.43}$$

i.e., (2.1.43) is a sufficient condition for the existence of positive solutions of (2.1.36).

Let $t_0 = \frac{1}{2}$, $\lambda(t) = 2t$. Then by Theorem 2.1.3, if

$$\tau(t) = t - \sqrt{t^2 - \ln 2t} \tag{2.1.44}$$

then Eq. (2.1.36) has a positive solution with respect to $t_0 = \frac{1}{2}$. In fact, $x(t) = e^{-t^2}$ is such a solution. We note that

$$\tau(\frac{e}{2}) = \frac{e}{2} - \sqrt{\left(\frac{e}{2}\right)^2 - 1} > \frac{1}{e} . \tag{2.1.45}$$

This example shows that (2.1.43) is not necessary for the existence of a positive solution of (2.1.36).

We now consider the linear equation of the form

$$x'(t) + p(t)x\big(t - \tau(t)\big) = 0 \tag{2.1.46}$$

where $p,\ \tau \in C([t_0, \infty), R_+)$, $\tau(t) \le t$ and $\lim_{t\to\infty} \big(t - \tau(t)\big) = \infty$.

Set $T_0 = \inf_{t \ge t_0} \{t - \tau(t)\}$.

Applying a similar idea to Theorem 2.1.3 we can prove the following results.

Theorem 2.1.4. *Eq. (2.1.46) has a positive solution with respect to t_0 if and only if there exists a continuous function $\lambda_0(t)$ on $[T_0, \infty)$ such that $\lambda_0(t) > 0$ for $t \ge t_0$ and*

$$\lambda_0(t) \ge p(t) \exp \int_{t-\tau(t)}^{t} \lambda_0(s)\, ds, \qquad t \ge t_0. \tag{2.1.47}$$

Remark 2.1.4. If $p(t) > 0$, then (2.1.47) can be replaced by

$$\tau(t) \le t - \Lambda_0^{-1}\left(\Lambda_0(t) - \ln \frac{\lambda_0(t)}{p(t)}\right), \quad t \ge t_0. \tag{2.1.48}$$

Corollary 2.1.1. *If*

$$\int_{t-\tau(t)}^{t} p(s)ds \le \tfrac{1}{e}\,, \quad t \ge t_0. \tag{2.1.49}$$

Then Eq. (2.1.46) has a positive solution with respect to t_0.

Proof: We take $\lambda_0(t) = ep(t)$, then (2.1.47) is satisfied. Then the corollary follows from Theorem 2.1.4. □

Theorem 2.1.5. *Assume that $\tau(t) \equiv \tau > 0$, $\int_{t_0}^{\infty} p(t)dt = \infty$. Then Eq. (2.1.46) has a positive solution with respect to t_0 if and only if there exists a continuous function $\lambda_0(t)$ on $[t_0 - \tau, \infty)$ such that*

$$\int_{t-\tau}^{t} p(u)du \leq t - \Lambda_0^{-1}\big(\Lambda_0(t) - \ln \lambda_0(t)\big), \quad t \geq t_0. \tag{2.1.50}$$

Proof: Set $u = \sigma(t) = \int_{t_0}^{t} p(s)ds, \quad t \geq t_0$. Then

$$t - \tau = \sigma^{-1}\left(u - \int_{\sigma^{-1}(u)-\tau}^{\sigma^{-1}(u)} p(s)ds\right).$$

Denote

$$z(u) = x\big(\sigma^{-1}(u)\big).$$

Then (2.1.46) becomes

$$z'(u) + z\left(u - \int_{\sigma^{-1}(u)-\tau}^{\sigma^{-1}(u)} p(s)\, ds\right) = 0. \tag{2.1.51}$$

By Theorem 2.1.3, (2.1.50) is a necessary and sufficient condition for Eq. (2.1.51) to have a positive solution with respect to 0. From the transform, it is equivalent to that Eq. (2.1.46) has a positive solution with respect to t_0. □

Remark 2.1.5. If we choose $\lambda_0(t) \equiv e$ in (2.1.50), then we obtain

$$\int_{t-\tau}^{t} p(u)\, du \leq \frac{1}{e}, \qquad t \geq t_0. \tag{2.1.52}$$

As we have mentioned, (2.1.52) is a sufficient condition and is not a necessary condition for the existence of a positive solution of (2.1.46).

Combining Theorem 2.1.1 and (2.1.52) we obtain the following corollary.

Corollary 2.1.2. *Let $p(t) \equiv p > 0$, $\tau(t) \equiv \tau > 0$. Then a necessary and sufficient condition for all solutions of (2.1.46) to be oscillatory is that $p\tau e > 1$.*

Remark 2.1.6. The above techniques can be used on the first order advanced type equations

$$x'(t) = x\big(t + \tau(t)\big) \tag{2.1.53}$$

and

$$x'(t) = p(t)x\big(t + \tau(t)\big) \tag{2.1.54}$$

where $p,\ \tau \in C([t_0, \infty), R_+)$. For example, (2.1.53) has a positive solution if and only if there exists a continuous function $\lambda \in C(T_0, \infty), R)$ such that

$$\tau(t) \le \Lambda^{-1}\big(\Lambda(t) + \ln\ \lambda(t)\big) - t \tag{2.1.55}$$

where $\Lambda(t) = \int_{t_0}^{t} \lambda(s)ds$. If we let $\lambda(t) \equiv \lambda > 0$, then (2.1.55) becomes $\tau(t) \le \frac{1}{e}$, which is a sufficient condition for the existence of positive solutions of (2.1.53).

For Eq. (2.1.54), assume

$$\liminf_{t\to\infty} \int_{t}^{t+\tau(t)} p(s)ds > \frac{1}{e}\,. \tag{2.1.56}$$

Then every solution of (2.1.54) oscillates.

If $\tau(t) \equiv \tau > 0$, then (2.1.54) has a positive solution if and only if there exists a continuous function $\lambda(t)$ such that

$$\int_{t}^{t+\tau} p(u)\,du \le \Lambda^{-1}\big(\Lambda(t) + \ln\ \lambda(t)\big) - t. \tag{2.1.57}$$

Corollary 2.1.2 is also true for Eq. (2.1.54).

2.2. The Distribution of Zeros of Oscillatory Solutions

In this section we will estimate the distances between adjacent zeros of oscillatory solutions of the first order delay differential equation (2.1.46).

Let us first consider the differential inequality

$$x'(t) + x(t - \tau) \le 0 \tag{2.2.1}$$

where $\tau > 0$ is a constant.

Lemma 2.2.1. *Assume that $\tau > \frac{1}{e}$. Let $x(t)$ be a solution of (2.2.1) on $[T_x, +\infty)$. If there exists $T > t_0 \geq T_x$ such that $x(t) > 0$ for $t \in [t_0, T]$, then*

$$\omega(T) > 1 + \frac{n}{\tau}(1 + \ln \tau), \quad \text{if} \quad T \geq t_0 + (n+2)\tau, \tag{2.2.2}$$

where $\omega(t) = \frac{x(t-\tau)}{x(t)}$, n is a nonnegative integer.

Proof: If $n = 0$, from (2.2.1), $x'(t) < 0$ for $t \in [t_0 + \tau, T]$. Therefore $\omega(t) = \frac{x(T-\tau)}{x(T)} > 1$ for $T \geq t_0 + 2\tau$.

If $n \geq 1$, dividing (2.2.1) by $x(t)$ and integrating it from $t - \tau$ to t_0 we have

$$\ln \frac{x(t-\tau)}{x(t)} \geq \int_{t-\tau}^{t} \frac{x(s-\tau)}{x(s)}\, ds. \tag{2.2.3}$$

Hence

$$\omega(t) \geq \exp \int_{t-\tau}^{t} \omega(s) ds \geq \exp\big(\tau \cdot \min_{t-\tau \leq s \leq t} \omega(s)\big). \tag{2.2.4}$$

Set $\omega(\xi_0) = \min_{T-t \leq s \leq T} \omega(s), \quad \xi_0 \in [T-\tau, T]$, and noting that

$$e^{rx} \geq x + \frac{1}{r}(1 + \ln r), \quad \text{for} \quad r > 0, \quad x \in R, \tag{2.2.5}$$

we have

$$\omega(T) \geq \exp\left[\tau \cdot \omega(\xi_0)\right] \geq \omega(\xi_0) + \frac{1}{\tau}(1 + \ln \tau). \tag{2.2.6}$$

Define a sequence $\{\xi_0, \xi_1, \ldots, \xi_{n-1}\}$ by

$$\omega(\xi_i) = \{\min \omega(s), \quad s \in [\xi_{i-1} - \tau, \xi_{i-1}]\}, \quad i = 1, 2, \ldots, n-1. \tag{2.2.7}$$

Clearly $\xi_i \in [\xi_{i-1} - \tau, \xi_{i-2}]$. In view of (2.2.4) and (2.2.5) we have

$$\omega(\xi_{i-1}) \geq \omega(\xi_i) + \frac{1}{\tau}(1 + \ln \tau). \tag{2.2.8}$$

From (2.2.6) and (2.2.8) we obtain

$$\omega(\tau) \geq \omega(\xi_0) + \frac{1}{\tau}(1+\ln\tau) \geq \omega(\xi_1) + \frac{2}{\tau}(1+\ln\,\tau)$$
$$\geq \cdots \geq \omega(\xi_{n-1}) + \frac{n}{\tau}(1+\ln\,\tau).$$

It is easy to see that $t_0 + 2\tau \leq \xi_{n-1} \leq t$ and hence $\omega(\xi_{n-1}) > 1$ by the first step. Therefore (2.2.2) holds. □

Theorem 2.2.1. *Assume that* $\tau > \frac{1}{e}$, $x(t)$ *is a solution of (2.2.1) on* $[T_x, +\infty)$, *and* I *is any closed interval in* $[T_x, +\infty)$ *with length* $n_0\tau$. *Then* $x(t)$ *cannot be always positive on* I, *where*

$$n_0 = \max\left\{4, 3 + \left[\frac{2-2\tau-\tau^2}{\tau(1+\ln\,\tau)}\right]\right\}, \tag{2.2.9}$$

and $[\cdot]$ *denotes the greatest integer function.*

Proof: Assume the contrary, then there is a $t_1 \geq T_x$, such that $x(t) > 0$ on $[t_1, t_1 + n_0\tau]$. Set $t^* = t_1 + n_0\tau$. Integrating (2.2.1) from $t^* - \tau$ to t^* we have

$$x(t^* - \tau) - x(t^*) \geq \int_{t^*-2\tau}^{t^*-\tau} x(s)ds.$$

Define a line ℓ as follows:

$$z(t) - x(t^* - \tau) = k(t - t^* + \tau)$$

where $k = -x(t^* - 2\tau)$. Hence

$$z(t^* + 2\tau) = x(t^* - \tau) + \tau \cdot x(t^* - 2\tau). \tag{2.2.10}$$

From (2.2.1), $x'(t^* - \tau) < k$, and so $x(t)$ lies above the line ℓ at a small left-neighborhood of the point $t = t^* - \tau$. If there is a point t on $[t^* - 2\tau, t^* - \tau]$ so that $x(t)$ lies below ℓ, then there exists $\xi \in [t^* - 2\tau, t^* - \tau)$ such that $x'(\xi) \geq k$. Hence

$$-x(\xi - \tau) > x'(\xi) \geq k = -x(t^* - 2\tau),$$

i.e.,

$$x(\xi - \tau) \le x(t^* - 2\tau). \tag{2.2.11}$$

Noting that $n_0 \ge 4$ and hence $t^* - 2\tau > \xi - \tau \ge t^* - 3\tau \ge t_1 + \tau$. On the other hand, $x(t)$ is decreasing for $t \ge t_1 + \tau$, so $x(\xi - \tau) > x(t^* - 2\tau)$, which contradicts (2.2.11). Therefore $x(t)$ lies above the line ℓ on $[t^* - 2\tau, t^* - \tau]$.

We see that

$$\int_{t^*-2\tau}^{t^*-\tau} x(s)ds > \frac{x(t^* - \tau) + z(t^* - 2\tau)}{2}\tau.$$

Using (2.2.10) we have

$$\int_{t^*-2\tau}^{t^*-\tau} x(s)ds > \frac{2x(t^* - \tau) + \tau x(t^* - 2\tau)}{2}\tau. \tag{2.2.12}$$

By Lemma 2.2.1

$$x(t^* - 2\tau) > \left\{1 + \frac{n_0 + 3}{\tau}(1 + \ln \tau)\right\} x(t^* - \tau). \tag{2.2.13}$$

Combining (2.2.12) and (2.2.13) we obtain

$$\int_{t^*-2\tau}^{t^*-\tau} x(s)ds > \frac{2 + \tau\left(1 + \frac{n_0+3}{\tau}(1 + \ln \tau)\right)}{2}\tau \cdot x(t^* - \tau). \tag{2.2.14}$$

Integrating (2.2.1) from $t^* - \tau$ to t^* and using (2.2.14) we obtain

$$\left\{1 - \frac{2 + \tau\left(1 + \frac{n_0-3}{\tau}(1 + \ln \tau)\right)}{2}\tau\right\} x(t^* - \tau) - x(t^*) > 0$$

and hence

$$\left\{1 - \frac{2 - \tau\left(1 + \frac{n_0-3}{\tau}(1 + \ln \tau)\right)}{2}\tau\right\} > 0$$

which contradicts the definition of n_0 as given in (2.2.9). □

From Theorem 2.2.1 and Lemma 2.2.1 we have the following corollary.

Corollary 2.2.1. *Assume that $\tau > \frac{1}{e}$, $x(t)$ is a solution of (2.2.1) and positive and decreasing on $[t_1, t_1 + \tau]$, then $x(t)$ can not always be positive on $[t_1 + \tau, t_1 + (n_0 - 1)\tau]$.*

Then we consider

$$x'(t) + p(t)x(t - \tau) = 0. \tag{2.2.15}$$

By the method of Theorem 2.1.3 and 2.1.5 the results of (2.2.1) can be extended to Eq. (2.2.15)

Theorem 2.2.1. *Assume that*

(i) $p \in C[t_0, \infty), (0, \infty))$, $p(t) \not\equiv 0$ *on any subinterval of* $[t_0, \infty)$,

(ii)
$$\liminf_{t\to\infty} \int_{t-\tau}^{t} p(s)ds = k > \frac{1}{e}\,. \tag{2.2.16}$$

Let $x(t)$ be a solution of (2.2.15) on $[T_x, +\infty)$, $T_x \geq t_0$. Let $\varepsilon > 0$ be a small number such that $k - \varepsilon > \frac{1}{e}$. Then there exists $t_\varepsilon \geq T_x$ such that the distances between adjacent zeros of $x(t)$ is less than $(1 + n_\varepsilon)\tau$ on $[t_\varepsilon, +\infty)$, where

$$n_\varepsilon = \max\left\{4, \quad 3 + \left[\frac{2 - 2(k-\varepsilon) - (k-\varepsilon)^2}{(k-\varepsilon)(1 + \ln(k-\varepsilon))}\right]\right\}. \tag{2.2.17}$$

Proof: In view of the assumptions there exists $t_\varepsilon \geq t_0$ such that

$$\int_{t-\tau}^{t} p(s)ds > k - \varepsilon \quad \text{for} \quad t \geq t_\varepsilon. \tag{2.2.18}$$

Set $u = \sigma(t) = \int_{t_0}^{t} p(s)ds$ and $z(u) = x(\sigma^{-1}(u))$. As in the proof of Theorem 2.1.4, (2.2.15) reduces to

$$z'(u)z\left(u - \int_{\sigma^{-1}(u)-\tau}^{\sigma^{-1}(u)} p(s)ds\right) = 0. \tag{2.2.19}$$

We shall show that $x(t)$ has zeros on any interval $[t_1, t_1 + (1 + n_\varepsilon)\tau]$, $t_1 \geq t_\varepsilon$. If not, $x(t) > 0$ on $[t_1, t_1 + (1 + n_\varepsilon)\tau]$, then $x'(t) < 0$ on $[t_1 + \tau, t_1 + (1 + n_\varepsilon)\tau]$. Set

$$u_1 = \sigma(t_1), \quad u_2 = \sigma(t_1 + \tau), \quad u_3 = \sigma(t_1 + 2\tau).$$

Then $z(u) > 0$ for $u \geq u_1$ and $z'(u) < 0$ for $u \geq u_2$ and

$$u - \int_{\sigma^{-1}(u)-\tau}^{\sigma^{-1}(u)} p(s)ds \geq u_2 \quad \text{for} \quad u \geq u_3.$$

Since

$$\int_{\sigma^{-1}(u)-\tau}^{\sigma^{-1}(u)} p(s)ds > k - \varepsilon, \quad \text{for} \quad u \geq u_1,$$

we have

$$z'(u) + z\big(u - (k-\varepsilon)\big) < 0, \quad \text{for} \quad u \geq u_3.$$

$z(u)$ is positive and decreasing for $u \geq u_3$. By Corollary 2.2.1, $z(u)$ has zeros on $[u_3, u_3 + (n_\varepsilon - 1)(k-\varepsilon)]$. Hence there exists $\widetilde{u} \in [u_3, u_3 + (n_\varepsilon - 1)(k-\varepsilon)]$ such that $z(\widetilde{u}) = 0$. Set $\widetilde{t} = \sigma^{-1}(\widetilde{u})$, then $x(\widetilde{t}) = 0$.

On the other hand, from (2.2.18) we have

$$\int_{t_1+2\tau}^{t_1+(n_\varepsilon+1)\tau} p(s)ds > (n_\varepsilon - 1)(k-\varepsilon).$$

Hence

$$\int_{t_0}^{t_1+(n_\varepsilon+1)\tau} p(s)ds > \int_{t_0}^{\widetilde{t}} p(s)ds > \int_{t_0}^{t_1+2\tau} p(s)ds,$$

which derives that

$$t_1 < t_1 + 2\tau \leq \widetilde{t} < t_1 + (1+n_\varepsilon)\tau.$$

This implies that $x(\widetilde{t}) > 0$, contradicting that $x(\widetilde{t}) = 0$.

Remark 2.2.1. Consider the autonomous equation

$$x'(t) + px(t-\tau) = 0, \quad p > 0, \quad \tau > 0, \tag{2.2.20}$$

where $p\tau > \frac{1}{e}$. Then according to Theorem 2.2.2 the distance between adjacent zeros of solutions of (2.2.20) is less than

$$\tau \max\left\{5, 4 + \left[\frac{2 - 2p\tau - (p\tau)^2}{p\tau(1+\ln(p\tau))}\right]\right\}. \tag{2.2.21}$$

2.3. Unstable Type Equations

We now consider the first order linear delay differential equation with unstable type of the form

$$x'(t) = p(t)x(\tau(t)) \tag{2.3.1}$$

where $p,\ \tau \in C([t_0,\infty), R_+)$, $\tau(t) \le t$ and $\lim_{t\to\infty} \tau(t) = \infty$.

Theorem 2.3.1. *Assume that*

$$\int_{t_0}^{\infty} p(t)dt = \infty. \tag{2.3.2}$$

Then Eq. (2.3.1) always has an unbounded positive solution, and every bounded solution of (2.3.1) is oscillatory.

Proof: The proof of the first part can be seen from Theorem 3.5.1. Now we will prove that every bounded solution of (2.3.1) is oscillatory. Let $x(t)$ be a bounded positive solution of (2.3.1), then $x(t)$ is nondecreasing and bounded. Hence $\lim\limits_{t\to\infty} x(t) = \ell > 0$ exists. From (2.3.1), we obtain

$$x'(t) \ge \frac{\ell}{2}\, p(t) \quad \text{for all large} \quad t. \tag{2.3.3}$$

(2.3.1) and (2.3.2) imply that $\lim x(t) = \infty$, contradicting the boundedness of $x(t)$. □

Example 2.3.1. Consider the equation

$$y'(t) = 2ty\,\Big(\sqrt{t^2 - \frac{3\pi}{2}}\,\Big). \tag{2.3.4}$$

By Theorem 2.3.1, every bounded solution of (2.3.4) is oscillatory. In fact, $y(t) = \cos t^2$ is such a solution. On the other hand, $y(t) = e^{\lambda t^2}$ is an unbounded positive solution of (2.3.4) where λ is the unique positive root of the equation

$$\lambda = \exp\,(-\frac{3}{2}\,\pi\lambda).$$

Theorem 2.3.2. *Assume that*

$$\int_{t_0}^{\infty} p(t)dt < \infty. \tag{2.3.5}$$

Then Eq. (2.3.1) has a bounded positive solution.

Proof: Let BC be a Banach space of bounded and continuous functions $x:\ [t_0, \infty) \to R$ with the super norm. Define a subset Ω of BC as follows:

$$\Omega = \{x \in BC: \ 0 \le x(t) \le 1, \quad t_0 \le t < \infty\}.$$

Clearly, Ω is a bounded, closed and convex subset of BC.

Define a mapping $\mathcal{T}: \Omega \to BC$ as follows:

$$(\mathcal{T}x)(t) = \begin{cases} \frac{1}{2} + \int_T^t p(s)x(\tau(s))ds, & t \ge T \\ (\mathcal{T}x)(T), & t_0 \le t \le T. \end{cases} \tag{2.3.6}$$

where T is chosen sufficiently large so that $\tau(t) \ge t_0$ for $t \ge T$ and $\int_T^\infty p(s)ds \le \frac{1}{2}$.

Clearly, for $x \in \Omega$, $0 \le (\mathcal{T}x)(t) \le 1$, i.e., $\mathcal{T}\Omega \subset \Omega$. Let $x_1, x_2 \in \Omega$, then

$$\begin{aligned} |\mathcal{T}x_1)(t) - (\mathcal{T}x_2)(t)| &\le \int_T^t p(s)|x_1(\tau(s)) - x_2(\tau(s))|ds \\ &\le \frac{1}{2}\,\|x_1 - x_2\|, \quad \text{for} \quad t \ge T \end{aligned}$$

and hence

$$\|\mathcal{T}x_1 - \mathcal{T}x_2\| \le \frac{1}{2}\,\|x_1 - x_2\|,$$

i.e., $\mathcal{T}$ is a contraction on Ω. Therefore by Banach fixed point theorem there is a $x \in \Omega$ so that

$$x(t) = \frac{1}{2} + \int_T^t p(s)x(\tau(s))ds, \quad t \ge T$$

and so $x'(t) = p(t)x(\tau(t)), \quad t \ge T$. Hence x is a bounded positive solution of (2.3.1). □

Corollary 2.3.1. *Every bounded solution of (2.3.1) oscillates if and only if (2.3.2) holds.*

Corollary 2.3.2. *Eq. (2.3.1) has an unbounded positive solution if and only if (2.3.2) holds.*

For the existence of oscillatory solutions of (2.3.1) we have the following result.

Theorem 2.3.3. *If $\tau(t) = t - \tau_1(t)$ in (2.3.1) and*

$$p(t_0)\tau_1(t_0) > 0 \tag{2.3.7}$$

then Eq. (2.3.1) has an oscillatory solution.

We consider (2.3.1) with $p(t) \equiv p > 0, \quad \tau(t) = t - \tau, \quad \tau > 0$.

By the analysis of the characteristic roots we have the following conclusion.

Theorem 2.3.4. *Assume that $p > 0, \quad \tau > 0$, then*

(i) *if*

$$p\tau < \frac{3}{2}\pi$$

then every oscillatory solution $x(t)$ of (2.3.1) tends to zero as $t \to \infty$;

(ii) *if*

$$p\tau = \frac{3}{2}\pi$$

then every oscillatory solution of (2.3.1) is bounded on R_+, and (2.3.1) has an oscillatory solution which does not tend to zero as $t \to \infty$.

(iii) *if*

$$p\tau > \frac{3}{2}\pi$$

then (2.3.1) has an unbounded oscillatory solution.

The proof of Theorem 2.3.3 and 2.3.4 can be seen from the original paper [72].

2.4. Equations with Oscillatory Coefficients

We consider the linear differential equation with deviating argument of the form

$$x'(t) + p(t)x(\tau(t)) = 0, \tag{2.4.1}$$

where $p,\ \tau \in C([t_0,\infty), R)$, τ is nondecreasing, $\lim_{t\to\infty} \tau(t) = \infty$, and p may change signs infinitely many times as $t \to \infty$.

We will use the following notations:

$$\tau^0(t) = t, \quad \tau^i(t) = \tau(\tau^{i-1}(t)), \quad i = 1, 2, \ldots \tag{2.4.2}$$

where $\tau^i(t)$ is defined on

$$E_i = \{t : \tau^{i-1}(t) \geq t_0\}, \quad i = 1, 2, \ldots . \tag{2.4.3}$$

Clearly, $\lim_{t\to\infty} \tau^i(t) = \infty, \quad i = 1, 2, \ldots .$

Lemma 2.4.1. *Assume that there exists a sequence $\{b_n\}$ and a positive integer $k \geq 3$ such that*

(i) $\lim_{n\to\infty} b_n = \infty, \quad \tau^k(b_n) < b_n, \quad n = 0, 1, \ldots;$

(ii) $p(t) \geq 0, \quad \tau(t) < t$ *for* $t \in [\tau^k(b_n), b_n], \quad n = 0, 1, \ldots;$

(iii) *there exists* $m_k \in [0, 1)$ *such that*

$$m_k \leq \int_{\tau(t)}^{t} p(s)ds, \quad t \in [\tau^{k-2}(b_n), b_n], \quad n = 0, 1, 2, \ldots .$$

Let $x(t)$ be a solution of the differential inequality

$$x'(t) + p(t)x(\tau(t)) \leq 0 \tag{2.4.4}$$

such that $x(t) > 0$ for $t \in [\tau^{(k+1)}(b_n), b_n]$. Then

$$\frac{x(t)}{x(\tau(t))} > \alpha_i \quad \text{for} \quad t \in [\tau^{k-2}(b_n), \tau^{i+1}(b_n)], \tag{2.4.5}$$

$i = 1, \ldots, k-3$, where

$$\alpha_0 = \frac{m_k^2}{2(1-m_k)}, \qquad \alpha_i = \alpha_{i-1}^2 + m_k \alpha_{i-1} + \tfrac{1}{2}\, m_k^2, \tag{2.4.6}$$

$i = 1, 2, \ldots, k-3$.

Proof: If $m_k = 0$, then $\alpha_i = 0, \quad i = 0, 1, \ldots, k-3$. Hence (2.4.5) holds. Now let us consider the case $0 < m_k < 1$. For any $t \in [\tau^{k-2}(b_n),\ \tau(b_n)]$, there exists $t^* \in [\tau^{k-2}(b_n), b_n]$ such that $\tau(t^*) \leq t < t^*$ and

$$\int_{\tau(t^*)}^{t^*} p(s)ds \geq m_k = \int_t^{t^*} p(s)ds. \tag{2.4.7}$$

From (2.4.4), we have

$$x'(t) \leq -p(t)x\big(\tau(t)\big) \leq 0, \quad t \in [\tau^k(b_n), b_n]. \tag{2.4.8}$$

Integrating (2.4.4) from t to t^* we have

$$x(t) \geq x(t^*) + \int_t^{t^*} p(s)x\big(\tau(s)\big)ds. \tag{2.4.9}$$

We note that $\tau^{k-1}(b_n) \leq \tau(t) \leq \tau(s) \leq \tau(t^*) \leq t \leq \tau(b_n)$ for $s \in [t, t^*]$ and integrating (2.4.4) from $\tau(s)$ to t we obtain

$$\begin{aligned}
x\big(\tau(s)\big) &\geq x(t) + \int_{\tau(s)}^{t} p(u)x\big(\tau(u)\big)du \\
&\geq x(t) + x\big(\tau(t)\big) \int_{\tau(s)}^{t} p(u)du \\
&= x(t) + x\big(\tau(t)\big)\left[\int_{\tau(u)}^{s} p(u)du - \int_t^s p(u)du\right] \\
&\geq x(t) + x\big(\tau(t)\big)\left[m_k - \int_t^s p(u)du\right].
\end{aligned} \tag{2.4.10}$$

Substituting (2.4.10) into (2.4.9) we have

$$\begin{aligned} x(t) &\geq x(t^*) + \int_t^{t^*} p(s)\Big\{x(t) + x\big(\tau(t)\big)\Big[m_k - \int_t^s p(u)du\Big]\Big\}ds \\ &= x(t^*) + m_k x(t) + x\big(\tau(t)\big)\Big[m_k^2 - \int_t^{t^*} ds \int_t^s p(s)p(u)du\Big]. \end{aligned} \tag{2.4.11}$$

By exchanging the order of the integral we have

$$\int_t^{t^*} ds \int_t^s p(s)p(u)du = \int_t^{t^*} du \int_u^{t^*} p(s)p(u)ds = \int_t^{t^*} ds \int_s^{t^*} p(s)p(u)du,$$

and hence

$$\begin{aligned} \int_t^{t^*} ds \int_t^s p(s)p(u)du &= \frac{1}{2}\Big[\int_t^{t^*} \int_t^s p(s)p(u)du + \int_t^{t^*} \int_s^{t^*} p(s)p(u)du\Big] \\ &= \tfrac{1}{2}\, m^2 k. \end{aligned}$$

Substituting this into (2.4.11) we obtain

$$x(t) \geq x(t^*) + m_k x(t) + \tfrac{1}{2}\, m_k^2 x\big(\tau(t)\big). \tag{2.4.12}$$

Thus

$$\frac{x(t)}{x\big(\tau(t)\big)} > \frac{m_k^2}{2(1 - m_k)} = \alpha_0, \quad t \in [\tau^{(k-2)}(b_n), \tau(b_n)], \tag{2.4.13}$$

i.e., (2.4.5) holds for $i = 0$. For $i = 1$, $t \in [\tau^{k-2}(b_n), \tau^2(b_n)]$, there exists $t^* \in [\tau^{k-2}(b_n), \tau(b_n)]$ such that $\tau(t^*) \leq t < t^*$ and (2.4.7) holds. By (2.4.13), we have

$$\frac{x(t^*)}{x\big(\tau(t^*)\big)} > \alpha_0. \tag{2.4.14}$$

By the same method as in the above proof we can show that (2.4.13) holds.

In view of the monotonic property of x we have

$$x(t^*) > \alpha_0 x(t) > \alpha_0^2 x\big(\tau(t)\big). \tag{2.4.15}$$

Substituting (2.4.15) into (2.5.12) we obtain

$$x(t) > \alpha_0^2 x\big(\tau(t)\big) + \alpha_0 m_k x\big(\tau(t)\big) + \tfrac{1}{2} m_k^2 x\big(\tau(t)\big) = \alpha_1 x\big(\tau(t)\big), \tag{2.4.16}$$

i.e., (2.4.5) holds for $i = 1$. By induction, we see (2.4.5) holds for $i = 0, 1, \ldots, k-3$.

□

Lemma 2.4.2. *In addition to the assumptions of Lemma 2.4.1, further assume that for some $i \in \{0, 1, \ldots, k-3\}$*

$$\int_{\tau(t)}^{t} p(s)ds \geq 1 - \alpha_i, \quad t \in [\tau^{k-2}(b_n), \tau^{i+1}(b_n)]. \tag{2.4.17}$$

Then (2.4.4) has no solution which is positive on $[\tau^{k+1}(b_n), b_n]$.

Proof: If not, there is a solution $x(t)$ of (2.4.4) with $x(t) > 0$ for $t \in [\tau^{k+1}(b_n), b_n]$. By Lemma 2.4.1 $x(t) > \alpha_i x\big(\tau(t)\big)$, $\quad t \in [\tau^{k-2}(b_n), \tau^{i+1}(b_n)]$ and $x'(t) \leq 0$, $t \in [\tau^k(b_n), b_n]$. Integrating (2.4.4) from $\tau(t)$ to t for $t \in [\tau^{k-2}(b_n), \tau^{i+1}(b_n)]$ we obtain

$$-x(t) + x\big(\tau(t)\big) \geq \int_{\tau(t)}^{t} p(s)x\big(\tau(s)\big)ds \geq x\big(\tau(t)\big) \int_{\tau(t)}^{t} p(s)ds.$$

Hence

$$\int_{\tau(t)}^{t} p(s)ds \leq 1 - \frac{x(t)}{x\big(\tau(t)\big)}, \quad t \in [\tau^{k-2}(b_n), \tau^{i+1}(b_n)].$$

In view of that $x(t) > \alpha_i x\big(\tau(t)\big)$ for $t \in [\tau^{k-2}(b_n), \tau^{i+1}(b_n)]$, we obtain

$$\int_{\tau(t)}^{t} p(s)ds < 1 - \alpha_i; \quad t \in [\tau^{k-2}(b_n), \tau^{i+1}(b_n)], \tag{2.4.18}$$

which contradicts (2.4.17). □

Similarly, we have the following lemma.

Lemma 2.4.3. *Assume the assumptions of Lemma 2.4.2 hold. Then the* *differential inequality*

$$x'(t) + p(t)x(\tau(t)) \geq 0 \tag{2.4.19}$$

has no solution $x(t)$ with $x(t) < 0$ on $[\tau^{k+1}(b_n), b_n]$.

Combining Lemma 2.4.2 and Lemma 2.4.3 we obtain the following oscillation criterion.

Theorem 2.4.1. *Under the assumptions of Lemma 2.4.2 every solution of Eq. (2.4.1) has at least one zero on $[\tau^{k+1}(b_n), b_n]$. Consequently, every solution of (2.4.1) is oscillatory.*

Remark 2.4.1. The first part of Theorem 2.1.2 can follow from Theorem 2.4.1. In fact, the assumptions of Lemma 2.4.2 are satisfied in this case.

Example 2.4.1. Consider the equation

$$x'(t) + 0.8x(t - \sin t) = 0 \tag{2.4.20}$$

where $\tau(t) = t - \sin t$ is nondecreasing. Let b be the real root of the equation

$$b = \tfrac{\pi}{2} + \sin b$$

which lies on the interval $(\frac{\pi}{2}, \pi)$. It is easy to see that $b \in (\frac{\pi}{2}, \frac{3}{4}\pi)$. Let $b_n = 2n\pi + b$, $n = 0, 1, 2, \ldots$. Then $\tau(b_n) = 2\pi + \frac{\pi}{2}$, $\tau^2(b_n) = 2n\pi + \frac{\pi}{2} - 1$, $\tau^3(b_n) = 2n\pi + \frac{\pi}{2} - 1 - \cos 1 > 2n\pi$. Clearly $\tau^3(b_n) < b_n$ and $\tau(t) < t$ for $t \in (\tau(b_n), b_n)$. Let $k = 3$, then the condition (i) of Lemma 2.4.1 is satisfied and

$$\int_{\tau(t)}^{t} 0.8ds = 0.8 \sin t \geq (0.4)\sqrt{2} = m_3, \quad t \in [\tau(b_n), b_n].$$

On the other hand, we see that

$$\int_{\tau^2(b_n)}^{\tau(b_n)} 0.8ds = 0.8 > 0.63161 \geq 1 - \alpha_0.$$

That is, all assumptions of Lemma 2.4.2 are satisfied. Therefore, by Theorem 2.5.1, every solution of (2.4.20) is oscillatory.

Example 2.4.2. Consider the equation

$$x'(t) + \alpha \sin\frac{t}{2}\, x(t - \frac{2\pi}{3}) = 0 \tag{2.4.21}$$

where $\alpha \in (\frac{3-\sqrt{3}}{3}, \frac{1}{2})$.

Set $k = 3$, $b_n = 2(2n-1)\pi$. Then the assumptions of Lemma 2.5.1 hold and

$$\alpha \le \int_{t-\frac{\pi}{2}}^{t} \alpha \sin\frac{s}{2}\, ds, \quad t \in [b_n - \frac{2\pi}{3}, b_n],$$

and

$$\int_{b_n - \frac{4\pi}{3}}^{b_n - \frac{2\pi}{3}} \alpha \sin\frac{s}{2}\, ds = 2\alpha > 1 - \frac{\alpha^2}{2(1-\alpha)},$$

i.e., all assumptions of Lemma 2.4.2 are satisfied. Therefore by Theorem 2.4.1 every solution of (2.4.21) is oscillatory.

2.5. Equations with Positive and Negative Coefficients

We consider the linear delay differential equation with nonpositive and nonnegative coefficients of the form

$$x'(t) + p(t)x(t-\tau) - Q(t)x(t-\sigma) = 0 \tag{2.5.1}$$

where $p;\ Q \in C([t_0, \infty), R_+)$, $\tau, \sigma \in (0, \infty)$.

The following is an oscillation criterion for (2.5.1).

Theorem 2.5.1. *Assume that* $\tau \ge \sigma \ge 0$ *and*

$$\bar{p}(t) = p(t) - Q(t + \sigma - \tau) \ge 0, \quad (\not\equiv 0), \quad t \ge t_1 \ge t_0,$$
$$\liminf_{t\to\infty} \int_{t-\tau}^{t} \bar{p}(s)ds > 0,$$

and for every $\lambda > 0$

$$\liminf_{t\to\infty} \left[\frac{1}{\lambda}\exp\left(\lambda \int_{t-\tau}^{t} \bar{p}(s)ds\right) + \int_{t-\tau}^{t-\sigma} Q(s+\sigma-\tau)\exp\left(\lambda \int_{s}^{t} \bar{p}(u)du\right)ds\right] > 1. \tag{2.5.2}$$

Then every solution of (2.5.1) is oscillatory.

In the next chapter we will prove a more general result, so we omit the proof here.

From (2.5.2) we can derive some explicit conditions easily.

Corollary 2.5.1. *If the condition (2.5.2) is replaced by the explicit condition*

$$\liminf_{t\to\infty}\left[e\int_{t-\tau}^{t}\bar{p}(s)ds+\int_{t-\tau}^{t-\sigma}Q(s+\sigma-\tau)ds\right]>1, \tag{2.5.3}$$

then the conclusion of Theorem 2.5.1 is true.

The following is a result for the existence of positive solutions of (2.5.1).

Theorem 2.5.2. *Assume that*

i) $\tau>\sigma\geq 0$ and

$$\bar{p}(t)=p(t)-Q(t+\sigma-\tau)\geq 0 \quad (\not\equiv 0), \quad t\geq t_1\geq t_0,$$

ii) $\int_{t_1}^{\infty}\bar{p}(t)dt=\infty$,

iii) *there exist* $T\geq t_1$ *and* $\lambda^*>0$ *such that*

$$\sup_{t\geq T}\left\{\frac{1}{\lambda^*}\exp\left(\lambda^*\int_{t-\tau}^{t}\bar{p}(u)du\right)\right.$$
$$\left.+\int_{t-\tau}^{t-\sigma}Q(s+\sigma-\tau)\exp\left(\lambda^*\int_{s}^{t}\bar{p}(u)du\right)ds\right\}\leq 1. \tag{2.5.4}$$

Then Eq. (2.5.1) has a positive solution which tends to zero as $t\to\infty$.

To prove Theorem 2.5.2, we first prove the following lemma.

Lemma 2.5.1. *Assume that condition i) in Theorem 2.5.2 holds. If the integral inequality*

$$\int_{t-\tau}^{t-\sigma}Q(s+\sigma)z(s)ds+\int_{t-\tau}^{\infty}\bar{p}(s+\tau)z(s)ds\leq z(t), \quad t\geq t_1 \tag{2.5.5}$$

has a continuous positive solution $z \in C[t_1 - \tau, \infty)$ *with* $\lim_{t\to\infty} z(t) = 0$. *Then there exists a positive solution* $x : [t_1 - \tau, \infty) \to (0, \infty)$ *of the integral equation*

$$\int_{t-\tau}^{t-\sigma} Q(s+\sigma)z(s)ds + \int_{t-\tau}^{\infty} \bar{p}(s+\tau)z(s)ds = z(t), \quad t \geq t_1. \tag{2.5.6}$$

Proof: Choose $T \geq t_1$ so large that $z(t) > z(T)$, for $t_1 - \tau \leq t < T$.

Define a set of functions as follows:

$$\Omega = \{\omega \in C([t_1 - \tau, \infty), R_+) : \quad 0 \leq \omega(t) \leq z(t), \quad t \geq t_1 - \tau\}$$

and a mapping $\mathcal{T}$ on Ω.

$$(\mathcal{T}\omega)(t) = \begin{cases} \int_{t-\tau}^{t-\sigma} Q(s+\sigma)\omega(s)ds + \int_{t-\tau}^{\infty} \bar{p}(s+\tau)\omega(s)ds, & t \geq T \\ (\mathcal{T}\omega)(T) + z(t) - z(T), & t_1 - \tau \leq t < T. \end{cases} \tag{2.5.7}$$

Clearly, $\mathcal{T}\Omega \subset \Omega$.

Define a sequence of functions as follows:

$$z_0 = z, \quad z_n = \mathcal{T} z_{n-1}, \quad n = 1, 2, \ldots .$$

It is easy to see that

$$0 \leq z_n(t) \leq z_{n-1}(t) \leq \cdots \leq z(t), \quad t \geq t_1 - \tau.$$

Hence $\lim_{n\to\infty} z_n(t) = x(t)$ exists and $x(t)$ is continuous and nonnegative for $t \geq t_1 - \tau$.

By Lebesgue's dominated convergence theorem $x(t)$ satisfies (2.5.6). Since $x(t) > 0$ for $t \in [t_1 - \tau, \tau)$, it follows from (2.5.6) that $x(t) > 0$ for all $t \geq t_1 - \tau$. The proof is complete.

Proof of Theorem 2.5.2: Set

$$z(t) = \exp\left(- \lambda^* \int_{t_1}^{t+\tau} \bar{p}(u)du \right). \tag{2.5.8}$$

We shall show that (2.5.8) satisfies integral inequality (2.5.5). In fact, (2.5.4) implies that

$$\frac{1}{\lambda^*}\exp\left(\lambda^* \int_{t-\tau}^{t} \bar{p}(u)du\right) + \int_{t-\tau}^{t-\sigma} Q(s+\sigma-\tau)\exp\left(\lambda^* \int_{s}^{t} \bar{p}(u)du\right)ds \le 1 \quad \text{for} \quad t \ge T. \tag{2.5.9}$$

It is not difficult to see the relations

$$\frac{1}{\lambda^*}\exp\left(\lambda^* \int_{t-\tau}^{t} \bar{p}(u)du\right) = \int_{t-2\tau}^{\infty} \bar{p}(s+\tau)\exp\left(\lambda^* \int_{s+\tau}^{t} \bar{p}(u)du\right)ds \tag{2.5.10}$$

and

$$\int_{t-\tau}^{t-\sigma} Q(s+\sigma-\tau)\exp\left(\lambda^* \int_{s}^{t} \bar{p}(u)du\right)ds = \int_{t-2\tau}^{t-\tau-\sigma} Q(s+\sigma)\exp\left(\lambda^* \int_{s+\tau}^{t} \bar{p}(u)du\right)ds. \tag{2.5.11}$$

Substituting (2.5.10) and (2.5.11) into (2.5.9) we obtain

$$\int_{t-2\tau}^{\infty} \bar{p}(s+\tau)\exp\left(\lambda^* \int_{s+\tau}^{t} \bar{p}(u)du\right)ds + \int_{t-2\tau}^{t-\tau-\sigma} Q(s+\sigma)\exp\left(\lambda^* \int_{s+\tau}^{t} \bar{p}(u)du\right)ds \le 1,$$

or

$$\int_{t-2\tau}^{\infty} \bar{p}(s+\tau)\exp\left(\lambda^* \int_{s+\tau}^{t_1} \bar{p}(u)du\right)ds + \int_{t-2\tau}^{t-\tau-\sigma} Q(s+\sigma)\exp\left(\lambda^* \int_{s+\tau}^{t_1} \bar{p}(u)du\right)ds \le \exp\left(-\lambda^* \int_{t_1}^{t} \bar{p}(u)du\right),$$

i.e.,

$$\int_{t-2\tau}^{\infty} \bar{p}(s+\tau)z(s)ds + \int_{t-2\tau}^{t-\tau-\sigma} Q(s+\sigma)z(s)ds \le z(t-\tau), \quad t \ge T.$$

The hypothesis of Lemma 2.5.1 is satisfied. Therefore the integral equation (2.5.6) has a positive solution $x(t)$ with $x(t) \le z(t)$. By differentiation we obtain

$$x'(t) = Q(t)x(t-\sigma) - p(t)x(t-\tau).$$

The proof is complete. □

From Theorem 2.5.1 and Theorem 2.5.2 we obtain the following corollary.

Corollary 2.5.2. *If $p > Q \ge 0$ are constants $\tau > \sigma \ge 0$, then every solution of Eq. (2.5.1) oscillates if and only if*

$$-\lambda\bar{p} + pe^{\lambda\bar{p}\tau} - Qe^{\lambda\bar{p}\sigma} > 0, \quad \textit{for all} \quad \lambda > 0 \tag{2.5.12}$$

where $\bar{p} = p - Q$.

Remark 2.5.1. If $Q(t) \equiv 0$, (2.5.4) becomes that there exists $\lambda^* > 0$ such that

$$\sup_{t\ge T}\left\{\frac{1}{\lambda^*}\exp\left(\lambda^*\int_{t-\tau}^{t} p(u)du\right)\right\} \le 1. \tag{2.5.13}$$

In particular, we take $\lambda^* = e$, then (2.5.4) leads to

$$\int_{t-\tau}^{t} p(u)du \le \frac{1}{e}, \quad \text{for} \quad t \ge T. \tag{2.5.14}$$

Example 2.5.1. Consider the equation

$$x'(t) + \frac{z(t-2)}{t^2}x(t-2) - \frac{1}{t}x(t) = 0 \tag{2.5.15}$$

where $p(t) = \frac{z(t-2)}{t^2}$, $Q(t) = \frac{1}{t}$, $\tau = 2$, $\sigma = 0$. Thus $\bar{p}(t) = p(t) - Q(t+\sigma-\tau) = \frac{z(t-2)}{t^2} - \frac{1}{t-2} > 0$ for $t \ge 10$ and $\int_{10}^{\infty}\bar{p}(t)dt = \infty$. It is easy to see that there exists $T \ge 10$ such that

$$\int_{t-\tau}^{t} \bar{p}(u)du \le \frac{\ln e}{2e}, \quad \int_{t-\tau}^{t} Q(s-\tau)ds \le \frac{1}{2e}, \quad \text{for} \quad t \ge T.$$

Choose $\lambda^* = 2e$, then (2.5.4) holds. Hence Eq. (2.5.15) has a positive solution which tends to zero as $t \to \infty$. In fact, $x(t) = \frac{1}{t}$ is such a solution of (2.5.15).

2.6. Equations with Several Delays

2.6.3. Equations with constant parameters

First we discuss the oscillation problem of the following equation

$$x'(t) + \sum_{i=1}^{n} p_i x(t - \tau_i) = 0 \tag{2.6.1}$$

where $p_i \in R$, $\tau_i \geq 0$ are constants. A known result for it is

Theorem 2.6.1. *Eq. (2.6.1) is oscillatory if and only if its characteristic equation*

$$f(\lambda) = \lambda + \sum_{i=1}^{n} p_i e^{-\lambda\tau_i} = 0 \tag{2.6.2}$$

has no real roots.

This theorem is of theoretical interest. But this condition is not easy to verify. In the following we give a new necessary and sufficient condition for oscillation of (2.6.1) which leads to some sharp explicit conditions for oscillation. Let

$$D = \{\ell = (\ell_1, \ldots, \ell_n) | \ell_i \geq 0, \quad i = 1, \ldots, n, \sum_{i=1}^{n} \ell_i = 1\}.$$

For any $\ell \in D$ define

$$f(\ell) = \sum_{i=1}^{n} \frac{\lambda_i}{\tau_i} \ln(ep_i\tau_i/\ell_i) \tag{2.6.3}$$

where $\ell_i = 0$ implies that the corresponding term vanishes.

Theorem 2.6.2. (i) *$f(\ell)$ has a maximum at a point $\ell^0 = (\ell_1^0, \ldots, \ell_n^0)$ on D which is uniquely determined by the condition that*

$$\lambda_i = \frac{1}{\tau_i} \ln(ep_i\tau_i/\ell_i), \quad i = 1, \ldots, n \tag{2.6.4}$$

have the same value if $\ell_i \neq 0$.

(ii) *Eq. (2.6.1) is oscillatory if and only if* $f(\ell^0) > 0$.

Another version of Theorem 2.6.2 is

Theorem 2.6.3. *Eq. (2.6.1) is oscillatory if and only if there exists an* $\ell^* = (\ell_1^*, \ldots, \ell_n^*)$ *such that* $\lambda_i \neq \lambda_j$ *for some* $i, j \in I_n = \{1, \ldots, n\}$ *and* $f(\ell^*) \geq 0$ *where* λ_i, $i \in I_n$ *is defined by (2.6.4).*

Corollary 2.6.1. *Assume that*

$$p_1\tau_1 > \tfrac{1}{e} \quad \text{for} \quad n = 1, \quad \text{and} \quad \sum_{i=1}^{n} p_i\tau_i \geq \tfrac{1}{e} \quad \text{for} \quad n \geq 2.$$

Then every solution of Eq. (2.6.1) is oscillatory.

Proof: We only prove it for the case that $n \geq 2$. Choose $0 < \alpha \leq 1$ such that $\sum_{i=1}^{n} \alpha e p_i \tau_i = 1$. Let $\ell_i^* = \alpha e p_i \tau_i$, $i = 1, \ldots, n$, and $\ell^* = (\ell_1^*, \ldots, \ell_n^*)$. Then $\ell^* \in D$, and λ_i are different for $i = 1, \ldots, n$. Furthermore,

$$f(\ell^*) = \alpha \ln \frac{1}{\alpha} \sum_{i=1}^{n} p_i \geq 0.$$

According to Theorem 2.6.3, Eq. (2.6.1) is oscillatory. □

From Theorems 2.6.2 and 2.6.3 we can derive some other explicit conditions for oscillation of (2.6.1).

In the next chapter we will prove more general results than Theorems 2.6.2 and 2.6.3. Therefore we omit the proofs here.

Remark 2.6.1. The following still is an open problem. Find an explicit necessary and sufficient condition for the oscillation of (2.6.1) which does not depend on solving an equation.

2.6.4. Equations with variable parameters

We consider the first order delay differential equations of the form

$$x'(t) + \sum_{i=1}^{n} p_i(t)x(t - \tau_i(t)) = 0 \tag{2.6.5}$$

where p_i, $\tau_i \in C([t_0, \infty), R_+)$, $t - \tau_i(t) \to \infty$ as $t \to \infty$, $i \in I_n = \{1, \dots, n\}$.

Let $f \in C(R, R)$ and define

$$\text{type } f = \limsup_{t\to\infty} \tfrac{1}{t} \ln |f(t)|. \tag{2.6.6}$$

Here $f(t)$ is allowed to vanish and we denote $\ln 0 = -\infty$. Clearly, a function f with type $-\infty$ tends to zero faster than any exponential functions as $t \to \infty$.

We start discussion with some basic lemmas.

Lemma 2.6.1. *If*

$$\liminf_{t\to\infty} \left\{ \inf_{\lambda>0} \frac{1}{\lambda} \sum_{i=1}^{n} p_i(t) e^{\lambda\tau_i(t)} \right\} > 1 \tag{2.6.7}$$

and Eq. (2.6.5) has nonoscillatory solutions. Then every nonoscillatory solution of (2.6.5) has type $-\infty$.

Proof: Without loss of generality assume Eq. (2.6.5) has a solution $x(t)$, $t \geq t_0 \geq 0$. From (2.6.7) we may assume that t_0 is so large that

$$\inf_{t\geq t_0, \lambda>0} \left\{ \frac{1}{\lambda} \sum_{i=1}^{n} p_i(t) e^{\lambda\tau_i(t)} \right\} > 1, \tag{2.6.8}$$

and hence $\sum_{i=1}^{n} p_i(t) > 0$ for $t \geq t_0$. We define a sequence $\{t_k\}$, $k = 0, 1, 2, \dots$ by $t_k = \sup\{t\colon \min_{1\leq i\leq n}(t - \tau_i(t)) \leq t_{k-1},\ i \in I_n\}$, $k \geq 1$, then $t_k \geq t_{k-1}$, $k \geq 1$. It is easy to see that $\{t_k\}$ is unbounded. Otherwise, there exists a constant A such that $\lim_{k\to\infty} t_k = A$ and hence $\tau_i(A_i) = 0$, $i \in I_n$. But then (2.6.8) and hence (2.6.7) is not true. Therefore $\lim_{k\to\infty} t_k = \infty$.

From (2.6.5) we have $x'(t) < 0$ for $t \geq t_1$. Define $u(t) = -\frac{x'(t)}{x(t)} > 0$, $t \geq t_1$.

Then

$$u(t) = \sum_{i=1}^{n} p_i(t) \exp \int_{t-\tau_i(t)}^{t} u(s)ds, \quad t \geq t_1. \tag{2.6.9}$$

We shall show that $\lim_{t\to\infty} u(t) = \infty$. To this end, define $\lambda_1 = \inf_{t \geq t_1} u(t)$, and

$$\lambda_{k+1} = \inf_{t \geq t_{k+1}} \sum_{i=1}^{n} p_i(t)\exp(\lambda_k \tau_i(t)). \tag{2.6.10}$$

Clearly, each λ_k is well defined. It is easy to see that

$$u(t) \geq \lambda_1, \quad t \geq t_1,$$

$$u(t) = \sum_{i=1}^{n} p_i(t)\exp \int_{t-\tau_i(t)}^{t} u(s)ds$$

$$\geq \sum_{i=1}^{n} p_i(t)\exp(\lambda_1 \tau_i(t)) \geq \lambda_2 \quad \text{for} \quad t \geq t_2.$$

By induction, we have

$$u(t) \geq \lambda_k \quad \text{for} \quad t \geq t_k, \quad k = 1, 2, \ldots .$$

Now we shall prove that $\lim_{k\to\infty} \lambda_k = \infty$.

We first claim that $\lambda_1 > 0$. Otherwise there is an arbitrarily large $t > t_0$ such that $u(s) \geq u(t) > 0$ for all $s \in [t_0, t]$. Then from (2.6.9)

$$u(t) = \sum_{i=1}^{n} p_i(t)\exp \int_{t-\tau_i(t)}^{t} u(s)ds \geq \sum_{i=1}^{n} p_i(t)\exp(u(t)\tau_i(t))$$

which contradicts (2.6.8). From (2.6.8) and (2.6.10) there is an $\alpha > 1$ such that $\lambda_{k+1} \geq \alpha\lambda_k$ for $k = 1, 2, \ldots$, and hence $\lambda_k \to \infty$ as $k \to \infty$. Since $x(t) = \exp(-\int_{t_0}^{t} u(s)ds)$ and hence $x(t)$ has type $-\infty$. The proof is complete. □

Lemma 2.6.2. *Let $f : R_+ \to (0, \infty)$ be continuous. If there exist b, $B > 0$ such that $f(t)/f(t-b) \geq B$ for $t \geq b$, then there exists $M > 0$ such that*

$$f(t) \geq M \exp\left(\tfrac{t}{b} \ln B\right), \quad \text{for} \quad t \geq b. \tag{2.6.11}$$

The proof is easy so we omit it here.

Lemma 2.6.3. *Let $p(t) \geq 0$ and $\tau(t)$ be continuous functions for $t \geq t_0 \geq 0$ and let $t - \tau(t)$ be nondecreasing such that $t - \tau(t) \to \infty$ as $t \to \infty$. Assume*

$$\liminf_{t\to\infty} \int_{t-\tau(t)}^{t} p(s)ds > 0 \tag{2.6.12}$$

and

$$\liminf_{t\to\infty} \tau(t) > 0. \tag{2.6.13}$$

Then the differential inequality

$$x(t)[x'(t) + p(t)x\big(t - \tau(t)\big)] \leq 0 \tag{2.6.14}$$

does not have any nonoscillatory solutions with type $-\infty$.

Proof: Let $x(t)$ be a nonoscillatory solution of (2.6.14). Without loss of generality, we assume $x(t) > 0$ for $t \geq t_0$. From (2.6.12) and (2.6.13) there are $b,\ B > 0$ such that for some $t_1 \geq t_0$

$$\int_{t-\tau(t)}^{t} p(s)ds \geq B, \quad \tau(t) \geq b, \quad t \geq t_1 \tag{2.6.15}$$

which implies that

$$x(t)/x\big(t - \tau(t)\big) \geq \frac{B^2}{4}, \quad t \geq t_1, \tag{2.6.16}$$

(see Lemma 2.1.3). Since $\tau(t) \geq b$ and $x(t)$ is nonincreasing we have

$$x(t)/x(t - b) \geq \frac{B^2}{4}, \quad t \geq t_1. \tag{2.6.17}$$

By Lemma 2.6.2, type $x > -\infty$. □

Remark 2.6.1. The following example shows that the conditions in Lemma 2.6.3 are sharp.

Example 2.6.1. The equation

$$x'(t) + e^{-1}t\exp(\tfrac{1}{2}t^{-2})x(t - t^{-1}) = 0 \tag{2.6.18}$$

has a solution $x(t) = \exp(-\,\frac{1}{2}t^2)$ with type $-\infty$. We see that (2.6.13) fails and (2.6.12) holds.

Theorem 2.6.4. *Assume that (2.6.7) holds and*

$$0 \le p_1(t) \le p_0, \quad 0 \le \tau_i(t) \le \tau_0, \quad i \in I_n. \tag{2.6.19}$$

Then every solution of (2.6.5) is oscillatory.

Proof: Let $x(t)$ be an eventually positive solution of (2.6.5). Then $x(t) > 0$ and $x'(t) \le 0$ for all large t. We first show that

$$\liminf_{t\to\infty} \max_{1\le i\le n} \{p_i(t)\tau_i(t)\} > 0. \tag{2.6.20}$$

If not, there exists a sequence $\{t_k\}$ such that $\lim_{k\to\infty} t_k = \infty$ and $p_i(t_k)\tau_i(t_k) \to 0$ as $k \to \infty$, $i \in I_n$. If $p_i(t_k)\tau_i(t_k) = 0$ for some t_k and all $i \in I_n$, then

$$\frac{1}{\lambda}\sum_{i=1}^{n} p_i(t_k)\exp\left[\tau_i(t_k)\right] \le np_0/\lambda \to 0, \quad \text{as} \quad \lambda \to \infty.$$

Hence we may assume that $\max\{p_i(t_k)\tau_i(t_k), \quad i \in I_n\} > 0$ for all $k = 1, 2, \ldots$. For each k, set $a_{ik} = \infty$ if $p_i(t_k)\tau_i(t_k) = 0$ and $a_{ik} = \frac{1}{\tau_i(t_k)} \ln\left(2p_0/p_i(t_k)\right)$ if $p_i(t_k)\tau_i(t_k) > 0$. From (2.6.19) we have

$$a_{ik} \ge \max\left\{\frac{1}{\tau_0}\ln\left(2p_0/p_i(t_k)\right), \quad \left(\tau_i(t_k)\right)^{-1}\ln^2\right\}.$$

In view of the fact that $p_i(t_k)\tau_i(t_k) \to 0$ as $k \to \infty$, it is not difficult to see that $a_{ik} \to \infty$ as $k \to \infty$ for each $i \in I_n$. Now let $\lambda(t_k) = \min\{a_{ik}, i = 1, 2, \ldots, n\}$. Then we have $0 < \lambda(t_k) < \infty$ and $\lambda(t_k) \to \infty$ as $k \to \infty$. Thus

$$\frac{1}{\lambda(t_k)}\sum_{i=1}^{n} p_i(t_k)\exp\left(\lambda(t_k), \tau_i(t_k)\right)$$

$$\leq \frac{1}{\lambda(t_k)}\Big[\sum_{a_{ik}<\infty} p_i(t_k)\exp\big(a_{ik}\tau_i(t_k)\big) + \sum_{a_{ik}=\infty} p_i(t_k)\Big]$$

$$\leq \frac{2np_0}{\lambda(t_k)} \to 0, \quad \text{as} \quad k \to \infty,$$

which contradicts (2.6.7) and hence (2.6.20) holds.

Next, if we let $p(t) = p_j(t)$, $\tau(t) = \tau_j(t)$ be such that $p_j(t)\tau_j(t) = \max\{p_i(t)\tau_i(t)$, $i \in I_n\}$ for each t, then $0 \leq p(t) \leq p_0$, $0 \leq \tau(t) \leq \tau_0$. From (2.6.20), there is a constant $c > 0$ such that $p(t)\tau(t) \geq c$ for all large t and so $p(t) \geq c/\tau_0$ and $\tau(t) \geq c/p_0$. From (2.6.5) we have

$$0 = x'(t) + \sum_{i=1}^{n} p_i(t)x\big(t - \tau_i(t)\big)$$

$$\geq x'(t) + p(t)x\big(t - \tau(t)\big) \geq x'(t) + \frac{c}{\tau_0}x(t - \frac{c}{p_0}),$$

and so, by Lemma 2.6.3, type $x > -\infty$.

On the other hand, by Lemma 2.6.1, under (2.6.1) the nonoscillatory solution x has type $-\infty$. This contradiction completes the proof of the theorem.

□

Considering that the functions $p_i(t)$ and $\tau_i(t)$ may be unbounded, we present the following oscillation criterion.

Theorem 2.6.5. *Suppose that there exists a nonempty subset I of the set $I_n = \{1, \ldots, n\}$ such that*

$$\liminf_{t\to\infty} \sigma(t) > 0 \tag{2.6.21}$$

and

$$\liminf_{t\to\infty} \int_{t-\sigma(t)}^{t} \sum_{i\in I} p_i(s)ds > 0 \tag{2.6.22}$$

where $\sigma(t) = \min\{\tau_i(t),\ i \in I\}$. If (2.6.7) holds, then every solution of Eq. (2.6.5) is oscillatory.

Proof: Assume that (2.6.5) has a solution $x(t) > 0$ for $t \geq t_0 \geq 0$, by Lemma 2.6.1, it is sufficient to show that type $x > -\infty$. From (2.6.7) we may assume that t_0 is sufficiently large so that (2.6.8) holds.

Then, from (2.6.5), $x'(t) < 0$ for $t \geq t_1 \geq t_0$, where $t_1 = \sup\{t : \min(t - \tau_i(t)) \leq t_0,\ i \in I_n\}$. Since

$$0 = x'(t) + \sum_{i=1}^{n} p_i(t)x\big(t - \tau_i(t)\big)$$

$$\geq x'(t) + x(t - \sigma(t)) \sum_{i \in I} p_i(t), \quad t \geq t_1.$$

By virtue of (2.6.2) and (2.6.22) and Lemma 2.6.3, we have type $x > -\infty$. The proof is complete. □

The following example shows that condition (2.6.19) or conditions (2.6.21) and (2.6.22) are important for Theorem 2.6.4 and Theorem 2.6.5 to hold.

Example 2.6.2. Consider the equation

$$x'(t) + p_1(t)x(t) + p_2(t)x(t-1) = 0. \tag{2.6.23}$$

Choose $a, b > 0$ so that $a \equiv b/(1 - e^{-b}) > 1$. Let t_0 be such that $t_0 > \frac{-1}{b} \ln(1 - e^{-b}) > 0$. $p_1(t) = a\big(g(t) - 1\big)$, $p_2(t) = a \exp(-g(t))$, $t \geq t_0$, where $g(t) = \big(1 - \exp(-b)\big)\exp(bt)$, $t \geq t_0$. For any $\lambda > 0$,

$$-\lambda + \inf_{t \geq t_0} \sum_{k=1}^{n} p_k(t)e^{\lambda \tau_k(\tau)} = -\lambda + \inf_{t \geq t_0} [p_1(t) + e^{\lambda} p_2(t)]$$

$$= \lambda \left[-1 + a \inf_{t \geq t_0} \frac{g(t) - 1 + \exp(\lambda - g(t))}{\lambda} \right].$$

The function

$$f(\lambda) = \frac{g(t) - 1 + \exp(\lambda - g(t))}{\lambda}, \quad \lambda > 0$$

minimized at $\lambda = g(t)$ and the minimum is 1. Hence

$$-\lambda + \inf_{t \geq t_0} \sum_{k=1}^{n} p_k(t)e^{\lambda \tau_k(t)} \geq \lambda(-1 + a) > 0$$

for all $\lambda > 0$, and so condition (2.6.7) holds.

On the other hand, it is easy to check that Eq. (2.6.23) has a positive solution

$$x(t) = \exp(-\exp(bt)), \quad t \geq t_0.$$

Furthermore, neither (2.6.19) nor (2.6.21) holds. Therefore nonoscillatory solutions with type $-\infty$ may exist (see Lemma 2.6.1).

Corollary 2.6.2. *Assume that*

i) *(2.6.19) holds or both (2.6.21) and (2.6.22) hold,*

ii)
$$\liminf_{t\to\infty} \sum_{i=1}^{n} p_i(t)\tau_i(t) > \tfrac{1}{e}\,. \tag{2.6.24}$$

Then every solution of (2.6.5) is oscillatory.

In order to present a nonoscillation criterion we need the following lemma.

Lemma 2.6.4. *Suppose that there exists a nonempty set $I \subseteq \{I_n\}$ such that*

i) $\tau_i(t) > 0$ *for* $t \geq t_0$, $\quad i \in I$, *and*

ii) $\sum_{i\in I} p_i(t) > 0$ *for* $t \geq t_0$.

If y is an eventually positive solution of the differential inequality

$$y'(t) + \sum_{k=1}^{n} p_k(t)y(t-\tau_k(t)) \leq 0, \tag{2.6.25}$$

then there exists an eventually positive solution $x(t)$ of (2.6.5) with

$$x(t) \leq y(t) \quad \textit{for all large} \quad t,$$

and $\lim_{t\to\infty} x(t) = 0$.

Proof: Since y is an eventually positive solution of (2.6.25) we have that eventually

$$y'(t) \leq -\sum_{i\in I} p_i(t)y(t-\tau_i(t)) \leq -\sum_{i\in I} p_i(t)y(t)$$

and hence, because of i), we have the $y'(t) < 0$ eventually.

Choose T sufficiently large. For any $\tilde{t} \geq t \geq T$ integrating (2.6.5) from t to $\tilde{t}$ we have

$$y(t) \geq y(\tilde{t}) + \int_t^{\tilde{t}} \sum_{k=1}^{n} p_k(s) y\big(s - \tau_k(s)\big) ds$$
$$> \int_t^{\tilde{t}} \sum_{k=1}^{n} p_k(s) y\big(s - \tau_k(s)\big) ds.$$

Because $\tilde{t}$ is arbitrary we get

$$y(t) \geq \int_t^{\infty} \sum_{k=1}^{n} p_k(s) y\big(s - \tau_k(s)\big) ds, \quad t \geq T.$$

Define a set

$$\Omega = \{x \in C([T,\infty), R_+) : \ 0 \leq x(t) \leq y(t), \quad t \geq T\}.$$

For any $x \in \Omega$, we define

$$\bar{x}(t) = \begin{cases} x(t), & t \geq T \\ x(T) + \big(y(t) - y(T)\big), & \tau \leq t \leq T. \end{cases}$$

Then define a mapping $\mathcal{T}$ on Ω as follows:

$$(\mathcal{T}x)(t) = \int_t^{\infty} \sum_{k=1}^{n} p_k(t) \bar{x}\big(s - T_k(s)\big) ds, \quad t \geq T.$$

Clearly, $\mathcal{T}$ is well-defined, and $(\mathcal{T}x)(t) \leq y(t)$ on $t \geq T$ and hence $\mathcal{T}\Omega \subset \Omega$.

Define a sequence $\{x_n(t)\}, \quad n = 0, 1, 2, \ldots$ as follows:

$$x_0 = y, \qquad t \geq T,$$
$$x_{n+1} = \mathcal{T}x_n, \quad t \geq T, \quad n = 0, 1, 2, \ldots .$$

By induction, we see that

$$0 \leq x_n(t) \leq x_{n-1}(t) \leq \cdots \leq y(t), \quad t \geq T,$$

and hence $x(t) = \lim_{n\to\infty} x_n(t)$ exists on $[T, \infty)$. Then we can apply the Lebesgue's dominated convergence theorem to show that $x = \mathcal{T}x$, i.e.,

$$x(t) = \int_t^\infty \sum_{k=1}^{n} p_k(s)\bar{x}\big(s - \tau_k(s)\big)ds, \quad t \geq T.$$

Obviously, $\lim_{t\to\infty} x(t) = 0$. Also, we have

$$x'(t) = -\sum_{k=1}^{n} p_k(t)\bar{x}\big(t - \tau_k(t)\big), \quad t \geq T,$$

which means that $\bar{x}$ or x is a solution on $[T, \infty)$ of Eq. (2.6.5). Moreover, it is easy to see that $\bar{x}(t) \leq y(t)$ for $t \geq T$.

It remains to prove that $\bar{x}$ is positive on $[T, \infty)$. Clearly, $\bar{x} > 0$ for $t \in [\tau, T)$ and $\bar{x} \geq 0$ for $t \geq T$. If there is a $\bar{t} \geq T$ such that $\bar{x}(\bar{t}) = 0$, set

$$t^* = \inf\{\bar{t} \mid \bar{x}(\bar{t}) = 0, \quad \bar{t} \geq T\},$$

then $\bar{x}(t^*) = 0, \quad \bar{x}(t) > 0$ for $t \in [\tau, t^*)$ and $\tau_i(t) > 0,\ i \in I,\ t \geq t_0$. Hence

$$x(t^*) = \int_{t^*}^\infty \sum_{k=1}^{\infty} p_k(s)\bar{x}\big(s - \tau_k(s)\big)ds > 0$$

contradicting that $\bar{x}(t^*) = 0$.

Theorem 2.6.6. *Suppose that there exists a nonempty set $I \subseteq I_n$ so that i) and ii) in Lemma 2.6.4 are satisfied. Moreover, assume that for a sufficiently large $T_0 \geq t_0$*

$$-\lambda + \sup_{t\geq T_0} \sum_{k=1}^{n} p_k(t)\exp\big(\lambda\tau_k(t)\big) \leq 0, \quad \text{for some} \quad \lambda > 0. \tag{2.6.26}$$

Then there exists a positive solution x of Eq. (2.6.5) satisfying

$$x(t) \leq \exp(-\lambda t) \quad \text{for all large} \quad t \tag{2.6.27}$$

Proof: The point T_0 is sufficiently large so that $t - \tau_k(t) \geq t_0$, for $t \geq T_0$,

$k \in I_n$. Set $y(t) = \exp(-\lambda t)$, $t \geq t_0$. Then

$$y'(t) + \sum_{k=1}^{n} p_k(t) y\big(t - \tau_k(t)\big) = -\lambda y(t) + \sum_{k=1}^{n} p_k(t) \exp\big(-\lambda(t - \tau_k(t))\big)$$

$$= y(t)\Big[-\lambda + \sum_{k=1}^{n} p_k(t) \exp\big(\lambda \tau_k(t)\big)\Big] \leq 0, \quad t \geq T_0.$$

This means that $y(t)$ is an eventually positive solution of (2.6.25). By Lemma 2.6.4, Eq. (2.6.5) has a solution $x(t)$ satisfying $x(t) \leq y(t) = \exp(-\lambda t)$ eventually. □

Corollary 2.6.3. *Consider the differential equation*

$$x'(t) + \sum_{k=1}^{n} p_k x(t - \tau_k) = 0 \tag{2.6.28}$$

where p_k, τ_k are positive constants, $k \in I_n$. Then every solution of (2.6.28) is oscillatory if and only if

$$-\lambda + \sum_{k=1}^{n} p_k \exp(\lambda \tau_k) > 0 \quad \textit{for all} \quad \lambda > 0. \tag{2.6.29}$$

Corollary 2.6.4. *Assume that the conditions of Lemma 2.6.4 hold. Moreover, assume that there exist $T \geq t_0$ and $\lambda > 0$ such that*

$$-\lambda + \sup_{t \geq T} \frac{\sum_{k=1}^{n} p_k(t) \exp\{\lambda \int_{t-\tau_k(t)}^{t} (\sum_{j=1}^{n} p_j(s)) ds\}}{\sum_{k=1}^{n} p_k(t)} \leq 0 \tag{2.6.30}$$

$$\textit{for a} \quad T_0 \geq t_0 \quad \textit{and} \quad \lambda > 0.$$

Then there exists a positive solution x of Eq. (2.6.5) satisfying $\lim_{t \to \infty} x(t) = 0$ and *eventually*

$$x(t) \leq \exp\left\{ -\lambda \int_{t_0}^{t} \left[\sum_{i=1}^{n} p_i(s) \right] ds \right\}. \tag{2.6.31}$$

Proof: Choose T_0 so large that $t - \tau_k(t) \geq t_0$ for $t \geq T_0$, $\quad k \in I_n$. Set

$$y(t) = \exp\left\{ -\lambda \int_{t_0}^{t} \left[\sum_{i=1}^{n} p_i(s) \right] ds \right\}.$$

It is easy to see that y is a positive solution of (2.6.25)

So the conclusion of the corollary follows from Lemma 2.6.4.

Remark 2.6.2. Suppose that $\sum_{k=1}^{n} p_k(t) > 0$, for $t \geq t_0$, and for a sufficiently large $T_0 \geq t_0$

$$\sup_{t \geq T_0} \left\{ \max_{k=1,2,\ldots,n} \int_{t-\tau_k(t)}^{t} \left(\sum_{j=1}^{n} p_j(s) \right) ds \right\} \leq \frac{1}{e}\,. \tag{2.6.32}$$

Then (2.6.30) holds. Consequently, the conclusion of Corollary 2.6.4 holds.

Corollary 2.6.5. *Assume that the assumptions i) and ii) in Lemma 2.6.4 hold. Then every solution of (2.6.5) is oscillatory if and only if differential inequality (2.6.25) has no eventually positive solutions.*

From this corollary we can derive some comparison results.

With Eq. (2.6.5) we associate the equation

$$x'(t) + \sum_{i=1}^{n} q_i(t) x\big(t - \sigma_i(t)\big) = 0. \tag{2.6.33}$$

Theorem 2.6.7. *Assume that the assumptions i) and ii) in Lemma 2.6.4 hold, and*

$$\sigma_k(t) \geq \tau_k(t), \quad q_k(t) \geq p_k(t), \quad k \in I_n \tag{2.6.34}$$

eventually. Then every solution of Eq. (2.6.5) is oscillatory implies the same for (2.6.33).

Proof: If not, let $x(t)$ be an eventually positive solution of (2.6.33). Then eventually

$$\begin{aligned} x'(t) &= -\sum_{i=1}^{n} q_i(t)x\big(t-\sigma_i(t)\big) \\ &\leq -\sum_{i=1}^{n} p_i(t)x\big(t-\sigma_i(t)\big) \\ &\leq -\sum_{i=1}^{n} p_i(t)x\big(t-\tau_i(t)\big). \end{aligned}$$

By Corollary 2.6.5, Eq. (2.6.5) has an eventually positive solution contradicting the assumption. □

Corollary 2.6.6. *If Eq. (2.6.5) has an eventually positive solution and*

$$0 \leq q_k(t) \leq p_k(t), \quad 0 \leq \sigma_k(t) \leq \tau_k(t), \quad k \in I_n \tag{2.6.35}$$

eventually, and there exists a nonempty set $I \in I_n$ such that $\sigma_i(t) > 0$ for $t \geq t_0$, $i \in I$ and $\sum_{i \in I} q_i(t) > 0$ for $t \geq t_0$. Then (2.6.33) has an eventually positive solution also.

Proof: Let $x(t)$ be an eventually positive solution of (2.6.5). Then eventually

$$x'(t) = -\sum_{i=1}^{n} p_i(t)x\big(t-\tau_i(t)\big) \leq -\sum_{i=1}^{n} q_i(t)x\big(t-\sigma_i(t)\big).$$

By Lemma 2.6.4, Eq. (2.6.33) has an eventually positive solution. □

We then consider (2.6.5) for the case that $\lim_{t\to\infty} \tau_i(t) = \widetilde{\tau}_i$ and $\lim_{t\to\infty} p_i(t) = \widetilde{p}_i$.

The "limiting equation" of (2.6.5) is

$$z'(t) + \sum_{i=1}^{n} \widetilde{p}_i z(t - \widetilde{\tau}_i) = 0. \tag{2.6.36}$$

Theorem 2.6.8. *Assume that τ_i, $p_i \in C([t_0,\infty), R_+)$, $\lim_{t\to\infty} \tau_i(t) = \widetilde{\tau}_i$, $\lim p_i(t) = \widetilde{p}_i$. Then (2.6.36) is oscillatory implies the same for (2.6.5).*

Proof: By Corollary 2.6.3 we have

$$-\lambda + \sum \widetilde{p}_i \exp(\lambda \widetilde{\tau}_i) > 0 \quad \text{for all} \quad \lambda > 0. \tag{2.6.37}$$

Assume the conclusion of the theorem is not true. Let $x(t)$ be a positive solution of (2.6.5). Then for any $\varepsilon > 0$ there exists a $T \geq t_0$ such that

$$x'(t) + \sum_{i=1}^{n} (\widetilde{p}_i - \varepsilon) x(t - \widetilde{\tau}_i + \varepsilon) \leq 0, \quad t \geq T.$$

In view of Corollary 2.6.5, it follows that

$$x'(t) + \sum_{i=1}^{n} (\widetilde{p}_i - \varepsilon) x(t - \widetilde{\tau}_i + \varepsilon) = 0$$

has a positive solution. Hence by Theorem 2.6.1 the characteristic equation

$$F(\mu) = -\mu + \sum_{i=1}^{n} (\widetilde{p}_i - \varepsilon) e^{\mu(\widetilde{\tau}_i - \varepsilon)} = 0$$

has a positive root μ^*. From (2.6.37) there exists an $m > 0$ such that

$$\sum_{i=1}^{n} \widetilde{p}_i \exp(\lambda \widetilde{\tau}_i) \geq \lambda + m \quad \text{for all} \quad \lambda > 0. \tag{2.6.38}$$

Choose $\varepsilon > 0$ such that

$$\mu^* \left(\exp(\mu^* \varepsilon) - 1\right) + \varepsilon \sum_{i=1}^{n} \exp(\mu^* \tau_i) < m.$$

Then

$$\sum_{i=1}^{n} \widetilde{p}_i \exp\left(\mu^* \widetilde{\tau}_i\right) < \mu^* + m$$

which contradicts (2.6.38). The proof is complete. □

2.6.5. Equations with oscillating coefficients

In the following we consider the equations of the form

$$x'(t) + \sum_{i=1}^{n} p_i(t) x\big(\tau_i(t)\big) = 0 \tag{2.6.39}$$

where $p_i : R_+ \to R$ are locally integrable, $\tau_i : R_+ \to R$ are measurable.

We have the following existence results for the positive solutions of (2.6.39) anywhere p_i and $t - \tau_i(t)$ may change signs.

Theorem 2.6.9. *Assume that either*

$$p_i(t) \geq 0, \quad \text{for} \quad t \geq 0, \quad i \in I_n \tag{2.6.40}$$

or

$$\tau_i(t) \leq t, \quad \text{for} \quad t \geq 0, \quad i \in I_n. \tag{2.6.41}$$

Further assume that there exist $\alpha_i \in R$, M_i, $i \in I_n$, λ, $t_0 \in R_+$ and locally integrable function $q : R_+ \to R_+$ such that

$$p_i(t) \leq M_i q(t), \quad \int_{\tau_i(t)}^{t} q(s)ds \leq \alpha_i, \quad t \geq t_0, \quad i \in I_n \tag{2.6.42}$$

and

$$\sum_{i=1}^{n} M_i e^{\alpha_i \lambda} - \lambda \leq 0. \tag{2.6.43}$$

Then Eq. (2.6.39) has a positive solution.

Proof: We will prove this theorem with condition (2.6.41). The theorem with condition (2.6.40) can be proved by a similar argument.

Set

$$\nu(t) = \inf\{s\colon s \geq t,\ \tau_i(\xi) \geq t,\ \text{for } \xi \geq s,\quad i \in I_n\}.$$

Let $t_1 \geq \nu(t_0)$.

Let $C([t_0,\infty),R)$ be a space of all continuous functions defined on $[t_0,\infty)$ with the Fréchet topology. Let $\Omega \subset C([t_0,\infty),R)$ be the subset satisfying:

a) $\dfrac{x\big(\tau_i(t)\big)}{x(t)} \leq e^{\alpha_i\lambda},\quad t \geq t_1,\ i \in I_n$

b) $x(t) = 1,\quad t \in [t_0,t_1]$

c) $\exp\big(-\lambda\int_{t_1}^t q(s)ds\big) \leq x(t) \leq \exp\big(\sum_{i=1}^n e^{\alpha_i\lambda}\int_{t_1}^t [p_i(s)]_- \,ds\big),\quad t \geq t_1$

where $[p_i(s)]_- = \max\big(-p(s),0\big)$.

Clearly, Ω is a nonempty, closed and convex subset of $C([t_0,\infty),R)$. Define an operator $\mathcal{T} : \Omega \to C([t_0,\infty),R)$ by

$$(\mathcal{T}x)(t) = \begin{cases} \exp\left(-\displaystyle\int_{t_1}^t \frac{\sum\limits_{i=1}^n p_i(s)x(T_i(s))}{x(s)}\right), & t \geq t_1 \\ 1, & t \in [t_0,t_1]. \end{cases}$$

It is easy to see that $\mathcal{T}\Omega \subseteq \Omega$ and $\mathcal{T}$ is continuous.

On every compact interval of $[t_0,\infty)$, $(\mathcal{T}x)(t)$, $\quad x \in \Omega$ is uniformly bounded and equicontinuous. By Ascoli theorem $\mathrm{cl}\{\mathcal{T}\Omega\}$ is compact in the topology of the Fréchet space $C([t_0,\infty),R)$. By the Schauder-Tychonoff fixed point theorem, there exists an $x \in \Omega$ such that $Tx = x$, i.e.,

$$x(t) = \exp\left(-\int_{t_1}^t \frac{\sum\limits_{i=1}^n p_i(s)x(\tau_i(s))ds}{x(s)}\right),\quad t \geq t_1$$

which implies that x is a positive solution of (2.6.39) on $[t_1,\infty)$.

Corollary 2.6.7. *Assume* $\sum\limits_{i=1}^n p_i(t) \geq 0,\ t \geq T$ *for a sufficiently large* $T \geq t_0$, *and*

$$\sup_{t\geq T}\left\{\max_{i\in I_n}\int_{t-T_i(t)}^t\left(\sum_{j=1}^n p_j(s)\right)ds\right\} \leq \frac{1}{e}\,.$$

Then Eq. (2.6.39) has a positive solution.

In fact, choose

$$q(t) = \sum_{i=1}^{n} p_i(t)$$

and $\lambda = e$. Then all assumptions of Theorem 2.6.9 are satisfied. This corollary follows from Theorem 2.6.9.

In particular, if $n = 1$, from Theorem 2.6.9 we obtain the following results.

Corollary 2.6.8. *Assume that either*

$$\tau(t) \le t, \quad \int_{\tau(t)}^{t} p_t(s)ds \le \frac{1}{e}\,, \quad t \ge t_0$$

or

$$p(t) \ge 0, \quad \int_{\tau(t)}^{t} p(s)ds \le \frac{1}{e}, \quad t \ge t_0.$$

Then (2.6.39) has a positive solution on $[t_0, \infty)$, *where* $p_t(s) = \max\big(p(s), 0\big)$.

Corollary 2.6.9. *Assume that:*

(i) *there exist* $\alpha_i \in R, M_i, i \in I_n, \lambda, t_0 \in R_+$ *such that one of the following holds:*

$$\begin{aligned} p_i(t) \le M_i, \quad & \tau_i(t) \ge t^{\alpha_i}, \quad t \ge t_0, \quad \text{or} \\ tp_i(t) \le M_i, \quad & \tau_i(t) \ge \alpha_i t, \quad t \ge t_0, \quad \text{or} \\ t\ \ln tp_i(t) \le M_i, \quad & \tau_i(t) \ge t^{\alpha_i}, \quad t \ge t_0; \end{aligned}$$

(ii) $\sum\limits_{i=1}^{n} M_i e^{\alpha_i \lambda} - \lambda \le 0$;

(iii) *one of (2.6.40) and (2.6.41) holds.*

Then Eq. (2.6.39) has a positive solution.

This corollary follows from Theorem 2.6.9 by taking $q(t) = 1,\ q(t) = \frac{1}{t}$ and $q(t) = \frac{1}{t \ln t}$ respectively.

Theorem 2.6.10. *Assume that there exist nonnegative numbers $M_i, \alpha_i, i \in I_n$, λ, $t_0 \in R_+$ and a locally integrable function $q : R_+ \to R_+$ such that*

$$|p_i(t)| \le M_i q(t), \quad \left| \int_{\tau_i(t)}^{t} q(s)ds \right| \le \alpha_i, \quad t \ge t_0, \quad i \in I_n,$$

and

$$\sum_{i=1}^{n} M_i \exp(\alpha_i \lambda) - \lambda \le 0.$$

Then Eq. (2.6.39) has a positive solution.

The proof is similar to the proof of Theorem 2.6.9. We only point out that the functions in the set Ω satisfy

$$\exp\left(-\lambda \int_{t_1}^{t} q(s)ds \right) \le x(t) \le \exp\left(\lambda \int_{t_1}^{t} q(s)ds \right), \quad t \ge t_1.$$

Corollary 2.6.10. *If for all sufficiently large t*

$$\int_{\tau(t)}^{t} |p(s)|ds \le \frac{1}{e}.$$

Then Eq. (2.6.39) has a positive solution.

Corollary 2.6.11. *If there exist constants $p_i, \tau_i, i \in I_n$ such that*

$$|p_i(t)| \le p_i, \quad |t - \tau_i(t)| \le \tau_i$$

and

$$\lambda = \sum_{i=1}^{n} p_i e^{\lambda \tau_i}$$

has a positive root, then Eq. (2.6.39) has a positive solution.

2.6.6. Advanced equations

Finally, we would like to mention that most oscillation criteria for Eq. (2.6.5) can be extended to the equations with advanced type of the form

$$x'(t) = \sum_{i=1}^{n} p_i(t)x\big(t + \tau_i(t)\big). \tag{2.6.44}$$

For instance, we present the following results.

Theorem 2.6.11. *Assume that p_i, $\tau_i \in C([t_0, \infty), R_+)$, with $\tau_i(t)$ uniformly bounded, $i \in I_n$ and such that*

$$\liminf_{t\to\infty}\left\{\inf_{\lambda>0}\frac{1}{\lambda}\sum_{i=1}^{n} p_i(t)\exp\big(\lambda\tau_i(t)\big)\right\} > 1. \tag{2.6.45}$$

Then every solution of (2.6.44) oscillates.

Proof: Suppose the contrary, assume $x(t) > 0$, $t \geq T$, is a solution of (2.6.44). Set

$$\lambda_0(t) = \frac{x'(t)}{x(t)}, \quad t \geq T. \tag{2.6.46}$$

Then

$$\lambda_0(t) = \sum_{i=1}^{n} p_i(t)\exp\left(\int_t^{t+\tau_i(t)} \lambda_0(s)ds\right), \quad t \geq T.$$

Define a set

$$\Omega = \{\lambda \in C([T, \infty), R) : \ \lambda(t) \geq 0, \ t \geq T\}$$

and a mapping $\mathcal{T}$ defined on Ω

$$(\mathcal{T}\lambda)(t) = \sum_{i=1}^{n} p_i(t)\exp\left(\int_t^{t+\tau_i(t)} \lambda(s)ds\right), \quad t \geq T.$$

Clearly, $\mathcal{T}\Omega \subset \Omega$, $\lambda_1(t) \geq \lambda_2(t) \geq 0$ implies $(\mathcal{T}\lambda_1)(t) \geq (\mathcal{T}\lambda_2)(t) \geq 0$ for $t \geq T$, and

$$\lambda_0(t) = (\mathcal{T}\lambda_0)(t), \quad \lambda_0(t) \geq 0, \quad t \geq T.$$

In view of the assumption, there exists $\tau_0 > 0$ such that $|\tau_i(t)| \leq \tau_0$, $i \in I_n$. Form (2.6.45) there exists $C > 1$ such that

$$\inf_{\lambda \in (0,\infty)} \frac{1}{\lambda} \sum_{i=1}^{n} p_i(t) \exp\bigl(\lambda \tau_i(t)\bigr) \geq C.$$

In particular, taking $\lambda = 1$, we have for $t \geq T$

$$C \leq \sum_{i=1}^{n} p_i(t) \exp\bigl(\tau_i(t)\bigr) \leq e^{\tau_0} \sum_{i=1}^{n} p_i(t).$$

Hence for $t \geq T$

$$\sum_{i=1}^{n} p_i(t) \geq \frac{C}{e^{\tau_0}} = M > 0,$$

$$\lambda_0(t) = (\mathcal{T}\lambda_0)(t) \geq \sum_{i=1}^{n} p_i(t) \geq M,$$

and so

$$\lambda_0(t) = (\mathcal{T}\lambda_0)(t) \geq \mathcal{T}M = \sum_{i=1}^{n} p_i(t) \exp\bigl(M\tau_i(t)\bigr) \geq MC, \quad t \geq T.$$

Repeating this procedure we obtain

$$\lambda_0(t) \geq C^k M \to \infty, \quad \text{as} \quad k \to \infty,$$

which contradicts (2.6.46). The proof is complete.

Corollary 2.6.12. *Assume that* $\tau_i(t) \leq \tau_0$, $i \in I_n$ *and*

$$\liminf_{t \to \infty} \sum_{i=1}^{n} p_i(t)\tau_i(t) > \frac{1}{e}\,.$$

Then every solution of (2.6.44) oscillates.

Theorem 2.6.12. *Assume that* $p_i,\quad \tau_i \in C([t_0,\infty),R_+)$, *and there exists* $\lambda^* > 0,\quad T \geq t_0$ *such that*

$$-\lambda^* + \sup_{t\geq T}\sum_{i=1}^{n} p_i(t)exp(\lambda^*\tau_i(t)) \leq 0.$$

Then (2.6.44) has a positive solution on (T,∞).

Remark 2.6.3. Finding conditions for the existence of oscillatory solutions of (2.6.5) is an open problem.

2.7. Equations with Forced Terms

We consider the forced differential equations of the form

$$x'(t) + f\big(t, x(\tau_1(t)),\ldots,x(\tau_n(t))\big) = q(t). \tag{2.7.1}$$

Assuming here $q \in C([t_0,\infty),R)$ and let $Q(t) \in C^1([t_0,\infty),R)$ such that $Q'(t) = q(t)$ for $t \geq t_0$.

The following result provides an oscillation criterion.

Theorem 2.7.1. *Assume that*

i) $f, q, \tau_i, i \in I_n$, *are continuous,*

ii) $\lim_{t\to\infty} \tau_i(t) = \infty, \quad i \in I_n$,

iii) *for* $t \geq T \geq t$.

$u_1 f(t,u_1,u_2,\ldots,u_n) > 0$ *whenever* $u_1 u_i > 0,\ i \in I_n$ *and for fixed* t, f *is nondecreasing with respect to* u_i, $i \in I_n$,

iv)

$$\int_T^\infty f\big(t, Q_+(\tau_1(t)),\ldots,Q_+(\tau_n(t))\big)dt = \infty$$
$$\int_T^\infty f\big(t, -q_-(\tau_1(t)),\ldots,-Q_-(\tau_n(t))\big)dt = -\infty \tag{2.7.2}$$

where $Q_+(t) = \frac{1}{2}\big(|Q(t)| + Q(t)\big)$, $\quad Q_-(t) = \frac{1}{2}\big(|Q(t)| - Q(t)\big)$. *Then every solution of (2.7.1) oscillates.*

Proof: Suppose the contrary, let $x(t)$, be an eventually positive solution of (2.7.1). Then

$$\big(x(t) - Q(t)\big)' = -f\big(t, x(\tau_1(t)), \ldots, x(\tau_n(t))\big). \tag{2.7.3}$$

In view of condition ii), $\big(x(t) - Q(t)\big)' < 0$ eventually. Condition iv) implies that $Q(t)$ always changes sign for sufficiently large t.

Therefore there exists a sequence $t_k \to \infty$ such that $x(t_k) > Q(t_k)$, and then

$$x(t) - Q(t) > 0 \quad \text{for} \quad t \geq T^*, \tag{2.7.4}$$

which implies that

$$\lim_{t\to\infty} \big(x(t) - Q(t)\big) = \ell$$

exists. From (2.7.3) we obtain that for sufficiently large T

$$\int_T^\infty f\big(t, x(\tau_1(t)), \ldots, x(\tau_n(t))\big)dt < \infty. \tag{2.7.5}$$

From (2.7.4), $x(t) > Q_+(t)$ for $t \geq T$. In view of the monotonic property of f and (2.7.5) we obtain

$$\int_T^\infty f\big(t, Q_+(\tau_1(t)), \ldots, Q_+(\tau_n(t))\big)dt < \infty \tag{2.7.6}$$

which contradicts (2.7.2). Similarly, we can derive a contradiction when Eq. (2.7.1) has an eventually negative solution. The proof is complete.

Example 2.7.1. We consider the equation

$$x'(t) + px(t - \pi) = \sin t + p\cos t \tag{2.7.7}$$

where p is a constant with $0 < p\pi e < 1$, then the homogeneous equation

$$x'(t) + px(t - \pi) = 0 \tag{2.7.8}$$

has a positive solution by Corollary 2.1.2. For (2.7.7) $Q(t) = -\cos t + p \sin t$, it is easy to see that

$$\int_T^\infty Q_\pm(t-\pi)dt = \infty. \tag{2.7.9}$$

By Theorem 2.7.1, every solution of (2.7.7) oscillates. In fact, $y(t) = -\cos t$ is such a solution.

Remark 2.7.1. From the proof of Theorem 2.7.1, it is easy to see that condition iv) can be replaced by iv),

$$\begin{aligned} \liminf_{t\to\infty} Q(t) &= -\infty \\ \limsup_{t\to\infty} Q(t) &= +\infty. \end{aligned} \tag{2.7.10}$$

We now consider the case that

$$\begin{aligned} \limsup_{t\to\infty} Q(t) &= M < \infty \\ \liminf_{t\to\infty} Q(t) &= m > -\infty. \end{aligned} \tag{2.7.11}$$

For the sake of simplicity we consider the equation of the form

$$x'(t) + p(t)x(\tau(t)) = q(t). \tag{2.7.12}$$

Theorem 2.7.2. *Assume that*

$$p \in C([t_0,\infty), R_+), \ \tau, \ q \in C([t_0,\infty), R), \ \lim_{t\to\infty} \tau(t) = \infty.$$

Set $Q'(t) = q(t)$, Q changes sign on (T,∞), where T is any number. If

$$\int_T^\infty p(t)\left[Q(\tau(t)) - m\right]_+ dt = \infty$$

and *(2.7.13)*

$$\int_T^\infty p(t)\left[M - Q(\tau(t))\right]_+ dt = \infty.$$

Then every solution of (2.7.12) oscillates.

Proof: Suppose the contrary, $x(t) > 0,\ t \geq t_0$, is a solution of (2.7.12), then there exists $t_1 \geq t_0$ such that $x\big(\tau(t)\big) > 0$ for $t \geq t_1$. As in the proof of Theorem 2.7.1 we know that $x(t) - Q(t) > 0$ and $\big(x(t) - Q(t)\big)' < 0$ for $t \geq t_1$, which implies that

$$\int_{t_1}^{\infty} p(s)x\big(\tau(s)\big)ds < \infty. \tag{2.7.14}$$

On the other hand (2.7.13) implies that

$$\int_{t_1}^{\infty} p(t)dt = \infty. \tag{2.7.15}$$

From (2.7.14) and (2.7.15) we obtain

$$\liminf_{t \to \infty} x(t) = 0. \tag{2.7.16}$$

Set

$$\lim_{t \to \infty} \big(x(t) - Q(t)\big) = \ell \geq 0. \tag{2.7.17}$$

We will show that $\ell = -m$. In fact, for any $\varepsilon > 0$ there exists $T \geq t_1$ such that

$$\ell < x(t) - Q(t) < \ell + \varepsilon,$$

and hence

$$-Q(t) < \ell + \varepsilon, \quad t \geq T.$$

Then

$$-m = -\liminf_{t \to \infty} Q(t) \leq \ell + \varepsilon. \tag{2.7.18}$$

On the other hand, there exists a sequence $\{t_n\}$ such that $\lim_{n \to \infty} t_n = \infty$ and $\lim_{n \to \infty} x(t_n) = 0$. From (2.7.17) we have

$$-Q(t_n) > \ell - x(t_n)$$

and hence

$$-\liminf_{n\to\infty} Q(t_n) \geq \ell$$

and so

$$-m \geq \ell. \tag{2.7.19}$$

(2.7.18) and (2.7.19) imply that $\ell = -m$. From (2.7.17),

$$x(t) - Q(t) > -m, \quad t \geq t_1$$

and so

$$x(t) > [Q(t) - m]_+, \quad t \geq t_1.$$

Choose $t_2 > t_1$ so large that $\tau(t) \geq t_1$ for $t \geq t_2$. Then

$$x\big(\tau(t)\big) > [Q(\tau(t)) - m]_+, \quad t \geq t_2.$$

Substituting this into (2.7.14) we obtain

$$\int_{t_2}^{\infty} p(t)[Q\big(\tau(t)\big) - m]_+ dt < \infty,$$

which contradicts (2.7.13). Similarly, we can prove that (2.7.11) has no eventually negative solutions. The proof is complete.

Example 2.7.2. Consider the equation

$$x'(t) + p(t)x(t - \pi) = -\cos t, \tag{2.7.20}$$

where

$$p(t) = \begin{cases} \frac{2}{\pi}t, & t \in [0, \frac{\pi}{2}] \\ \frac{2}{\pi}(\pi - t), & t \in (\frac{\pi}{2}, \pi] \end{cases}$$

and $p(t + 2\pi) = p(t)$ for all $t \geq 0$.

In this case $Q(t) = -\sin t$ and $M = 1$, $m = -1$.

Clearly, we have

$$\int_{\pi}^{\infty} p(t)[1 - Q(t - \pi)]_{+} dt = \infty,$$
$$\int_{\pi}^{\infty} p(t)[1 + Q(t - \pi)]_{+} dt = \infty. \tag{2.7.21}$$

By Theorem 2.7.2, every solution of (2.7.20) oscillates. We note that condition (2.7.2) does not hold for (2.7.20).

In the next chapter we will allow a result for the existence of nonoscillatory solutions of (2.7.1).

2.8. Single Population Models with Delays

2.8.7. Delay logistic equation

We consider the delay logistic equation with a constant carrying capacity k

$$\frac{dx(t)}{dt} = r(t)x(t)\big(1 - \frac{x\big(t - \tau(t)\big)}{k}\big), \tag{2.8.1}$$

where r, τ are positive continuous functions defined on $[0, \infty)$ and k is a positive constant. Denote $\tau^* = \sup_{t \geq 0} \tau(t)$.

Motivated by the application of (2.8.1), we consider solutions of (2.8.1) with initial conditions of the type

$$x(s) = \varphi(s) \quad \text{where} \quad \varphi \geq 0$$
$$\text{is continuous on} \quad [-\tau^*, 0] \quad \text{and} \quad \varphi(0) > 0. \tag{2.8.2}$$

It is not difficult to show that solutions of (2.8.1) and (2.8.2) are defined for all $t \geq 0$ and remain positive for $t \geq 0$.

A solution of (2.8.1) and (2.8.2) is said to be oscillatory about k if there exists a sequence $\{t_n\}$ such that $\lim_{n \to \infty} t_n = \infty$ and $k - x(t_n) = 0, \quad n = 1, 2, \ldots$.

Set

$$y(t) = \frac{x(t)}{k} - 1, \quad t \geq -\tau^*. \tag{2.8.3}$$

Then (2.8.1) becomes

$$y'(t) = -r(t)\big(1 + y(t)\big)y\big(t - \tau(t)\big), \quad t > 0, \tag{2.8.4}$$

whose initial conditions are inherited from (2.8.2).

By way of (2.8.3), the oscillation of x about k is equivalent to that of y (about zero) as in the usual sense. We note from the positivity of $x(t)$ that $1 + y(t) > 0$ for $t \geq 0$.

We state some results on oscillation of Eq. (2.8.4) without proofs.

Theorem 2.8.1. *Assume that*

i) $r,\ \tau \in C\big(R_+, (0,\infty)\big)$, *and* $\lim_{t\to\infty}\big(t - \tau(t)\big) = \infty$,

ii) $\int_0^\infty r(s)ds = \infty$.

Then every solution of Eq. (2.8.4) is either oscillatory or converges to zero, eventually monotonically, as $t \to \infty$.

Theorem 2.8.2. *In addition to the assumptions of Theorem 2.8.1 assume*

$$\liminf_{t\to\infty} \int_{t-\tau(t)}^{t} r(s)ds > \frac{1}{e}\,.$$

Then every solution of Eq. (2.8.4) is oscillatory.

Theorem 2.8.3. *Assume there exists a* $t_0 \geq 0$ *such that*

$$\int_{t-\tau(t)}^{t} r(s)ds \leq \frac{1}{e}\,, \qquad t \geq t_0.$$

Then Eq. (2.8.4) has a nonoscillatory solution.

Corollary 2.8.1. *If* $r(t) \equiv r > 0,\quad \tau(t) \equiv \tau > 0$, *then every solution of Eq. (2.8.4) is oscillatory if and only if* $r\tau e > 1$.

Let us consider now the delay logistic equation of the form

$$\frac{dx(t)}{dt} = r(t)x(t)\left[1 - \frac{x(t - n\tau)}{k(t)}\right] \tag{2.8.5}$$

where n is a positive integer, $\tau > 0$, $r,\ k \in C\big(R_+, (0,\infty)\big)$.

The following result is concerned with the oscillation of solution about k.

Theorem 2.8.4. *Assume that*

i) *k is τ-periodic and positive and*

$$a = \max_{t\geq 0} k(t), \quad b = \min_{t\geq 0} k(t), \quad a > b, \tag{2.8.6}$$

ii)

$$\int_0^\infty r(t)[a - k(t)]dt = \infty \tag{2.8.7}$$

and

$$\int_0^\infty r(t)[k(t) - b]dt = \infty. \tag{2.8.8}$$

Then every positive solution x of Eq. (2.8.5) is oscillatory about $k(t)$, i.e., there exists a sequence $\{t_n\}$ such that $\lim_{n\to\infty} t_n = \infty$ and $x(t_n) = \beta(t_n)$, $n = 1, 2, \ldots$.

Proof: Assume the contrary, and let $x(t)$ be a solution of (2.8.5) such that $x(t) > k(t)$ for $t \geq t_0 > 0$. Then $x'(t) < 0$, $t \geq t_0$, and hence $\lim_{t\to\infty} x(t) = \ell \geq a$ exists. We claim $\ell = a$. Otherwise, from (2.8.5) we see that for $t \geq t_1 \geq t_0$

$$x'(t) \leq \frac{\ell}{2}\left(1 - \frac{\ell}{a}\right) r(t).$$

Noting that (2.8.7) and (2.8.8) imply that $\int_0^\infty r(t)dt = \infty$. We have

$$x(t) \leq x(t_1) + \frac{\ell}{2}\left(1 - \frac{\ell}{a}\right)\int_{t_1}^t r(s)ds \to -\infty, \quad \text{as} \quad t \to \infty$$

contradicting that $x(t) > k(t) \geq b$.

From (2.8.5)

$$x'(t) \leq \frac{a}{2}\, r(t)\left[1 - \frac{a}{k(t)}\right] \leq -\frac{1}{2}\, r(t)(a - k(t))$$

for all large t, which leads to a contradiction.

Similarly, (2.8.5) has no positive solution $x(t)$ with $x(t) < k(t)$ eventually. The proof is complete.

The following provides a sufficient condition for the existence of a solution of (2.8.5) with $x(t) > k(t)$ eventually.

Theorem 2.8.5. *Assume* $0 < b < a < \infty$ and $\int_{t_0}^{\infty} r(t)dt < \infty$. *Then Eq. (2.8.5) has a solution $x(t)$ satisfying that $x(t) > k(t)$ eventually.*

Proof: Choose T so large that

$$\frac{b}{2} = b^* < k(t) < a^* = a + 1, \quad t \geq T$$

and

$$\int_T^{\infty} r(t)dt \leq \varepsilon = \min\left\{\frac{b^*}{2(b^* + 2(a^* + 1))}, \frac{b^*}{(1 + a^*)(a^* + 1 - b^*)}\right\}. \tag{2.8.9}$$

Let CB be the Banach space of all bounded continuous functions defined on R_+ with the sup norm. Define a subset Ω of CB as follows:

$$\Omega = \{x \in X : \; a^* \leq x(t) \leq a^* + 1, \quad t \geq 0\}.$$

Clearly Ω is a nonempty, bounded, closed and convex subset of CB.
Define a mapping $\mathcal{T} : \Omega \to CB$ as follows:

$$(\mathcal{T}x)(t) = \begin{cases} a^* + \int_t^{\infty} r(s)x(s)\left[\frac{x(s-n\tau)}{k(s)} - 1\right]ds, & t \geq T \\ (\mathcal{T}x)(T), & 0 \leq t \leq T. \end{cases} \tag{2.8.10}$$

For any $x \in \Omega$ and $t \geq T$

$$a^* \leq (\mathcal{T}x)(t) \leq a^* + (a^* + 1)\left(\frac{a^* + 1}{b^*} - 1\right)\int_T^{\infty} r(s)ds \leq a^* + 1.$$

Hence $\mathcal{T}\Omega \subset \Omega$.

For any $x_1, x_2 \in \Omega$, $t \geq T$ we see that

$$|(\mathcal{T}x_1)(t) - (\mathcal{T}x_2)(t)| \leq \int_t^{\infty} r(s)\Big[|x_1(s) - x_2(s)|$$
$$+ \frac{1}{k(s)}|x_1(s)x_1(s - n\tau) - x_2(s)x_2(s - n\tau)|\Big]ds$$

$$\leq \int_T^\infty r(s)\Big[\|x_1 - x_2\| + \frac{1}{b^*}x_1(s)|x_1(s-n\tau) - x_2(s-n\tau)|$$

$$+ \frac{1}{b^*}x_2(s-n\tau)|x_1(s) - x_2(s)|\Big]\, ds$$

$$\leq \left[1 + \frac{2}{b^*}(a^*+1)\right]\int_T^\infty r(s)ds\|x_1 - x_2\|$$

$$\leq \frac{1}{2}\|x_1 - x_2\|.$$

Hence

$$\|\mathcal{T}x_1 - \mathcal{T}x_2\| = \sup_{t\geq 0}|(\mathcal{T}x_1)(t) - (\mathcal{T}x_2)(t)|$$

$$= \sup_{t\geq T}|(\mathcal{T}x_1)(t) - (\mathcal{T}x_2)(t)|$$

$$\leq \frac{1}{2}\|x_1 - x_2\|,$$

i.e., $\mathcal{T}$ is a contraction on Ω. Therefore there is an $x \in \Omega$ so that $\mathcal{T}x = x$. It is easy to see that x is a solution of Eq. (2.8.5) and $x(t) > k(t)$ for $t \geq T$. The proof is complete. □

2.8.8. Generalized delay logistic equation (I)

We now consider the generalized delay logistic equation

$$\frac{dx(t)}{dt} = r(t)\,x(t)\left(1 - \frac{x(\tau(t))}{k}\right)\left|1 - \frac{x(\tau(t))}{k}\right|^{\alpha-1} \tag{2.8.11}$$

where $r,\ \tau \in C(R_+, (0,\infty))$, $\tau(t) \leq t$, $\lim_{t\to\infty}\tau(t) = \infty$, $k > 0$, $\alpha > 0$, $\alpha \neq 1$.

Make a change of variables:

$$y(t) = \frac{x(t)}{k} - 1. \tag{2.8.12}$$

Then (2.8.11) becomes

$$y'(t) = -r(t)(1 + y(t))y(\tau(t))|y(\tau(t))|^{\alpha-1}. \tag{2.8.13}$$

We are concerned with the existence of nonoscillatory solutions of (2.8.11) about k or the existence of nonoscillatory solutions of (2.8.13).

First we establish a basic lemma.

Lemma 2.8.1. *A sufficient and necessary condition for the existence of a nonoscillatory solution of Eq. (2.8.13) is that there exist constant t_1 and C_α such that the integral equation*

$$\lambda(t) = \left(1 + C_\alpha \exp\left(-\int_{t_0}^{t} r(s)\lambda(s)ds\right)\right)|C_\alpha|^{\alpha-1}$$
$$\times \exp\left(\int_{\tau(t)}^{t} r(s)\lambda(s)ds\right) + (1-\alpha)\int_{t_1}^{\tau(t)} r(s)\lambda(s)ds\Big). \tag{2.8.14}$$

has a continuous positive solution.

Proof: If Eq. (2.8.13) has an eventually positive solution $y(t)$, then there exists $t_1 > 0$ such that

$$y(\tau(t)) > 0, \quad y'(t) < 0, \quad t \geq t_1.$$

Let $\lambda(t) = -\frac{y'(t)}{r(t)y(t)}$, $y(t_1) = C_\alpha > 0$, then $\lambda(t)$ is a positive continuous solution of (2.8.14).

Similarly, if Eq. (2.8.13) has an eventually negative solution x which is larger than -1 then $-1 < C_\alpha < 0$.

Conversely, if (2.8.14) has a positive continuous solution $\lambda(t)$, then

$$y(t) = C_\alpha \exp\left(-\int_{t_1}^{t} \lambda(s)r(s)ds\right) \tag{2.8.15}$$

is a nonoscillatory solution of Eq. (2.8.13). The proof is complete. □

We discuss the following two cases for α respectively

i) $0 < \alpha < 1$

ii) $\alpha > 1$.

Theorem 2.8.6. *Assume that $\alpha \in (0,1)$. Then a necessary and sufficient* *condition for every solution of (2.8.13) to be oscillatory is*

$$\int_0^\infty r(t)dt = \infty. \tag{2.8.16}$$

Proof: i) *Necessity.* If

$$\int_0^\infty r(t)dt < \infty, \tag{2.8.17}$$

then there exists $t_0 \geq 0$ such that

$$\int_{t_0}^\infty r(s)ds \leq \frac{1}{(2-\alpha)e(1+C_\alpha)C_\alpha^{\alpha-1}} := k,$$

where C_α is a positive number.

Set $T_0 = \inf_{t \geq t_0} \tau(t)$. Let $C([T_0, \infty), R)$ denote a locally convex of all continuous functions defined on $[\tau_0, +\infty)$ endowed with the topology of uniform convergence on compact subsets of $[T_0, \infty)$. Define a subset Ω of $([T_0, \infty), R)$ as follows:

$$\Omega = \{x \in C([T_0, \infty), R) : \; x(t) \geq 0, \quad |x(t)| \leq e(1+C_\alpha)C_\alpha^{\alpha-1}, \; t \geq T_0\}.$$

Clearly, Ω is a nonempty, bounded, closed and convex subset of $C([T_0, \infty), R)$.

Define a mapping $\mathcal{T}$ on Ω as follows:

$$(\mathcal{T}x)(t) = \begin{cases} C_\alpha^{\alpha-1}\left(1 + C_\alpha \exp(-\int_{\tau_0}^t r(s)x(s)ds)\right) \\ \quad \times \exp\left((1-\alpha)\int_{T_0}^{\tau(t)} r(s)x(s)ds + \int_{\tau(t)}^t r(s)x(s)ds\right), & t \geq t_0 \\ (\mathcal{T}x)(T), & t_0 \geq t \geq T_0. \end{cases}$$

It is easy to see that $\mathcal{T}\Omega \subset \Omega$.

To prove $\mathcal{T}$ is a completely continuous operator on Ω, it is sufficient to prove that the integral operator

$$(\mathcal{H}x)(t) = \exp\left((1-\alpha)\int_{T_0}^{\tau(t)} r(s)x(s)ds + \int_{\tau(t)}^t r(s)x(s)ds\right)$$

is continuous on compact subsets of $[T_0, +\infty)$ and $\mathcal{H}\Omega$ is relatively compact.

Assume $\{x_i\}$, $x \in \Omega$ and $x_i \to x$ as $i \to \infty$ which means that $\lim_{i\to\infty} x_i(t) = x(t)$ uniformly on any compact subset of $[T_0, \infty)$. In view of

$$\int_{T_0}^{t} r(s)x(s)dx \leq eC_\alpha^{\alpha-1}(1+C_\alpha)\int_{T_0}^{t} r(s)ds < \infty$$

and using the Lebesgue's bounded convergence theorem we obtain

$$(\mathcal{H}x_i)(t) \to (\mathcal{H}x)(t), \quad \text{as} \quad i \to \infty,$$

i.e., $\mathcal{T}$ is continuous.

From the integrability of $\int_{T_0}^{\infty} r(s)x(s)ds$, it is easy to see that $\mathcal{H}\Omega$ is equicontinuous and uniformly bounded. Then $\mathcal{H}\Omega$ is relatively compact. It follows that $\mathcal{T}$ is a completely continuous operator on Ω. Using the Schauder-Tychonov fixed point theorem, there exists a $\lambda \in \Omega$ such that

$$\lambda(t) = \left(1 + C_\alpha \exp\left(-\int_{T_0}^{t} r(s)\lambda(s)ds\right)\right)$$
$$\times C_\alpha^{\alpha-1}\exp\left((1-\alpha)\int_{T_0}^{\tau(t)} r(s)\lambda(s)ds + \int_{\tau(t)}^{t} r(s)x(s)ds\right), \quad t \geq T.$$

By Lemma 2.8.1, Eq. (2.8.13) has a nonoscillatory solution.

ii) *Sufficiency.* If (2.8.13) has an eventually positive solution, by Lemma 2.8.1 there exist t_1, C_α, and a continuous function $\lambda(t)$ satisfying

$$\lambda(t) = \left(1 + C_\alpha \exp(-\int_{t_1}^{t} \lambda(s)r(s)ds)\right)$$
$$\times C_\alpha^{\alpha-1}\exp\left((1-\alpha)\int_{t_1}^{\tau(t)} \lambda(s)r(s)ds + \int_{\tau(t)}^{t} \lambda(s)r(s)ds\right)$$
$$\geq C_\alpha^{\alpha-1}\exp\left((1-\alpha)\int_{t_1}^{\tau(t)} \lambda(s)r(s)ds + \int_{\tau(t)}^{t} \lambda(s)r(s)ds\right)$$
$$\geq C_\alpha^{\alpha-1}\exp\left((1-\alpha)\int_{t_1}^{t} r(s)\lambda(s)ds\right).$$

Set

$$z(t) = \exp\left(-\int_{t_1}^{t}(1-\alpha)r(s)\lambda(s)ds\right),$$

then

$$z'(t) \le -|C_\alpha|^{\alpha-1}(1-\alpha)r(t)z(t_1).$$

Integrating it we have

$$z(t) \le z(t_1) - |C_\alpha|^{\alpha-1}(1-\alpha)z(t_1)\int_0^t r(s)ds$$

which implies that $\lim_{t\to\infty} z(t) = -\infty$, a contradiction.

Similarly, we can show that Eq. (2.8.13) has no eventually negative solution $y(t)$ with $1 + y(t) > 0$. The proof is complete.

Theorem 2.8.7. *Assume that $\alpha > 1$. Then a necessary and sufficient condition for the existence of a nonoscillatory solution of (2.8.13) is that there exists a nonnegative continuous function $\lambda(t)$ such that*

$$\exp\left((1-\alpha)\int_T^{\tau(t)} \lambda(s)r(s)ds + \int_{\tau(t)}^t \lambda(s)r(s)ds\right) \le m\lambda(t), \quad t \ge T \qquad (2.8.18)$$

where m and T are some positive constants.

Proof: i) *Sufficiency.* First we consider the case that $\int_{t_0}^\infty r(s)\lambda(s)ds < +\infty$. Then there exist ℓ, T and $C_\alpha > 0$ such that

$$\int_T^\infty \lambda(s)r(s)ds < \ell$$

$$(1 + C_\alpha)C_\alpha^{\alpha-1} < \frac{1}{m_\ell}.$$

Define a mapping $\mathcal{T}$ on $C([T_0, \infty), R_+)$ as follows

$$(\mathcal{T}y)(t) = \begin{cases} \int_t^\infty r(s)(1 + y(x))y^\alpha(\tau(s))ds, & t \ge T \\ (\mathcal{T}y)(T) + C_\alpha \exp\left(-\int_{T_0}^t r(s)\lambda(s)ds\right) & \\ \quad - C_\alpha \exp\left(-\int_{T_0}^T \lambda(s)r(s)ds\right), & T_0 \le t \le T. \end{cases}$$

Obviously, $\mathcal{T}$ is an increasing operator.

Set

$$y_0 = C_\alpha \exp\Big(- \int_T^t \lambda(s) r(s) ds \Big)$$

$$y_{n+1} = T y_n, \quad n = 1, 2, \ldots .$$

Then we see that

$$y_0(t) \geq y_1(t) \geq \cdots \geq y_n(t) \geq \cdots \geq 0, \quad t \geq T_0. \tag{2.8.19}$$

In fact,

$$\begin{aligned} y_1(t) = (T y_0)(t) &\leq \int_t^\infty r(s) \Big(1 + C_\alpha \exp\Big(- \int_T^s \lambda(u) r(u) du \Big)\Big) \\ &\quad \times C_\alpha^\alpha \exp\Big(- \alpha \int_T^{\tau(s)} \lambda(u) r(u) du \Big) ds \\ &\leq C_\alpha^\alpha (1 + C_\alpha) m \int_t^\infty r(s) \lambda(s) ds \, \exp\Big(- \int_T^t \lambda(s) r(s) ds \Big) \\ &\leq C_\alpha \exp\Big(- \int_T^t r(s) \lambda(s) ds \Big) = y_0(t), \quad t \geq T. \end{aligned}$$

Combining this together with the increment of T we obtain (2.8.19). Then $\lim_{n \to \infty} y_n(t) = y(t) \geq 0$, $t \geq T_0$ exists.

According to the Lebesgue's convergence theorem

$$y(t) = \begin{cases} \int_t^\infty r(s)\big(1 + y(s)\big) y^\alpha\big(\tau(s)\big) ds, & t \geq T \\ y(T) + C_\alpha \exp\big(- \int_{T_0}^t \lambda(s) r(s) ds \big) & \\ \quad - C_\alpha \exp\big(- \int_{T_0}^T \lambda(s) r(s) ds \big) & T_0 \leq t \leq T. \end{cases}$$

It is easy to see that $y(t) > 0$ on $[T_0, T)$ and hence $y(t) > 0$ for all $t \geq T_0$. Therefore y is a positive solution of Eq. (2.8.13) on $[T, \infty)$.

We now consider the case: $\int_{t_0}^\infty r(s) \lambda(s) ds = \infty$. Using (2.8.18) we see that for sufficiently large $t \geq T$

$$\frac{\int_t^\infty r(u)(1+C_\alpha \exp(-\int_T^u r(s)\lambda(s)ds)C_\alpha^{\alpha-1}\exp(-\alpha\int_T^{\tau(t)} r(s)\lambda(s)ds))du}{\exp(-\int_T^t r(s)\lambda(s)ds)}$$

$$\leq (1+C_\alpha)C_\alpha^{\alpha-1}m < \frac{1}{\ell} < 1.$$

Hence there exists a sufficiently large T_1 such that

$$\int_t^\infty r(u)\Big(1+C_\alpha \exp\Big(-\int_T^u r(s)\lambda(s)ds\Big)C_\alpha^{\alpha-1}\exp\Big(-\alpha\int_T^{\tau(u)} r(s)\lambda(s)ds\Big)\Big)du$$
$$\leq \exp\Big(-\int_T^t r(s)\lambda(s)ds\Big), \quad t \geq T_1.$$

By using the similar method to the previous one, we can show that (2.8.13) has a nonoscillatory solution.

ii) *Necessity.* If (2.8.13) has an eventually positive solution then from Lemma 2.8.1 there exists a continuous positive function $\lambda(t)$ such that

$$\lambda(t) = \Big(1+C_\alpha \exp\Big(-\int_{t_1}^t \lambda(s)r(s)ds\Big)\Big)$$
$$\times C_\alpha^{\alpha-1}\exp\Big((1-\alpha)\int_{t_1}^{\tau(t)} r(s)\lambda(s)ds + \int_{\tau(t)}^t r(s)\lambda(s)ds\Big)$$
$$\geq C_\alpha^{\alpha-1}\exp\Big((1-\alpha)\int_{t_1}^{\tau(t)} r(s)\lambda(s)ds + \int_{\tau(t)}^t r(s)\lambda(s)ds\Big). \tag{2.8.20}$$

Let $m = \frac{1}{C_\alpha^{\alpha-1}}$. Then (2.8.20) implies (2.8.18).

If (2.8.13) has an eventually negative solution, then

$$\lambda(t) \geq (1-|C_\alpha|)|C_\alpha^{\alpha-1}|\exp\Big((1-\alpha)\int_{t_1}^{\tau(t)} \lambda(s)r(s)ds + \int_{\tau(t)}^t \lambda(s)r(s)ds\Big),$$

where $|C_\alpha| < 1$. Thus (2.8.18) is true also. The proof is complete.

In the following we derive some explicit sufficient conditions for the existence of nonoscillatory solutions of (2.8.13).

Corollary 2.8.1. *If $\alpha > 1$ and*

$$(1-\alpha)\int_{t_1}^{\tau(t)} r(s)ds + \int_{\tau(t)}^{t} r(s)ds \leq A, \quad t \geq t_0 \geq 0, \tag{2.8.21}$$

where A is some positive number. Then Eq. (2.8.13) has a nonoscillatory solution.

In fact, (2.8.21) follows from (2.8.18) by taking $\lambda(t) \equiv \lambda > 0$.

Remark 2.8.1. If

$$\int_0^\infty r(t)dt < \infty$$

or

$$\int_0^\infty r(t)dt = \infty, \quad \text{and} \quad \int_{\tau(t)}^{t} r(s)ds \leq A, \quad t \geq t_0 \geq 0.$$

Then (2.8.21) is satisfied.

Example 2.8.1. Consider

$$y'(t) = -\frac{1}{t+1}\,(1+y(t))y^2(\sqrt{t}). \tag{2.8.22}$$

Since

$$-\int_{t_0}^{\sqrt{t}} \frac{ds}{1+s} + \int_{\sqrt{t}}^{t} \frac{ds}{1+s} = \ln(1+t) - \ln(1+t+\sqrt[2]{t}) + \ln(\sqrt{t_0}+1)$$
$$\leq \ln(\sqrt{t_0}+1),$$

the condition (2.8.21) is satisfied. By Corollary 2.8.1, Eq. (2.8.22) has a nonoscillatory solution. In fact, $y(t) = \frac{1}{t}$ is such a solution.

2.8.9. Generalized delay logistic equation (II)

Here we discuss the equation

$$x'(t) = r(t)x(t)\,\frac{1-x(t-\tau(t))}{1-cx(t-\tau(t))} \tag{2.8.23}$$

where $c \in (0,1)$, r, $\tau \in C(R_+, R_+)$, $\tau(t) \leq t$, $\lim_{t\to\infty}(t-\tau(t)) = \infty$. When $c = 0$, (2.8.23) reduces to (2.8.1).

From an ecological point of view, we will restrict our attention to the bounded positive solutions of (2.8.23). Denote

$$E_0 = \{t-\tau(t) : t-\tau(t) \leq 0,\ t \geq 0\} \cup \{0\}.$$

The initial conditions for (2.8.23) which we consider are

$$c^{-1} > x(\theta) = \varphi(\theta) \geq 0, \quad \theta \in E_0, \quad \varphi(0) > 0 \tag{2.8.24}$$

where $\varphi(\theta)$ is continuous on E_0. Assume that $\tau(t) \geq \tau > 0$ for all $t \geq 0$. It is known, by the method of steps, the initial problem (2.8.23) and (2.8.24) always (locally) has a unique positive solution.

Make the change of variables

$$y(t) = \frac{1-x(t)}{1-cx(t)}\,.$$

Then (2.8.23) and (2.8.24) reduce to

$$y'(t) = -\frac{r(t)}{1-c}\,(1-y(t))(1-cy(t))y(t-\tau(t)), \tag{2.8.25}$$

and

$$y(\theta) = [1-\varphi(\theta)]/[1-c\varphi(\theta)], \quad \theta \in E_0. \tag{2.8.26}$$

Obviously, $-\infty < y(\theta) \leq 1$, $y(0) < 1$. Thus, the study of the oscillatory behavior of solutions of (2.8.23) about 1 is equivalent to the study of the oscillatory behavior of solutions of (2.8.25) with respect to zero.

It is easy to see that the solution of (2.8.23) and (2.8.24) is positive, which implies that the solution of (2.8.25) and (2.8.26) satisfies $1-y(t) > 0$ and $1-cy(t) > 0$.

We call a solution of a differential equation global, if it exists for all $t \geq 0$. In the following, we will consider global solutions of (2.8.23). For this we first give a sufficient condition to guarantee the existence of global solutions of (2.8.23).

Theorem 2.8.8. *Assume that $t-\tau(t)$ is continuous and nondecreasing, $\lim_{t\to\infty}(t-\tau(t) = \infty$, and $\tau(t) \geq \tau > 0$ for $t \geq 0$. Assume further that*

$r(t) \geq r_0 > 0$ is continuous,

$$\sigma = \sup_{t \geq 0} \left\{ \int_{t-\tau(t)}^{t} r(\theta)\, d\theta \right\}$$

is finite, $e^{\sigma} < c^{-1}$, and that $0 < \varphi(\theta) < c^{-1}$ for $\theta \in [-T(0), 0)$ and $\varphi(0) < c^{-1}e^{-\sigma}$. Then the solutions $x(t)$ of (2.8.23) are bounded, and $\limsup_{t\to\infty} x(t) \leq e^{\sigma}$.

Proof: Since $\tau(t) \geq \tau > 0$ for $t \geq 0$, there exists a $\bar{t} > 0$ such that $\bar{t} = \tau(\bar{t})$. For $0 \leq t \leq \bar{t}$, we have

$$x(t) = x(0)\exp\left(\int_0^t r(\theta)\, \frac{1 - x(\theta - \tau(\theta))}{1 - cx(\theta - \tau(\theta))} d\theta \right). \tag{2.8.27}$$

Since $0 \leq \varphi(\theta) < c^{-1}$, $\quad 1 - u(\theta - \tau(\theta))/(1 - cu(\theta - \tau(\theta))) \leq 1$, thus for $0 \leq t \leq \bar{t}$

$$\int_0^t r(\theta)\, \frac{1 - x(\theta - \tau(\theta))}{1 - cx(\theta - \tau(\theta))}\, d\theta \leq \int_0^t r(\theta) d\theta \leq \sigma. \tag{2.8.28}$$

Obviously, (2.8.27) and (2.8.28) imply that

$$x(t) \leq x(0)e^{\sigma} < c^{-1}, \quad \text{for} \quad 0 \leq t \leq \bar{t}.$$

Hence we have shown $0 \leq x(t) < c^{-1}$ for $-\tau(0) \leq t \leq \bar{t}$. Assume that $t_1 - \tau(t_1) = \bar{t}$. Then for $t \leq t_1$, we have

$$x'(t) = r(t)x(t)\left(1 - \frac{(1-c)x(t - \tau(t))}{1 - cx(t - \tau(t))}\right) \leq r(t)x(t).$$

Thus

$$x(t) \leq x(t_0)\exp \int_{t_0}^{t} r(\theta) d\theta.$$

From this we have

$$x\big(t - \tau(t)\big) \geq x(t)\exp\left(- \int_{t-\tau(t)}^{t} r(\theta) d\theta \right)$$

and hence

$$x(t-\tau(t)) \geq x(t)e^{-\sigma}.$$

This implies for $\bar{t} \leq t \leq t_1$

$$x'(t) \leq r(t)x(t)\left(\frac{1-x(t)e^{-\sigma}}{1-cx(t)e^{-\sigma}}\right).$$

Since $r(t) \geq r_0 > 0$, we see that all positive solutions of

$$x'(t) = r(t)x(t)\left(\frac{1-x(t)e^{-\sigma}}{1-cx(t)e^{-\sigma}}\right) \tag{2.8.29}$$

with initial values less than δ, for $\delta > 0$, $e^{\sigma} < \delta < c^{-1}e^{\sigma}$, will be bounded by δ and have e^{σ} as their limit. Therefore, for $E \leq t \leq t_1$, $0 \leq x(t) < c^{-1} < c^{-1}e^{\sigma}$. By repeating this argument (assume that $t_2 > t_1$ is such that $t_2 - \tau(t_2) = t_1$, we can prove that (2.8.29) holds for $t_1 \leq t \leq t_2$. Define $t_{i+1} - \tau(t_{i+1}) = t_i$, then $t_i \to +\infty$ as $i \to +\infty$), we see $x(t)$ is bounded and satisfies (2.8.29) for all $t \geq \bar{t}$, thus the solution $x(t)$ of (2.8.23) is bounded by $x(0)e^{\sigma} < c^{-1}$ and

$$\limsup_{t\to\infty} x(t) \leq e^{\sigma}.$$

This completes the proof of the theorem.

The corresponding result for Eq. (2.8.25) is the following.

Theorem 2.8.9. *Assume that $r(t)$ and $\tau(t)$ are the same as described in Theorem 2.8.8, and the initial function $y(\theta)$, $\theta \in [-\tau(0), 0]$ satisfies $y(\theta) \leq 1$, $y(0) > (1-c^{-1}e^{-\sigma})/(1-e^{-\sigma})$. Then the solution $y(t)$ of (2.8.25) is bounded, and*

$$\liminf_{t\to\infty} y(t) \geq (1-e^{\sigma})/(1-ce^{\sigma}).$$

Remark 2.8.2. As we mentioned before, when $c = 0$ (2.8.23) reduces to the logistic equation. In general, from the ecological point of view c is small, so that equation (2.8.23) is close to the logistic equation, and hence $e^{\sigma} < c^{-1}$ holds naturally.

In the following we will consider only global solutions of (2.8.23).

Theorem 2.8.10. *Assume that $r(t)$ and $\tau(t)$ are positive and continuous, $\lim_{t\to\infty}\big(t-\tau(t)\big)=\infty$, $\int_0^\infty r(t)dt=\infty$. Then every global solution of (2.8.23) is either oscillatory with respect to 1 or tends to 1 as $t\to\infty$.*

Proof: By Theorem 2.8.8, a global solution $x(t)$ of Eq. (2.8.23) satisfies $0<x(t)<C^{-1}$. Then $x'(t)<0$ for $t>T^*$, where $T^*-\tau(T^*)>T$, and hence

$$\lim_{t\to\infty} x(t)=\mu\geq 1 \quad \text{exists.} \tag{2.8.30}$$

If $\mu>1$, then we have

$$x'(t)\leq r(t)\mu\,\frac{1-\mu}{1-c\mu}, \quad \text{for} \quad t\geq T^*,$$

which leads to $\lim x(t)=-\infty$ which contradicts (2.8.30). Therefore $\mu=1$.

Similarly, we can prove if $0<x(t)<1$, for $t\geq T$, then $\lim_{t\to\infty} x(t)=1$.

Next, we would like to discuss oscillation for Eq. (2.8.25).

Theorem 2.8.11. *Let $r,\ \tau\in C(R_+,(0,\infty))$, with $\lim_{t\to\infty}\big(t-\tau(t)\big)=\infty$,*

$$\liminf_{t\to\infty}\int_{t-\tau(t)}^{t} r(\theta)d\theta>\frac{1-c}{e}\,.$$

Then every solution of Eq. (2.8.25) oscillates.

Proof: Let $y(t)$ be a negative solution of (2.8.25), then

$$y'(t)+\frac{r(t)}{1-C}\,y\big(t-\tau(t)\big)\geq 0. \tag{2.8.31}$$

By Theorem 2.1.1 this is impossible under the assumptions of the theorem.

Now let $y(t)$ be a positive solution of (2.8.25). Then $y'(t)<0$ and $y(t)\to 0$ as $t\to\infty$ by Theorem 2.8.10.

Dividing (2.8.25) by $y(t)$ and integrating from $t-\tau(t)$ to t we have

$$\ln\frac{y(t-\tau(t))}{y(t)}=\frac{1}{1-c}\int_{t-\tau(t)}^{t} r(s)\big(1-y(s)\big)\big(1-cy(s)\big)\,\frac{y(s-\tau(s))}{y(s)}\,ds \tag{2.8.32}$$

Let w be defined by

$$w(t) = \frac{y(t-\tau(t))}{y(t)} .$$

Clearly $w(t) \geq 1$. From (2.8.32) we have

$$\ln\, w(t) = \frac{w(\xi)}{1-c} \int_{t-\tau(t)}^{t} r(s)\big(1-y(s)\big)\big(1-cy(s)\big)ds \tag{2.8.33}$$

where $\xi \in \big(t-\tau(t), t\big)$. Similar to Lemma 2.1.3 we can prove that w is bounded. Taking the lim inf on both sides of (2.8.33) we obtain

$$\frac{\ln\, \ell}{\ell} \geq \liminf_{t\to\infty} \frac{1}{1-c} \int_{t-\tau(t)}^{t} r(s)ds \tag{2.8.34}$$

where $\ell = \liminf_{t\to\infty} w(t)$. Since $\frac{\ln\, \ell}{\ell} \leq \frac{1}{e}$, (2.8.34) contradicts the assumptions of the theorem. □

Theorem 2.8.12. *Suppose $r(t)$ is bounded above, and there is a $t_1 > 0$, such that*

$$\int_{t-\tau(t)}^{t} r(\theta)d\theta \leq \frac{1-c}{e} \quad \textit{for} \quad t \geq t_1.$$

Then Eq. (2.8.25) has a bounded nonoscillatory solution on $[t, \infty)$.

Proof: Assume $r(t) \leq M$ for all $t \geq 0$. Let BC denote the space of all bounded continuous functions defined on R_+, with the supernorm. Let $\Omega \subset BC$ denote the subset satisfying:

(i) $y(t)$ is Lipschitzian and nonincreasing on R_+

(ii) $\text{lip}\;\; y(t)|_{R_+} \leq Me(1-a)/(1-c)$

(iii) $y(t) \equiv 1-a, \quad t \in [0,t], \quad a \in (0,1)$

(iv) $e\, y\big(t-\tau(t)\big) \geq y(t)e \geq y\big(t-\tau(t)\big), \quad t \geq t_1$

(v) $(1-a)\exp\big(-\frac{e}{1-c}\int_{t_1}^{t} r(\theta)d\theta\big) \leq y(t) \leq 1-a, \quad t \geq t_1$

$$\text{(vi)} \quad |y'(t)| \leq Me\,\frac{1-a}{1-c}, \quad t \in R_+, \quad t \geq t_1$$

where lip $y(t)|_{R_+}$ denotes the Lipschitzian constant of y over R_+. Let

$$y_0(t) = \begin{cases} 1-a, & t \in [0, t_1] \\ (1-a)\exp\big(-\frac{e}{1-c}\int_{t_1}^{t} r(\theta)d\theta\big), & t \geq t_1. \end{cases}$$

Then $y_0(t) \in \Omega$ and S is nonempty. It is not difficult to see that Ω is convex and compact.

Define a mapping $\mathcal{T} : \Omega \to BC$ as follows:

$$(\mathcal{T}y)(t) = \begin{cases} 1-a, & t \in [0, t_1] \\ (1-a)\exp\big(-\frac{1}{1-c}\int_{t_1}^{t} r(\theta) & \\ \quad \times (1-y(\theta))(1-cy(\theta))\,\frac{y(\theta-\tau(\theta))}{y(\theta)}\,d\theta\big), & t \geq t_1. \end{cases}$$

It is easy to see that $\mathcal{T}\Omega \subset \Omega$, and $\mathcal{T}$ is completely continuous. Hence, by the Schauder-Tychonov fixed point theorem, we conclude that $\mathcal{T}$ has a fixed point y in S. Obviously, y^* is a positive, bounded solution of Eq. (2.8.25) on $[t, \infty)$. The proof is complete.

Corollary 2.8.2. *If $r(t) \equiv r$, $\tau(t) \equiv \tau$, r and τ are positive constants. Then every solution of (2.8.25) is oscillatory if and only if $r\tau e > 1 - c$.*

2.9. Notes

Lemma 2.1.1 is obtained by Koplatadze and Canturija [97]. Lemma 2.1.2 is due to Kwong [101]. Lemma 2.1.3 is taken from Yu, Wang, Zhang and Qian [184], also see Jian [92], Erbe and Zhang [44].

Lemma 2.1.4 is adopted from [92], also see Elbert and Stavroulakis [33]. Theorem 2.1.1 is obtained by Koplatadze and Canturiza [97], a special case is by Ladas and Ladde, see [110]. Theorem 2.1.2 is based on [92] and [101], and related work can be found also in [184]. Theorem 2.1.3 and Theorem 2.1.4 are taken from Zhou [219], Theorem 2.1.5 is new. The results in Section 2.2 are taken from Li [124]. Theorem 2.3.1 and 2.3.2 are new, Theorem 2.3.3 and 2.3.4 are taken from Györi [72]. The materials of the Section 5.4 are based on [184]. Theorem 2.5.1

and 2.5.2 are taken from Zhang and Gopalsamy [203], also studied by Ladas and Qian [106]. Theorem 2.6.1 is obtained by Tramov and Ladas, etc., by a different method, see [110 or 76]. Theorems 2.6.2 and 2.6.3 are extracted from Huang [84]. The proposition is established by many authors such as Györi [74], Kwong [101], Li [126], Philos [150], Chen and Huang [15] and Chuanxi, Ladas, Zhang and Zhao [25]. Theorems 2.6.4 and 2.6.5 are based on Chen and Huang [15]. Corollary 2.6.2 is obtained by Hunt and Yorke [85] using a different method. Lemma 2.6.4, Theorem 2.6.6, and Corollary 2.6.4 are taken from Philos [150]. Corollary 2.6.5 is obtained by several authors, Wei [170], Györi and Ladas [76]. Theorem 2.6.7 is based on Lalli and Zhang [117]. Theorem 2.6.8 is an improved form of a result by Györi and Ladas [76]. Theorems 2.6.9, 2.6.10 and Corollaries 2.6.7-2.6.10 are taken from Nadareishvili [139]. An oscillation criterion for Eq. (2.6.5) with oscillating coefficients can be found in Yu, Wang, Zhang and Qian [184]. The first study of the oscillation of differential equations with advanced type is due to Zhang and Ding [199], and Zhang [202]. The corresponding results for differential inequalities with advanced type is due to Onose, see [76]. Theorems 2.6.11 and 2.6.12 are found in Zhou [217]. Theorem 2.7.1 is from Zhang [191], also see Erbe and Zhang [44]. Theorem 2.7.2 is a new result by Yu and Zhang. Theorems 2.8.1-2.8.3 are taken from Zhang and Gopalsamy [208]. The global attractivity of the delay logistic equation is studied by Zhang and Gopalsamy [207]. Theorems 2.8.4 and 2.8.5 are obtained by J.S. Yu and B.G. Zhang. Theorems 2.8.6 and 2.8.7 are due to Li [125], also see Aiello [4]. Eq. (2.8.23) without delay as an ecological model of the single species is posed by Cui and Law [27]. Theorems 2.8.8 - 2.8.12 are taken from Kuang, Zhang and Zhao [70], in which some results for the existence of periodic solutions and the global attractivity of solutions are obtained.

There are many important results for oscillation and nonoscillation of nonlinear delay differential equations in the literature. Some of then can be found in the monographs by Ladde, V. Lakshmikantham and Zhang [110], Györi and Ladas [76].

Some further results for delay ecological equations can also be found in the monograph by Gopalsamy [51].

3

Oscillation of First Order Neutral Differential Equations

3.0. Introduction

In general, the theory of neutral delay differential equations is more complicated than the theory of delay differential equations without neutral terms. For example, Snow (see also Gyori and Ladas, Chapter 6, [76]) has shown that even though the characteristic roots of a neutral differential equation may all have negative real parts, it is still possible for some solutions to be unbounded.

In this chapter, we will present a systematic study for the oscillation theory of first order NDDEs, which contains some recent results and consequently is a useful source for researchers in this field.

In Section 3.1 we consider the linear neutral delay differential equations with constant parameters, where the "characteristic equation" method plays an important role. Some necessary and sufficient conditions for the characteristic equations to have no real roots are presented.

In Sections 3.2 and 3.3 we consider the neutral delay differential equations with variable coefficients of the form

$$\big(x(t) + p(t)x(\tau(t))\big)' + q(t)x\big(t - \sigma(t)\big) = 0 \tag{3.0.1}$$

where $q(t) \geq 0$ and $-1 \leq p(t) \leq 0$, $p(t) \equiv -1$, $p(t) \leq -1$, or $p(t) > 0$, respectively.

Since the behavior of solutions of Eq. (3.0.1) is closely dependent on the range of $p(t)$, we deal with the case that $-1 \leq p(t) \leq 0$ in Section 3.2, and the other cases in Section 3.3. Some sharp conditions for all solutions to be oscillatory and for the existence of positive solutions are presented.

In Section 3.4, we present a comparison result for oscillation. In Section 3.5, we consider Eq. (3.0.1) in the unstable case. In Section 3.6, we discuss a class of sublinear neutral delay differential equations.

In Sections 3.7 and 3.8, we deal with the neutral delay differential equations with positive and negative coefficients of the form

$$\big(x(t) - c(t)x(t - r)\big)' + p(t)x(t - \tau) - q(t)x(t - \sigma) = 0.$$

Oscillation criteria, criteria for the existence of positive solutions and results of linearized oscillation are given.

In Section 3.9, we are concerned with the neutral delay differential equations with a nonlinear neutral term. In particular, the equation

$$\big(x(t) + px^{\alpha}(t - \tau)\big)' + q(t)x^{\beta}(t - \sigma) = 0$$

is considered, where $\alpha \neq 1$, $\beta \neq 1$.

The deviating arguments in the neutral differential equation (3.0.1) can be of various types. Usually, we investigate the following cases:

1) $\tau(t) \leq t, \quad \sigma(t) \geq 0,$ 2) $\tau(t) \leq t, \quad \sigma(t) \leq 0,$

3) $\tau(t) \geq t, \quad \sigma(t) \geq 0,$ 4) $\tau(t) \geq t, \quad \sigma(t) \leq 0,$

and the mixed type.

In this chapter we mainly discuss case 1), which corresponds to the neutral delay differential equations. But the methods and techniques employed here can be used for equations of other types.

3.1. Characteristic Equations

We first state a basic result on the oscillation of the neutral delay differential equation with constant parameters

$$\left(x(t)+\sum_{j=1}^{m} p_j x(t-\tau_j)\right)' + \sum_{i=1}^{k} q_i x(t-\sigma_i), \quad t \geq t_0 \tag{3.1.1}$$

where p_j, τ_j, and q_i, σ_i are given real numbers, and $\tau_j,\ \sigma_i \geq 0,\ j \in I_m,\ i \in I_k$.

Theorem 3.1.1. *Every solution of Eq. (3.1.1) is oscillatory if and only if its characteristic equation*

$$\lambda + \lambda \sum_{j=1}^{m} p_j exp(-\lambda\tau_j) + \sum_{i=1}^{k} q_i\ exp(-\lambda\sigma_i) = 0, \tag{3.1.2}$$

has no real roots.

The above proposition can be extended to higher order NDEs.

From Theorem 3.1.1 the characteristic equation (3.1.2) plays an important role in the investigation of the oscillation and asymptotic behavior of solutions of Eq. (3.1.1). But to determine if (3.1.2) has a real root is quite a problem itself. In the following we derive other necessary and sufficient conditions for the oscillation of Eq. (3.1.1) which can be easily applied to get some explicit sufficient conditions.

For sake of convenience, we consider

$$\left(x(t) - p_1 x(t-\tau_1) - p_2 x(t-\tau_2)\right)' + \sum_{i=1}^{m} q_i x(t-\sigma_i) = 0 \tag{3.1.3}$$

where $p_j\ (j=1,2,),\quad q_i,\ \sigma_i\ (i=1,2,\ldots,m) \in R_+,\quad \tau_i > 0\ (i=1,2)$.

From Theorem 3.1.1 Eq. (3.1.3) is oscillatory if and only if its characteristic equation

$$F(\lambda) = \lambda(1 - p_1 e^{-\lambda\tau_1} - p_2 e^{-\lambda\tau_2}) + \sum_{i=1}^{m} q_i e^{-\lambda\sigma_i} = 0 \tag{3.1.4}$$

has no real roots, or, $F(\lambda) > 0$ for all $\lambda \in R$.

From (3.1.4) we see that if $p_1 + p_2 = 1$ then every solution of Eq. (3.1.3) is oscillatory.

For the case that $0 < p_1 + p_2 < 1$, we define a subset D of the space ℓ^1 and a function f on D as follows:

$$D = \{t = (t_{ijk}) : (ijk) \in I_1,\ t_{ijk} \geq 0,\ \Sigma_1\ t_{ijk} = 1\}$$

and

$$f(t) = \Sigma_1\ \frac{t_{ijk}}{i\tau_1 + (k-i)\tau_2 + \sigma_j} \ln\left(eC_k^i p_1^i p_2^{k-i} q_j [i\tau_1 + (k-i)\tau_2 + \sigma_j]/t_{ijk}\right) \tag{3.1.5}$$

where $I_1 = \{(ijk) :\ j = 1, \ldots, m;\ k = 0, 1, \ldots;\ i = 0, \ldots, k\}$, $\Sigma_1 = \sum_{(ijk)\in I_1}$, and in (3.1.5) $t_{ijk} = 0$ implies the corresponding terms vanish. Here $C_k^i = \frac{k!}{i!(k-i)!}$.

It is easy to see that the series in (3.1.5) is convergent in D and then $f(t)$ is well-defined on D. In fact, for $t \in D$

$$\Sigma_1\ \frac{t_{ijk}}{i\tau_1 + (k-i)\tau_2 + \sigma_j} \ln\left(eC_k^i p_1^i p_2^{k-i} q_j [i\tau_1 + (k-i)\tau_2 + \sigma_j]\right) < \infty$$

by the comparison test and the definition of D. Also

$$\left|\Sigma_1\ \frac{t_{ijk}}{i\tau_1 + (k-i)\tau_2 + \sigma_j} \ln\ t_{ijk}\right|$$

$$\leq \left(\Sigma_1\ \frac{1}{(i\tau_1 + (k-i)\tau_2 + \sigma_j)^2}\right)^{1/2} \left(\Sigma_1 t_{ijk}^2 (\ln t_{ijk})^2\right)^{1/2},$$

the right hand side is convergent for $t \in D$.

Theorem 3.1.2. *Assume* $0 < p_1 + p_2 < 1$. *Then*

i) $f(t)$ *achieves its maximum at a point* $t^0 = (t_{ijk}^0)$ *on* D *which is completely determined by the condition that*

$$\frac{1}{i\tau_1 + (k-i)\tau_2 + \sigma_j} \ln\left(C_k^i p_1^i p_2^{k-i} q_j [i\tau_1 + (k-i)\tau_2 + \sigma_j]/t_{ijk}^0\right) \tag{3.1.6}$$

has the same value for all $(ijk) \in I_1,\ t_{ijk}^0 \neq 0$.

ii) *Eq. (3.1.3) is oscillatory if and only if* $f(t^0) > 0$.

For the case that $p_1 + p_2 > 1$, we define the following three subsets D_b, $b = 1, 2, 3$, of the space ℓ^1, and three functions $h_b(t)$ defined on D_b, $b = 1, 2, 3$, respectively: (We have in the expansions below $C^i_{-(k+1)} = \frac{(-k-1)(-k-2)\dots(-k-i)}{i!}$ $i, k = 1, 2, \dots$.)

$$D_1 = \{S = (S_{ijk}) : (ijk) \in I_2,\ S_{ijk}C^i_{-(k+1)}[(k+i+1)\tau_1 - i\tau_2 - \sigma_j] \geq 0,$$

$$\Sigma_2 S_{ijk} = 1\},$$

$$D_2 = \{S = (S_{ijk}) : (ijk) \in I_2,\ S_{ijk}C^i_{-(k+1)}[(k+i+1)\tau_2 - i\tau_1 - \sigma_j] \geq 0,$$

$$\Sigma_2 S_{ijk} = 1\},$$

$$D_3 = \{S = (S_{jk}) : (j, k) \in I_3,\ S_{jk}[(k+1)\tau_1 - \sigma_j] \geq 0,\ \Sigma_3 S_{jk} = 1\},$$

$$h_1(S) = \Sigma_2 \frac{S_{ijk}}{(k+i+1)\tau_1 - i\tau_2 - \sigma_j} \times \ln\left(eC^i_{-(k+1)}p_1^{-(k+i+1)}p_2^i q_j[(k+i+1)\tau_1 - i\tau_2 - \sigma_j]/S_{ijk}\right), \tag{3.1.7}$$

$$h_2(S) = \Sigma_2 \frac{S_{ijk}}{(k+i+1)\tau_2 - i\tau_1 - \sigma_j} \times \ln\left(eC^i_{-(k+1)}p_1^i p_2^{-(k+i+1)} q_j[(k+i+1)\tau_2 - i\tau_1 - \sigma_j]/S_{ijk}\right), \tag{3.1.8}$$

$$h_3(S) = \Sigma_3 \frac{S_{jk}}{(k+1)\tau_1 - \sigma_j} \ln\left(e(2p_1)^{-(k+1)}q_j[(k+1)\tau_1 - \sigma_j]/S_{jk}\right); \tag{3.1.9}$$

where $I_2 = \{(ijk) :\ j = 1, \dots, m;\ k,\ i = 1, 2, \dots\}$, $I_3 = \{(jk) : j = 1, \dots, m,$ $k = 0, 1, \dots\}$, $\Sigma_2 = \sum_{(ijk)\in I_2}$, $\Sigma_3 = \sum_{(jk)\in I_3}$, where $S_{ijk} = 0$ or $S_{jk} = 0$ imply that the corresponding terms vanish. Different from the series in (3.1.5), we cannot guarantee that all the series in (3.1.7) – (3.1.9) are convergent in D_b, instead, as shown in the Remark 3.1.1, we see there exists at least one series which is convergent on D_b.

Theorem 3.1.3. *Assume $p_1 + p_2 > 1$. Then*

i) *for some $b = 1, 2$, or 3, $h_b(t)$ has a maximum at a point $S^{(b)} = (S^{(b)}_{ijk})$, $b = 1, 2$, or $S^{(3)} = (S^{(3)}_{ijk})$ on D_b, which is completely determined by one of* *the conditions that*

$$\frac{1}{(k+i+1)\tau_1 - i\tau_2 - \sigma_j}$$
$$\times \ln\left(C^i_{-(k+1)} p_1^{-(k+i+1)} p_2^i q_j[(k+i+1)\tau_1 - i\tau_2 - \sigma_j]/S^{(1)}_{ijk}\right), \tag{3.1.10}$$

$$\frac{1}{(k+i+1)\tau_2 - i\tau_1 - \sigma_j}$$
$$\times \ln\left(C^i_{-(k+1)} p_1^i p_2^{-(k+i+1)} q_j[(k+i+1)\tau_2 - i\tau_1 - \sigma_j]/S^{(2)}_{ijk}\right), \tag{3.1.11}$$

or

$$\frac{1}{(k+1)\tau_1 - \sigma_j} \ln\left((2p_1)^{-(k+1)} q_j[(k+1)\tau_1 - \sigma_j]/S^{(3)}_{jk}\right) \tag{3.1.12}$$

has the same value for all $(ijk) \in I_2$ *or* $(jk) \in I_3$, *such that* $S^{(d)}_{ijk} \neq 0,\ d = 1, 2$, *or* $S^{(3)}_{jk} \neq 0$, *and* $(k+i+1)\tau_1 - i\tau_2 - \sigma_j \neq 0$, *or* $(k+1)\tau_1 - \sigma_j \neq 0$.

ii) *Eq. (3.1.3) is oscillatory if and only if* $0 < h_b(S^{(b)}) < \infty$ *for some* $b = 1, 2$, *or* 3.

Theorems 3.1.2 and 3.1.3 can be applied to get a series of sufficient conditions for oscillation of Eq. (3.1.3). To use Theorem 3.1.2 we need to find a $t^* \in D$ such that $f(t^*) > 0$, whereas, to use Theorem 3.1.3, we need firstly to choose a suitable function h_b which is convergent on D_b, and then find a $S^* \in D_b$ such that $h_b(S^*) > 0$. The following corollaries are derived from Theorems 3.1.1 and 3.1.3, where

$$q = \left(\prod_{j=1}^m q_j\right)^{1/m}, \quad \sigma = \frac{1}{m}\sum_{j=1}^m \sigma_j.$$

Corollary 3.1.3. *Assume* $p_1 + p_2 < 1$. *Then each one of the following is sufficient for Eq. (3.1.3) to be oscillatory:*

i) $\Sigma_1 C^i_k p_1^i p_2^{k-i} q_j[i\tau_1 + (k-i)\tau_2 + \sigma_j] \geq \frac{1}{e}$, $\quad$ (3.1.13)

ii) $mq \sum_{k=0}^{\infty} \sum_{i=0}^{k} C_k^i p_1^i p_2^{k-i}[i\tau_1 + (k-i)\tau_2 + \sigma] \geq \frac{1}{e}$,

iii) *let* $a = \Sigma_1 C_k^i p_1^i p_2^{k-i} q_j$, $\alpha_{ijk} = \frac{C_k^i p_1^i p_2^{k-i} q_j}{i\tau_1 + (k-i)\tau_2 + \tau_j}$, $(ijk) \in I_1$, and $\Sigma_1 \frac{\alpha_{ijk}}{a} \ln(ea[i\tau_1 + (k-i)\tau_2 + \sigma_j]) \geq 0$.

Corollary 3.1.4. *Assume* $p_1 + p_2 > 1$.

i) *Let* $\tau_1 \neq \tau_2$, and

$$\Sigma_3 (2p_1)^{-(k+1)} q_j [(k+1)\tau_1 - \sigma_j] e^{\left(\frac{1}{\tau_1 - \tau_2} \ln \frac{p_1}{p_2}\right)[(k+1)\tau_1 - \sigma_j]} = 1. \quad (3.1.14)$$

Then Eq. (3.1.3) is oscillatory if and only if

$$\Sigma_3 (2p_1)^{-(k+1)} q_j e^{\left(\frac{1}{\tau_1 - \tau_2} \ln \frac{p_1}{p_2}\right)[(k+1)\tau_1 - \sigma_j]} - \frac{1}{\tau_1 - \tau_2} \ln \frac{p_1}{p_2} > 0. \quad (3.1.15)$$

Otherwise, each one of the following is sufficient for Eq. (3.1.3) to be oscillatory:

ii)
$$\sum_{(ijk) \in I_2^*} C_{-(k+1)}^i p_1^{-(k+i+1)} p_2^i q_j [(k+i+1)\tau_1 - i\tau_2 - \sigma_j] \geq \frac{1}{e}, \quad (3.1.16)$$

or

$$mq \sum_{(ij) \in J_2^*} C_{-(k+1)}^i p_1^{-(k+i+1)} p_2^i q_j [(k+i+1)\tau_1 - i\tau_2 - \sigma] \geq \frac{1}{e} \quad (3.1.17)$$

where

$$I_2^* = \Big\{(ijk) : C_{-(k+1)}^i > 0, \quad (k+i+1)\tau_1 - i\tau_2 - \sigma_j > 0,$$

$$(k+i+1)\tau_2 - i\tau_1 - \sigma_j > 0, \quad \text{and} \quad \left(\frac{p_1}{p_2}\right)^{k+2i+1} \frac{(k+i+1)\tau_2 - i\tau_1 - \sigma_j}{(k+i+1)\tau_1 - i\tau_2 - \sigma_j} > 1\Big\}$$

$$J_2^* = \Big\{(ik) : C_{-(k+1)}^i > 0, \quad (k+i+1)\tau_1 - i\tau_2 - \sigma > 0,$$

$$(k+i+1)\tau_2 - i\tau_1 - \sigma > 0 \quad \text{and} \quad \left(\frac{p_1}{p_2}\right)^{k+2i+1} \frac{(k+i+1)\tau_2 - i\tau_1 - \sigma}{(k+i+1)\tau_1 - i\tau_2 - \sigma} > 1\Big\}$$

iii) *(3.1.16) and (3.1.17) hold if* p_1 *and* p_2, τ_1 *and* τ_2 *exchange their positions respectively.*

It is easy to see that $I_2^* \neq \emptyset$ (empty set) and has infinitely many terms provided $p_1 > p_2$, or $p_1 = p_2$ and $\tau_1 > \tau_2$.

To prove Theorem 3.1.2 we need the following lemmas.

Lemma 3.1.1. *Assume $0 < p_1 + p_2 < 1$. Define*

$$G_1(\lambda) = \lambda + \sum_{j=1}^{m} q_j e^{-\lambda\sigma_j}(1 - p_1 e^{-\lambda\tau_1} - p_2 e^{-\lambda\tau_2})^{-1}, \tag{3.1.18}$$

and let $\lambda^* < 0$ be such that $p_1 e^{-\lambda^*\tau_1} + p_2 e^{-\lambda^*\tau_2} = 1$. Then there exists a $\lambda_0 \in (\lambda^*, \infty)$ such that

$$0 < p_1 e^{-\lambda_0\tau_1} + p_2 e^{-\lambda_0\tau_2} < 1, \tag{3.1.19}$$

and $G_1(\lambda_0)$ is the minimal value of $G_1(\lambda)$ in (λ^*, ∞).

Proof: Noting that $G_1(\infty) = \infty$ and $G_1(\lambda^* + 0) = \infty$, the conclusion is obvious by the continuity of $G_1(\lambda)$. □

Lemma 3.1.2. *Assume $0 < p_1 + p_2 < 1$, λ_0 is defined by Lemma 3.1.1, and let*

$$t_{ijk}^0 = C_k^i p_1^i p_2^{k-i} q_j [i\tau_1 + (k-i)\tau_2 + \sigma_j] e^{-\lambda_0[i\tau_1 + (k-i)\tau_2 + \sigma_j]}, \quad (ijk) \in I_1. \tag{3.1.20}$$

Then

i) $t^0 = (t_{ijk}^0) \in D$,

ii) *(3.1.6) has the same value for all $(ijk) \in I_1$, $t_{ijk}^0 \neq 0$,*

iii) $f(t^0) = G_1(\lambda_0)$,

iv) *$f(t^0)$ is the maximum value of $f(t)$ on D.*

Proof: By (3.1.19) there exists a neighborhood Λ on λ_0 such that

$$0 < p_1 e^{-\lambda\tau_1} + p_2 e^{-\lambda\tau_2} < 1$$

for all $\lambda \in \Lambda$. Expanding $(1 - p_1 e^{-\lambda\tau_1} - p_2 e^{-\lambda\tau_2})^{-1}$ we get for $\lambda \in \Lambda$

$$G_1(\lambda) = \lambda + \sum_{i=1}^{m} q_j e^{-\lambda\sigma_j} \sum_{k=0}^{\infty} (p_1 e^{-\lambda\tau_1} + p_2 e^{-\lambda\tau_2})^k$$
$$= \lambda + \Sigma_1 C_k^i p_1^i p_2^{k-i} q_j e^{-\lambda[i\tau_1 + (k-i)\tau_2 + \sigma_j]}.$$

Thus

$$G_1'(\lambda) = 1 - \Sigma_1 C_k^i p_1^i p_2^{k-i} q_j [i\tau_1 + (k-i)\tau_2 + \sigma_j] e^{-\lambda[i\tau_1 + (k-i)\tau_2 + \sigma_j]}.$$

Since $G'(\lambda_0) = 0$ we have

$$\Sigma_1 C_k^i p_1^i p_2^{k-i} q_j [i\tau_1 + (k-i)\tau_2 + \sigma_j] e^{-\lambda_0[i\tau_1 + (k-i)\tau_2 + \sigma_j]} = 1,$$

i.e., $\Sigma_1 t_{ijk}^0 = 1$, or $t^0 \in D$. From (3.1.20)

$$\frac{1}{i\tau_1 + (k-i)\tau_2 + \sigma_j} \ln \left(C_k^i p_1^i p_2^{k-i} q_j [i\tau_1 + (k-i)\tau_2 + \sigma_j] / t_{ijk}^0 \right) = \lambda_0 \qquad (3.1.21)$$

for $\ell_{ijk}^0 \neq 0$. Furthermore by (3.1.5) and (3.1.21)

$$f(t^0) = \Sigma_1 \Big\{ \frac{\ell_{ijk}^0}{i\tau_1 + (k-i)\tau_2 + \sigma_j} + \frac{\ell_{ijk}^0}{i\tau_1 + (k-i)\tau_2 + \sigma_j}$$
$$\times \ln(C_k^i p_1^i p_2^{k-i} q_j [i\tau_1 + (k-i)\tau_2 + \sigma_j] / \ell_{ijk}^0) \Big\} \qquad (3.1.22)$$
$$= \Sigma_1 \{ C_k^i p_1^i p_2^{k-i} q_j e^{-\lambda_0[i\tau_1 + (k-i)\tau_2 + \sigma_j]} + \lambda_0 \ell_{ijk}^0 \}$$
$$= G_1(\lambda_0).$$

For $t = (t_{ijk}) \in D$ from (3.1.22) we have

$$f(t^0) = \Sigma_1 \{ C_k^i p_1^i p_2^{k-i} q_j e^{-\lambda_0[i\tau_1 + (k-i)\tau_2 + \sigma_j]} + \lambda_0 t_{ijk}^0 \}$$
$$\geq \Sigma_1 \frac{t_{ijk}^0}{i\tau_1 + (k-i)\tau_2 + \sigma_j} \ln \left(e C_k^i p_1^i p_2^{k-i} q_j [i\tau_1 + (k-i)\tau_2 + \sigma_j] / t_{ijk}^0 \right)$$
$$= f(t).$$

□

Proof of Theorem 3.1.2: The first part is shown by Lemma 3.1.2. Assume Eq. (3.1.3) is oscillatory. Thus $F(\lambda) > 0$ for F defined by (3.1.4) and all $\lambda \in R$. In particular,

$$F(\lambda_0) = \lambda_0\left(1 - p_1 e^{-\lambda_0 \tau_1} - p_2 e^{-\lambda_0 \tau_2}\right) + \sum_{j=1}^{m} q_j e^{-\lambda_0 \sigma_j} > 0$$

where λ_0 satisfies $0 < p_1 e^{-\lambda_0 \tau_1} + p_2 e^{-\lambda_0 \tau_2} < 1$ by Lemma 3.1.1. Hence

$$G_1(\lambda_0) = \lambda_0 + \sum_{j=1}^{m} q_j e^{-\lambda_0 \sigma_j}\left(1 - p_1 e^{-\lambda_0 \tau_1} - p_2 e^{-\lambda_0 \tau_2}\right)^{-1} > 0.$$

From Lemma 3.1.2 we have $f(t^0) = G_1(\lambda_0) > 0$.

On the other hand, if $f(t^0) > 0$, by Lemma 3.1.2, $G_1(\lambda_0) > 0$. Hence $G_1(\lambda) > 0$ for all $\lambda \in (\lambda^*, \infty)$, where λ^* is given by Lemma 3.1.1. If Eq. (3.1.3) is not oscillatory, then there exists a $\lambda_1 \in R$ such that

$$F(\lambda_1) = \lambda_1\left(1 - p_1 e^{-\lambda_1 \tau_1} - p_2 e^{-\lambda_1 \tau_2}\right) + \sum_{j=1}^{m} q_j e^{-\lambda_1 \sigma_j} = 0. \tag{3.1.23}$$

Obviously $\lambda_1 < 0$ and thus $0 < p_1 e^{-\lambda_1 \tau_1} + p_2 e^{-\lambda_1 \tau_2} < 1$, i.e., $\lambda_1 \in (\lambda^*, \infty)$. (3.1.23) gives us that

$$0 = \lambda_1 + \sum_{j=1}^{m} q_j e^{-\lambda_1 \sigma_j}\left(1 - p_1 e^{-\lambda_1 \tau_1} - p_2 e^{-\lambda_1 \tau_2}\right)^{-1} = G_1(\lambda_1)$$

which contradicts $G_1(\lambda) > 0$ on (λ^*, ∞). □

To prove Theorem 3.1.3 we need the following lemmas.

Lemma 3.1.3. *Assume $p_1 + p_2 > 1$. Define*

$$G_2(\mu) = -\mu + \sum_{j=1}^{m} q_j e^{-\mu \sigma_j}\left(p_1 e^{-\mu \tau_1} + p_2 e^{-\mu \tau_2} - 1\right)^{-1},$$

and let $\mu^* > 0$ be such that $p_1 e^{-\mu^* \tau_1} + p_2 e^{-\mu^* \tau_2} = 1$. Then there exists a $\mu_0 \in (-\infty, \mu^*)$ such that

$$p_1 e^{-\mu_0 \tau_1} + p_2 e^{-\mu_0 \tau_2} > 1, \tag{3.1.24}$$

and $G_2(\mu_0)$ is the minimum value of $G_2(\mu)$ in $(-\infty, \mu^*)$.

Proof: Similar to Lemma 3.1.1. □

Lemma 3.1.4. *Assume $p_1 + p_2 > 1$, μ_0 is defined by Lemma 3.1.3. Let*

i) $\frac{p_2}{p_1} e^{-\mu_0(\tau_2-\tau_1)} < 1$, *and*

ii) $S^{(1)}_{ijk} = C^i_{-(k+1)} p_1^{-(k+i+1)} p_2^i q_j [(k+i+1)\tau_1 - i\tau_2 - \sigma_j] e^{\mu_0[(k+i+1)\tau_1 - i\tau_2 - \sigma_j]}$, $(ijk) \in I_2$.

Then

i) $S^{(1)} = (S^{(1)}_{ijk}) \in D_1$,

ii) *(3.1.10) has the same value for all $(ijk) \in I_2$, $S^{(1)}_{ijk} \neq 0$,*

iii) $h_1(S^{(1)}) = G_2(\mu_0)$,

iv) *$h_1(S^{(1)})$ is the maximal value of $h_1(S)$ on D_1.*

Proof: By (3.1.24) and condition i), there exists a neighborhood $\mathcal{U}$ of μ_0 such that
$0 < (p_1 e^{-\mu\tau_1} + p_2 e^{-\mu\tau_2})^{-1} < 1$, and $\frac{p_2}{p_1} e^{-\mu(\tau_2-\tau_1)} < 1$ for $\mu \in \mathcal{U}$.

$$G_2(\mu) = -\mu + \sum_{j=1}^{m} q_j e^{-\mu\sigma_j} (p_1 e^{-\mu\tau_1} + p_2 e^{-\mu\tau_2})^{-1} \left[1 - (p_1 e^{-\mu\tau_1} + p_2 e^{-\mu\tau_2})^{-1}\right]^{-1}$$

$$= -\mu + \sum_{j=1}^{m} \sum_{k=0}^{\infty} q_j e^{-\mu\sigma_j} (p_1 e^{-\mu\tau_1} + p_2 e^{-\mu\tau_2})^{-(k+1)}$$

$$= -\mu + \sum_{j=1}^{m} \sum_{k=0}^{\infty} p_1^{(k+1)} q_j e^{\mu[(k+1)\tau_1 - \sigma_j]} \left(1 + \frac{p_2}{p_1} e^{-\mu(\tau_2-\tau_1)}\right)^{-(k+1)} \tag{3.1.25}$$

$$= -\mu + \Sigma_2 C^i_{-(k+1)} p_1^{-(k+1)} p_2^i q_j e^{\mu[(k+1)\tau_1 - i\tau_2 - \sigma_j]}. \tag{3.1.26}$$

The rest of the proof is similar to that of Lemma 3.1.2 and hence is omitted. □

Lemma 3.1.5. *Let the conditions of Lemma 3.1.4 hold if p_1 and p_2, τ_1 and τ_2 exchange their positions, respectively, and $S^{(1)}_{ijk}$ are replaced by $S^{(2)}_{ijk}$. Then the* *conclusion of Lemma 3.1.4 holds if $S^{(1)}$, D_1, h_1, and (3.1.10) are replaced by $S^{(2)}$, D_2, h_2, and (3.1.11), respectively.*

Lemma 3.1.6. *Assume $p_1 + p_2 > 1$, μ_0 is defined by Lemma 3.1.3. Let*

i) $\frac{p_2}{p_1} e^{-\mu_0(\tau_2 - \tau_1)} = 1$, *and*

ii) $S_{jk}^{(3)} = (2p_1)^{-(k+1)} q_j[(k+1)\tau_1 - \sigma_j] e^{\mu_0[(k+1)\tau_1 - \sigma_j]}, \ (jk) \in I_3.$ (3.1.27)

Then

i) $S^{(3)} = (S_{jk}^{(3)}) \in D_3$,

ii) *(3.1.12) has the same value for all $(jk) \in I_3$, $S_{jk}^{(3)} \neq 0$,*

iii) $h_3(S^{(3)}) = G_2(\mu_0)$,

iv) *$h_3(S^{(3)})$ is the maximum value of $h_3(S)$ on D_3.*

Proof: From (3.1.25) and condition i)

$$G_2(\mu) = -\mu + \Sigma_3 (2p_1)^{-(k+1)} q_j e^{\mu[(k+1)\tau_1 - \sigma_j]}. \tag{3.1.28}$$

The rest of the proof is omitted.

Remark 3.1.1. There exists at least one of h_b which is well-defined on D_b $(b = 1, 2, 3)$. In fact, there is at least one expansion of (3.1.26), its dual form, and (3.1.28), of $G_2(\mu)$, which converges at $\mu = \mu_0$. Therefore, at least one of $h_b(S^{(b)})$, $(b = 1, 2, 3)$, has a finite value and satisfies $h_b(S^{(b)}) = G_2(\mu_0)$.

Proof of Theorem 3.1.3: The first part is shown by Lemmas 3.1.4 – 3.1.6.

Assume Eq. (3.1.3) is oscillatory. Then $F(\mu) > 0$ for all $\mu \in R$. In particular,

$$F(\mu_0) = \mu_0 (1 - p_1 e^{-\mu_0 \tau_1} - p_2 e^{-\mu_0 \tau_2}) + \sum_{j=1}^{m} q_j e^{-\mu_0 \sigma_j} > 0$$

where μ_0 satisfies (3.1.24). Hence

$$G_2(\mu_0) = -\mu_0 + \sum_{j=1}^{m} q_j e^{-\mu_0 \sigma_j} (p_1 e^{-\mu_0 \tau_1} + p_2 e^{-\mu_0 \tau_2} - 1)^{-1} > 0.$$

By Remark 3.1.1 there exists $b = 1, 2$, or 3, such that $h_b(S^{(b)}) = G_2(\mu_0)$. Therefore, $0 < h_b(S^{(b)}) < \infty$ for some $b = 1, 2$, or 3.

On the other hand, if $0 < h_b(S^{(b)}) < \infty$ for some $b = 1, 2$, or 3, by Remark 3.1.1, $G_2(\mu_0) > 0$. Hence $G_2(\mu) > 0$ for all $\mu \in (-\infty, \mu^*)$. Assume Eq. (3.1.3) is not oscillatory. Then there exists a $\mu_1 \in R$ such that

$$F(\mu_1) = \mu_1(1 - p_1 e^{-\mu_1 \tau_1} - p_2 e^{-\mu_1 \tau_2}) + \sum_{j=1}^{m} q_j e^{-\mu_1 \sigma_j} = 0.$$

Obviously, $\mu_1 > 0$, and thus $p_1 e^{-\mu_1 \tau_1} + p_2 e^{-\mu_1 \tau_2} > 1$, i.e., $\mu_1 \in (-\infty,\ \mu^*)$, and

$$G_2(\mu_1) = -\mu_1 + \sum_{j=1}^{m} q_j e^{-\mu_1 \sigma_j} \left(p_1 e^{-\mu_1 \tau_1} + p_2 e^{-\mu_1 \tau_2} - 1\right)^{-1} = 0,$$

contradicting that $G_2(\mu) > 0$ on $(-\infty, \mu^*)$. □

Proof of Corollary 3.1.3:

i) Choose

$$t^*_{ijk} = eC^i_k p^i_1 p^{k-i}_2 q_j [i\tau_1 + (k-i)\tau_2 + \sigma_j]c, \ (ijk) \in I_1,$$

such that $\Sigma_1 t^*_{ijk} = 1$. By (3.1.13) we have $0 < c \leq 1$. Hence

$$f(t^*) = \Sigma_1 e C^i_k p^i_1 p^{k-i}_2 q_j c \ln \frac{1}{c} \geq 0.$$

Noting that t^* does not make (3.1.6) have the same value for $(ijk) \in I$, we see $t^* \neq t^0$. So $f(t^0) > 0$. By Theorem 3.1.2, Eq. (3.1.3) is oscillatory. The proofs for ii) and iii) are similar. In fact, for ii) we choose

$$t^*_{ijk} = eC^i_k p^i_1 p^{k-i}_2 q [i\tau_1 + (k-i)\tau_2 + \sigma_j]c,$$

for iii) we choose $t^*_{ijk} = C^i_k p^i_1 p^{k-i}_2 q_j / a$. □

Proof of Corollary 3.1.4:

i) The conditions of Lemma 3.1.6 are satisfied, where $\mu_0 = \frac{1}{r_1 - r_2} \ln \frac{p_1}{p_2}$.

Hence from (3.1.9) and (3.1.27)

$$h_3(S^{(3)}) = \Sigma_3 \frac{S^{(3)}_{jk}}{(k+1)\tau_1 - \sigma_j}\left(1 - \ln e^{-\mu_0[(k+1)\tau_1 - \sigma_j]}\right)$$

$$= \Sigma_3 (2p_1)^{-(k+1)} q_j e^{\left(\frac{1}{\tau_1-\tau_2}\ln\frac{p_1}{p_2}\right)[(k+1)\tau_1-\sigma_j]} - \frac{1}{\tau_1-\tau_2}\ln\frac{p_1}{p_2}.$$

Thus $h_3(S^{(3)}) > 0$ if and only if (3.1.15) holds.

ii) Without loss of generality we may assume the left hand side of (3.1.16) is finite, for otherwise we can replace I_2^* by its suitable subset in (3.1.16) such that the above assumption holds. Choose

$$S^*_{ijk} = \begin{cases} eC^i_{-(k+1)} p_1^{-(k+i+1)} p_2^i q_j [(k+i+1)\tau_1 - i\tau_2 - \sigma_j]c, & (ijk) \in I_2^* \\ 0, & (ijk) \in I_2 \backslash I_2^* \end{cases}$$

such that $\Sigma_2 S^*_{ijk} = 1$, i.e., $S^* = (S^*_{ijk}) \in D_2$. By (3.1.16) we have $0 < c \leq 1$. From (3.1.7)

$$h_1(S^*) = \sum_{(ijk)\in I_2^*} eC^i_{-(k+1)} p_1^{-(k+i+1)} p_2^i q_j c \ln \frac{1}{c} \geq 0.$$

From (3.1.8) and the definition of I_2^*

$$h_2(S^*) = \sum_{(ijk)\in I_2^*} eC^i_{-(k+1)} p_1^{-(k+i+1)} p_2^i q_j \frac{(k+i+1)\tau_1 - i\tau_2 - \sigma_j}{(k+i+1)\tau_2 - i\tau_1 - \sigma_j} c$$

$$\ln\left[\left(\frac{p_1}{p_2}\right)^{k+2i+1} \frac{(k+i+1)\tau_2 - i\tau_1 - \sigma_j}{(k+i+1)\tau_1 - i\tau_2 - \sigma_j}\right] > 0.$$

By Remark 3.1.1 we get that there exists a $b = 1$ or 2 such that $h_b(S^{(b)}) < \infty$. Noting that S^* does not make (3.1.10) have the same value for $(ijk) \in I_2$, hence $h_b(S^{(b)}) > h_b(S^*) \geq 0$. By Theorem 3.1.3, Eq. (3.1.3) is oscillatory.

iii) The proof here is similar and so is omitted. □

If we let $p_1 = p,\ p_2 = 0,\ \tau_1 = \tau$, then Eq. (3.1.3) becomes

$$\frac{d}{dx}[x(t) - px(t-\tau)] + \sum_{j=1}^{m} q_j x(t - \sigma_j) = 0, \tag{3.1.29}$$

and the above results can be reformulated as follows.

Define two sets

$$D_1 = \left\{ t = (t_{jk}) : t_{jk} \geq 0,\ j = 1, \ldots, m;\ k = 0, 1, \ldots ; \sum_{j=1}^{m} \sum_{k=0}^{\infty} t_{jk} = 1 \right\},$$

$$D_2 = \left\{ S = (S_{jk}) : S_{jk}[(k+1)\tau - \sigma_j] \geq 0,\ j = 1, \ldots, m;\ k = 0, 1, \ldots ; \right.$$

$$\left. \sum_{j=1}^{m} \sum_{k=0}^{\infty} S_{jk} = 1 \right\}.$$

On D_1 we define a function

$$f(t) = \sum_{j=1}^{m} \sum_{k=0}^{\infty} \frac{t_{jk}}{k\tau + \sigma_j} \ln \left(e p^k q_j (k\tau + \sigma_j)/t_{jk} \right)$$

on D_2 we define a function

$$h(s) = \sum_{j=1}^{m} \sum_{k=0}^{\infty} \frac{S_{jk}}{(k+1)\tau_1 - \sigma_j} \ln \left(e p^{-(k+1)} q_j [(k+1)\tau - \sigma_j]/S_{jk} \right).$$

Theorem 3.1.4. *Assume $0 < p < 1$. Then*

i) *$f(t)$ has a maximum at a point $t^0 = (t^0_{jk})$ on D_1 which is determined by the condition that*

$$\frac{1}{k\tau + \sigma_j} \ln \left(p^k q_j [k\tau + \sigma_j]/t^0_{jk} \right)$$

has the same value for all $j = 1, \ldots, m,\ k = 0, 1, \ldots, t^0_{jk} \neq 0$.

ii) *Eq. (3.1.29) is oscillatory if and only if $f(t^0) > 0$.*

Theorem 3.1.5. *Assume $p > 1$. Then*

i) *$h(S)$ has a maximum value at a point $S^0 = (S^0_{jk})$ on D_2, which is determined by the condition that*

$$\frac{1}{(k+1)\tau - \sigma_j} \ln \left(p^{-(k+1)} q_j [(k+1)\tau - \sigma_j)/S^0_{jk} \right)$$

has the same value for all $j = 1, \ldots, m$, $k = 0, 1, \ldots, S^0_{jk} \neq 0$, $(k+1)\tau - \sigma_j \neq 0$.

ii) *Eq. (3.1.29) is oscillatory if and only if* $h(S^0) > 0$.

Corollary 3.1.5. *Assume* $0 < p < 1$. *Then each of the following is sufficient for Eq. (3.1.29) to be oscillatory:*

i) $\sum_{j=1}^{m} \sum_{k=0}^{\infty} p^k q_j (k\tau + \sigma_j) \geq \frac{1}{e}$,

ii) $mq \sum_{k=0}^{\infty} p^k (k\tau + \sigma) \geq \frac{1}{e}$,

iii) *let* $a = \sum_{j=1}^{m} \sum_{k=0}^{\infty} p^k q_j$, $\alpha_{jk} = \frac{p^k q_j}{k\tau + \sigma_j}$, $j = 1, \ldots, m$, $k = 0, 1, \ldots$, *and*

$$\sum_{j=1}^{m} \sum_{k=0}^{\infty} \frac{\alpha_{jk}}{a} \ln[ea(k\tau + \sigma_j)] \geq 0.$$

Corollary 3.1.6. *Assume* $p > 1$. *Then each one of the following is sufficient for Eq. (3.1.29) to be oscillatory:*

i) *there is a* k_j *for each* $j = 1, \ldots, m$, *such that* $(k_j + 1)\tau - \sigma_j > 0$, *and*

$$\sum_{j=1}^{m} \sum_{k=k_j}^{\infty} p^{-(k+1)} q_j [(k+1)\tau - \sigma_j] \geq \frac{1}{e},$$

ii) *there is a* k_0 *such that* $(k_0 + 1)\tau - \sigma > 0$, *and*

$$mq \sum_{k=k_0}^{\infty} p^{-(k+1)} [(k+1)\tau - \sigma] \geq \frac{1}{e}.$$

3.2. Equations with Variable Coefficients (I)

3.2.1. We consider the linear neutral differential equations of the form

$$\big(x(t) - p(t)x(t-\tau)\big)' + q(t)x\big(t - \sigma(t)\big) = 0, \quad t \geq t_0. \tag{3.2.1}$$

Clearly, the oscillatory behavior of solutions of (3.2.1) depends on the range of p. In this section, we discuss the case that $0 \leq p \leq 1$.

The following result reduces the oscillation problem of NDE (3.2.1) to a corresponding problem of delay differential inequality of type (2.1.1).

Theorem 3.2.1. *Assume that*

i) $\tau \in (0, \infty)$, $p, q, \sigma \in C([t_0, \infty), R_+)$ *such that* $0 \leq p(t) \leq 1$,

$$\int_{t_0}^{\infty} q(t)dt = \infty \quad \text{and} \quad \lim_{t\to\infty} (t - \sigma(t)) = \infty;$$

ii) *there exists a positive integer* N *such that the differential inequality*

$$y'(t)+q(t)[1+p(t-\sigma(t))+\cdots+\prod_{i=0}^{N-1} p(t-\sigma(t)-i\tau)]y(t-\sigma(t)) \leq 0 \quad (3.2.2)$$

has no eventually positive solutions.

Then every solution of (3.2.1) is oscillatory.

Proof: Assume the contrary, and let $x(t)$ be an eventually positive solution of (3.2.1). Set

$$z(t) = x(t) - p(t)x(t - \tau). \quad (3.2.3)$$

Then $z'(t) \leq 0$. If $z(t) < 0$ eventually, then $x(t) \leq p(t)x(t-\tau) \leq x(t-\tau)$ eventually, which implies that $x(t)$ is bounded. Hence $z(t)$ is bounded, and $\lim_{t\to\infty} z(t) = \ell$ exists. From (3.2.1), we obtain that

$$\int_{t_0}^{\infty} q(t)x(t - \sigma(t))dt < \infty.$$

This together with condition i) derives that $\liminf_{t\to\infty} x(t) = 0$. Hence there exists a sequence $\{t_n\}$ such that $\lim_{n\to\infty} t_n = \infty$ and $\lim_{n\to\infty} x(t_n) = 0$. Since

$$\begin{aligned} 0 &= \lim_{n\to\infty} (z(t_n + \tau) - z(t_n)) \\ &= \lim_{n\to\infty} (x(t_n + \tau) - (p(t_n + \tau) + 1)x(t_n) + p(t_n)x(t_n - \tau)) \end{aligned}$$

we see that

$$\lim_{n\to\infty} (x(t_n + \tau) + p(t_n)x(t_n - \tau)) = 0$$

which implies that $\lim_{n\to\infty} p(t_n)x(t_n - \tau) = 0$ and hence

$$\ell = \lim_{n\to\infty} \big(x(t_n) - p(t_n)x(t_n - \tau)\big) = 0.$$

This contradicts the assumption. Therefore $z(t) > 0$ eventually. Obviously, $x(t) > z(t)$ for all large t. Considering that $z(t)$ is decreasing we have

$$\begin{aligned} x(t) &= z(t) + p(t)x(t-\tau) \\ &\geq \left[1 + p(t) + \cdots + \prod_{i=0}^{N-1} p(t - i\tau)\right] z(t) \end{aligned} \tag{3.2.4}$$

for all large t. Substituting (3.2.4) into (3.2.1) we obtain

$$z'(t) + q(t)\left[1 + p(t - \sigma(t)) + \cdots + \prod_{i=0}^{N-1} p(t - \sigma(t) - i\tau)\right] z(t - \sigma(t)) \leq 0$$

which contradicts assumption ii). □

Remark 3.2.1. Some sufficient conditions for Eq. (3.2.2) to have no eventually positive solutions have been presented in Theorems 2.1.1 and 2.1.2.

In the following we shall give a nonoscillation result for Eq. (3.2.1).

Theorem 3.2.2. *Assume that* $p \in C([t_0,\infty), R_+)$, $q \in C([t_0,\infty),(0,\infty))$, *and*
$\tau \in (0,\infty)$, $\sigma(t) \equiv \sigma \in [0,\infty)$. *Further assume that there exist* $\lambda^* \in (0,\infty)$ *and* $T \geq t_0$ *such that*

$$\sup_{t\geq T} \left[p(t-\sigma)\frac{q(t)}{q(t-\tau)} e^{\lambda^*\tau} + \frac{1}{\lambda^*}\, q(t)e^{\lambda^*\sigma}\right] \leq 1. \tag{3.2.5}$$

Then Eq. (3.2.1) has a positive solution on $[T,\infty)$.

Proof: First we claim that the integral equation

$$w(t) = p(t-\sigma)\frac{q(t)}{q(t-\tau)}w(t-\tau)\exp\Big(\int_{t-\tau}^{t} w(s)ds\Big)$$
$$+ q(t)\exp\int_{t-\sigma}^{t} w(s)ds, \quad t \geq T+m, \tag{3.2.6}$$

possesses a positive continuous solution on $[T,\infty)$, where $m = \max\{\tau,\sigma\}$. To this end, set $w_1(t) = 0, \quad t \geq T$, and for $k = 1, 2, \ldots$

$$w_{k+1}(t) = \begin{cases} p(t-\sigma)\frac{q(t)}{q(t-\tau)}w_k(t-\tau)\exp\big(\int_{t-\tau}^{t} w_k(s)ds\big) \\ \qquad +q(t)\exp\big(\int_{t-\sigma}^{t} w_k(s)ds\big), \qquad t \geq T+m \\ w_{k+1}(T+m), \qquad T \leq t < T+m. \end{cases} \tag{3.2.7}$$

Clearly, by (3.2.7) and (3.2.5), $w_2(t) > 0 = w_1(t)$ for $t \geq T$ and $w_2(t) \leq \lambda^*$ for $t \geq T$.

Assume for some positive integer k, $w_{k-1}(t) \leq w_k(t)$ for $t \geq T$. Then by (3.2.7) and (3.2.5), $w_{k+1}(t) \geq w_k(t)$ for $t \geq T$, and

$$w_{k+1}(t) \leq p(t-\sigma)\frac{q(t)}{q(t-\tau)}\lambda^* e^{\lambda^*\tau} + q(t)e^{\lambda^*\sigma}$$
$$\leq \lambda^* \sup_{t\geq T}\left[p(t-\sigma)\frac{q(t)}{q(t-\tau)}e^{\lambda^*\tau} + \frac{1}{\lambda^*}q(t)e^{\lambda^*\sigma}\right] \leq \lambda^*.$$

By induction we see that for $t \geq T$

$$w_1(t) < w_2(t) \leq w_3(t) \leq \cdots \leq w_k(t) \leq \cdots \leq \lambda^*.$$

Then it follows that $\lim_{k\to\infty} w_k(t) = w(t)$ exists and is positive on $[T,\infty)$. By taking limits on both sides of (3.2.7) and by using the Lebesgue's monotone convergence theorem we see that $w(t)$ is a solution of (3.2.6) on $[T,\infty)$. We also see that the sequences $\{w_k(t)\}_{k=1}^{\infty}$ converge uniformly on $[T, T+m]$. Then it follows by (3.2.6) that $w(t)$ is a continuous function on $[T,\infty)$. Set $z(t) = \exp\big(-\int_T^t w(s)ds\big)$. Then $w(t) = -\frac{z'(t)}{z(t)}$ and $z'(t) < 0$ for $t \geq T$. Thus (3.2.6) reduces to

$$z'(t) = p(t-\sigma)\ \frac{q(t)}{q(t-\tau)}z'(t-\tau) - q(t)z(t-\sigma). \tag{3.2.8}$$

Now define $x(t) = -z'(t+\sigma)/q(t+\sigma)$. It is easy to see that $x(t)$ is a positive solution of Eq. (3.2.1) on $[T,\infty)$. □

Theorem 3.2.3. *Assume that*

i) $p,\ q \in C([t_0,\infty),R)$, $0 \le p(t) \le p < 1$, $\tau \in (0,\infty)$ *and* $\sigma(t) \equiv \sigma \in [0,\infty)$;

ii) *there exists* $\mu > 0$ *such that*

$$p(t)\mu e^{\mu\tau} + |q(t)|e^{\mu\sigma} \le \mu, \quad t \ge t_0. \tag{3.2.9}$$

Then Eq. (3.2.1) has a positive solution on $[t_0,\infty)$.

Proof: Set $p(t) \equiv p(t_0)$, $t \le t_0$, and

$$q(t) = \begin{cases} \dfrac{t+m-t_0}{m}\, q(t_0), & t_0 - m \le t \le t_0 \\ 0, & t \le t_0 - m \end{cases}$$

where $m = \max\{\tau,\sigma\}$, then p, q are well defined on R and i) and ii) hold on R. Define a set of functions on R as:

$$X = \{\lambda\colon \lambda \in C(R,R),\ |\lambda(t)| \le \mu \ \text{ for } t \ge t_0 - m,\ \lambda(t) \equiv 0 \ \text{ for } t \le t_0 - m\}$$

where the norm of λ is $\|\lambda\| = \sup_{t \ge t_0 - m} |\lambda(t)|e^{-2\mu t}$. Then X is a Banach space. Define a mapping on X as follows:

$$(T\lambda)(t) = q(t)\left[\exp \int_{t-\sigma}^{t} \lambda(s)ds + \sum_{i=1}^{\infty}\prod_{j=0}^{i-1} p(t-\sigma-j\tau)\exp\int_{t-\sigma-i\tau}^{t}\lambda(s)ds\right].$$

Clearly, $(T\lambda)(t) \equiv 0$ for $t \le t_0 - m$. $(T\lambda)(t)$ is continuous and

$$\begin{aligned} |(T\lambda)(t)| &\le |q(t)|\left[\exp\int_{t-\sigma}^{t}|\lambda(s)|ds \right.\\ &\qquad \left. + \sum_{i=1}^{\infty}\prod_{j=0}^{i-1} p(t-\sigma-j\tau)\exp\int_{t-\sigma-i\tau}^{t}|\lambda(s)|ds\right] \\ &\le |q(t)|\left(e^{\mu\sigma} + \sum_{i=1}^{\infty} p^i e^{\mu(\sigma+i\tau)}\right) \\ &\le |q(t)|e^{\mu\sigma}/(1-pe^{\mu\tau}) \le \mu, \end{aligned}$$

i.e., $\mathcal{T}X \subset X$. We show that $\mathcal{T}$ is a contraction on X. In fact, for $\lambda_1, \lambda_2 \in X$, and $t \geq t_0 - m$

$$\begin{aligned}
|(\mathcal{T}\lambda_1)(t) - (\mathcal{T}\lambda_2)(t)| &\leq q(t)\Bigg[\left|\exp\int_{t-\sigma}^{t}\lambda_1(s)ds - \exp\int_{t-\sigma}^{t}\lambda_2(s)ds\right| \\
&\quad + \sum_{i=1}^{\infty} p^i\left|\exp\Big(\int_{t-\sigma-i\tau}^{t}\lambda_1(s)ds\Big) - \exp\Big(\int_{t-\sigma-i\tau}^{t}\lambda_2(s)ds\Big)\right|\Bigg] \\
&\leq |q(t)|e^{\mu\sigma}\Bigg\{\int_{t-\sigma}^{t}|\lambda_2(s) - \lambda_1(s)|ds \\
&\quad + \sum_{i=1}^{\infty} p^i e^{i\mu\tau}\int_{t-\sigma-i\tau}^{t}|\lambda_1(s) - \lambda_2(s)|ds\Bigg\} \\
&\leq |q(t)|\ \|\lambda_1 - \lambda_2\|e^{\mu\sigma}\Bigg\{\int_{t-\sigma}^{t}e^{2\mu s}ds + \sum_{i=1}^{\infty}p^i e^{i\mu\tau}\int_{t-\sigma-i\tau}^{t}e^{2\mu s}ds\Bigg\} \\
&= \frac{e^{2\mu t}}{2\mu}|q(t)|\ \|\lambda_1 - \lambda_2\|e^{\mu\sigma}\Bigg[1 - e^{-2\mu\sigma} + \sum_{i=1}^{\infty}p^i e^{i\mu\tau}(1 - e^{-2\mu(\sigma+i\tau)})\Bigg] \\
&\leq \frac{e^{2\mu t}}{2\mu}|q(t)|\ \|\lambda_1 - \lambda_2\|e^{\mu\sigma}/(1 - pe^{\mu\tau}) \\
&\leq \frac{1}{2}e^{2\mu t}\|\lambda_1 - \lambda_2\|.
\end{aligned}$$

Hence

$$\|\mathcal{T}\lambda_1 - \mathcal{T}\lambda_2\| = \sup_{t\geq t_0-m}|(\mathcal{T}\lambda_1)(t) - (\mathcal{T}\lambda_2)(t)|e^{-2\mu t} \leq \tfrac{1}{2}\|\lambda_1 - \lambda_2\|.$$

By Banach contraction theorem, there is a $\lambda \in X$ such that $\mathcal{T}\lambda = \lambda$, i.e.,

$$\lambda(t) = q(t)\Bigg[\exp\int_{t-\sigma}^{t}\lambda(s)ds + \sum_{i=1}^{\infty}\prod_{j=0}^{i-1}p(t-\sigma-j\tau)\exp\int_{t-\sigma-i\tau}^{t}\lambda(s)ds\Bigg].$$

Set $y(t) = \exp\big(-\int_{t_0-m}^{t}\lambda(s)ds\big)$. Then $y(t) > 0$ and

$$y'(t) = -q(t)\Bigg[y(t-\sigma) + \sum_{i=1}^{\infty}\prod_{j=0}^{i-1}p(t-\sigma-j\tau)y(t-\sigma-i\tau)\Bigg]. \qquad (3.2.10)$$

Denote

$$x(t) = y(t) + \sum_{i=1}^{\infty} \prod_{j=0}^{i-1} p(t - j\tau) y(t - i\tau).$$

Then $x(t) > 0$ and $y(t) = x(t) - p(t)x(t - \tau)$. Hence $x(t)$ is a positive solution of (3.2.1). □

Corollary 3.2.1. *Assume that $p,\ q \in C([t_0, \infty), R)$,*

$$0 \le p(t) \le p < 1, \quad |q(t)| \le q, \quad \sigma(t) \equiv \sigma > 0, \quad t \ge t_0, \tag{3.2.11}$$

and the equation

$$\big(x(t) - px(t - \tau)\big)' + qx(t - \sigma) = 0 \tag{3.2.12}$$

has an eventually positive solution. Then Eq. (3.2.1) also has an eventually positive solution.

In fact, (3.2.12) having an eventually positive solution implies that the characteristic equation of (3.2.12)

$$\lambda(1 - pe^{-\lambda\tau}) + qe^{-\lambda\sigma} = 0$$

has a real root λ^*. It is easy to see that λ^* must be negative. Set $\mu = -\lambda^*$. Then $\mu pe^{\mu\tau} + qe^{\mu\sigma} = \mu$, and hence (3.2.9) holds. Therefore, the conclusion of Corollary 3.2.1 follows.

3.2.2. We consider the neutral differential equation

$$\big(x(t) - x(t - \tau)\big)' + q(t)x(t - \sigma) = 0, \quad t \ge t_0, \tag{3.2.13}$$

where $q \in C([t_0, \infty), R_+)$, $\tau, \sigma \in (0, \infty)$, $m = \max\{\tau, \sigma\}$. Set

$$q_0(t) \equiv q(t),$$
$$q_1(t) = tq(t) \int_t^\infty q(u)du,$$
$$\dots$$
$$q_n(t) = tq(t) \int_t^\infty q_{n-1}(u)du, \quad n = 1, 2, \dots \tag{3.2.14}$$

provided the right hand side of (3.2.14) exists.

Theorem 3.2.4. *Assume that either*

$$\int_{t_0}^\infty q(t)dt = \infty, \tag{3.2.15}$$

or there exists an integer $\ell > 0$ such that $q_1, \dots, q_{\ell-1}$ are defined, and

$$\int_{t_0}^\infty sq(s) \int_s^\infty q_{\ell-1}(u)duds = \infty. \tag{3.2.16}$$

Then every solution of (3.2.13) is oscillatory.

Proof: Assume the contrary, and let $x(t)$ be an eventually positive solution of (3.2.13). Set

$$z(t) = x(t) - x(t - \tau). \tag{3.2.17}$$

Then it is easy to see that $z(t) > 0$ eventually. Hence there exists $M > 0$ such that

$$x(t) \geq M, \quad t \geq T \geq t_0. \tag{3.2.18}$$

Substituting (3.2.18) into (3.2.13) we have

$$z'(t) + Mq(t) \leq 0. \tag{3.2.19}$$

If (3.2.15) holds, (3.2.19) leads to

$$\lim_{t \to \infty} z(t) = -\infty, \tag{3.2.20}$$

a contradiction. If

$$\int_{t_0}^{\infty} q(t)dt < \infty, \tag{3.2.21}$$

from (3.2.19) we obtain

$$z(t) \geq M \int_t^{\infty} q(s)ds, \quad t \geq T. \tag{3.2.22}$$

Hence

$$\begin{aligned} x(t) &\geq x(t-\tau) + M \int_t^{\infty} q(s)ds \\ &\geq M\Big(\int_t^{\infty} q(s)ds + \int_{t-\tau}^{\infty} q(s)ds + \cdots + \int_{t-\{[\frac{t-T}{\tau}]-1\}\tau}^{\infty} q(s)ds\Big) \\ &\geq M\left[\frac{t-T}{\tau}\right] \int_t^{\infty} q(s)ds, \quad t \geq T, \end{aligned}$$

where $[\cdot]$ denotes the greatest integer function. Hence there exist $M_1 > 0$ and $T_1 \geq T$ such that

$$x(t-\sigma) \geq M_1 t \int_t^{\infty} q(s)ds, \quad t \geq T_1. \tag{3.2.23}$$

Substituting (3.2.23) into (3.2.13) we have

$$z'(t) + M_1 t q(t) \int_t^{\infty} q(s)ds \leq 0, \quad t \geq T_1.$$

If

$$\lim_{t\to\infty} \int_T^{\infty} sq(s) \int_s^{\infty} q(u)duds = \infty, \tag{3.2.24}$$

we obtain (3.2.20). Otherwise

$$z(t) \geq M_1 \int_t^{\infty} q_1(s)ds.$$

Repeating above procedures $\ell-1$ times we get a contradiciton with (3.2.16).

□

As an example we consider the equation

$$\big(x(t) - x(t-\tau)\big)' + t^{\alpha} x(t-\sigma) = 0. \tag{3.2.25}$$

From Theorem 3.2.4 we have the following result.

Corollary 3.2.2. *If $\alpha > -2$, then every solution of (3.2.25) is oscillatory.*

In fact, there exists an integer $n > 2$ such that

$$\alpha \geq -2 + \frac{1}{n}\,. \tag{3.2.26}$$

Let $\ell = n$, then (3.2.16) holds. Corollary 3.3.2 follows from Theorem 3.2.4.

Remark 3.2.2. We will see from Theorem 3.2.6 that if $\alpha < -2$, then (3.2.25) has a bounded positive solution. This shows that condition (3.2.26) is sharp.

Example 3.2.1. Consider the equation

$$\big(x(t) - x(t-1)\big)' + \frac{1}{2(t-1)\sqrt{t}\,(\sqrt{t} + \sqrt{t-1})}\, x(t-1) = 0. \tag{3.2.27}$$

$x(t) = \sqrt{t}$ is a solution of (3.2.27), which implies $\alpha > -2$ is the best possibility for oscillation of (3.2.25).

Example 3.2.2. Consider the equation

$$\big(x(t) - x(t-1)\big)' + \frac{1}{t \ln^2 t}\, x(t-1) = 0, \quad t \geq 2. \tag{3.2.28}$$

It is easy to see that (3.2.16) with $\ell = 1$ is satisfied. Therefore every solution of (3.2.28) oscillates by Theorem 3.2.4. We note that (3.2.15) does not hold here.

In the following we present a necessary and sufficient condition for the existence of a bounded positive solution of Eq. (3.2.13).

Theorem 3.2.5. *Eq. (3.2.13) possesses a bounded positive solution if and* *only if*

$$\sum_{i=0}^{\infty} \int_{t_0+i\tau}^{\infty} q(t)dt < \infty. \tag{3.2.29}$$

Proof: *Sufficiency.* By (3.2.29) there exists $T \geq t_0$ such that for $t \geq T$

$$\sum_{i=0}^{\infty} \int_{t+i\tau}^{\infty} q(t)dt \leq 1. \tag{3.2.30}$$

Define a function

$$H(t) = \begin{cases} \int_t^{\infty} q(s)ds, & t \geq T \\ (t - T + \tau)H(T)/\tau, & T - \tau \leq t < T \\ 0, & t < T - \tau. \end{cases} \tag{3.2.31}$$

It is easy to see that $H \in C(R, R_+)$. Let

$$y(t) = \sum_{i=0}^{\infty} H(t - i\tau), \quad t \geq T. \tag{3.2.32}$$

Then $y \in C([T, \infty), R_+)$. Condition (3.2.30) implies that $0 < y(t) \leq 1$ for $t \geq T$. From (3.2.31) we have

$$y(t) = y(t - \tau) + H(t), \quad t \geq T + \tau.$$

Define a set X as follows:

$$X = \{x : \ x \in C([T, \infty), R), \quad 0 \leq x(t) \leq y(t), \quad t \geq T\}.$$

The set X is considered to be endowed with the usual pointwise ordering $\leq$:

$$x_1 \leq x_2 \Longleftrightarrow x_1(t) \leq x_2(t) \quad \text{for all} \quad t \geq T.$$

Then for any $A \subset X$ there exists $\inf A$ and $\sup A$. Define a mapping $\mathcal{T}$ on X as follows:

$$(\mathcal{T}x)(t) = \begin{cases} x(t-\tau) + \int_t^\infty q(s)x(s-\sigma)ds, & t \geq T+m \\ (\mathcal{T}x)(T+m)\frac{ty(t)}{(T+m)y(T+m)} + y(t)(1-\frac{t}{T+m}), & T \leq t \leq T+m. \end{cases}$$

It is easy to see that

$$\begin{aligned}(\mathcal{T}x)(t) &\leq y(t-\tau) + \int_t^\infty q(s)y(s-\sigma)ds \\ &\leq y(t-\tau) + H(t) = y(t), \quad t \geq T+m\end{aligned}$$

and

$$(\mathcal{T}x)(t) \leq y(t) \quad \text{for} \quad t \in [T, T+m],$$

i.e., $\mathcal{T}X \subseteq X$. It is obvious that $\mathcal{T}$ is continuous and nondecreasing. Therefore, by Knaster's fixed point theorem there exists an $x \in X$ such that $\mathcal{T}x = x$. That is,

$$x(t) = x(t-\tau) + \int_t^\infty q(s)x(s-\sigma)ds, \quad t \geq T+m$$

which implies that $x(t)$ is a bounded positive solution of (3.2.13) on $[T+m, \infty)$.

Necessity. Let $x(t)$ be a bounded and positive solution of Eq. (3.2.13). Set

$$z(t) = x(t) - x(t-\tau).$$

Then $z'(t) \leq 0$. As in the proof of Theorem 3.2.4, $z(t) > 0$ for $t \geq t_0$. Hence $x(t) > x(t-\tau)$, $t \geq t_0$, and there exist $M > 0$ and $t_1 \geq t_0$ such that $x(t) \geq M$, $t \geq t_1$. From (3.2.13), $z'(t) \leq -Mq(t)$, $t \geq t_1 + \sigma$. Integrating it from t to ∞ for $t \geq t_1 + \sigma$ we have

$$x(t) \geq x(t-\tau) + M\int_t^\infty q(s)ds.$$

Then

$$x(t_1+\sigma+n\tau) \geq x(t_1+\sigma) + M\sum_{i=1}^n \int_{t_1+\sigma+i\tau}^\infty q(t)dt.$$

Since $x(t)$ is bounded, from the preceding inequality we obtain that

$$\sum_{i=1}^{\infty}\int_{t_1+\sigma+i\tau}^{\infty} q(s)ds < \infty. \qquad \square$$

The following result will lead to an equivalent version to condition (3.2.29). To this end we consider the second order delay equation

$$x''(t) + q(t)x(t-\tau) = 0. \tag{3.2.33}$$

Theorem 3.2.6. *Assume that $q \in C([t_0,\infty), R_+)$, $\tau > 0$. Then the following three propositions are equivalent*

(i) *Every bounded solution of Eq. (3.2.33) is oscillatory;*

(ii) $\int_{t_0}^{\infty} tq(t)dt = \infty$;

(iii) $\sum\limits_{i=0}^{\infty} \int_{t_0+i\tau}^{\infty} q(t)dt = \infty$.

Proof: (i) $\Longrightarrow$ (ii). It is sufficient to show that Eq. (3.2.33) has a bounded positive solution if

$$\int_{t_0}^{\infty} tq(t)dt < \infty. \tag{3.2.34}$$

In fact, from (3.2.34), there exists $T > t_0$ such that $\int_T^{\infty} tq(t)dt \le \frac{1}{4}$. We introduce the Banach space X of all bounded continuous functions $x : [t_0,\infty) \to R$ with the sup norm $\|x\| = \sup\{|x(t)| : t \in [t_0,\infty)\}$. We consider the subset Ω of X defined by

$$\Omega = \{x \in X : 1 \le x(t) \le 2, \quad t \ge t_0\}. \tag{3.2.35}$$

It is obvious that Ω is a bounded, closed and convex subset of X. We define the operator $\mathcal{T}$ on Ω by

$$(\mathcal{T}x)(t) = \begin{cases} 1 + \int_T^t (s-T)q(s)x(s-\tau)ds & \\ \quad + (t-T)\int_t^{\infty} q(s)x(s-\tau)ds, & t \ge T \\ (\mathcal{T}x)(T), & t_0 \le t \le T. \end{cases} \tag{3.2.36}$$

In view of (3.2.35) we have

$$1 \leq (\mathcal{T}x)(t) \leq 1 + 2\Big(\int_T^\infty (s-T)q(s)ds\Big) \leq 1 + \tfrac{1}{2} \leq 2,$$

i.e., $\mathcal{T}\Omega \subseteq \Omega$.

Let x_1, x_2 be elements of Ω. Then

$$\begin{aligned}
|(\mathcal{T}x_1)(t) - (\mathcal{T}x_2)(t)| \leq & \int_T^t (s-T)q(s)|x_1(s-\tau) - x_2(s-\tau)|ds \\
& + (t-\tau)\int_t^\infty q(s)|x_1(x-\tau) - x_2(s-\tau)|ds \\
\leq & \|x_1 - x_2\|\left(\int_T^t (s-T)q(s)ds + (t-T)\int_t^\infty q(s)ds\right) \\
\leq & \tfrac{1}{2}\|x_1 - x_2\|, \quad t \geq T.
\end{aligned}$$

Hence

$$\|\mathcal{T}x_1 - \mathcal{T}x_2\| = \sup_{t \geq t_0} |(\mathcal{T}x_1)(t) - (\mathcal{T}x_2)(t)| \leq \tfrac{1}{2}\|x_1 - x_2\|.$$

That is, $\mathcal{T}$ is a contraction operator on Ω. Then there exists an $x \in \Omega$ such that $\mathcal{T}x = x$, or

$$x(t) = \begin{cases} 1 + \int_T^t (s-T)q(s)x(s-\tau)ds & \\ \quad +(t-T)\int_t^\infty q(s)x(s-\tau)ds, & t \geq T, \\ x(T_1), & t_0 \leq t \leq T. \end{cases} \tag{3.2.37}$$

which implies that $x(t)$ is a bounded positive solution of (3.2.33).

(ii) $\Longrightarrow$ (i). If not, let $x(t)$ be a bounded positive solution of Eq. (3.2.33). Then $x''(t) \leq 0$, $x'(t) > 0$ and $x(t) > 0$ eventually. Thus $\lim_{t\to\infty} x(t) = \alpha > 0$ exists, because $x(t)$ is bounded. Integrating (3.2.33) twice we have

$$x(t) \geq \int_T^t (s-T)q(s)x(s-\tau)ds + (t-T)\int_t^\infty q(s)x(s-\tau)ds \tag{3.2.38}$$

where T is a sufficiently large number. Then condition (ii), implies that $\lim_{t\to\infty} x(t) = \infty$ which contradicts the boundedness of x.

(i) $\Longrightarrow$ (iii). It is sufficient to show that Eq. (3.2.33) has a bounded positive solution if

$$\sum_{i=0}^{\infty} \int_{t_0+i\tau}^{\infty} q(t)dt < \infty. \tag{3.2.39}$$

From (3.2.39), there exist $T \geq t_0$ such that

$$\sum_{i=0}^{\infty} \int_{T+i\tau}^{\infty} q(s)ds \leq \frac{1}{2\tau}. \tag{3.2.40}$$

Let X and Ω be the same as before. Define an operator $\mathcal{T}$ on Ω as follows:

$$(\mathcal{T}x)(t) = \begin{cases} 1 + \int_T^t \int_s^\infty q(u)x(u-\tau)duds, & t \geq T \\ 1, & t_0 \leq t \leq T. \end{cases}$$

For any $x \in \Omega$ we have

$$\begin{aligned} 1 \leq (\mathcal{T}x)(t) &\leq 1 + 2\int_T^t \int_s^\infty q(u)duds \\ &< 1 + 2\int_T^\infty \int_s^\infty q(u)duds \\ &= 1 + 2\sum_{i=0}^{\infty} \int_{T+i\tau}^{T+(i+1)\tau} \int_s^\infty q(u)duds \\ &\leq 1 + 2\tau \sum_{i=0}^{\infty} \int_{T+i\tau}^{\infty} q(u)du \leq 2, \quad t \geq T. \end{aligned}$$

Therefore $\mathcal{T}\Omega \subseteq \Omega$. For any $x_1, x_2 \in \Omega$, it is not difficult to see that $\|\mathcal{T}x_1 - \mathcal{T}x_2\| \leq \frac{1}{2}\|x_1 - x_2\|$, i.e., $\mathcal{T}$ is a contraction. Hence there exists a fixed point $x \in \Omega$ such that $\mathcal{T}x = x$. It is easy to see that $x(t)$ is a bounded positive solution of Eq. (3.2.33).

(iii) $\Longrightarrow$ (i). If not, let $x(t)$ be a bounded positive solution of Eq. (3.2.33). Then $x''(t) \leq 0$, $x'(t) > 0$, $x(t) > 0$ eventually. Integrating (3.2.33) we have

$$x'(t) \geq \int_t^\infty q(u)x(u-\tau)du, \quad t \geq T, \tag{3.2.41}$$

where T is a sufficiently large number. Integrating (3.2.41) from $t-\tau$ to t we have

$$\begin{aligned} x(t) &\geq x(t-\tau) + \int_{t-\tau}^{t} \int_{s}^{\infty} q(u)x(u-\tau)duds \\ &\geq x(t-\tau) + \tau \int_{t}^{\infty} q(u)x(u-\tau)du, \quad t \geq T+\tau. \end{aligned} \tag{3.2.42}$$

Hence there exists $\alpha > 0$ such that $x(t) \geq \alpha$ for $t \geq T$. From (3.2.42) we have

$$x(t) \geq x(t-\tau) + \alpha\tau \int_{t}^{\infty} q(u)du, \quad t \geq T+\tau.$$

By induction

$$x(T+(n+1)\tau) \geq x(T) + \alpha\tau \sum_{i=1}^{n+1} \int_{T+i\tau}^{\infty} q(u)du.$$

Since $x(t)$ is bounded, it follows that

$$\sum_{i=1}^{\infty} \int_{T+i\tau}^{\infty} q(u)du < \infty$$

which contradicts (iii). □

Combining Theorems 3.2.5 and 3.2.6 we obtain the following result.

Theorem 3.2.7. *Eq. (3.2.13) possesses a bounded positive solution if and only if*

$$\int_{t_0}^{\infty} tq(t)dt < \infty. \tag{3.2.43}$$

Example 3.3.2. Consider

$$\left(x(t) - x(t-2)\right)' + \frac{4(t-1)}{t^2(t-3)(t-2)} x(t-2) = 0, \quad t \geq 4. \tag{3.2.44}$$

It is easy to see that (3.2.43) holds. Therefore, by Theorem 3.2.5, Eq. (3.2.44) has a bounded positive solution. In fact, $x(t) = \frac{t-1}{t}$ is such a solution.

We have a similar result for the more general form of neutral differential equations

$$(x(t) - x(\tau(t)))' + \sum_{i=1}^{m} q_i(t) f_i(\sigma(t)) = 0, \quad t \geq t_0. \tag{3.2.45}$$

Theorem 3.2.8. *Assume that*

i) $\tau,\ \sigma_i \in C([t_0, \infty), R)$, *$\tau$ is increasing,* $t - \tau^* \leq \tau(t) < t$ *for* $t \geq t_0$, *and* $\lim_{t\to\infty} \sigma_i(t) = \infty$, $i = 1, \ldots, m$, *where* $\tau^* > 0$;

ii) $q_i \in C([t_0, \infty), R_+)$, $\quad i = 1, \ldots, m$;

iii) $f_i \in C(R, R)$, *f_i is nondecreasing,* $x f_i(x) > 0$ *for* $x \neq 0$, $\quad i = 1, \ldots, m$.
Then Eq. (3.2.45) has a bounded nonoscillatory solution if and only if

$$\int_{t_0}^{\infty} t \sum_{i=1}^{m} q_i(t) dt < \infty. \tag{3.2.46}$$

3.2.3. We return to the equation (3.2.1).

Theorem 3.2.9. *Assume that*

i) $p(t) \geq 1, \quad q(t) > 0, \quad \int_T^\infty q(t)dt = \infty$ *and* $\tau > \sigma$;

ii) $\liminf_{t\to\infty} \int_t^{t+\tau-\sigma} \big(q(s)/p(s + \tau - \sigma)\big) ds > 0$;

iii) *there exists a positive and continuous function $H(t)$ such that*

$$\liminf_{t\to\infty} \int_t^{t+\tau-\sigma} H(s)ds > 0, \quad \liminf_{t\to\infty} \{q(t)/p(t + \tau - \sigma)H(t)\} > 0;$$

iv) *there exists $T \geq t_0 + \tau$ such that*

$$\inf_{t \geq T, \lambda > 0} \left\{ \frac{q(t)H(t+\tau)}{p(t+\tau-\sigma)q(t+\tau)H(t)} \exp\left(\lambda \int_t^{t+\tau} H(s)ds\right) + \frac{q(t)}{\lambda p(t+\tau-\sigma)H(t)} \exp\left(\lambda \int_t^{t+\tau-\sigma} H(s)ds\right) \right\} > 1. \tag{3.2.47}$$

Then every solution of (3.2.1) is oscillatory.

Proof: Assume the contrary, and let $x(t)$ be an eventually positive solution of Eq. (3.2.1). Set $z(t) = x(t) - p(t)x(t-\tau)$. From condition (i) we have $z(t) < 0$ and $z'(t) < 0$ for $t \geq t_1 \geq t_0$. Eq. (3.2.1) becomes

$$z'(t) - p(t-\sigma)\,\frac{q(t)}{q(t-\tau)} z'(t-\tau) + q(t)z(t-\sigma) = 0, \quad t \geq t_0,$$

or equivalently

$$z'(t) = \frac{q(t)}{p(t+\tau-\sigma)q(t+\tau)} z'(t+\tau) + \frac{q(t)}{p(t+\tau-\sigma)}\, z(t+\tau-\sigma), \quad t \geq t_0 - \tau. \tag{3.2.48}$$

Set $\lambda(t)H(t) = z'(t)/z(t)$, for $t \geq t_1$. Then

$$z(t) = z(t_1)\exp\left(\int_{t_1}^{t} \lambda(s)H(s)ds\right), \quad t \geq t_1, \tag{3.2.49}$$

and $\lambda(t) > 0$ for $t \geq t_1$. Substituting (3.2.49) into (3.2.48) we have

$$\begin{aligned} \lambda(t)H(t) &= \frac{q(t)}{p(t+\tau-\sigma)q(t+\tau)} \lambda(t+\tau)H(t+\tau)\exp\left(\int_t^{t+\tau} \lambda(s)H(s)ds\right) \\ &\quad + \frac{q(t)}{p(t+\tau-\sigma)}\exp\left(\int_t^{t+\tau-\sigma} \lambda(s)H(s)ds\right), \qquad (3.2.50) \\ &> \frac{q(t)}{p(t+\tau-\sigma)}\exp\left(\int_t^{t+\tau-\sigma} \lambda(s)H(s)ds\right), \quad t \geq t_1 - \tau. \end{aligned}$$

By Lemma 3.3.2 (see Section 3.3) we have

$$\liminf_{t\to\infty} \int_t^{t+\tau-\sigma} \lambda(s)H(s)ds < \infty.$$

It is not difficult to see that $\liminf_{t\to\infty} \lambda(t) = \lambda_0 \in (0,\infty)$.

From (3.2.47) there exists a $\alpha \in (0,1)$ such that

$$\alpha \inf_{t\geq T,\lambda>0} \left\{ \frac{q(t)H(t+\tau)}{p(t+\tau-\sigma)q(t+\tau)H(t)} \exp\left(\lambda \int_t^{t+\tau} H(s)ds\right) + \frac{1}{\lambda} \frac{q(t)}{p(t+\tau-\sigma)H(t)} \exp\left(\lambda \int_t^{t+\tau-\sigma} H(s)ds\right)\right\} > 1. \tag{3.2.51}$$

There exists a $t_2 \geq \max\{t_1, T\}$ such that $\lambda(t) > \alpha\lambda_0, \quad t \geq t_2$. Substituting it into (3.2.50) we have

$$\lambda(t)H(t) > \alpha\lambda_0 \frac{q(t)H(t+\tau)}{p(t+\tau-\sigma)q(t+\tau)} \exp\left(\alpha\lambda_0 \int_t^{t+\tau} H(s)ds\right) + \frac{q(t)}{p(t+\tau-\sigma)} \exp\left(\alpha\lambda_0 \int_t^{t+\tau-\sigma} H(s)ds\right), \quad t \geq t_2. \tag{3.2.52}$$

Thus, for $s \geq t_2$

$$\lambda(s) > \inf_{t\geq t_2} \left\{ \alpha\lambda_0 \frac{q(t)H(t+\tau)}{p(t+\tau-\sigma)q(t+\tau)H(t)} \exp\left(\alpha\lambda_0 \int_t^{t+\tau} H(s)ds\right) + \frac{q(t)}{p(t+\tau-\sigma)H(t)} \exp\left(\alpha\lambda_0 \int_t^{t+\tau-\sigma} H(s)ds\right)\right\}.$$

Taking inferior limits in s we have

$$\lambda_0 \geq \inf_{t\geq t_2} \left\{ \alpha\lambda_0 \frac{q(t)H(t+\tau)}{p(t+\tau-\sigma)q(t+\tau)H(t)} \exp\left(\alpha\lambda_0 \int_t^{t+\tau} H(s)ds\right) + \frac{q(t)}{p(t+\tau-\sigma)H(t)} \exp\left(\alpha\lambda_0 \int_t^{t+\tau-\sigma} H(s)ds\right)\right\}. \tag{3.2.53}$$

Letting $\lambda_1 = \alpha\lambda_0$ in (3.2.53) we have

$$\alpha \inf_{t \geq t_2} \left\{ \frac{q(t)H(t+\tau)}{p(t+\tau-\sigma)q(t+\tau)H(t)} \exp\left(\lambda_1 \int_t^{t+\tau} H(s)ds \right) + \frac{1}{\lambda_1} \frac{q(t)}{p(t+\tau-\sigma)H(t)} \exp\left(\lambda_1 \int_t^{t+\tau-\sigma} H(s)ds \right) \right\} \leq 1. \tag{3.2.54}$$

Since $t_2 \geq T$ and $\lambda_1 > 0$. (3.2.54) contradicts (3.2.51). □

From Theorem 3.2.9 we can obtain different sufficient conditions for oscillation of Eq. (3.2.1) by different choices of $H(t)$. For instance, if we choose $H(t) = q(t)/p(t+\tau-\sigma)$ or $H(t) = 1$, then (3.2.47) becomes

$$\inf_{t \geq T, \lambda > 0} \left\{ \frac{1}{p(t+2\tau-\sigma)} \exp\left(\lambda \int_t^{t+\tau} \frac{q(s)}{p(s+\tau-\sigma)} ds \right) + \frac{1}{\lambda} \exp\left(g\lambda \int_t^{t+\tau-\sigma} \frac{q(s)}{p(s+\tau-\sigma)} ds \right) \right\} > 1, \tag{3.2.55}$$

or

$$\inf_{t \geq T, \lambda > 0} \left\{ \frac{q(t)}{p(t+\tau-\sigma)q(t+\tau)} e^{\lambda\tau} + \frac{q(t)}{\lambda p(t+\tau-\sigma)} e^{\lambda(\tau-\sigma)} \right\} > 1. \tag{3.2.56}$$

Since $e^x \geq ex$ for $x > 0$, (3.2.55) and (3.2.56) lead to the following corollaries.

Corollary 3.2.6. *In addition to condition (i) of Theorem 3.2.9 further assume that*

$$\liminf_{t \to \infty} \{ q(t)/p(t+\tau-\sigma) \} > 0 \tag{3.2.57}$$

and

$$\liminf_{t \to \infty} \left\{ \frac{1}{p(t+2\tau-\sigma)} + e \int_t^{t+\tau-\sigma} \frac{q(s)}{p(s+\tau-\sigma)} ds \right\} > 1. \tag{3.2.58}$$

Then every solution of Eq. (3.2.1) is oscillatory.

Corollary 3.2.7. *In addition to condition (i) of Theorem 3.2.9 and (3.2.57) further assume that*

$$\liminf_{t\to\infty}\left\{\frac{q(t)}{p(t+\tau-\sigma)q(t+\tau)}+\frac{q(t)}{p(t+\tau-\sigma)}\,e(\tau-\sigma)\right\}>1. \tag{3.2.59}$$

Then every solution of Eq. (3.2.1) is oscillatory.

Theorem 3.2.10. *Assume that*

(i) $\tau>\sigma\geq 0,\quad p(t)\geq 1,\quad q(t)>0,\quad \int_{t_0}^{\infty}q(t)dt=\infty$, and

$$\liminf_{t\to\infty}\{q(t)/p(t+\tau-\sigma)\}>0;$$

(ii) *there exists* $\lambda^*>0$ *and* $T\geq t_0$ *such that*

$$\sup_{t\geq T}\left\{\frac{q(t)}{p(t+\tau-\sigma)q(t+\tau)}e^{\lambda^*\tau}+\frac{1}{\lambda^*}\,\frac{q(t)}{p(t+\tau-\sigma)}e^{\lambda^*(\tau-\sigma)}\right\}\leq 1. \tag{3.2.60}$$

Then Eq. (3.2.1) has a positive solution on $[T+\tau-\sigma,\infty)$.

Proof: At first, we show that the integral equation

$$\lambda(t)=\frac{q(t)}{p(t+\tau-\sigma)q(t+\tau)}\lambda(t+\tau)\exp\int_t^{t+\tau}\lambda(s)ds$$
$$+\frac{q(t)}{p(t+\tau-\sigma)}\exp\int_t^{t+\tau-\sigma}\lambda(s)ds \tag{3.2.61}$$

has a positive solution.

To this end we define a sequence $\{\lambda_k(t)\}$ as follows: $\lambda_1(t)=0,\quad t\geq T$, and

$$\lambda_{k+1}(t)=\frac{q(t)}{p(t+\tau-\sigma)q(t+\tau)}\lambda_k(t)\exp\left(\int_t^{t+\tau}\lambda_k(s)ds\right)$$
$$+\frac{q(t)}{p(t+\tau-\sigma)}\exp\left(\int_t^{t+\tau-\sigma}\lambda_k(s)ds\right),\quad t\geq T. \tag{3.2.62}$$

Noting (3.2.60) we have

$$\lambda_1(t) = 0 < \lambda_2(t) = \frac{q(t)}{p(t+\tau-\sigma)} \leq \lambda^*, \quad t \geq T.$$

Assume that $\lambda_{k-1}(t) \leq \lambda_k(t) \leq \lambda^*$ for $t \geq T$ and some positive integer k. From (3.2.62) we have $\lambda_k(t) \leq \lambda_{k+1}(t)$ for $t \geq T$, and

$$\lambda_{k+1}(t) \leq \frac{q(t)}{p(t+\tau-\sigma)q(t+\tau)}\lambda^* \exp(\lambda^*\tau) + \frac{q(t)}{p(t+\tau-\sigma)}\exp\big(\lambda^*(\tau-\sigma)\big) \leq \lambda^*$$

for $t \geq T$. Therefore $\lim_{k\to\infty} \lambda_k(t) = \lambda(t)$ exists and $\lambda^* \geq \lambda(t) > 0$ for $t \geq T$. Taking the limits in (3.2.62) as $k \to \infty$ and using the Lebesgue's convergence theorem we obtain that $\lambda(t)$ is a solution of (3.2.61). Set

$$z(t) = -\exp \int_T^t \lambda(s)ds < 0, \quad t \geq T. \tag{3.2.63}$$

Then $z'(t) = \lambda(t)z(t)$, and (3.2.61) becomes

$$z'(t) = \frac{q(t)}{p(t+\tau-\sigma)q(t+\tau)}z'(t+z) + \frac{q(t)}{p(t+\tau-\sigma)}z(t+\tau-\sigma). \tag{3.2.64}$$

Let

$$x(t) = -\frac{z'(t+\sigma)}{q(t+\sigma)} > 0, \quad t \geq T-\sigma. \tag{3.2.65}$$

Then (3.2.64) becomes

$$\big(x(t) - p(t)x(t-\tau)\big)' + q(t)x(t-\sigma) = 0.$$

That is, we have found a positive solution $x(t)$ of Eq. (3.2.1). □

3.2.4.

Theorem 3.2.11. *Assume that $p(t) \equiv p \neq 1$, $\sigma(t) \equiv \sigma > 0$, $\tau > 0$, $q \in C([t_0,\infty), R_+)$ and $\int_0^\infty q(t)dt < \infty$. Then Eq. (3.2.1) has a positive solution.*

Proof: Let BC be the Banach space of all bounded continuous functions on $[t_0, \infty)$ with the super norm.

i) Consider the case that $0 < p < 1$.

Let t^* be so large that $t^* - \tau \geq t_0$, $t^* - \sigma \geq t_0$ and $\int_{t^*}^{\infty} q(s)ds \leq \frac{1-p}{2}$. Set $\Omega = \{x \in BC : 1 \leq x(t) \leq 2, \ t \geq t_0\}$. Then Ω is a bounded, closed and convex subset of BC. Define a mapping $\mathcal{T} : \Omega \to BC$ as follows:

$$(\mathcal{T}x)(t) = \begin{cases} 1 - p + px(t-\tau) + \int_t^{\infty} q(s)x(s-\sigma)ds, & t \geq t^* \\ (\mathcal{T}x)(t^*), & t_0 \leq t \leq t^*. \end{cases}$$

Clearly, $\mathcal{T}$ is continuous and $\mathcal{T}\Omega \subset \Omega$. For $x_1, \ x_2 \in \Omega$ and $t \geq t^*$

$$\begin{aligned} |(\mathcal{T}x_1)(t) - (\mathcal{T}x_2)(t)| &\leq p|x_1(t-\tau) - x_2(t-\tau)| \\ &\qquad + \int_t^{\infty} q(s)|x_1(s-\sigma) - x_2(s-\sigma)|ds \\ &\leq \|x_1 - x_2\|(p + \frac{1-p}{2}) = \frac{1+p}{2}\|x_1 - x_2\|. \end{aligned}$$

Hence

$$\begin{aligned} \|\mathcal{T}x_1 - \mathcal{T}x_2\| &= \sup_{t \geq t_0} |(\mathcal{T}x_1)(t) - (\mathcal{T}x_2)(t)| \\ &= \sup_{t \geq t^*} |(\mathcal{T}x_1)(t) - (\mathcal{T}x_2)(t)| = \frac{1+p}{2}\|x_1 - x_2\|. \end{aligned}$$

$1 + p/2 < 1$ implies that $\mathcal{T}$ is a contraction. Then there is a $x \in \Omega$ that $\mathcal{T}x = x$. It is easy to see that $x(t)$ is a solution of (3.2.1) on $[t^*, \infty)$.

ii) Consider the case that $p > 1$.

Let t^* be so large that $t^* - \sigma \geq t_0$ and $\int_{t^*+\tau}^{\infty} q(s)ds \leq \frac{p-1}{2}$. Define a subset Ω of BC and a mapping $\mathcal{T}$ on Ω as follows:

$$\Omega = \{x \in BC : \frac{p-1}{2} \leq x(t) \leq p, \quad t \geq t_0\}$$

and

$$(\mathcal{T}x)(t) = \begin{cases} p - 1 + \frac{1}{p}\, x(t+\tau) - \frac{1}{p}\int_{t+\tau}^{\infty} q(s)x(s-\sigma)ds, & t \geq t^* \\ (\mathcal{T}x)(t^*), & t_0 \leq t \leq t^*. \end{cases}$$

It is easy to show that $\mathcal{T}\Omega \subset \Omega$ and

$$\|\mathcal{T}x_1 - \mathcal{T}x_2\| \le \frac{1+p}{2p}\,\|x_1 - x_2\|$$

for $x_1,\ x_2 \in \Omega$, i.e., $\mathcal{T}$ is a contraction on Ω. Then there is a $x \in \Omega$ such that $\mathcal{T}x = x$, or $x(t)$ is a positive solution of (3.2.1) on $t \ge t^*$.

The rest can be discussed similarly. We only give an outline for those.

iii) The case that $-1 < p \le 0$.
Choose t^* so that $t^* - \tau \ge t_0$, $t^* - \sigma \ge t_0$ and $\int_{t^*}^{\infty} q(s)ds \le \frac{1+p}{2}$. Define

$$\Omega = \left\{x \in BC :\ \frac{(1+p)^2}{1-p} \le x(t) \le \frac{2(1+p)}{1-p},\ t \ge t_0\right\}$$

and

$$(\mathcal{T}x)(t) = \begin{cases} 1 + p + px(t-\tau) + \int_t^{\infty} q(s)x(s-\sigma)ds, & t \ge t^* \\ (\mathcal{T}x)(t^*), & t_0 \le t \le t^* \end{cases}$$

iv) The case that $p = -1$.
Choose t^* so that $t^* + \tau - \sigma \ge t_0$ and $\int_{t^*+\tau}^{\infty} q(s)ds \le \frac{1}{2}$. Define

$$\Omega = \{x \in BC :\ 1 \le x(t) \le 2,\ t \ge t_0\}$$

and

$$(\mathcal{T}x)(t) = \begin{cases} 1 + \sum_{i=1}^{\infty} \int_{t+(2i-1)\tau}^{t+2i\tau} q(s)x(s-\sigma)ds, & t \ge t^* \\ (\mathcal{T}x)(t^*), & t_0 \le t \le t^*. \end{cases}$$

v) The case that $p < -1$.
Choose t^* so that $t^* + \tau - \sigma \ge t_0$ and $\int_{t^*+\tau}^{\infty} q(s)ds \le -\frac{p+1}{2}$. Define

$$\Omega = \left\{x \in CB :\ \frac{(1+p)^2}{(p-1)p} \le x(t) \le \frac{2(1+p)}{p-1},\quad t \ge t_0\right\}$$

and

$$(\mathcal{T}x)(t) = \begin{cases} 1 + \frac{1}{p} + \frac{1}{p}x(t+\tau) - \frac{1}{p}\int_{t+\tau}^{\infty} q(s)x(s-\sigma)ds, & t \geq t^* \\ (\mathcal{T}x)(t^*), & t_0 \leq t \leq t^*. \end{cases}$$

Theorem 3.2.12. *Assume that $\sigma > \tau > 0$, $p(t) \equiv p < 0$, $q \in C([t_0, \infty), R)$ and the equation*

$$y'(t) + \frac{Q(t)}{1-p}\, y\big(t - (\sigma - \tau)\big) = 0 \tag{3.2.66}$$

is oscillatory. Then Eq. (3.2.1) is oscillatory, where $Q(t) = \min\{q(t),\ q(t-\tau)\}$.

Proof: Otherwise, there is an eventually positive solution $x(t)$ of (3.2.1). Set

$$z(t) = x(t) - px(t-\tau),$$

and

$$y(t) = z(t) - pz(t-\tau).$$

Then eventually $z(t) > 0$, $z'(t) < 0$, and $y(t) > 0$, $y'(t) < 0$. We see that

$$\begin{aligned} y'(t) &= z'(t) - pz'(t-\tau) \\ &= -q(t)x(t-\sigma) + pq(t-\tau)x(t-\tau-\sigma) \\ &\leq -Q(t)z(t-\sigma). \end{aligned}$$

On the other hand

$$\begin{aligned} y\big(t - (\sigma - \tau)\big) &= z\big(t - (\sigma - \tau)\big) - pz(t-\sigma) \\ &\leq (1-p)z(t-\sigma). \end{aligned}$$

Thus the inequality

$$y'(t) \leq -\frac{Q(t)}{1-p}\, y\big(t - (\sigma - \tau)\big)$$

has a positive solution, which implies that Eq. (3.2.66) has a nonoscillatory solution. This contradicts the assumption. □

3.3. Equations with Variable Coefficients (II)

3.3.5. We discuss the neutral delay differential equations of the form

$$\big(x(t) - p(t)x(t-\tau)\big)' + q(t)\prod_{i=1}^{n} |x(t-\sigma_i)|^{\alpha_i} \operatorname{sign} x(t-\sigma_i) = 0 \tag{3.3.1}$$

where $p,\ q \in C([t_0,\infty), R)$, $\tau,\ \sigma_i \in (0,\infty)$, $\alpha_i \geq 0$, and $\sum_{i=1}^{n} \alpha_i = 1$, $i = 1, 2, \ldots, n$. Denote $m = \max\{\tau, \sigma_i,\ i = 1, \ldots, n\}$. If $n = 1$, (3.3.1) reduces to

$$\big(x(t) - p(t)x(t-\tau)\big)' + q(t)x(t-\sigma) = 0. \tag{3.3.2}$$

Lemma 3.3.1. *Assume q is positive and p is bounded and nonnegative, and there exists a $t^* \geq t_0$ such that*

$$p(t^* + n\tau) \leq 1, \quad n = 0, 1, 2, \ldots . \tag{3.3.3}$$

Let $x(t)$ be an eventually positive solution of Eq. (3.3.1) and set

$$z(t) = x(t) - p(t)x(t-\tau). \tag{3.3.4}$$

Then eventually $z(t) > 0$ and $z'(t) < 0$.

Proof: It follows from (3.3.1) that $z'(t) < 0$ eventually. It remains to show that $z(t) > 0$ eventually. Otherwise, $z(t)$ is eventually negative. Thus there exists a sufficiently large T such that $z(t) \leq -d < 0$ for $t \geq T$, where d is a positive constant. Hence

$$x(t) \leq -d + p(t)x(t-\tau) \quad \text{for} \quad t \geq T.$$

In particular,

$$x\big(t^* + (n+N)\tau\big) \leq -nd + x\big(t^* + (N-1)\tau\big), \quad n = 1, 2, \ldots$$

if $t^* + N\tau \geq T$. Hence $x(t)$ can not be eventually positive. This contradiction proves the lemma. □

Lemma 3.3.2. *Assume* $\sigma > 0$, $q \in C([t_0, \infty), (0, \infty))$, $\lambda \in C([t_0 - \sigma, \infty), (0, \infty))$ *satisfying*

$$\liminf_{t \to \infty} \int_{t-\sigma}^{t} q(s)ds > 0 \tag{3.3.5}$$

and

$$\lambda(t) \geq q(t)\exp\Big(\int_{t-\sigma}^{t} \lambda(s)ds\Big), \quad t \geq t_0. \tag{3.3.6}$$

Then

$$\liminf_{t \to \infty} \int_{t-\sigma}^{t} \lambda(s)ds < \infty. \tag{3.3.7}$$

Proof: Define

$$Q(t) = \int_{t_0}^{t} q(s)ds, \quad t \geq t_0.$$

(3.3.5) implies that $\lim_{t \to \infty} Q(t) = +\infty$, and $Q(t)$ is strictly increasing. Then $Q^{-1}(t)$ is well defined, strictly increasing, and $\lim_{t \to \infty} Q^{-1}(t) = +\infty$.

(3.3.5) implies that there exist $c > 0$ and $T_1 \geq t_0$ such that

$$Q(t) - Q(t - \sigma) \geq \frac{c}{2} \quad \text{for} \quad t \geq T_1$$

and thus

$$Q^{-1}\big(Q(t) - \frac{c}{2}\big) \geq t - \sigma, \quad t \geq T_1.$$

Set

$$\Lambda(t) = \exp\Big(-\int_{T}^{t} \lambda(s)ds\Big). \tag{3.3.8}$$

(3.3.6) implies that

$$\Lambda'(t) \le -q(t)\Lambda(t-\sigma), \quad t \ge t_0. \tag{3.3.9}$$

By Lemma 2.1.3 $\frac{\Lambda(t-\sigma)}{\Lambda(t)}$ is bounded above under the condition (3.3.5). Then (3.3.7) is true. □

We are now ready to prove the following result.

Theorem 3.3.1. *In addition to the assumptions of Lemma 3.3.1 assume (3.3.5) hold, and either*

$$\liminf_{t\to\infty}\left\{\inf_{\lambda>0}\left[\prod_{i=1}^{n} p^{\alpha_i}(t-\sigma_i)\frac{q(t)}{q(t-\tau)}e^{\lambda\tau} + \frac{q(t)}{\lambda}exp\left(\lambda\sum_{i=1}^{n}\alpha_i\sigma_i\right)\right]\right\} > 1, \tag{3.3.10}$$

or

$$\liminf_{t\to\infty}\left\{\inf_{\lambda>0}\left[\prod_{i=1}^{n} p^{\alpha_i}(t-\sigma_i)\exp\left(\lambda\int_{t-\tau}^{t} q(s)ds\right) + \frac{1}{\lambda}\exp\left(\lambda\sum_{i=1}^{n}\alpha_i\int_{t-\sigma_i}^{t} q(s)ds\right)\right]\right\} > 1. \tag{3.3.11}$$

Then every solution of (3.3.1) is oscillatory.

Proof: First we assume (3.3.10) holds. Without loss of generality, assume that Eq. (3.3.1) has an eventually positive solution $x(t)$. Let $x(t) > 0$, $x(t-m) > 0$, for $t \ge T_1 \ge t_0$. Then by Lemma 3.3.1, $z(t) > 0$, $z'(t) < 0$ for $t \ge T_1$, where $z(t)$ is defined by (3.3.4). From (3.3.1), we have

$$\begin{aligned} z'(t) &= -q(t)\prod_{i=1}^{n} x^{\alpha_i}(t-\sigma_i) \\ &= -q(t)\prod_{i=1}^{n}\left[z(t-\sigma_i) + p(t-\sigma_i)x(t-\sigma_i-\tau)\right]^{\alpha_i} \\ &\le -q(t)\left[\prod_{i=1}^{n} z^{\alpha_i}(t-\sigma_i) + \prod_{i=1}^{n} p^{\alpha_i}(t-\sigma_i)\prod_{i=1}^{n} x^{\alpha_i}(t-\sigma_i-\tau)\right] \end{aligned}$$

$$= -q(t)\prod_{i=1}^{n} z^{\alpha_i}(t-\sigma_i) + \frac{q(t)}{q(t-\tau)}\prod_{i=1}^{n} p^{\alpha_i}(t-\sigma_i)z'(t-\tau). \tag{3.3.12}$$

Set $\lambda(t) = -\frac{z'(t)}{z(t)}$. Then (3.3.12) reduces to

$$\lambda(t) \geq \lambda(t-\tau)\frac{q(t)}{q(t-\tau)}\prod_{i=1}^{n} p^{\alpha_i}(t-\sigma_i)\exp\Big(\int_{t-\tau}^{t}\lambda(s)ds\Big) + q(t)\exp\Big(\sum_{i=1}^{n}\alpha_i\int_{t-\sigma_i}^{t}\lambda(s)ds\Big). \tag{3.3.13}$$

It is obvious that $\lambda(t) > 0$ for $t \geq T_1$. From (3.3.13) we have

$$\lambda(t) \geq q(t)\exp\Big(\overline{\alpha}\int_{t-\sigma_*}^{t}\lambda(s)ds\Big)$$

where $\sigma_* = \min_{1\leq i\leq n}\{\sigma_i\}$, $\overline{\alpha} = \min_{1\leq i\leq n}\{\alpha_i\}$. In view of Lemma 3.3.2, we have

$$\liminf_{t\to\infty}\int_{t-\sigma_*}^{t}\lambda(s)ds < \infty$$

which implies that $\liminf_{t\to\infty}\lambda(t) < \infty$. Now we show that $\liminf_{t\to\infty}\lambda(t) > 0$. In fact, if $\liminf_{t\to\infty}\lambda(t) = 0$, then there exists a sequence $\{t_n\}$ such that $t_n \geq T_1$, $\lim_{n\to\infty} t_n = \infty$ and $\lambda(t_n) \leq \lambda(t)$, for $t \in [T_1, t_n]$. From (3.3.13), we have

$$\lambda(t_n) \geq \lambda(t_n)\frac{q(t_n)}{q(t_n-\tau)}\prod_{i=1}^{n} p^{\alpha_i}(t_n-\sigma_i)\exp(\lambda(t_n)\tau) + q(t_n)\exp\Big(\lambda(t_n)\sum_{i=1}^{n}\alpha_i\sigma_i\Big).$$

Hence

$$\frac{q(t_n)}{q(t_n-\tau)}\prod_{i=1}^{n} p^{\alpha_i}(t_n-\sigma_i)\exp(\lambda(t_n)\tau) + \frac{1}{\lambda(t_n)}Q(t_n)\exp(\lambda(t_n)\sum_{i=1}^{n}\alpha_i\sigma_i) \leq 1$$

which contradicts (3.3.10), and therefore,

$$0 < \liminf_{t \to \infty} \lambda(t) = h < \infty. \tag{3.3.14}$$

From (3.3.10), there exist $\alpha \in (0,1)$ such that

$$\alpha \liminf_{t\to\infty} \left\{ \inf_{\lambda>0} \left[\prod_{i=1}^{n} p^{\alpha_i}(t-\sigma_i) \frac{q(t)}{q(t-\tau)} e^{\lambda\tau} + \frac{q(t)}{\lambda} \exp\left(\lambda \sum_{i=1}^{n} \alpha_i \sigma_i\right)\right]\right\} > 1. \tag{3.3.15}$$

In view of (3.3.14) we assume that

$$\lambda(t) > \alpha h, \quad t \geq T_2. \tag{3.3.16}$$

Substituting (3.3.16) into (3.3.13) we obtain

$$\lambda(t) > \alpha h \frac{q(t)}{q(t-\tau)} \prod_{i=1}^{n} p^{\alpha_i}(t-\sigma_i) \exp(\alpha h \tau) + q(t) \exp\left(\alpha h \sum_{i=1}^{n} \alpha_i \sigma_i\right)$$

for $t \geq T_2 + m$. Hence

$$h \geq \liminf_{t\to\infty} \left\{ \alpha h \, \frac{q(t)}{q(t-\tau)} \prod_{i=1}^{n} p^{\alpha_i}(t-\sigma_i) \exp(\alpha h \tau) + q(t) \exp\left(\alpha h \sum_{i=1}^{n} \alpha_i \sigma_i\right)\right\}.$$

Set $\lambda^* = \alpha h$. Then

$$\lambda^* \geq \alpha \liminf_{t\to\infty} \left\{ \lambda^* \frac{q(t)}{q(t-\tau)} \prod_{i=1}^{n} p^{\alpha_i}(t-\sigma_i) \exp(\lambda^* \tau) + q(t) \exp\left(\lambda^* \sum_{i=1}^{n} \alpha_i \sigma_i\right)\right\}$$

which contradicts (3.3.15) and completes the proof of this theorem under condition (3.3.10).

If (3.3.11) holds, we let $\lambda(t)q(t) = -\frac{z'(t)}{z(t)}$. Then (3.3.12) becomes

$$\lambda(t) \geq \lambda(t-\tau)\prod_{i=1}^{n} p^{\alpha_i}(t-\sigma_i)\exp\Big(\int_{t-\tau}^{t} \lambda(s)q(s)ds\Big)$$
$$+ \exp\Big(\sum_{i=1}^{n} \alpha_i \int_{t-\sigma_i}^{t} \lambda(s)q(s)ds\Big). \tag{3.3.17}$$

By Lemma 3.3.2, we know that

$$\liminf_{t\to\infty} \int_{t-\sigma}^{t} \lambda(s)q(s)ds < \infty. \tag{3.3.18}$$

From (3.3.5) and (3.3.18), we conclude that $\liminf_{t\to\infty} \lambda(t) < \infty$. From (3.3.17) $\lambda(t) \geq 1$ and so $0 < \liminf_{t\to\infty} \lambda(t) = h < \infty$.

From (3.3.11), there exists $\alpha \in (0,1)$ such that

$$\alpha \liminf_{t\to\infty} \left\{ \inf_{\lambda>0} \left[\prod_{i=1}^{n} p^{\alpha_i}(t-\sigma_i)\exp\Big(\lambda \int_{t-\tau}^{t} q(s)ds\Big) \right.\right.$$
$$\left.\left. + \frac{1}{\lambda} \exp\left(\lambda \sum_{i=1}^{n} \alpha_i \int_{t-\sigma_i}^{t} q(s)ds\right)\right]\right\} > 1.$$

By a similar argument to the first part of the proof, we reach a contradiction.

□

Corollary 3.3.1. *In addition to the assumptions of Lemma 3.3.1 assume (3.3.5) holds and*

$$\liminf_{t\to\infty}\left\{\inf_{\lambda>0}\left[p(t-\sigma)\frac{q(t)}{q(t-\tau)}e^{\lambda\tau} + \frac{1}{\lambda}q(t)e^{\lambda\sigma}\right]\right\} > 1. \tag{3.3.19}$$

Then every solution of (3.3.2) is oscillatory.

Example 3.3.1. Consider

$$\big(x(t) - (2+\sin t)x(t-\pi)\big)' + (3+2\sin t)x(t-\frac{\pi}{2}) = 0, \quad t \geq \pi. \tag{3.3.20}$$

It is easy to see that (3.3.19) holds. Therefore, every solution of (3.3.20) is oscillatory. In fact, $x(t) = \sin t$ is such a solution.

It is easy to obtain explicit conditions for oscillation of Eq. (3.3.1) from Theorem 3.3.1. The following is an example.

Corollary 3.3.2. *In addition to the assumptions of Lemma 3.3.1 assume (3.3.5) holds and*

$$\liminf_{t\to\infty}\left\{\prod_{i=1}^{n} p^{\alpha_i}(t-\sigma_i)+e\sum_{i=1}^{n}\alpha_i\int_{t-\sigma_i}^{t} q(s)ds\right\}>1. \tag{3.3.21}$$

Then every solution of (3.3.1) is oscillatory.

The following is a nonoscillation result for Eq. (3.3.1).

Theorem 3.3.2. *Assume that*

(i) $p, q \in C([t_0,\infty), R)$, $p(t)\geq 0$, *and* $0\leq q(t)\leq M$, *where* M *is a constant;*

(ii) *There exists a positive number* α *such that for* $t\geq t_0$

$$0\leq p(t)e^{\alpha t}\leq c<1 \tag{3.3.22}$$

and

$$p(t)e^{\alpha\tau}+\exp\left(\alpha\sum_{i=1}^{n}\alpha_i\sigma_i\right)\int_t^\infty q(s)\exp[-\alpha(s-t)]ds\leq 1. \tag{3.3.23}$$

Then Eq. (3.3.1) has an eventually positive solution which approaches zero exponentially.

Proof: Let BC denote the Banach space of all bounded and continuous real valued functions defined on $[t_0-m,\infty)$ with the sup norm. Let Ω be the subset of BC defined by

$$\Omega=\{y\in BC: 0\leq y(t)\leq 1 \quad \text{for} \quad t\geq t_0-m\}.$$

Now we define operators $\mathcal{T}_1$ and $\mathcal{T}_2$ on Ω as follows:

$$(\mathcal{T}_1 y)(t) = \begin{cases} p(t)e^{\alpha\tau}y(t-\tau), & t \geq t_0 \\ \frac{t}{t_0}(\mathcal{T}_1 y)(t_0) + (1 - \frac{t}{t_0}), & t_0 - m \leq t \leq t_0; \end{cases}$$

$$(\mathcal{T}_2 y)(t) = \begin{cases} \exp(\alpha \sum_{i=1}^{n} \alpha_i \sigma_i) \int_t^\infty q(s) & \\ \quad \times \exp[-\alpha(s-t)] \prod_{i=1}^{n} y^{\alpha_i}(s-\sigma_i)ds, & t \geq t_0 \\ \frac{t}{t_0}(\mathcal{T}_2 y)(t_0), & t_0 - m \leq t \leq t_0. \end{cases}$$

In view of (3.3.22) and (3.3.23) we have $\mathcal{T}_1 y_1 + \mathcal{T}_2 y_2 \in \Omega$ for every pair $y_1, y_2 \in \Omega$, and $\mathcal{T}_1$ is a contraction on Ω. It is easy to see that $|\frac{d}{dt}(\mathcal{T}_2 y)(t)| \leq N$ for some positive constant N. Hence $\mathcal{T}_2$ is completely continuous on Ω. By Krasnoselskii's fixed point theorem, there exists a $y \in \Omega$ such that $\mathcal{T}_1 y + \mathcal{T}_2 y = y$. That is, for $t \geq t_0$

$$\begin{aligned} y(t) &= p(t)e^{\alpha\tau}y(t-\tau) \\ &\quad + \exp\left(\alpha \sum_{i=1}^{n} \alpha_i \sigma_i\right) \int_t^\infty q(s)\exp[-\alpha(s-t)] \prod_{i=1}^{n} y^{\alpha_i}(s-\sigma_i)ds, \end{aligned}$$

and for $t_0 - m \leq t < t_0$

$$y(t) = \frac{t}{t_0}\, y(t_0) + (1 - \frac{t}{t_0}) > 0.$$

It follows that $y(t) > 0$ for all $t \geq t_0 - m$. Set $x(t) = y(t)e^{-\alpha t}$. Then

$$x(t) = p(t)x(t-\tau) + \int_t^\infty q(s) \prod_{i=1}^{n} x^{\alpha_i}(s-\sigma_i)ds, \quad t \geq t_0.$$

Consequently,

$$\big(x(t) - p(t)x(t-\tau)\big)' + q(t) \prod_{i=1}^{n} x^{\alpha_i}(t-\sigma_i) = 0, \quad t \geq t_0$$

i.e. $x(t)$ is a positive solution of Eq. (3.3.1) and $x(t)$ approaches zero exponentially. □

Remark 3.3.1. When $p(t) \equiv p$, $q(t) \equiv q$ are both positive constants, condition (3.3.23) becomes

$$pe^{\alpha\tau} + \frac{q}{\alpha}\exp\left(\alpha \sum_{i=1}^{n} \alpha_i \sigma_i\right) \le 1 \tag{3.3.24}$$

for some $\alpha > 0$. From Theorems 3.3.1 and 3.3.2 condition (3.3.24) provides a sufficient and necessary condition for the existence of a nonoscillatory solution of Eq. (3.3.1).

Example 3.3.2. Consider the equation

$$\left(x(t) - \frac{1}{2e}x(t-1)\right)' + q(t)x^{1/3}(t)x^{1/3}(t-1)x^{1/3}(t-2) = 0 \tag{3.3.25}$$

where $q(t) = \frac{t}{2et^{1/3}(t-1)^{1/3}(t-2)^{1/3}}$, $t \ge 3$. It is not difficult to see that all assumptions of Theorem 3.3.21 are satisfied. Therefore, Eq. (3.3.25) has a nonoscillatory solution which tends to zero as $t \to \infty$. In fact, $x(t) = te^{-t}$ is such a solution.

Corollary 3.3.3. *Assume that there exist positive constants p and q such that*

$$0 \le p(t) \le p < 1, \quad 0 \le q(t) \le q, \tag{3.3.26}$$

and that the linear equation

$$(x(t) - px(t-\tau))' + qx\left(t - \sum_{i=1}^{n} \alpha_i \sigma_i\right) = 0 \tag{3.3.27}$$

has a nonoscillatory solution. Then Eq. (3.3.1) also has a nonoscillatory solution.

In fact, from Theorem 3.1.1, (3.3.27) having a nonoscillatory solution implies that the characteristic equation of (3.3.27)

$$\lambda - \lambda pe^{-\lambda\tau} + q \exp\left(-\lambda \sum_{i=1}^{n} \alpha_i \sigma_i\right) = 0 \tag{3.3.28}$$

has a negative real root λ. Let $\alpha = -\lambda > 0$, then (3.3.24) is satisfied. The conclusion of the corollary follows from Theorem 3.3.2.

The following result is for Eq. (3.3.1) with oscillating coefficients $p(t)$ and $q(t)$.

Theorem 3.3.3. *Assume that $p \in C^1([t_0,\infty),R)$, $q \in C([t_0,\infty),R)$ and there exists a positive number μ such that*

$$\mu|p(t)|e^{\mu\tau} + |p'(t)|e^{\mu\tau} + |q(t)|\exp\left(\mu\sum_{i=1}^{n}\alpha_i\sigma_i\right) \le \mu, \quad t \ge t_0. \tag{3.3.29}$$

Then Eq. (3.3.1) has a positive solution for $t \ge t_0$.

Proof: Set

$$p_1(t) = \begin{cases} p(t) & t \ge t_0 \\ \frac{t-t_0+\tau}{\tau}\, p(t_0), & t_0-\tau \le t \le t_0 \\ 0, & t_0-m-\tau \le t \le t_0-\tau, \end{cases}$$

$$p_2(t) = \begin{cases} p'(t), & t \ge t_0 \\ \frac{t-t_0+\tau}{\tau}\, p'(t_0), & t_0-\tau \le t \le t_0 \\ 0, & t_0-m-\tau \le t \le t_0-\tau, \end{cases}$$

and

$$p_3(t) = \begin{cases} q(t), & t \ge t_0 \\ \frac{t-t_0+\tau}{\tau}\, q(t_0), & t_0-\tau \le t \le t_0 \\ 0, & t_0-m-\tau \le t \le t_0-\tau. \end{cases}$$

Then p_i, $i=1,2,3$, are continuous on $[t_0-m-\tau,\infty)$. From (3.3.29) we have

$$\mu|p_1(t)|e^{\mu\tau} + |p_2(t)|e^{\mu\tau} + |p_3(t)|\exp\left(\mu\sum_{i=1}^{n}\alpha_i\sigma_i\right) \le \mu, \quad t \ge t_0-m-\tau. \tag{3.3.30}$$

We introduce the Banach space X of all bounded continuous functions $x : [t_0-m-\tau,\infty) \to R$ with the norm $\|x\| = \sup_{t \ge t_0-m-\tau} |x(t)|e^{-\eta t}$ where $\eta > 0$

satisfies the following inequality

$$|p_1(t)|e^{\mu\tau}(e^{-\eta\tau}+\frac{\mu}{\eta})+\frac{e^{\mu\tau}}{\eta}|p_2(t)|+\frac{1}{\eta}|p_3(t)|\exp\left(\mu\sum_{i=1}^{n}\alpha_i\sigma_i\right)\le\frac{1}{2}$$

$$\text{for}\quad t\ge t_0-\tau. \tag{3.3.31}$$

We consider the subset Ω of X as follows

$$\Omega=\{\lambda\in X:\ |\lambda(t)|\le\mu,\quad t\ge t_0-m-\tau\}.$$

Clearly, Ω is a bounded, closed and convex subset of X. Define an operator $\mathcal{T}$ on Ω as follows:

$$(\mathcal{T}\lambda)(t)=\begin{cases}\lambda(t-\tau)p_1(t)\exp(\int_{t-\tau}^{t}\lambda(s)ds)+p_2(t)\exp(\int_{t-\tau}^{t}\lambda(s)ds)\\ \quad+p_3(t)\exp(\sum_{i=1}^{n}\alpha_i\int_{t-\sigma_i}^{t}\lambda(s)ds),\quad t\ge t_0-\tau\\ 0,\qquad t_0-m-\tau\le t\le t_0-\tau.\end{cases}$$

In view of (3.2.41), we have

$$\begin{aligned}|(\mathcal{T}\lambda)(t)|&\le\mu|p_1(t)|e^{\mu\tau}+|p_2(t)|e^{\mu\tau}\\ &\quad+|p_3(t)|\exp\left(\mu\sum_{i=1}^{n}\alpha_i\sigma_i\right)\le\mu\quad\text{for}\quad t\ge t_0-\tau\end{aligned}$$

which shows that $\mathcal{T}$ maps Ω into itself.

Next, we show that $\mathcal{T}$ is a contraction on Ω. In fact, for any $\lambda_1,\ \lambda_2\in\Omega$ and $t\ge t_0-\tau$

$$\begin{aligned}&|(\mathcal{T}\lambda_1)(t)-(\mathcal{T}\lambda_2)(t)|\\ &\le|p_1(t)|\left|\lambda_1(t-\tau)\exp\left(\int_{t-\tau}^{t}\lambda_1(s)ds\right)-\lambda_2(t-\tau)\exp\left(\int_{t-\tau}^{t}\lambda_2(s)ds\right)\right|\\ &\quad+|p_2(t)|\left|\exp\left(\int_{t-\tau}^{t}\lambda_1(s)ds\right)-\exp\left(\int_{t-\tau}^{t}\lambda_2(s)ds\right)\right|\\ &\quad+|p_3(t)|\left|\exp\left(\sum_{i=1}^{n}\alpha_i\int_{t-\sigma_i}^{t}\lambda_1(s)ds\right)-\exp\left(\sum_{i=1}^{n}\alpha_i\int_{t-\sigma_i}^{t}\lambda_2(s)ds\right)\right|\end{aligned}$$

$$\leq |p_1(t)|\left(e^{\mu\tau}|\lambda_1(t-\tau)-\lambda_2(t-\tau)|+\mu e^{\mu\tau}\int_{t-\tau}^{t}|\lambda_1(s)-\lambda_2(s)|ds\right)$$

$$+\,e^{\mu\tau}|p_2(t)|\int_{t-\tau}^{t}|\lambda_1(s)-\lambda_2(s)|ds$$

$$+\,|p_3(t)|\exp\left(\mu\sum_{i=1}^{n}\alpha_i\sigma_i\right)\sum_{i=1}^{n}\alpha_i\int_{t-\sigma_i}^{t}|\lambda_1(s)-\lambda_2(s)|ds$$

$$\leq\left\{e^{\mu\tau}|p_1(t)|e^{\eta\tau}\left[e^{-\eta\tau}+\frac{\mu}{\eta}(1-e^{-\eta\tau})\right]+\frac{1}{\eta}e^{\mu\tau}|p_2(t)|(1-e^{-\eta\tau})e^{\eta\tau}\right.$$

$$\left.+\,\frac{1}{\eta}|p_3(t)|\exp\left(\mu\sum_{i=1}^{n}\alpha_i\sigma_i\right)\sum_{i=1}^{n}\alpha_i(1-e^{-\eta\sigma_i})e^{\eta t}\right\}\|\lambda_1-\lambda_2\|$$

$$\leq e^{\eta t}\left\{|p_1(t)|e^{\mu\tau}\left(e^{-\eta\tau}+\frac{\mu}{\eta}\right)+\frac{e^{\mu\tau}}{\eta}|p_2(t)|\right.$$

$$\left.+\,\frac{1}{\eta}|p_3(t)|\exp\left(\mu\sum_{i=1}^{n}\alpha_i\sigma_i\right)\right\}\|\lambda_1-\lambda_2\|$$

$$\leq\frac{1}{2}e^{\eta t}\|\lambda_1-\lambda_2\|.$$

Hence

$$\|\mathcal{T}\lambda_1-\mathcal{T}\lambda_2\|=\sup_{t\geq t_0-m-\tau}|(\mathcal{T}\lambda_1)(t)-(\mathcal{T}\lambda_2)(t)|e^{-\eta t}$$

$$=\sup_{t\geq t_0-\tau}|(\mathcal{T}\lambda_1)(t)-(\mathcal{T}\lambda_2)(t)|e^{-\eta t}\leq\frac{1}{2}\|\lambda_1-\lambda_2\|,$$

i.e., $\mathcal{T}$ is a contraction on Ω. Therefore there exists a $\lambda\in\Omega$ such that $\mathcal{T}\lambda=\lambda$. That is,

$$\lambda(t)=\begin{cases}\lambda(t-\tau)p_1(t)\exp\left(\int_{t-\tau}^{t}\lambda(s)ds\right)+p_2(s)\exp\left(\int_{t-\tau}^{t}\lambda(s)ds\right)\\ \quad+\,p_3(t)\exp\left(\sum\limits_{i=1}^{n}\alpha_i\int_{t-\sigma_i}^{t}\lambda(s)ds\right),\quad t\geq t_0-\tau,\\ 0,\qquad t_0-m-\tau\leq t\leq t_0-\tau.\end{cases}$$

According to the definition of p_i, we have

$$\lambda(t) = \lambda(t-\tau)p(t)\exp\Big(\int_{t-\tau}^{t}\lambda(s)ds\Big) + p(t)\exp\Big(\int_{t-\tau}^{t}\lambda(s)ds\Big)$$
$$+ q(t)\exp\Big(\sum_{i=1}^{n}\alpha_i\int_{t-\sigma_1}^{t}\lambda(s)ds\Big), \quad t \geq t_0.$$

Set $x(t) = \exp(-\int_{t_0}^{t}\lambda(s)ds)$. Then we can find that $x(t)$ is a positive solution of Eq. (3.3.1) on $[t_0, \infty)$. □

Example 3.3.3. Consider

$$\Big(x(t) - \frac{2\sin t}{t-1}x(t-1)\Big)' + \frac{1-2\cos t}{1-t}x(t-1) = 0. \tag{3.3.32}$$

It is easy to see that (3.3.29) with $\mu = 1$ is satisfied for $t \geq T$, where T is a sufficiently large number. Then Eq. (3.3.32) has a positive solution on $[T, \infty)$. In fact, $x(t) = t$ is such a solution of Eq. (3.3.32).

3.3.6. We now consider the neutral delay differential equations with several delays of the form

$$(x(t) - px(t-\tau))' + \sum_{i=1}^{n} q_i(t)x(t-\sigma_i) = 0, \quad t \geq t_0 \tag{3.3.33}$$

where $p \in [0,1]$, $q_i \quad (i = 1,2,\ldots,n) \in C([t_0,\infty), R_+)$ and $\tau_i, \sigma_i \ (i = 1,2,\ldots,n) \in (0,\infty)$. Denote $m = \max\{\tau, \sigma_1, \ldots, \sigma_n\}$.

Lemma 3.3.3. *Assume that* $\liminf_{t\to\infty}\int_{t-\sigma_i}^{t} q_1(s)ds > 0$ *and*

$$q_j(t-\tau) \leq q_j(t), \quad t \geq t_0 + m, \quad j = 1,2,\ldots,n. \tag{3.3.34}$$

Let $x(t)$ be an eventually positive solution of (3.3.33) and set

$$z(t) = x(t) - px(t-\tau). \tag{3.3.35}$$

Then eventually $z(t) > 0, \quad z'(t) < 0$, *and*

$$z'(t) - pz'(t-\tau) + \sum_{i=1}^{n} q_i(t)z(t-\sigma_i) \le 0. \tag{3.3.36}$$

Proof: As in the proof of Theorem 3.2.1, it is easy to see that $z(t) > 0, \; z'(t) < 0$ eventually. From (3.3.34) and (3.3.35) we have

$$\begin{aligned} z'(t) - pz'(t-\tau) &= -\sum_{i=1}^{n}[q_i(t)x(t-\sigma_i) - pq_i(t-\tau)x(t-\tau-\sigma_i)] \\ &\le -\sum_{i=1}^{n} q_i(t)z(t-\sigma_i) \end{aligned}$$

eventually.

Theorem 3.3.4. *Assume that the assumptions of Lemma 3.3.3 hold. Further assume that for all* $\mu > 0$, *and* $\ell = \tau, \sigma_1, \ldots, \sigma_n$

$$\liminf_{t\to\infty}\left\{pe^{\mu\tau} + \frac{1}{\ell\mu}\sum_{i=1}^{n} e^{\mu\sigma_i}\int_t^{t+\ell} q_i(s)ds\right\} > 1. \tag{3.3.37}$$

Then every solution of Eq. (3.3.33) is oscillatory.

Proof: Assume (3.3.33) has an eventually positive solution $x(t)$. By Lemma 3.3.3 there exists a $T \ge t_0$ such that $z(t) > 0, \; z'(t) < 0$, and (3.3.36) holds for $t \ge T$. Let $w(t) = -\frac{z'(t)}{z(t)}$, $t \ge T$. Then $w(t) > 0, \; t \ge T$, and (3.3.36) becomes

$$w(t) \ge pw(t-\tau)\exp\left(\int_{t-\tau}^{t} w(s)ds\right) + \sum_{i=1}^{n} q_i(t)\exp\left(\int_{t-\sigma_i}^{t} w(s)ds\right) \tag{3.3.38}$$

for $t \ge \tau + m$.

We now define a sequence of functions $\{w_k(t)\}$ for $k = 1, 2, \ldots$ and $t \ge T$, and a sequence of numbers $\{\mu_k\}$ for $k = 1, 2, \ldots,$ as follows:

$$w_1(t) \equiv 0, \quad t \ge \tau$$

and for $k = 1, 2, \ldots, t \geq \tau + km$

$$w_{k+1}(t) = pw_k(t-\tau)\exp\left(\int_{t-\tau}^{t} w_k(s)ds\right) + \sum_{i=1}^{n} q_i(t)\exp\left(\int_{t-\sigma_i}^{t} w_k(s)ds\right) \tag{3.3.39}$$

and $\mu_1 = 0$, and for $k = 1, 2, \ldots$

$$\mu_{k+1} = \inf_{t \geq T} \min_{\ell=\tau,\sigma_1,\ldots,\sigma_n} \left\{ p\mu_k e^{\mu_k \tau} + \frac{1}{\ell} \sum_{i=1}^{n} e^{\mu_k \sigma_k} \int_t^{t+\ell} q_i(s)ds \right\}. \tag{3.3.40}$$

We claim that

i) $0 = \mu_1 < \mu_2 < \ldots,$
ii) $w_k(t) \leq w(t)$ for $t \geq T + (k-1)m$ and $k = 1, 2, \ldots,$
iii) $\frac{1}{\ell}\int_t^{t+\ell} w_k(s)ds \geq \mu_k$ for $t \geq T + (k+1)m, \quad l = 1, 2, \ldots,$ and $\ell = \tau, \sigma_1, \ldots, \sigma_n$.

In fact, i) and ii) follow from (3.3.39) and (3.3.40) by induction. Clearly iii) is true for $k = 1$. Assume iii) is true for some k. Then (3.3.39) and (3.3.40) imply that for $t \geq T + km, \quad \ell = \tau, \sigma_1, \ldots, \sigma_n$

$$\begin{aligned}\frac{1}{\ell}\int_t^{t+\ell} w_{k+1}(s)ds &= \frac{1}{\ell}\int_t^{t+\ell} w_k(s-\tau)\exp\left(\int_{s-\tau}^{s} w_k(\theta)d\theta\right)ds \\ &\quad + \frac{1}{\ell}\sum_{i=1}^{n}\int_t^{t+\ell} q_i(s)\exp\left(\int_{s-\sigma_i}^{s} w_k(\theta)d\theta\right)ds \\ &\geq p\mu_k e^{\mu_k\tau} + \frac{1}{\ell}\sum_{i=1}^{n} e^{\mu_k\sigma_i}\int_t^{t+\ell} q_i(s)ds \\ &\geq \mu_{k+1}.\end{aligned}$$

Hence iii) holds.

Let $\mu^* = \lim_{n\to\infty} \mu_n$. From (3.3.37) and (3.3.40) there exists an $\alpha > 1$ such that $\mu_{k+1} \geq \alpha\mu_k, \ k = 1, 2, \ldots,$ which implies that $\mu^* = \infty$.

In view of ii) and iii) $\lim_{t\to\infty}\int_t^{t+\sigma_1} w(s)ds = \infty$. From the definition of w, we have

$$\frac{z(t)}{z(t+\sigma_1)} = \exp\int_t^{t+\sigma_1} w(s)ds.$$

Thus

$$\lim_{t\to\infty}\frac{z(t)}{z(t+\sigma_1)}=\infty. \tag{3.3.41}$$

From (3.3.33)

$$\begin{aligned} z'(t) &= -\sum_{i=1}^{n} q_i(t)x(t-\sigma_i) \le -q_1(t)x(t-\sigma_1) \\ &\le -q_1(t)z(t-\sigma_1). \end{aligned}$$

By Lemma 2.1.3, $\frac{z(t-\sigma_1)}{z(t)}$ is bounded above which contradicts (3.3.41). □

Theorem 3.3.5. *Assume there exists a $\mu^* > 0$ and a $T \ge t_0$ such that for $\ell = \tau, \sigma, \ldots, \sigma_n$*

$$\sup_{t\ge T}\{pe^{\mu^*\tau} + \frac{1}{\ell\mu^*}\sum_{i=1}^{n} e^{\mu^*\sigma_i}\int_t^{t+\ell} q_i(s)ds\} \le 1. \tag{3.3.42}$$

Then (3.3.33) has a positive solution on $[T+m,\infty)$.

Proof: First we claim that the integral equation

$$v(t) = pv(t-\tau)\exp\int_{t-\tau}^{t} v(s)ds + \sum_{i=1}^{n} q_i(s)\exp\int_{t-\sigma_i}^{t} v(s)ds \tag{3.3.43}$$

has a positive solution on $[T+m,\infty)$. To this end set $v_1(t)\equiv 0,\quad t\ge T$, and for $k = 1,2,\ldots$

$$v_{k+1}(t) = \begin{cases} pv_k(t-\tau)\exp(\int_{t-\tau}^{t} v_k(s)ds) \\ \quad + \sum_{i=1}^{n} q_i(t)\exp(\int_{t-\sigma_i}^{t} v_k(s)ds), & t\ge T+m \\ \beta_{k+1}(t) & T\le t< T+m \end{cases} \tag{3.3.44}$$

where $\{\beta_k\}$ are given function sequences satisfying

i) $\beta_k \in C^2([T,T+m),[0,\infty))$ with $\beta_k' \ge 0$ and $\beta_k'' \ge 0,\quad t\in[T,T+m)$, $k=1,2,\ldots$;

ii) $\beta_k(t) = 0, \quad t \in [T, T+m-\tau], \quad \beta_k(T+m) = v_k(T+m)$, and $\beta_k(t)$ are increasing in k for $t \in [T+m-\tau, T+m)$ and $k = 1, 2, \ldots$;

iii) for every $\ell = \tau, \sigma_1, \ldots, \sigma_n$ and $t \in [T+m-\ell, T+m), \quad k = 1, 2, \ldots$

$$\int_t^{T+m} \beta_k(s)ds \le \int_{t+\ell}^{T+m+\ell} v_k(s)ds.$$

Clearly, $v_1(t) \le v_2(t) \le \ldots$. By induction we will show that for $k = 1, 2, \ldots$ and $\ell = \tau, \sigma_1, \ldots, \sigma_n$

$$\frac{1}{\ell}\int_t^{t+\ell} v_k(s)ds \le \mu^*, \quad t \ge T. \tag{3.3.45}$$

In fact, (3.3.45) is true for $k = 1$. Assume (3.3.45) holds for some k. Then from (3.3.42) and (3.3.44) we have for $t \ge T+m, \quad \ell = \tau, \sigma_1, \ldots, \sigma_n$

$$\begin{aligned} &\frac{p}{\ell}\int_t^{t+\ell} v_k(s-\tau)\exp\left(\int_{s-\tau}^{s} v_k(\theta)d\theta\right)ds \\ &\qquad + \frac{1}{\ell}\sum_{i=1}^{r}\int_t^{t+\ell} q_i(s)\exp\left(\int_{s-\sigma_i}^{s} v_k(\theta)d\theta\right)ds \\ &\qquad \le p\mu^* e^{\mu^*\tau} + \frac{1}{\ell}\sum_{\ell=1}^{n} e^{\mu^*\sigma}\int_t^{t+\ell} q_i(s)ds \\ &\qquad \le \mu^*. \end{aligned} \tag{3.3.46}$$

For every $\ell = \tau, \sigma_1, \ldots, \sigma_n$ and $t \in [T+m-\ell, T+m)$ from (3.3.46) and condition iii) for $\{\beta_k\}$ we have

$$\begin{aligned} \frac{1}{\ell}\int_t^{t+\ell} v_{k+1}(s)ds &= \frac{1}{\ell}\left[\int_t^{T+m} \beta_{k+1}(s)ds + \int_{T+m}^{t+\ell} v_{k+1}(s)ds\right] \\ &\le \frac{1}{\ell}\int_{T+m}^{T+m+\ell} v_{k+1}(s)ds \le \mu^*. \end{aligned}$$

From the monotonic property of $\beta_{k+1}(t)$ with respect to t we see that (3.3.45) also holds for $t \in [T, T+m-\ell), \quad \ell = \tau, \sigma_1, \ldots, \sigma_n$.

Let $v(t) = \lim_{k\to\infty} v_k(t)$. Then $v(t) \equiv 0, \quad t \in [T, T+m-\tau], \quad v(t)$ is increasing on $[T+m-\tau, T+m)$, and for $t \geq T$ and $\ell = \tau, \sigma_1, \ldots, \sigma_n$

$$\frac{1}{\ell}\int_t^{t+\ell} v(s)ds \leq \mu^*.$$

Taking limits as $k \to \infty$ on both sides of (3.3.44), from the Lebesgue's monotone convergence theorem, we see that $v(t)$ satisfies (3.3.43) for $t \geq T+m$. It is easy to see that $v(t)$ is well defined on $[T, \infty)$. Furthermore, from condition (i) of $\{\beta_k\}$ we know that $v(t)$ is continuous on $[T, T+m]$, and from (3.3.40) $v(t)$ is continuous on $[T, \infty)$. Thus $v(t)$ is a continuous positive solution of (3.3.43) on $[T+m, \infty)$. Set

$$x(t) = \exp\left(-\int_{\tau+m}^{t} v(s)ds\right), \quad t \geq \tau + m.$$

Then $x(t)$ is a positive solution of (3.3.33). □

Corollary 3.3.4. *If*

$$\sum_{i=1}^{n} e^{\mu\sigma_i}\int_t^{t+\ell} q_i(s)ds \tag{3.3.47}$$

is nondecreasing in t for $\ell = \tau, \sigma_1, \ldots, \sigma_n$. Then (3.3.33) is oscillatory if and only if for all $\mu > 0$ and $\ell = \tau, \sigma_1, \ldots, \sigma_n$

$$\lim_{t\to\infty}\left\{pe^{\mu\tau} + \frac{1}{\ell\mu}\sum_{i=1}^{n} e^{\mu\sigma_i}\int_t^{t+\ell} q_i(s)ds\right\} > 1. \tag{3.3.48}$$

Corollary 3.3.5. *If there exist $i \in I_n = \{1, \ldots, n\}$ and $\ell > 0$ such that $\liminf_{t\to\infty}\int_{t-\sigma_i}^{t} q_i(s)ds > 0$ and $\tau = m_0\ell, \quad \sigma_i = m_i\ell$ for some integers m_i, $i = 0, 1, \ldots, n$. Furthermore*

$$q_i(t) = g_i(t) + h_i(t), \quad i = 1, 2, \ldots, n,$$

where $g_i(t)$ are ℓ-periodic functions with $\frac{1}{\ell}\int_t^{t+\ell} g_i(s)ds = g_i^*, \quad h_i(t)$ are nondecreasing with $\lim_{t\to\infty} h_i(t) = h_i^*, \quad i = 1, 2, \ldots, n$. Then (3.3.33) is oscillatory if and *only if for all* $\mu > 0$

$$pe^{\mu\tau} + \frac{1}{\mu}\sum_{i=1}^{n} e^{\mu\sigma_i}(g_i^* + h_i^*) > 1. \tag{3.3.49}$$

For the case that $q_i(t) \equiv 0$ and $h_i(t) \equiv h_i^*$, $i = 1, 2, \ldots, n$, (3.3.49) becomes

$$pe^{\mu\tau} + \frac{1}{\mu}\sum_{i=1}^{n} e^{\mu\sigma_i} h_i^* > 1. \tag{3.3.50}$$

this implies that Eq. (3.3.33) with constants parameters is oscillatory if and only if its characteristic equation has no real roots.

If $h_i(t) \equiv 0$, $i = 1, 2, \ldots, n$. Eq. (3.3.33) becomes

$$\big(x(t) - px(t-\tau)\big)' + \sum_{i=1}^{n} g_i(t)x(t-\sigma_i) = 0 \tag{3.3.51}$$

where $g_i(t)$ are ℓ-periodic functions. Then (3.3.51) is oscillatory if and only if its "characteristic equation"

$$\lambda(1 - pe^{-\lambda\tau}) + \sum_{i=1}^{n} g_i^* e^{-\lambda\sigma_i} = 0 \tag{3.3.52}$$

has no real roots, where $g_i^* = \frac{1}{\ell}\int_t^{t+\ell} g_i(s)ds$, $i = 1, 2, \ldots, n$.

By employing the above technique in Eq. (2.7.1) the condition (2.7.3) in Theorem 2.7.1 can be improved to assuming that for all $\lambda > 0$ and $j = 1, 2, \ldots, n$

$$\liminf_{t\to\infty}\left\{\frac{1}{\lambda\tau_j(t)}\sum_{i=1}^{n}\int_t^{t+\tau_j(t)} p_i(s)e^{\lambda\tau_i(s)}ds\right\} > 1. \tag{3.3.53}$$

Corollary 3.3.6. *If* $\tau_i(t)$ *are nondecreasing for* $i = 1, 2, \ldots, n$, *and*

$$\liminf_{t\to\infty}\sum_{i=1}^{n}\int_t^{t+\tau_i(t)} p_i(s)ds > \frac{1}{e}\,. \tag{3.3.54}$$

Then Eq. (2.7.1) is oscillatory.

From Theorem 3.3.4 we derive some explicit conditions for oscillation of Eq. (3.3.33).

Theorem 3.3.6. *Assume that the assumptions of Lemma 3.3.3 hold. Furthermore, for* $\ell = \tau, \sigma_1, \ldots, \sigma_n$

$$\liminf_{t\to\infty} \sum_{i=1}^{n}\sum_{k=0}^{\infty}\left(\frac{1}{\ell}\int_t^{t+\ell} q_i(s)ds\right)p^k(k\tau+\sigma_i) > \frac{1}{e}\,. \tag{3.3.55}$$

Then every solution of Eq. (3.3.33) is oscillatory.

Proof: We shall show that (3.3.37) is true for all $\mu > 0$, and then Eq. (3.3.33) is oscillatory by Theorem 3.3.4. In fact

$$\begin{aligned}
&\frac{1}{\mu}\sum_{i=1}^{n}\left(\frac{1}{\ell}\int_t^{t+\ell} q_i(s)ds\right)e^{\mu\sigma_i}(1-pe^{\mu\tau})^{-1}\\
&\quad=\frac{1}{\mu}\sum_{i=1}^{n}\sum_{k=0}^{\infty}\left(\frac{1}{\ell}\int_t^{t+\ell} q_i(s)ds\right)p^k e^{\mu(k\tau+\sigma_i)}\\
&\quad\geq\sum_{i=1}^{n}\sum_{k=0}^{\infty}\left(\frac{1}{\ell}\int_t^{t+\ell} q_i(s)ds\right)p^k e(k\tau+\sigma_i).
\end{aligned}$$

Thus (3.3.55) implies that

$$\liminf_{t\to\infty}\frac{1}{\mu}\sum_{i=1}^{n}\left(\frac{1}{\ell}\int_t^{t+\ell} q_i(s)ds\right)e^{\mu\sigma_i}(1-pe^{\mu\tau})^{-1} > 1.$$

Hence (3.3.37) holds for all $\mu > 0$ and $\ell = \tau, \sigma_1, \ldots, \sigma_n$. The proof is complete. □

For the case $p \geq 1$, we can prove the similar results using above techniques.

Theorem 3.3.7. *Assume that*

i) $p \geq 1, \quad q_i(t) \leq q_i(t-\tau)$, *for* $i = 1, 2, \ldots, n$ *and all large* T;

ii) $\tau > \sigma_i, \quad i = 1, 2, \ldots, n, \ \liminf_{t\to\infty}\int_t^{t+\frac{\tau-\max\sigma_i}{2}}\sum_{i=1}^{n} q_i(s)ds > 0$;

iii) *for all* $\mu > 0$, $\ell = \tau, \tau - \sigma_i$, *and* $i = 1, 2, \ldots, n$

$$\liminf_{t\to\infty}\left\{\frac{1}{p}e^{\mu\tau} + \frac{1}{\ell p\mu}\sum_{i=1}^{n} e^{\mu(\tau-\sigma_i)}\int_t^{t+\ell} q_i(s)ds\right\} > 1. \tag{3.3.56}$$

Then every solution of (3.3.33) is oscillatory.

Theorem 3.3.8. *Assume that*

i) $p \geq 1$, $\liminf_{t\to\infty} \sum_{i=1}^{n} q_i(t) > 0$, $\tau > \sigma_i$, $i = 1, 2, \ldots, n$;

ii) *there exist* $\lambda^* > 0$ *and* $T \geq t_0$, *for* $\ell = \tau, \tau - \sigma_i$, $i = 1, 2, \ldots, n$

$$\sup_{t\geq T}\left\{\frac{1}{p}e^{\lambda^*\tau} + \frac{1}{\ell p\lambda^*}\sum_{i=1}^{n} e^{\lambda^*(\tau-\sigma_i)}\int_t^{t+\ell} q_i(s)ds\right\} \leq 1. \tag{3.3.57}$$

Then Eq. (3.3.33) has a positive solution on $[T, \infty)$.

Corollary 3.3.4. *Assume that there exist* $\ell > 0$ *and positive integers* $m_0, m_1, \ldots, m_n$ *such that* $\tau = m_0\ell$, $\sigma_i = m_i\ell$ *and* $m_0 > m_i$, $i = 1, 2, \ldots, n$, *and* $q_i(t)$ *are* ℓ*-periodic functions. Then every solution of (3.3.33) is oscillatory if and only if its "characteristic equation"*

$$\lambda(1 - pe^{-\lambda\tau}) + \sum_{i=1}^{n} q_i^* e^{-\lambda\sigma_i} = 0 \tag{3.3.58}$$

has no real roots, where

$$q_i^* = \frac{1}{\ell}\int_t^{t+\ell} q_i(s)ds, \quad i = 1, 2, \ldots, n.$$

Remark 3.3.2. The following condition is sufficient for (3.3.56) to hold: for $\ell = \tau,\ \tau - \sigma_i,\ i = 1, 2, \ldots, n$

$$\liminf_{t\to\infty}\sum_{i=1}^{n}\sum_{k=0}^{\infty}\left(\frac{1}{\ell}\int_t^{t+\ell} q_i(s)ds\right)\frac{1}{p^{k+1}}((k+1)\tau - \sigma_i) > \frac{1}{e}\,. \tag{3.3.59}$$

3.4. Comparison Results

We consider neutral delay differential equations of the form

$$\big(x(t) - px(t-\tau)\big)' + \sum_{i=1}^{n} q_i(t)x\big(t - \sigma_i(t)\big) = 0, \quad t \geq t_0. \tag{3.4.1}$$

In the following we present a necessary and sufficient condition for oscillation of all solutions of Eq. (3.4.1) by a comparison method.

Theorem 3.4.1. *Assume that*

(i) p, τ *are constants,* $0 \leq p < 1, \quad \tau > 0$;
(ii) $\sigma_i \in C(R_+, R_+), \quad \lim_{t\to\infty} \big(t - \sigma_i(t)\big) = \infty, \quad i = 1, \ldots, n$;
(iii) $q_i \in C(R_+, R_+), \quad i = 1, \ldots, n$. *Then every solution of Eq. (3.4.1) is oscillatory if and only if the differential inequality*

$$\big(x(t) - px(t-\tau)\big)' + \sum_{i=1}^{n} q_i(t)x\big(t - \sigma_i(t)\big) \leq 0 \tag{3.4.2}$$

has no eventually positive solutions.

Let us now compare Eq. (3.4.1) with the equation

$$\big(x(t) - \overline{p}x(t-\tau)\big)' + \sum_{i=1}^{n} \overline{q}_i(t)x\big(t - \sigma_i(t)\big) = 0, \quad t \geq t_0. \tag{3.4.3}$$

Theorem 3.4.2. *Assume that all the assumptions of Theorem 3.4.1 hold. Further assume that* $0 \leq p \leq \overline{p} < 1, \quad \overline{q}_i(t) \geq q_i(t) \geq 0, \quad i = 1, \ldots, n$. *Then the oscillation of Eq. (3.4.1) implies the oscillation of Eq. (3.4.3).*

Since we will establish more general results in Chapter 5, we omit the proofs here.

We now consider a more general neutral differential equation

$$\big(x(t)-px(t-\tau)\big)'+\sum_{i=1}^{n} q_i(t)x\big(t-\sigma_i(t)\big)+F\big(t, x(g_1(t)), \ldots, x\big(g_m(t)\big)\big) = 0. \tag{3.4.4}$$

Theorem 3.4.3. *In addition to all assumptions of Theorem 3.4.1 we assume* $g_i \in C(R_+, R)$, $F \in C(R_+ \times R^m, R)$ *and* $F(t, y_1, \ldots, y_m)y_1 \geq 0$ *whenever* $y_1 y_j > 0$, $j = 1, 2, \ldots, m$. *Then the oscillation of Eq. (3.4.1) implies the oscillation of Eq. (3.4.4).*

Proof: If not, assume Eq. (3.4.1) is oscillatory, and Eq. (3.4.4) has an eventually positive solution $x(t)$. Then $x(t)$ satisfies the differential inequality

$$\big(x(t) - px(t-\tau)\big)' + \sum_{i=1}^{n} q_i(t)x\big(t - \sigma_i(t)\big) \leq 0. \tag{3.4.5}$$

By Theorem 3.4.1, Eq. (3.4.1) has a nonoscillatory solution, which contradicts the assumption. □

For example, we consider the mixed neutral differential equation

$$\big(x(t) - px(t-\tau)\big)' + \sum_{i=1}^{n} q_i x(t - \sigma_i) + \sum_{j=1}^{m} r_j x(t + \delta_j) = 0, \tag{3.4.6}$$

where $0 < p < 1$, τ, q_i, σ_i are positive constants, r_j, δ_j are nonnegative constants, $j = 1, 2, \ldots, m$.

According to Theorem 3.4.3 we have the following conclusion.

If every solution of

$$\big(x(t) - px(t-\tau)\big)' + \sum_{i=1}^{n} q_i x(t - \sigma_i) = 0 \tag{3.4.7}$$

is oscillatory, then every solution of Eq. (3.4.6) is also oscillatory.

Using Theorem 3.4.1 we obtain the following "Linearized oscillation" result for the nonlinear equation

$$\big(x(t) - px(t-\tau)\big)' + \sum_{i=1}^{n} q_i f_i\big(x(t - \sigma_i)\big) = 0. \tag{3.4.8}$$

Theorem 3.4.4. *Assume*

(i) $p \in (0,1)$, τ, $q_i \in (0, \infty)$, $\sigma_i \in [0, \infty)$, $i = 1, \ldots, n$;
(ii) $f_i \in C(R, R)$, $xf_i(x) > 0$ *for* $x \neq 0$, $i = 1, \ldots, n$;

(iii) *there exists an* $\varepsilon > 0$ *such that* $\frac{f_i(u)}{u} \geq 1$ *for* $|u| \leq \varepsilon, \quad i = 1, \ldots, n.$

Then every solution of (3.4.8) is oscillatory if every solution of the "linearized equation" (3.4.7) is oscillatory.

Proof: Let $x(t)$ be a positive solution of (3.4.8). Then we have $\lim_{t\to\infty} x(t) = 0$, and

$$\left(x(t) - px(t-\tau)\right)' + \sum_{i=1}^{n} q_i x(t-\sigma_i) \leq 0.$$

By Theorem 3.4.1, (3.4.7) has a nonoscillatory solution. This contradicts our assumption.

Remark 3.4.2. Theorem 3.4.4 is true for (3.4.8) with variable coefficients $q_i(t) \in C(R_+, R_+)$, $i = 1, \ldots, n$, satisfying $\int_{t_0}^{\infty} \sum_{i=1}^{n} q_i(t)dt = \infty$.

We then compare the oscillation of two neutral delay differential equations

$$\left(x(t) - p(t)x(t-\tau)\right)' + q(t)x(t-\sigma) = 0 \tag{3.4.9}$$

and

$$\left(x(t) - \overline{p}(t)x(t-\tau)\right)' + \overline{q}(t)x(t-\sigma) = 0. \tag{3.4.10}$$

Theorem 3.4.5. *Assume that* $p, \ \overline{p}, \ q, \ \overline{q} \in C([t_0, \infty), R_+), \quad \tau > \sigma \geq 0; \quad \overline{p}(t) \geq 1, \ \overline{q}(t) > 0, \quad \int_{t_0}^{\infty} \overline{q}(t)dt = \infty$; *and*

$$\frac{\overline{p}(t+\tau-\sigma)}{p(t+\tau-\sigma)} \leq \frac{\overline{q}(t)}{q(t)} \leq 1, \quad t \geq T \geq t.$$

Then (3.4.9) is oscillatory implies that (3.4.10) is oscillatory.

Proof: If not, let $x(t)$ be an eventually positive solution of (3.4.10). Set $z(t) = x(t) - \overline{p}(t)x(t-\tau)$. Then $z(t) < 0, \quad z'(t) < 0$ eventually, and

$$z'(t) = \frac{\overline{q}(t)}{\overline{p}(t+\tau-\sigma)\overline{q}(t+\tau)} z'(t+\tau) + \frac{\overline{q}(t)}{\overline{p}(t+\tau-\sigma)} z(t+\tau-\sigma).$$

Set $\lambda(t) = \frac{z'(t)}{z(t)}$. Then $\lambda(t) > 0$ for $t \geq T$ and

$$\begin{aligned}\lambda(t) &= \frac{\overline{q}(t)}{\overline{p}(t+\tau-\sigma)\overline{q}(t+\tau)}\lambda(t+\tau)\exp\int_t^{t+\tau}\lambda(s)ds \\ &\quad + \frac{\overline{q}(t)}{\overline{p}(t+\tau-\sigma)}\exp\int_t^{t+\tau-\sigma}\lambda(s)ds \\ &\geq \frac{q(t)}{p(t+\tau-s)q(t+\tau)}\lambda(t+\tau)\exp\int_t^{t+\tau}\lambda(s)ds \\ &\quad + \frac{q(t)}{p(t+\tau-\sigma)}\exp\int_t^{t+\tau-\sigma}\lambda(s)ds. \end{aligned} \tag{3.4.11}$$

Define $\lambda_0(t) = \lambda(t)$, $\quad t \geq T$, and for $n = 0, 1, 2, \ldots$ and $t \geq T$

$$\begin{aligned}\lambda_{n+1}(t) &= \frac{q(t)}{p(t+\tau-\sigma)q(t+\tau)}\lambda_n(t+\tau)\exp\int_t^{t+\tau}\lambda_n(s)ds \\ &\quad + \frac{q(t)}{p(t+\tau-\sigma)}\exp\int_t^{t+\tau-\sigma}\lambda_n(s)ds. \end{aligned} \tag{3.4.12}$$

Clearly,

$$0 \leq \lambda_n(t) \leq \lambda_{n-1}(t) \leq \cdots \leq \lambda_0(t) = \lambda(t), \quad t \geq T.$$

Hence $\lim_{n\to\infty}\lambda_n(t) = \lambda^*(t)$ exists and $\lambda^*(t) \geq 0$, $t \geq T$. By the Lebesgue's dominated convergence theorem to (3.4.12) we obtain that

$$\begin{aligned}\lambda^*(t) &= \frac{q(t)}{p(t+\tau-\sigma)q(t+\tau)}\lambda^*(t+\tau)\exp\int_t^{t+\tau}\lambda^*(s)ds \\ &\quad + \frac{q(t)}{p(t+\tau-\sigma)}\exp\int_t^{t+\tau-\sigma}\lambda^*(s)ds. \end{aligned} \tag{3.4.13}$$

Set

$$y(t) = -\exp\int_T^t \lambda^*(s)ds < 0. \tag{3.4.14}$$

Then $y'(t) = \lambda^*(t)y(t)$, and (3.4.13) becomes

$$y'(t) = \frac{q(t)}{p(t+\tau-\sigma)q(t+\tau)}y'(t+\tau) + \frac{q(t)}{p(t+\tau-\sigma)}y(t+\tau-\sigma). \tag{3.4.15}$$

Let $x(t) = -\frac{y'(t+\sigma)}{q(t+\sigma)} > 0$. Thus (3.4.15) reduces to

$$\big(x(t) - p(t)x(t-\tau)\big)' + q(t)x(t-\sigma) = 0.$$

This means that Eq. (3.4.15) has a positive solution $x(t)$, contradicting the assumption. □

3.5. Unstable Type Equations

We consider neutral differential equations of the forms

$$\big(x(t) - px(t-\tau)\big)' - q(t)x\big(g(t)\big) = 0, \quad t \geq t_0 \tag{3.5.1}$$

and

$$\big(x(t) - px(t-\tau)\big)' - q(t)x\big(g(t)\big) = F(t), \quad t \geq t_0. \tag{3.5.2}$$

where $p \geq 0$, $\tau > 0$, $q \in C([t_0, \infty), R_+)$, g, $F \in C([t_0, \infty), R)$, $g(t) \leq t$ and $\lim_{t\to\infty} g(t) = \infty$.

We first show some properties of nonoscillatory solutions of Eq. (3.5.1).

Lemma 3.5.1. *Let $x(t)$ be a positive solution of (3.5.1). Then $x(t)$ is of one of the following types of asymptotic behavior:*

(a) $\lim_{t\to\infty} x(t) = 0$, (b) $\lim_{t\to\infty} x(t) = \ell \neq 0$, (c) $\lim_{t\to\infty} x(t) = \infty$.

To prove this lemma we shall use the following facts:

Lemma 3.5.2. *Let x, $z \in C([t, \infty), R)$ satisfy*

$$z(t) = x(t) - px(t-\tau), \quad t \geq t_0 + \max\{0, \tau\},$$

where p, $\tau \in R$. Assume that x is bounded on $[t_0, \infty)$ and $\lim_{t\to\infty} z(t) = \ell$ exists. Then the following statements hold:

(i) *If $p = 1$, then $\ell = 0$;*

(ii) *If $p \neq \pm 1$, then $\lim_{t\to\infty} x(t)$ exists.*

The next lemma is for the first order linear difference equation

$$x(t+1) = p(t)x(t) + y(t), \quad x(t_0) = x_0 \tag{3.5.3}$$

where t_0 and t are integer-valued, p and y are given functions with $p(t) \neq 0$ for all t. The corresponding homogeneous equation to (3.5.3) is

$$z(t+1) = p(t)z(t), \quad z(t_0) = z_0. \tag{3.5.3$'$}$$

Lemma 3.5.3. *Let $p(t) \neq 0$ and $y(t)$ be given for $t = t_0, t_0+1, \dots$. Then the solution of Eq. (3.5.3)$'$ is*

$$z(t) = z(t_0) \prod_{s=t_0}^{t-1} p(s), \quad t = t_0+1, t_0+2, \dots$$

where $\prod_{s=t_0}^{t_0-1} p(s) \equiv 1$; and the solution of Eq. (3.5.3) is

$$x(t) = \sum_{s=t_0}^{t-1} \left(y(s) \prod_{u=s+1}^{t-1} p(u) \right) + x_0 \prod_{s=t_0}^{t-1} p(s) \tag{3.5.4}$$

where $\sum_{s=t_0}^{t_0-1} p(s) \equiv 1$.

Proof of Lemma 3.5.1: Set $z(t) = x(t) - px(t-\tau)$, then $z'(t) \geq 0$.

First we assume that $z(t) > 0$ eventually. Then $\lim_{t\to\infty} z(t) = \ell \leq \infty$. If $\ell = \infty$, then $x(t) \geq z(t) \to \infty$, as $t \to \infty$, i.e., (c) holds.

If $\ell < \infty$ and $x(t)$ is bounded, Lemma 3.5.2 implies that $\lim_{t\to\infty} x(t)$ exists when $p \neq 1$, i.e., either (a) or (b) holds. When $p = 1$, $x(t) \geq x(t-\tau) + \frac{\ell}{2} \geq \cdots \geq x(t-n\tau) + n \cdot \frac{\ell}{2} \to \infty$ as $n \to \infty$. This is impossible since $x(t)$ is bounded.

If $\ell < \infty$ and $x(t)$ is unbounded, then there exists $\{t_n\}$ such that $x(t_n) = \max_{t \leq t_n} x(t) \to \infty$ as $n \to \infty$. For $p \in (0,1)$, $z(t_n) \geq x(t_n)(1-p) \to \infty$ as $n \to \infty$, which contradicts the boundedness of z. For $p = 1$, we have the same argument as the above. For $p > 1$, we have

$$x(t) \geq px(t-\tau) \geq \cdots \geq p^n x(t-n\tau)$$

which implies that (c) holds.

Next we consider the case that $z(t) < 0$ eventually. Then $\lim_{t\to\infty} z(t) = \ell \le 0$. If $p \ne 1$ and x is bounded, by Lemma 3.5.2 either (a) or (b) holds. If $p = 1$ and $z(t) \to \ell < 0$ as $t \to \infty$, then

$$x(t) \le x(t-\tau) + \frac{\ell}{2}\,.$$

Applying this inequality repeatedly we obtain that $\lim_{t\to\infty} x(t) = -\infty$. This is impossible. If $p = 1$ and $\lim_{t\to\infty} z(t) = 0$, we define

$$\overline{z}(t) = \begin{cases} z(t), & t \ge t_0 \\ (t - t_0 + \tau)\overline{z}(t_0)/\tau, & t_0 - \tau \le t \le t_0 \\ 0, & t \le t_0 - \tau. \end{cases}$$

Clearly, $\overline{z}$ is continuous on R and

$$x(t) = \sum_{i=0}^{[\frac{t-t_0}{\tau}+1]} \overline{z}(t - i\tau). \tag{3.5.5}$$

Hence $x(t)$ is monotonic, and the conclusion of Lemma 3.5.1 holds.

We now consider the case that $x(t)$ is unbounded and $z(t) \to \ell \le 0$ as $t \to \infty$.

If $p \in (0,1]$, then $x(t_n) = \max_{t \le t_n} x(t) \to \infty$ as $n \to \infty$. This is impossible since $z(t_n) \ge x(t_n)(1-p)$ and $z(t)$ is bounded.

If $p > 1$, from (3.5.4), let $\overline{x}(t)$ be an unbounded solution, then every solution of $x(t) = px(t-\tau) + w(t)$ takes the form $x(t) = \overline{x}(t) + w(t)$, where $w(t)$ is some solution of $w(t) = pw(t-\tau)$. Clearly, $w(t) \to \infty$ as $t \to \infty$ and hence $x(t) \to \infty$ as $t \to \infty$.

Theorem 3.5.1. *Based on the range of p we have the conclusions:*

(i) *If $p \ge 0$, Eq. (3.5.1) has a positive solution $x(t)$ satisfying (b) or (c);*

(ii) *If $p = 1$, Eq. (3.5.1) has an unbounded solution $x(t)$ satisfying (c);*

(iii) *If $p > 1$, Eq. (3.5.1) has an unbounded solution $x(t)$ which tends to infinity exponentially;*

(iv) *If $0 \le p < 1$, and $\int_{t_0}^{\infty} q(t)dt = \infty$, Eq. (3.5.1) has an unbounded positive solution, and every bounded solution of Eq. (3.5.1) is either oscillatory or tends to zero as $t \to \infty$.*

Proof: Choose a positive continuous function $H(t)$ such that

$$\int_{t_0}^{\infty} q(t)H(t)dt = \infty \tag{3.5.6}$$

and

$$\lim_{t\to\infty}\left\{\frac{q(t)}{\exp(\int_{t_0}^{t} q(s)H(s)ds)}\right\} = 0. \tag{3.5.7}$$

Define a function v by

$$v(t) = \exp\left[\int_{t_0}^{t} \exp\left(\int_{t_0}^{s} q(u)H(u)du\right)ds\right]. \tag{3.5.8}$$

Let BC be the Banach space of all bounded and continuous functions $y : [t_0, \infty) \to R$ with the sup norm. Define a subset Ω of BC as follows:

$$\Omega = \{y \in BC : \ 0 \le y(t) \le 1, \quad t_0 \le t < \infty\}.$$

Clearly Ω is a bounded, closed and convex subset of BC. Now we define a mapping $\mathcal{T}$ on Ω as follows:

$$(\mathcal{T}y)(t) = \begin{cases} p\,\frac{v(t-\tau)}{v(t)} + \frac{1}{v(t)}\int_T^t q(s)v\big(g(s)\big)y\big(g(s)\big)ds + \frac{1}{2v(t)}, & t \ge T \\ \frac{t}{T}\,(\mathcal{T}y)(T) + (1-\frac{t}{T}), & t_0 \le t \le T, \end{cases} \tag{3.5.9}$$

where T is sufficiently large so that $t-\tau \ge t_0,\ g(t) \ge t_0,\ v(t) \ge 1$, and

$$p\,\frac{v(t-\tau)}{v(t)} + \frac{1}{v(t)}\int_T^t q(s)v\big(g(s)\big)ds \le \frac{1}{2} \tag{3.5.10}$$

for $t \ge T$. In fact, from (3.5.7) and (3.5.8) it is easy to see that

$$\frac{v(t-\tau)}{v(t)} \to 0 \quad \text{and} \quad \frac{\int_{t_0}^t q(s)v\big(g(s)\big)ds}{v(t)} \to 0 \quad \text{as} \quad t \to \infty$$

which shows that (3.5.10) is true for large t. Thus we have $\mathcal{T}\Omega \subset \Omega$. Let y_1 and

y_2 be two functions in Ω. Then

$$|(\mathcal{T}y_2)(t) - (\mathcal{T}y_1)(t)| \le p\frac{v(t-\tau)}{v(t)}|y_2(t-\tau) - y_1(t-\tau)|$$

$$+ \frac{1}{v(t)}\int_T^t q(s)v(g(s))|y_2(g(s)) - y_1(g(s))|ds$$

$$\le \frac{1}{2}\|y_2 - y_1\|, \quad t \ge T,$$

and

$$\|\mathcal{T}y_2 - \mathcal{T}y_2\| = \sup_{t\ge t_0} |(\mathcal{T}y_2)(t) - (\mathcal{T}y_1)(t)|$$

$$= \sup_{t\ge T} |(\mathcal{T}y_2)(t) - (\mathcal{T}y_1)(t)| \le \frac{1}{2}\|y_2 - y_1\|,$$

which shows that $\mathcal{T}$ is a contraction on Ω. Hence there is a function $y \in \Omega$ such that $\mathcal{T}y = y$. That is

$$y(t) = \begin{cases} p\frac{v(t-\tau)}{v(t)} + \frac{1}{v(t)}\int_T^t q(s)v(g(s))y(g(s))ds + \frac{1}{2v(t)}, & t \ge T \\ \frac{t}{T}y(T) + (1 - \frac{t}{T}), & t_0 \le t \le T. \end{cases}$$

Obviously $y(t) > 0$ for $t \ge t_0$. Set $x(t) = y(t)v(t)$. Then

$$x(t) - px(t-\tau) = \int_T^t q(s)x(g(s))ds + \frac{1}{2}, \quad t \ge T. \tag{3.5.11}$$

Therefore, $x(t)$ is a positive solution of (3.5.1) for $t \ge T$. This proves (i), (ii) and the first part of (iv). In the case that $p > 1$ we have

$$x(t) \ge px(t-\tau) \ge \cdots \ge p^n x(t - n\tau),$$

or

$$x(t) \ge x(t_0)\exp\left(\mu(t - t_0)\right), \quad \text{for} \quad t \ge t_0,$$

where $\mu = \frac{\ln p}{\tau} > 0$, which shows that (iii) is true.

In order to prove the second part of (iv), we let $x(t)$ be a bounded positive solution of Eq. (3.5.1) and define $z(t) = x(t) - px(t-\tau)$. Then $\lim_{t\to\infty} z(t) = \ell$

exists. If $\ell > 0$, then $x(g(t)) \geq \ell$ for $t \geq T_1 \geq t_0$. Hence

$$z(t) - z(T_1) = \int_{T_1}^{t} q(s)x(g(s))ds \geq \ell \int_{T_1}^{t} q(s)ds.$$

Since $\int_{T_1}^{t} q(s)ds \to \infty$ as $t \to \infty$, we have a contradiction. In view of (3.5.11) we assume that $p \neq 0$. Now for $p \in (0,1)$ it is impossible that $\ell < 0$, thus $\lim_{t\to\infty} z(t) = 0$ and hence $\lim_{t\to\infty} x(t) = 0$. This completes the proof. □

Theorem 3.5.2. *Assume that* $p \in [0,1)$, $\int_{t_0}^{\infty} q(t)dt = \infty$, *and there exists a positive integer* k *such that*

$$u'(t) + q(t)\sum_{i=0}^{k} \frac{1}{p^{i+1}} u(g(t) + (i+1)\tau) \geq 0 \tag{3.5.12}$$

has no eventually negative solutions. Then every bounded solution of (3.5.1) is oscillatory.

Proof: From the proof of Theorem 3.5.1, we see that if $x(t)$ is a bounded positive solution of (3.5.1), then $z(t) < 0$ eventually. Also by induction we see that

$$x(t-\tau) > -\sum_{i=0}^{k} \frac{1}{p^{i+1}} z(t + i\tau).$$

Substituting it into (3.5.1) we obtain

$$z'(t) \geq -q(t)\sum_{i=0}^{k} \frac{1}{p^{i+1}} z[g(t) + (i+1)\tau]$$

which contradicts the fact that (3.5.2) has no eventually negative solutions.

Example 3.5.1. Consider

$$\left(x(t) - \tfrac{1}{2}x(t-1)\right)' = q(t)x(t-10) \tag{3.5.13}$$

where $q \in C([t_0,\infty), R_+)$. If the delay differential equation

$$y'(t) + q(t)\sum_{i=0}^{8} 2^{i+1} y(t-(9-i)) = 0$$

is oscillatory, then every bounded solution of (3.5.13) is oscillatory. In fact, we may choose $k = 8$. Then all assumptions of Theorem 3.5.2 hold.

Theorem 3.5.2. *Assume that $p < 0$, $p \neq -1$, and $\int_{t_0}^{\infty} q(t)dt = \infty$. Then every bounded solution of (3.5.1) is oscillatory.*

Proof: Let $x(t) > 0$ be a bounded solution. Set $z(t) = x(t) - px(t-\tau)$, then $z'(t) \geq 0$. Thus $\lim_{t\to\infty} z(t) = \ell$ exists and $\ell > 0$. From (3.5.1) we have

$$\int_{t_0}^{\infty} q(t)x\big(g(t)\big)dt < \infty.$$

In view of the assumption we have that $\liminf_{t\to\infty} x\big(g(t)\big) = 0$. Then there exists a sequence $\{t_n\}$ such that $\lim_{n\to\infty} t_n = \infty$ and $\lim_{n\to\infty} x\big(g(t_n) - 2\tau\big) = 0$. Noting that

$$\begin{aligned} &z\big(g(t_n)\big) - z\big(g(t_n) - \tau\big) \\ &\qquad = x\big(g(t_n)\big) - px\big(g(t_n) - \tau\big) - \big(x(g(t_n) - \tau) - px(g(t_n) - 2\tau)\big), \end{aligned}$$

and letting $n \to \infty$ we have

$$\liminf_{n\to\infty} \big(x(g(t_n)) - (p+1)x(g(t_n) - \tau)\big) = 0.$$

If $p < -1$, then

$$\liminf_{n\to\infty} x\big(g(t_n)\big) = 0 \quad \text{and} \quad \liminf_{n\to\infty} x\big(g(t_n) - \tau\big) = 0,$$

and so $\liminf_{n\to\infty} z(t_n) = \lim_{t\to\infty} z(t) = 0$, contradicting that $\ell > 0$.

If $-1 < p < 0$, we have

$$\liminf_{n\to\infty} \big((1+p)x(g(t_n) - \tau) - px(g(t_n) - 2\tau)\big) = 0$$

which implies that $\lim_{n\to\infty} z\big(g(t_n) - \tau\big) = 0$, contradicting that $\ell > 0$ again.

Example 3.5.2. Consider

$$\big(x(t) + \tfrac{1}{2}x(t - 2\pi)\big)' = \tfrac{3}{2}x(t - \tfrac{3}{2}\pi). \tag{3.5.14}$$

It is easy to see that the conditions in Theorem 3.5.2 are satisfied. Therefore, every bounded solution of (3.5.14) is oscillatory. In fact $x(t) = \sin t$ is such a solution.

For the case that $p = -1$, we need the following lemma.

Lemma 3.5.1. *Let $p < 0$. Assume that $x \in C([t_0, \infty), R_+)$, $x(t) - px(t-\tau)$ is increasing, and $x(t) - px(t-\tau) \geq \ell > 0$, for $t \geq T \geq t_0$. Then for every $t_1 \in [T, \infty)$ the set*

$$E = \{t \mid t_1 \leq t \leq t_1 + 2\tau,\ -px(t-\tau) \geq \beta\}$$

satisfies mes $(E) \geq \tau$, where $\beta = \min\{\frac{\ell}{2}, -\frac{p\ell}{2}\}$, and mes (E) denotes Lebesgue measure.

In Chapter 4, we will prove a more general result, so we omit the proof here.

Theorem 3.5.4. *Let $p = -1$. Assume $\int_E q(t)dt = \infty$ for every closed subset E of $[t_0, \infty)$ whose intersection with every interval of the form $[t, t+2\tau]$, $t_0 \leq t < \infty$, has a measure not less than τ. Then every bounded solution of (3.5.1) is oscillatory.*

Proof: Let $x(t)$ be a bounded positive solution of (3.5.1). As shown before,

$$\int_{t_0}^{\infty} q(t)x(g(t))dt < \infty.$$

Obviously, the conditions of Lemma 3.5.1 are satisfied. The intersection of the set
$E = \{t \mid T \leq t < \infty,\quad x(t-\tau) \geq \frac{\ell}{2}\}$ with the interval $[s, s+2\tau]$, $s \in [T, \infty)$, has a measure not less than τ. Therefore,

$$\int_T^{\infty} q(t)x(g(t))dt \geq \int_E q(t)x(g(t))dt \geq \frac{\ell}{2}\int_E q(t)dt = \infty.$$

This contradiction proves the theorem. □

Now we briefly discuss an oscillatory property of the forced equation (3.5.2).

Theorem 3.5.5. *Assume that there exists a function f such that $F(t) = f'(t)$, and*

$$\liminf_{t\to\infty} f(t) = -\infty \quad \textit{and} \quad \limsup_{t\to\infty} f(t) = \infty. \tag{3.5.15}$$

Then every bounded solution of (3.5.2) is oscillatory.

Proof: Let $x(t)$ be an eventually positive and bounded solution of (3.5.2) and set $z(t) = x(t) - px(t-\tau)$. Then

$$\big(z(t) - f(t)\big)' = q(t)x\big(g(t)\big) \geq 0.$$

If $z(t) - f(t) \leq 0$ eventually, then $z(t) \leq f(t)$ eventually, which contradicts (3.5.15) and the boundedness of z. Hence $z(t) - f(t) > 0$ eventually, which is impossible since $z(t)$ is bounded. □

Example 3.5.2. The equation

$$\big(x(t) + x(t-\pi)\big)' - tx(t-2\pi) = -t\sin t \tag{3.5.16}$$

satisfies the assumptions of Theorem 3.5.5. Hence every bounded solution of (3.5.16) is oscillatory. In fact, $x(t) = \sin t$ is such a solution.

3.6. Sublinear Equations

Consider the nonlinear neutral differential equations of the form

$$\big(x(t) + p(t)x(\tau(t))\big)' + f\big(t, x(g_1(t)), \ldots, x(g_N(t))\big) = 0. \tag{3.6.1}$$

Definition 3.6.1. $f(t, y_1, \ldots, y_N)$ is said to be strongly sublinear if there exist constants $\beta \in (0,1)$ and $m > 0$ such that $|z|^{-\beta}|f(t,z,\ldots,z)|$ is nonincreasing in $|z|$ for $0 < |z| \leq m$.

The following notations will be employed:

$$g^*(t) = \max_{1\leq i\leq N} g_i(t)$$
$$R(g^*, \tau) = \{t \in [t_0, \infty) : \ t_0 \leq g^*(t) < \tau(t)\}.$$

Theorem 3.6.1. *Assume that*

(i) $p \in C([t_0, \infty), (0, \infty))$, $1 < \mu \leq p(t) \leq \nu$ *where* μ *and* ν *are constants;*

(ii) $\tau \in C([t_0, \infty), R)$ *is strictly increasing, and satisfies that* $\tau(t) < t$, *and* $\lim_{t\to\infty} \tau(t) = \infty$;

(iii) $g_i \in C([t_0, \infty), R)$, $\lim_{t\to\infty} g_i(t) = \infty$, $i = 1, 2, \ldots, N$,

(iv) $f \in C([t_0, \infty) \times R^N, R)$ *is nondecreasing in each* y_i *and strongly sublinear, and* $y_i f(t_1, y_1, \ldots, y_N) \geq 0$, $\not\equiv 0$ *for* $y_1 y_i > 0$, $1 \leq i \leq N$;

(v) $$\int_{R(g^*,\tau)} |f(t, a, \ldots, a)| dt = \infty \quad \text{for every constant} \quad a \neq 0. \tag{3.6.2}$$

Then every solution of Eq. (3.6.1) is oscillatory.

Proof: Assume the contrary and let $x(t)$ be a nonoscillatory solution of Eq. (3.6.1). Without loss of generality we may assume that $x(t)$ is eventually positive. Set

$$z(t) = x(t) + p(t)x\big(\tau(t)\big). \tag{3.6.3}$$

Then $z(t) > 0$ and $z'(t) \leq 0$ for all large t. From (3.6.3) and using (i), we have

$$\begin{aligned} x(t) &= \frac{z\big(\tau^{-1}(t)\big) - x\big(\tau^{-1}(t)\big)}{p\big(\tau^{-1}(t)\big)} \\ &= \frac{z\big(\tau^{-1}(t)\big)}{p\big(\tau^{-1}(t)\big)} - \frac{1}{p\big(\tau^{-1}(t)\big)}\left[\frac{z\big(\tau^{-2}(t)\big)}{p\big(\tau^{-2}(t)\big)} - \frac{x\big(\tau^{-2}(t)\big)}{p\big(\tau^{-2}(t)\big)}\right] \\ &\geq \frac{z\big(\tau^{-1}(t)\big)}{p\big(\tau^{-1}(t)\big)} - \frac{z\big(\tau^{-2}(t)\big)}{p\big(\tau^{-1}(t)\big)p\big(\tau^{-2}(t)\big)} \\ &\geq \frac{\mu - 1}{\mu\nu} z\big(\tau^{-1}(t)\big) \end{aligned} \tag{3.6.4}$$

for all large t, where τ^{-1} is the inverse function of τ and $\tau^{-2}(t) \equiv \tau^{-1}\big(\tau^{-1}(t)\big)$. Setting $y(t) = \frac{\mu-1}{\mu\nu} z(t)$ and noting that f is sublinear, from (3.6.1) and (3.6.4) we have

$$y'(t) + \frac{\mu - 1}{\mu\nu} f\big(t, y(\tau^{-1}(g_1(t))), \ldots, y(\tau^{-1}(g_N(t)))\big) \leq 0, \quad t \geq T \tag{3.6.5}$$

provided $T \geq t_0$ is sufficiently large. From (3.6.2) and (3.6.5), it is easy to show that $\lim_{t\to\infty} z(t) = 0$. Thus there exists $T_1 \geq T$ such that $y\big(\tau^{-1}(g^*(t))\big) \leq m$ for

$t \geq T_1$. In view of (iv), we have

$$\begin{aligned} &f\big(t, y(\tau^{-1}(g_1(t))), \ldots, y(\tau^{-1}(g_N(t)))\big) \\ &\quad \geq f\big(t, y(\tau^{-1}(g^*(t))), \ldots, y(\tau^{-1}(g^*(t)))\big) \\ &\quad \geq m^{-\beta}\big(y(\tau^{-1}(g^*(t)))\big)^{\beta} f(t, m, \ldots, m), \quad t \geq T_1. \end{aligned} \tag{3.6.6}$$

From (3.6.5) and (3.6.6) we have

$$-y'(t) \geq \frac{\mu - 1}{\mu\nu} m^{-\beta}\big(y(\tau^{-1}(g^*(t)))\big)^{\beta} f(t, m, \ldots, m), \quad t \geq T_1. \tag{3.6.7}$$

Dividing (3.6.7) by $\big(y(t)\big)^{\beta}$ and integrating it over $R(g^*, \tau) \cap [T_1, \infty)$, we have

$$\int_{R(g^*,\tau)\cap[T_1,\infty)} f(s, m, \ldots, m)ds \leq \frac{\mu\nu m^{\beta}\big(y(T_1)\big)^{1-\beta}}{(\mu - 1)(1 - \beta)} < \infty$$

which contradicts (3.6.2). □

Theorem 3.6.2. *Assume that (i) – (iv) in Theorem 3.6.1 hold excepting the assumption of sublinearity of function f. Then Eq. (3.6.1) has a bounded nonoscillatory solution which is bounded away from zero if and only if*

$$\int_{t_0}^{\infty} |f(t, b, \ldots, b)|dt < \infty \quad \textit{for some constant} \quad b \neq 0. \tag{3.6.8}$$

Proof: *Necessity.* Let $x(t)$ be a bounded positive solution satisfying $x\big(g_i(t)\big) \geq c > 0$ for $t \geq T$, $1 \leq i \leq N$. Then $z(t)$ defined by (3.6.3) is bounded and nonincreasing. Integrating (3.6.1) we have

$$\int_{T}^{\infty} f(t, c, \ldots, c)dt \leq z(T) - z(\infty) < z(T) < \infty$$

which contradicts (3.6.2).

Sufficiency. Let $c > 0$ and $T > t_0$ be such that $\mu c \leq b$ and

$$T_0 = \min\ \big\{\tau(T),\ \inf_{t\geq T}\ \{g_1(t)\}, \ldots, \inf_{t\geq T}\ \{g_N(t)\}\big\} \geq t_0$$

and

$$\int_T^\infty f(t, \mu c, \ldots, \mu c) dt \le (\mu - 1)c. \tag{3.6.9}$$

Define a set $X \subset C[T_0, \infty)$ as follows:

$$X = \{x \in C[T_0, \infty) : \ c \le x(t) \le \mu c \quad \text{for } t \ge T,\ x(t) = x(T) \quad \text{for } T_0 \le t \le T\}. \tag{3.6.10}$$

For every $x \in X$ we define

$$\widehat{x}(t) = \begin{cases} \sum_{i=1}^{\infty} \frac{(-1)^{i-1} x\left(\tau^{-i}(t)\right)}{H_i\left(\tau^{-i}(t)\right)}, & t \ge T \\ \widehat{x}(T), & T_0 \le t \le T \end{cases} \tag{3.6.11}$$

where $\tau^0(t) \equiv t, \quad \tau^i(t) = \tau\left(\tau^{i-1}(t)\right), \quad \tau^{-i}(t) = \tau^{-1}\left(\tau^{-(i-1)}(t)\right)$ and

$$H_0(t) \equiv 1, \quad H_i(t) = \prod_{j=0}^{i-1} p\left(\tau^j(t)\right), \quad i = 1, 2, \ldots . \tag{3.6.12}$$

Then $0 < \widehat{x}(t) \le \mu c, \quad t \ge T$. Define a mapping $\mathcal{T}$ on X as follows:

$$(\mathcal{T}x)(t) = \begin{cases} c + \int_t^\infty f\left(s, \widehat{x}(g_1(s)), \ldots, \widehat{x}(g_N(s))\right) ds, & t \ge T \\ \mathcal{T}x(t), & T_0 \le t \le T. \end{cases} \tag{3.6.13}$$

In view of (3.6.9), $\mathcal{T}X \subset X$. It is not difficult to prove that $\mathcal{T}$ is continuous and $\mathcal{T}(X)$ is relatively compact in the topology of the Fréchet space $C[T_0, \infty)$. Therefore, by the Schauder Tychonoff fixed point theorem, there exists an $x \in X$ such that $x = \mathcal{T}x$. That is

$$x(t) = c + \int_t^\infty f\left(x, \widehat{x}(g_1(s)), \ldots, \widehat{x}(g_N(s))\right) ds, \quad t \ge T. \tag{3.6.14}$$

From (3.6.11) we have

$$\widehat{x}(t) + p(t)\widehat{x}\left(\tau(t)\right) = x(t),$$

and thus

$$\widehat{x}(t) + p(t)\widehat{x}\left(\tau(t)\right) = c + \int_t^\infty f\left(s, \widehat{x}(g_1(s)), \ldots, \widehat{x}(g_N(s))\right) ds, \quad t \ge T. \tag{3.6.15}$$

Differentiating (3.6.15) we have that the function $\widehat{x}(t)$ defined by (3.6.11) satisfies Eq. (3.6.1) for $t \geq T$ and

$$\frac{\mu-1}{\mu\nu}x\big(\tau^{-1}(t)\big) \leq \widehat{x}(t) \leq x(t), \quad t \geq T. \tag{3.6.16}$$

Therefore, $\widehat{x}(t)$ is a bounded nonoscillatory solution of Eq. (3.6.1) which is bounded away from zero. □

Combining Theorems 3.6.1 and 3.6.2, we obtain the following result.

Corollary 3.6.1. *Let $g^*(t) < \tau(t)$ and (i)–(iv) in Theorem 3.6.1 hold. Then, (3.6.2) is a necessary and sufficient condition for oscillation of all solutions of Eq. (3.6.1).*

Example 3.6.1. Consider the neutral equation

$$\big(x(t) + \mu x(t-2\pi)\big)' + (1+\mu)\cos^{\frac{2}{3}} t\, x^{\frac{1}{3}}\big(t - \tfrac{5}{2}\,\pi\big) = 0 \tag{3.6.17}$$

where $\mu > 1$ is a constant.

It is easy to see that all the assumptions of Theorem 3.6.1 are satisfied. Therefore every solution of (3.6.17) oscillates. In fact, $x(t) = \sin t$ is such a solution. The following example shows that condition (3.6.2) is not sufficient for oscillation of linear equations.

Example 3.6.2. Consider the linear equation

$$\big(x(t) + 2x(t-\tau)\big)' + \frac{3t^2 - 4t + 4}{t(t-2)^2}x(t) = 0. \tag{3.6.18}$$

It is easy to see that all assumptions of Theorem 3.6.1 are satisfied. But (3.6.18) has a positive solution $x(t) = \frac{1}{t}$.

We now consider the nonlinear neutral differential equation described by

$$\big(x(t) - px(t-\tau)\big)' + \sum_{i=1}^{n} q_i(t) f_i\big(x(g_i(t))\big) = 0 \tag{3.6.19}$$

where $p \geq 0, \quad \tau > 0, \quad q_i \in C([t_0, \infty), R_+), \quad g_i \in C([t_0, \infty), R), \quad \lim_{t\to\infty} g_i(t) = \infty,$
$f_i \in C(R, R), \quad x f_i(x) > 0$ for $x \neq 0, \quad i = 1, 2, \ldots, n.$

f is said to be locally sublinear near $x = 0$ if there exists $\alpha > 0$ such that $f(x)$ is increasing in $(0, \alpha)$ and $(-\alpha, 0)$ and

$$\int_0^\alpha \frac{dx}{f(x)} < \infty \quad \text{and} \quad \int_{-\alpha}^0 \frac{dx}{f(x)} < \infty. \tag{3.6.20}$$

Theorem 3.6.3. *Assume that $p \in [0, 1)$, there exists $i_0 \in I_n = \{1, \ldots, n\}$ such that $f_{i_0}(x)$ is locally sublinear near $x = 0$, and $\int_{A_{i_0}} q_{i_0}(t)dt = \infty$. Then (3.6.19) is oscillatory, where*

$$A_{i_0} = \{t \in [t_0, \infty) : \ t_0 \leq g_{i_0}(t) \leq t\}. \tag{3.6.21}$$

Proof: Let $x(t)$ be an eventually positive solution of (3.6.19), say, $x(t) > 0$, $x(g_i(t)) > 0$ for $t \geq t_1 \geq t_0$, $i = 1, \ldots, n$. Set $z(t) = x(t) - px(t - \tau)$, then $z'(t) \leq 0$. If $z(t) > 0$, $t \geq t_2 \geq t_1$, we have $\lim_{t\to\infty} z(t) = \ell_1 \geq 0$. If $\ell_1 > 0$, since $\ell_1 \leq z(t) \leq z(t_2)$ for $t \geq t_2$ and

$$x(t) = \sum_{i=1}^{N(t)-1} p^{i-1} z(t - (i-1)\tau) + p^{N(t)-1} x(t - (N(t) - 1)\tau)$$

where $N(t)$ is a positive integer such that $T - \tau < t - (N(t) - 1)\tau \leq T$, we have

$$\ell_1 \leq x(t) \leq \frac{z(t_2)}{1 - p} + \max_{t_2 - \tau \leq t \leq t_2} x(t) = N_1,$$

and

$$f_{i_0}(x(g_{i_0}(t))) \geq \min_{\ell_1 \leq u \leq N_1} f_{i_0}(u) + L > 0, \quad t \geq t_2.$$

From (3.6.19) we see

$$-z'(t) \geq q_{i_0}(t) f_{i_0}(x(g_{i_0}(t))) \geq q_{i_0}(t) L.$$

Integrating it from t_2 to t we have

$$z(t_2) - z(t) \geq L \int_{t_2}^{t} q_{i_0}(s)ds.$$

Letting $t \to \infty$ we obtain

$$z(t_2) - \ell_1 \geq L \int_{t_2}^{\infty} q_{i_0}(s)ds \geq L \int_{A_{i_0} \cap [t_2, \infty)} q_{i_0}(t)dt$$

which contradicts that $\int_{A_{i0}} q_{i_0}(t)dt = \infty$.

If $z(t) \to 0$ as $t \to \infty$, then $x(t) \to 0$ as $t \to \infty$. Then there exists $t_3 \geq t_2$ such that $0 \leq x\big(g_{i_0}(t)\big) \leq \alpha$ and $z(t) \leq \alpha$ for $t \geq t_3$. Since f is nondecreasing for $t \in A_{i_0} \cap [t_3, \infty)$

$$f_{i_0}\big(x(g_{i_0}(t))\big) \geq f_{i_0}\big(x(g_{i_0}(t)) - px(g_{i_0}(t) - \tau)\big) \geq f_{i_0}\big(z(t)\big).$$

From (3.6.19)

$$-z'(t) \geq q_{i_0}(t) f_{i_0}\big(x(g_{i_0}(t))\big) \geq q_{i_0}(t) f_{i_0}\big(z(t)\big).$$

Then we have

$$\begin{aligned}\int_{z(t)}^{z(t_3)} \frac{du}{f_{i_0}(u)} &= \int_{t_3}^{t} \frac{-z'(s)}{f_{i_0}\big(z(s)\big)} ds \\ &\geq \int_{A_{i_0} \cap [t_3, t]} \frac{-z'(s)}{f_{i_0}\big(z(s)\big)} ds \geq \int_{A_{i_0} \cap [t_3, t]} q_{i_0}(t)dt.\end{aligned}$$

Letting $t \to \infty$ we have

$$\infty > \int_0^{\alpha} \frac{du}{f_{i_0}(u)} \geq \int_{A_{i_0} \cap [t_3, \infty)} q_{i_0}(t)dt$$

which contradicts $\int_{A_{i_0}} q_{i_0}(t)dt = \infty$ again.

The last possibility is that $z(t) < 0, \quad z'(t) \leq 0, \quad t \geq t_2$. Then $\lim_{t \to \infty} z(t) = -\ell_2, \quad \ell_2 > 0$ or $\ell_2 = \infty$. If $\ell_2 = \infty$, then $x(t)$ is unbounded, so there exists a sequence $\{t_n\}$ satisfying $\lim_{n \to \infty} t_n = \infty$, $x(t_n) = \max_{t_2 \leq t \leq t_n} x(t)$, and $\lim_{n \to \infty} x(t_n) = \infty$. Thus $z(t_n) = x(t_n) - px(t_n - \tau) \geq (1-p)x(t_n) > 0$, which contradicts the negativity of z. If $z(t) \to -\ell_2, \quad \ell_2 \in (0, \infty)$, then $\lim_{t \to \infty} x(t) = -\frac{\ell_2}{1-p} < 0$ contradicting the positivity of x. □

Remark 3.6.1. The local sublinear condition for f_i is important here. In fact, Theorem 3.6.3 is not true for linear equations. For example, consider the equation

$$\left(x(t) - \tfrac{1}{2}x(t-1)\right)' + \frac{t^2 - 4t + 4}{4t^2(t-1)}\, x(t-1) = 0 \tag{3.6.22}$$

where $q(t) = \frac{t^2-4t+4}{4t^2(t-1)}$. Clearly, $\int_{t_0}^{\infty} q(t)dt = \infty$. But (3.6.22) has a nonoscillatory solution $x(t) = \frac{1}{t}$.

Using Tychonoff-Schauder fixed point theorem we can prove the following result for the existence of nonoscillatory solutions. We omit the proof here.

Theorem 3.6.4. *Let $0 \le p < 1$ and $\sum_{i=1}^{n} \int_{t_0}^{\infty} q_i(t)dt < \infty$. Then (3.6.19) has a bounded nonoscillatory solution x which does not tend to zero as $t \to \infty$.*

Corollary 3.6.2. *Assume that $0 \le p < 1$, $g_i(t) \le t$, $f_i(x)$ is locally sublinear near $x = 0$, $i = 1, 2, \ldots, n$. Then every solution of Eq. (3.6.19) is oscillatory if and only if*

$$\sum_{i=1}^{n} \int_{t_0}^{\infty} q_i(t)dt = \infty.$$

3.7. Equations with Mixed Coefficients

In this section we consider neutral delay differential equations with positive and negative coefficients of the form

$$\left(x(t) - c(t)x(t-r)\right)' + p(t)x(t-\tau) - q(t)x(t-\sigma) = 0. \tag{3.7.1}$$

The following is an oscillation criterion for (3.7.1).

Theorem 3.7.1. *Assume that*

(i) *$r \in (0, \infty)$, and τ, $\sigma \in [0, \infty)$, and $c, p, q \in C([t_0, \infty), R_+)$;*

(ii) *$\tau \ge \sigma$;*

(iii) *$p(t) \ge q(t - \tau + \sigma)$ and $p(t) \not\equiv q(t - \tau + \sigma)$ for $t \ge t_0 + \tau - \sigma$;*

(iv) $1 - c(t) - \int_{t-\tau+\sigma}^{t} q(s)ds \geq 0$ *for all sufficiently large* t;

(v) *every solution of the delay differential equation*

$$y'(t) + \big(p(t) - q(t - \tau + \sigma)\big)\left(1 + c(t - \tau) + \int_{t-\tau+\sigma}^{t} q(s - \tau)ds\right) y(t - \tau) = 0 \tag{3.7.2}$$

is oscillatory.

Then every solution of (3.7.1) is oscillatory.

Proof: Assume the contrary, and let $x(t)$ be an eventually positive solution of Eq. (3.7.1). Denote

$$y(t) = x(t) - c(t)x(t - r) - \int_{t-\tau+\sigma}^{t} q(s)x(s - \sigma)ds. \tag{3.7.3}$$

Then it is easy to see that $y'(t) \leq 0$ and $y(t) > 0$, eventually. In view of (3.7.3) and (3.7.1),

$$y'(t) = -\big(p(t) - q(t - \tau + \sigma)\big)x(t - \tau). \tag{3.7.4}$$

From (3.7.3) we have

$$\begin{aligned} x(t) &\geq y(t) + c(t)y(t - r) + \int_{t-\tau+\sigma}^{t} q(s)y(s - \sigma)ds \\ &\geq \left(1 + c(t) + \int_{t-\tau+\sigma}^{t} q(s)\, ds\right) y(t). \end{aligned} \tag{3.7.5}$$

Substituting (3.7.5) into (3.7.4) and noting condition (iii), we have

$$y'(t) + \big(p(t) - q(t - \tau + \sigma)\big)\left(1 + c(t - \tau) + \int_{t-\tau+\sigma}^{t} q(s - \tau)ds\right) y(t - \tau) \leq 0. \tag{3.7.6}$$

Hence, (3.7.2) has a positive solution, which contradicts (v). □

Theorem 3.7.2. *Assume that*

i) $r > 0, \quad \tau > \sigma \geq 0, \quad c, p, q \in C([t_0, \infty), R_+)$;

ii) *there exists* $t_1 \geq t_0$ *such that for* $t \geq t_1$

$$0 \leq c(t) + \int_{t-\tau}^{t-\sigma} q(s+\sigma)ds \leq 1,$$

$$\overline{p}(t) \equiv p(t) - q(t-\tau+\sigma) \geq 0; \tag{3.7.7}$$

iii) $\liminf\limits_{t\to\infty} \int_{t-\tau}^{t} \overline{p}(s)ds > 0$;

iv) *there exists* $T \geq t_1 + \max\{r, \tau\}$ *such that*

$$\inf_{t\geq T, \lambda>0} \left\{ \frac{1}{\lambda}\exp\left(\lambda \int_{t-\tau}^{t} \overline{p}(s)ds\right) + c(t-\tau)\exp\left(\lambda \int_{t-\tau}^{t} \overline{p}(s)ds\right) \right.$$
$$\left. + \int_{t-\tau}^{t-\sigma} q(s+\sigma-\tau)\exp\left(\lambda \int_{s}^{t} \overline{p}(u)du\right) \right\} > 1. \tag{3.7.8}$$

Then every solution of (3.7.1) is oscillatory.

To prove this theorem we need the following lemmas.

Lemma 3.7.1. *Assume that conditions i)-iii) in Theorem 3.7.2 hold. Let* $x(t)$ *be an eventually positive solution of (3.7.1). Set*

$$u(t) = x(t) - c(t)x(t-r) - \int_{t-\tau}^{t-\sigma} q(s+\sigma)x(s)ds. \tag{3.7.9}$$

Then $u'(t) \leq 0, \quad u(t) > 0$ *eventually.*

Proof: From (3.7.1), (3.7.9), and condition ii) in Theorem 3.7.2, we see that

$$u'(t) = -\overline{p}(t)x(t-\tau) \leq 0, \quad t \geq t_2 \geq t_1. \tag{3.7.10}$$

Denote $\lim\limits_{t\to\infty} u(t) = \ell$. We show that $\ell \geq 0$.

First we consider the case that $x(t)$ is unbounded, i.e., $\limsup\limits_{t\to\infty} x(t) = \infty$. Then there exists a sequence $\{s_n\}$ such that $\lim\limits_{n\to\infty} s_n = \infty, \quad \lim\limits_{n\to\infty} x(s_n) = \infty$

and $x(s_n) = \max\{x(t) : t_2 \le t \le s_n\}, \quad n = 1, 2, \ldots$. Hence from ii)

$$u(s_n) = x(s_n) - c(s_n)x(s_n - r) - \int_{s_n-\tau}^{s_n-\sigma} q(s+\sigma)x(s)ds$$

$$\ge x(s_n)\left[1 - c(s_n) - \int_{s_n-\tau}^{s_n-\sigma} q(s+\sigma)ds\right] \ge 0, \quad n = 1, 2, \ldots .$$

Therefore $\ell = \lim_{t\to\infty} u(t) = \lim_{n\to\infty} u(s_n) \ge 0$.

Next we consider the case that $x(t)$ is bounded. Set $\limsup_{t\to\infty} x(t) = \bar{\ell} < \infty$. Then there exists a sequence $\{\bar{s}_n\}$ such that $\lim_{n\to\infty} \bar{s}_n = \infty$ and $\lim_{n\to\infty} x(\bar{s}_n) = \bar{\ell}$. Denote $r_1 = \min\{r, \sigma\}$, $r_2 = \max\{r, \tau\}$, and

$$x(\xi_n) = \max\{x(s) : \bar{s}_n - r_2 \le s \le \bar{s}_n - r_1\}, \quad n = 1, 2, \ldots .$$

Clearly $\lim_{n\to\infty} \xi_n = \infty$ and $\limsup_{n\to\infty} x(\xi_n) \le \bar{\ell}$. From ii)

$$x(\bar{s}_n) - u(\bar{s}_n) = c(\bar{s}_n)x(\bar{s}_n - r) + \int_{\bar{s}_n-\tau}^{\bar{s}_n-\sigma} q(s+\sigma)x(s)ds$$

$$\le x(\xi_n)[c(\bar{s}_n) + \int_{\bar{s}_n-\tau}^{\bar{s}_n-\sigma} q(s+\sigma)ds] \le x(\xi_n).$$

Taking superior limits on both sides as $n \to \infty$ we obtain that $\bar{\ell} - \ell \le \bar{\ell}$, and so $\ell \ge 0$. Then $u(t) > \ell \ge 0$ for $t \ge t_2$. □

Lemma 3.7.2. *Assume that i)-iii) in Theorem 3.7.2 hold. Let $x(t)$ be an eventually positive solution and let u be defined by (3.7.9). Define a set*

$$\Lambda = \{\lambda > 0 : \ u'(t) + \lambda\bar{p}(t)u(t) \le 0 \quad \textit{holds eventually}\}.$$

Then $\Lambda \ne \emptyset$, and has an uniform upper bound which is independent of $x(t)$.

Proof: From Lemma 3.7.1 and its proof, there exists $t_2 \ge t_1$ such that $u'(t) \le 0$, $u(t) > 0$, for $t \ge t_2$, and

$$u'(t) = -\bar{p}(t)x(t-\tau) \le -\bar{p}(t)u(t-\tau) \le -\bar{p}(t)u(t), \quad t \ge t_2 + \tau, \tag{3.7.11}$$

i.e., $1 \in \Lambda$. Now we show that Λ has an upper bound which does not depend on $x(t)$. Set

$$\liminf_{t\to\infty} \int_{t-\tau}^{t} \overline{p}(s)ds = 3k.$$

From condition iii) in Theorem 3.7.2 we see $k > 0$. Hence there exists $t_3 \geq t_2 + \tau$ such that

$$\int_{t-\tau}^{t} \overline{p}(s)ds > 2k, \quad t \geq t_3.$$

By Lemma 2.1.3 this together with (3.7.11) implies that $\frac{u(t-\tau)}{u(t)} \leq \frac{1}{k^2}$, $t \geq t_3$. From (3.7.1) and condition (iii) in Theorem 3.7.2, it is easy to see that $\liminf\limits_{t\to\infty} x(t) = 0$. Hence there exists a sequence $\{s_n\}$ such that $s_n \geq t_3 + 2\tau$, $\lim\limits_{n\to\infty} s_n = \infty$, and $x(s_n - \tau) = \min\{x(s-\tau) : t_3 \leq s \leq s_n\}$, $n = 1, 2, \ldots$. Integrating (3.7.10) from $s_n - \tau$ to s_n we have

$$\begin{aligned} u(s_n) - u(s_n - \tau) &= -\int_{s_n-\tau}^{s} \overline{p}(s)x(s-\tau)ds \\ &\leq -x(s_n - \tau)\int_{s_n-\tau}^{s_n} \overline{p}(s)ds \leq -2kx(s_n - \tau), \quad n = 1, 2, \ldots, \end{aligned}$$

and hence $u(s_n - \tau) > 2kx(s_n - \tau)$. Substituting it into (3.7.10) we obtain

$$\begin{aligned} u'(s_n) &= -\overline{p}(s_n)x(s_n - \tau) \\ &> -\frac{1}{2k}\,\overline{p}(s_n)u(s_n - \tau) > -\frac{1}{2k^3}\,\overline{p}(s_n)u(s_n) \end{aligned}$$

which means that $\frac{1}{2k^3} \overline{\in} \Lambda$, i.e., $\frac{1}{2k^3}$ is an upper bound of Λ which does not depend on $x(t)$. □

Proof of Theorem 3.7.2: Let $x(t)$ be an eventually positive solution. By Lemmas 3.7.1 and 3.7.2 $u'(t) \leq 0$, $u(t) > 0$, $t \geq t_2$. We claim that there exists an $\alpha > 1$ such that $\lambda_0 \in \Lambda$ implies that $\alpha\lambda_0 \in \Lambda$.

In fact, from condition (iv), there exists $\alpha > 1$ such that

$$\inf_{\lambda>0}\inf_{t\geq T}\left\{\frac{1}{\lambda}\exp\left(\lambda\int_{t-\tau}^{t}\overline{p}(s)ds\right)+c(t-\tau)\exp\left(\lambda\int_{t-r}^{t}\overline{p}(s)ds\right)\right.$$
$$\left.+\int_{t-\tau}^{t-\sigma}q(s-\tau+\sigma)\exp\left(\lambda\int_{s}^{t}\overline{p}(u)du\right)\right\}\geq\alpha>1. \tag{3.7.12}$$

By the definition of Λ we see that eventually

$$u'(t)+\lambda_0\overline{p}(t)u(t)\leq 0. \tag{3.7.13}$$

Set $z(t)=u(t)\exp\left(\lambda_0\int_{t_2}^{t}\overline{p}(s)ds\right)$. Then

$$z'(t)=[u'(t)+\lambda_0\overline{p}(t)u(t)]\ \exp\ (\lambda_0\int_{t_2}^{t}\overline{p}(s)ds)\leq 0.$$

From (3.7.10) and (3.7.13) we have $x(t-\tau)\geq\lambda_0 u(t)$, and then from (3.7.9) and (3.7.12)

$$\begin{aligned}
u'(t)&=-\overline{p}(t)x(t-\tau)\\
&=-\overline{p}(t)\left[u(t-\tau)+c(t-\tau)x(t-\tau-r)+\int_{t-2\tau}^{t-\tau-\sigma}q(s+\sigma)x(s)ds\right]\\
&\leq-\overline{p}(t)\left[u(t-\tau)+\lambda_0 c(t-\tau)u(t-r)+\lambda_0\int_{t-2\tau}^{t-\tau-\sigma}q(s+\sigma)u(s+\tau)ds\right]\\
&=-\overline{p}(t)\left[u(t-\tau)+\lambda_0 c(t-\tau)u(t-r)+\lambda_0\int_{t-\tau}^{t-\sigma}q(s+\sigma-\tau)u(s)ds\right]\\
&=-\overline{p}(t)\left[z(t-\tau)\exp\left(-\lambda_0\int_{t_2}^{t-\tau}\overline{p}(s)ds\right)\right.\\
&\qquad+\lambda_0 c(t-\tau)z(t-r)\exp\left(-\lambda_0\int_{t_2}^{t-r}\overline{p}(s)ds\right)\\
&\qquad\left.+\lambda_0\int_{t-\tau}^{t-\sigma}q(s+\sigma-\tau)z(s)\exp\left(-\lambda_0\int_{t_2}^{s}\overline{p}(u)du\right)ds\right]\\
&\leq-\overline{p}(t)\left[\exp\left(\lambda_0\int_{t-\tau}^{t}\overline{p}(s)ds\right)+\lambda_0 c(t-\tau)\exp\left(\lambda_0\int_{t-r}^{t}\overline{p}(s)ds\right)\right.
\end{aligned}$$

$$+ \lambda_0 \int_{t-r}^{t-\sigma} q(s+\sigma-\tau)\exp\Big(\lambda_0 \int_s^t \overline{p}(u)du\Big)ds\Big] u(t)$$

$$\leq - \inf_{t\geq T} \exp\Big(\lambda_0 \int_{t-\tau}^t \overline{p}(s)ds\Big) + \lambda_0 c(t-\tau)\exp\Big(\lambda_0 \int_{t-r}^t \overline{p}(s)ds\Big)$$

$$+ \lambda_0 \int_{t-r}^{t-\sigma} q(s+\sigma-\tau)\exp\Big(\lambda_0 \int_s^t \overline{p}(u)du\Big)ds\Big] \overline{p}(t)u(t)$$

$$\leq -\alpha\lambda_0\overline{p}(t)u(t).$$

This implies that $\alpha\lambda_0 \in \Lambda$. Repeating this we obtain that $\alpha^m\lambda_0 \in \Lambda$ for any positive integer m, which contradicts the boundedness of Λ. □

Theorem 3.7.3. *Assume that*

i) $r > 0, \quad \tau > \sigma \geq 0, \quad c, p, q \in C([t_0, \infty), R_+)$;

ii) $\overline{p}(t) \geq 0$, *for* $t \geq t_1 \geq t_0$ *and* $\int_{t_0}^{\infty} \overline{p}(t)dt = \infty$, *where* $\overline{p}$ *is defined in Theorem 3.7.1;*

iii) *For some* $T \geq t_1 + \max\{r, \tau\}$, *there exists* $\lambda^* > 0$ *such that*

$$\sup_{t\geq T} \Big\{ \frac{1}{\lambda^*} \exp\Big(\lambda^* \int_{t-\tau}^t \overline{p}(u)du\Big) + c(t-\tau)\exp\Big(\lambda^* \int_{t-r}^t \overline{p}(s)ds\Big)$$

$$+ \int_{t-r}^{t-\sigma} q(s-\tau+\sigma)\exp\Big(\lambda^* \int_s^t \overline{p}(u)du\Big)ds\Big\} \leq 1. \tag{3.7.14}$$

Then Eq. (3.7.1) has a positive solution x satisfying $\lim_{t\to\infty} x(t) = 0$ and

$$x(t) \leq \exp\Big(-\lambda^* \int_{t_1}^{t+\tau} \overline{p}(u)du\Big) \quad \text{for all large} \quad t. \tag{3.7.15}$$

To prove this theorem we first show the following lemma.

Lemma 3.7.3. *Let conditions i) and ii) in Theorem 3.7.2 hold. Further assume* *that the integral inequality*

$$c(t)z(t-r) + \int_{t-\tau}^{t-\sigma} q(s+\sigma)z(s)ds + \int_{t-\tau}^{\infty} [p(s+\tau) - q(s+\sigma)]z(s)ds \le z(t), \quad t \ge t_1, \tag{3.7.16}$$

has a positive solution $z : [t_1 - m, \infty) \to (0,\infty)$ *such that* $\lim_{t\to\infty} z(t) = 0$ *where* $m = \max\{\tau, r\}$. *Then there exists a positive function* $x : [t_1 - m, \infty) \to (0,\infty)$ *satisfying the integral equation*

$$c(t)x(t-r) + \int_{t-\tau}^{t-\sigma} q(s+\sigma)x(s)\,ds + \int_{t-\tau}^{\infty} [p(s+\tau) - q(s+\sigma)]x(s)ds = x(t), \quad t \ge t_1 \tag{3.7.17}$$

and $0 < x(t) \le z(t), \quad t \ge t_1$.

Proof: Choose $T \ge t_1$ such that $z(t) > z(T)$ for $t_1 - m \le t < T$. Define a set of functions as follows:

$$\Omega = \{w \in C([t_1 - m, \infty), R_+) : \; 0 \le w(t) \le z(t), \quad t \ge t_1 - m\} \tag{3.7.18}$$

and define a mapping $\mathcal{T}$ on Ω by

$$(\mathcal{T}w)(t) = \begin{cases} c(t)w(t-r) + \int_{t-\tau}^{t-\sigma} q(s+\sigma)w(s)ds \\ \quad + \int_{t-\tau}^{\infty} (p(s+\tau) - q(s+\sigma))w(s)ds, & t \ge T \\ (\mathcal{T}w)(T) + z(t) - z(T), & t_1 - m \le t < T. \end{cases} \tag{3.7.19}$$

By (3.7.16) $\mathcal{T}\Omega \subset \Omega$. Define a sequence $\{x_n\}$ on Ω by $x_0 = z, \quad x_n = \mathcal{T}x_{n-1}, \quad n = 1, 2, \ldots$. It is easy to see that

$$0 \le x_n(t) \le x_{n-1}(t) \le \cdots \le x_1(t) \le z(t), \quad t \ge t_1 - m.$$

Then $\lim_{n\to\infty} x_n(t) = x(t)$ exists for $t \ge t_1 - m$.

It follows from the Lebesgue's dominated convergence theorem that $x(t)$ satisfies (3.7.17). Clearly, $x(t)$ is continuous for $t \geq t_1 - m$. Since $x(t) > 0$ on $[t_1 - m, T)$, it follows that $x(t) > 0$ for all $t \geq t_1 - m$.

Proof of Theorem 3.7.3: Set

$$z(t) = \exp\left(- \lambda^* \int_T^{t+\tau} \overline{p}(s)ds \right), \quad t \geq T - \tau. \tag{3.7.20}$$

It is obvious that $z(t)$ is continuous for $t \geq T - \tau$ and $\lim_{t\to\infty} z(t) = 0$. We shall show that $z(t)$ satisfies the integral inequality (3.7.16) for $t \geq T$. From (3.7.14), we have

$$\frac{1}{\lambda^*}\exp\left(\lambda^* \int_t^{t+\tau} \overline{p}(u)du\right) + c(t)\exp\left(\lambda^* \int_{t+\tau-r}^{t+\tau} \overline{p}(s)ds\right) \\ + \int_t^{t+\tau-\sigma} q(s-\tau+\sigma)\exp\left(\lambda^* \int_s^{t+\tau} \overline{p}(u)du\right) ds \leq 1, \quad t \geq T - \tau. \tag{3.7.21}$$

We observe that

$$\begin{aligned} &\int_{t-\tau}^{\infty} \overline{p}(s+\tau)\exp\left(\lambda^* \int_{s+\tau}^{t+\tau} \overline{p}(u)du\right) ds \\ &\qquad = \exp\left(\lambda^* \int_0^{t+\tau} \overline{p}(u)du\right) \int_{t-\tau}^{\infty} \overline{p}(s+\tau)\exp\left(\lambda^* \int_{s+\tau}^{0} \overline{p}(u)du\right) ds \\ &\qquad = \frac{1}{\lambda^*}\exp\left(\lambda^* \int_0^{t+\tau} \overline{p}(u)du\right) \exp\left(\lambda^* \int_t^{0} \overline{p}(u)du\right) \\ &\qquad = \frac{1}{\lambda^*}\exp\left(\lambda^* \int_t^{t+\tau} \overline{p}(u)du\right) \end{aligned} \tag{3.7.22}$$

and

$$\begin{aligned} &\int_t^{t+\tau-\sigma} q(s-\tau+\sigma)\exp\left(\lambda^* \int_s^{t+\tau} \overline{p}(u)du\right) ds \\ &\qquad = \int_{t-\tau}^{t-\sigma} q(s+\sigma)\exp\left(\lambda^* \int_{s+\tau}^{t+\tau} \overline{p}(u)du\right) ds. \end{aligned} \tag{3.7.23}$$

Substituting (3.7.22) and (3.7.23) into (3.7.21) we have

$$c(t)\exp\left(\lambda^* \int_{t+\tau-r}^{t+\tau} \overline{p}(u)du\right) + \int_{t-r}^{t-\sigma} q(s+\sigma)\exp\left(\lambda^* \int_{s+\tau}^{t+\tau} \overline{p}(u)du\right)ds$$
$$+ \int_{t-\tau}^{\infty} \overline{p}(s+\tau)\exp\left(\lambda^* \int_{s+\tau}^{t+\tau} \overline{p}(u)du\right)ds \le 1, \quad t \ge T-\tau. \tag{3.7.24}$$

From (3.7.24) and (3.7.20) we obtain

$$c(t)z(t-r) + \int_{t-\tau}^{t-\sigma} q(s+\sigma)z(s)ds + \int_{t-\tau}^{\infty} \overline{p}(s+\tau)z(s)ds \le z(t), \quad t \ge T-\tau,$$

i.e., (3.7.16) has a continuous positive solution $z(t)$ defined by (3.7.20). By Lemma 3.7.3, (3.7.17) has a positive solution $x(t)$ on $[T,\infty)$ satisfying $x(t) \le z(t)$. Thus x is a positive solution of (3.7.1) satisfying $\lim_{t\to\infty} x(t) = 0$. □

Remark 3.7.1. Condition (3.7.14) is sharp in the following sense: if c, p, q are constants with $c > 0$ and $p > q$, condition (3.7.14) is not only a sufficient condition but also a necessary condition for the existence of a positive solution which tends to zero as $t \to \infty$. This is based on the fact that for the autonomous case, Eq. (3.7.1) has a positive solution which tends to zero if and only if its characteristic equation

$$f(\lambda) = \lambda - c\lambda e^{-\lambda r} + pe^{-\lambda\tau} - qe^{-\lambda\sigma} = 0 \tag{3.7.25}$$

has a negative real root.

In the same sense, condition (3.7.8) is sharp.

Corollary 3.7.1. *Assume that conditions i) and ii) in Theorem 3.7.2 hold, and*

$$\liminf_{t\to\infty}\left\{e\int_{t-\tau}^{t} \overline{p}(s)ds + c(t-\tau) + \int_{t-\tau}^{t-\sigma} q(s-\tau+\sigma)ds\right\} > 1. \tag{3.7.26}$$

Then every solution of (3.7.1) is oscillatory.

Example 3.7.1. Consider

$$\left(x(t) - \tfrac{1}{2}\, x(t-2)\right)' + \left(1 + \tfrac{1}{\pi} + \sin t\right)x(t-\pi) - \tfrac{1}{\pi}\, x\left(t - \tfrac{\pi}{2}\right) = 0 \tag{3.7.27}$$

and

$$\left(x(t) - \tfrac{1}{3}\, x(t-1)\right)' + \frac{2(1+e)}{3e} x(t-1) - \tfrac{2}{3}\, x(t - \tfrac{1}{2}) = 0. \tag{3.7.28}$$

It is easy to see that for (3.7.27) and (3.7.28) all assumptions of Corollary 3.7.1 are satisfied. Therefore, these equations are oscillatory.

3.8. Linearized Oscillation

The following linearized oscillation result reveals that, under some assumptions, certain nonlinear differential equations have the same oscillatory behavior as an associated linear equation with constant coefficients.

Consider the neutral differential equation with positive and negative coefficients

$$\left(x(t) - P(t)G(x(t-\tau))\right)' + Q_1(t)H_1\left(x(t-\sigma_1)\right) - Q_2(t)H_2\left(x(t-\sigma_2)\right) = 0 \tag{3.8.1}$$

with the following assumptions

(1) $P, Q_1, Q_2 \in C([t_0, \infty), R_+), \quad G, H_1, H_2 \in C(R, R)$;

(2) $\tau \in (0, \infty), \quad \sigma_1, \sigma_2 \in [0, \infty), \quad \sigma_1 \geq \sigma_2$;

(3) $\limsup_{t \to \infty} P(t) = p_2 \in (0,1); \quad \liminf_{t\to\infty} P(t) = p_1 \in (0,1)$;

(4) $\liminf_{t\to\infty} Q_1(t) = q_1, \quad \limsup_{t\to\infty} Q_2(t) = q_2$;

(5) $0 \leq \frac{G(u)}{u} \leq 1$ for $u \neq 0, \quad \lim_{u\to\infty} \frac{G(u)}{u} = 1$;

(6) $\frac{H_1(u)}{H_2(u)} \geq 1$ and $0 < \frac{H_2(u)}{u} \leq M$ for $u \neq 0, \quad \lim_{u\to 0} \frac{H_2(u)}{u} = 1$;

(7) $1 - p_2 - Mq_2(\sigma_1 - \sigma_2) > 0$.

We associate Eq. (3.8.1) with the linear equation with constant coefficients

$$\left(y(t) - p_2 y(t-\tau)\right)' + q_1 y(t-\sigma_1) - q_2 y(t-\sigma_2) = 0. \tag{3.8.2}$$

The following is the linearized oscillation result for Eq. (3.8.1).

Theorem 3.8.1. *Assume that every solution of Eq. (3.8.2) is oscillatory. Then every solution of Eq. (3.8.1) is also oscillatory.*

To prove this theorem we first show some related lemmas.

Lemma 3.8.1. *Consider the neutral delay differential equation*

$$\left(y(t)-py(t-\tau)\right)'+q_1y(t-\sigma_1)-q_2y(t-\sigma_2)=0 \tag{3.8.3}$$

where $p\in(0,1]$, $\tau,q_1,q_2\in(0,\infty)$, $\sigma_1,\sigma_2\in[0,\infty)$. *Suppose that every solution of Eq. (3.8.3) is oscillatory. Then there exists an* $\varepsilon_0>0$ *such that for any* $\varepsilon,\varepsilon_1,\varepsilon_2\in[0,\varepsilon_0)$ *every solution of the differential equation*

$$\left(y(t)-(p-\varepsilon)y(t-\tau)\right)'+(q_1-\varepsilon_1)y(t-\sigma_1)-(q_2+\varepsilon_2)y(t-\sigma_2)=0 \tag{3.8.4}$$

is also oscillatory.

Proof: By Theorem 2.1.1 it suffices to show that the characteristic equation of (3.8.4) has no real roots. The assumption that every solution of Eq. (3.8.3) is oscillatory implies that the characteristic equation

$$f(\lambda)=\lambda-\lambda pe^{-\lambda\tau}+q_1e^{-\lambda\sigma_1}-q_2e^{-\lambda\sigma_2}=0 \tag{3.8.5}$$

of Eq. (3.8.3) has no real roots. In view of $f(\infty)=\infty$ we see that $f(\lambda)>0$ for all $\lambda\in R$. This means that $f(0)=q_1-q_2>0$ and either $\sigma_1\geq\sigma_2$ or $\sigma_1<\sigma_2\leq\tau$. Then $f(-\infty)=\infty$ and so $m:=\min_{\lambda\in R}f(\lambda)>0$ exists. Therefore,

$$\lambda-\lambda pe^{-\lambda\tau}+q_1e^{-\lambda\sigma_1}-q_2e^{-\lambda\sigma_2}\geq m,\quad \lambda\in R.$$

Set

$$\delta=\tfrac{1}{3}\min\{p,q_1-q_2\}\quad\text{and}\quad g(\lambda)=\delta(|\lambda|e^{-\lambda\tau}+e^{-\lambda\sigma_1}+e^{-\lambda\sigma_2}).$$

Observing that

$$f(\lambda)-g(\lambda)=\lambda-(p\lambda+\delta|\lambda|)e^{-\lambda\tau}+(q_1-\delta)e^{-\lambda\sigma_1}-(q_2+\delta)e^{-\lambda\sigma_2},$$

we have $\lim_{|\lambda|\to\infty}(f(\lambda)-g(\lambda))=\infty$. In particular, there exists a $\lambda_0>0$ such that $f(\lambda)-g(\lambda)>\frac{m}{2}$ for $|\lambda|\geq\lambda_0$. Let $\eta=|\lambda_0|e^{\lambda_0\tau}+e^{\lambda_0\sigma_1}+e^{\lambda_0\sigma_2}$ and set $\varepsilon_0=\min\{\delta,m/2\eta\}$.

To complete the proof, it suffices to show that for any $\varepsilon, \varepsilon_1, \varepsilon_2 \in [0, \varepsilon_0)$ the characteristic equation

$$\lambda - \lambda(p-\varepsilon)e^{-\lambda\tau} + (q_1 - \varepsilon_1)e^{-\lambda\sigma_1} - (q_2+\varepsilon_2)e^{-\lambda\sigma_2} = 0$$

of Eq. (3.8.4) has no real roots. In fact, for $|\lambda| \geq \lambda_0$

$$\lambda - \lambda(p-\varepsilon)e^{-\lambda\tau} + (q_1 - \varepsilon_1)e^{-\lambda\sigma_1} - (q_2+\varepsilon_2)e^{-\lambda\sigma_2}$$
$$= f(\lambda) - [-\varepsilon\lambda e^{-\lambda\tau} + \varepsilon_1 e^{-\lambda\sigma_1} + \varepsilon_2 e^{-\lambda\sigma_2}]$$
$$\geq f(\lambda) - g(\lambda) > \tfrac{m}{2} > 0.$$

Also, for $|\lambda| \leq \lambda_0$

$$\lambda - \lambda(p-\varepsilon)e^{-\lambda\tau} + (q_1 - \varepsilon_1)e^{-\lambda\sigma_1} - (q_2+\varepsilon_2)e^{-\lambda\sigma_2}$$
$$\geq f(\lambda) - \varepsilon_0[|\lambda_0|e^{\lambda_0\tau} + e^{\lambda_0\sigma_1} + e^{\lambda_0\sigma_2}]$$
$$\geq m - \tfrac{m}{2} > 0. \qquad \square$$

Lemma 3.8.2. *Assume that $F \in C([T,\infty),(0,\infty))$, $H \in C(R_+, R_+)$, $\tau \in (0,\infty)$, $\sigma_1, \sigma_2 \in [0,\infty)$, $c_1, c_2 \in [0,\infty)$; $H(u)$ is nondecreasing in a neighborhood of the origin and $\sigma_1 \geq \sigma_2$. Let $m = \max\{\tau, \sigma_1\}$ and suppose that the integral inequality*

$$F(t)z(t-\tau) + c_1\int_{t-\sigma_1}^{t-\sigma_2} H\big(z(s)\big)ds + c_2\int_{t-\sigma_1}^{\infty} H\big(z(s)\big)ds \leq z(t), \quad t \geq T \quad (3.8.6)$$

has a continuous positive solution $z : [T-m,\infty) \to (0,\infty)$ such that $\lim_{t\to\infty} z(t) = 0$.

Then there exists a positive solution $x : [T-m,\infty) \to (0,\infty)$ of the corresponding integral equation

$$F(t)x(t-\tau) + c_1\int_{t-\sigma_1}^{t-\sigma_2} H\big(x(s)\big)ds + c_2\int_{t-\sigma_1}^{\infty} H\big(x(s)\big)ds = x(t), \quad t \geq T. \quad (3.8.7)$$

Proof: Choose a $T_1 \geq T$ and a $\delta > 0$ such that $z(t) > z(T_1)$ for $T-m \leq t < T_1$, $0 < z(t) < \delta$ for $t \geq T_1 - m$ and $H(u)$ is nondecreasing in $[0,\delta]$. Define a set of functions

$$\Omega = \{w \in C([T-m,\infty), R_+) : \ 0 \le w(t) \le z(t), \quad t \ge T-m\}$$

and define a mapping $\mathcal{T}$ on Ω as follows:

$$(\mathcal{T}w)(t) = \begin{cases} F(t)w(t-\tau) + c_1 \int_{t-\sigma_1}^{t-\sigma_2} H\big(w(s)\big)ds & \\ \quad + c_2 \int_{t-\sigma_1}^{\infty} H\big(w(s)\big)ds & t \ge T_1 \\ (\mathcal{T}w)(T_1) + z(t) - z(T_1), & T-m \le t < T_1. \end{cases} \tag{3.8.8}$$

It is obvious that $\mathcal{T}$ is continuous. Also $w_1, w_2 \in \Omega$ with $w_1 \le w_2$ implies $\mathcal{T}w_1 \le \mathcal{T}w_2$. From (3.8.6), $\mathcal{T}z \le z$ and so $w \in \Omega$ implies $0 \le \mathcal{T}w \le \mathcal{T}z \le z$. Thus $\mathcal{T} : \Omega \to \Omega$.

Define a sequence $\{x_n\}$ on Ω as follows:

$$x_0 = z, \quad \text{and} \quad x_n = \mathcal{T}x_{n-1}, \quad n = 1, 2, \dots . \tag{3.8.9}$$

By induction, we have that

$$0 \le x_n(t) \le x_{n-1}(t) \le z(t) \quad \text{for} \quad t \ge T-m, \quad n = 1, 2, \dots . \tag{3.8.10}$$

Set $x(t) = \lim_{n\to\infty} x_n(t)$ for $t \ge T-m$. It follows from the Lebesgue's dominated convergence theorem that $x(t)$ satisfies (3.8.6). From (3.8.8) $x(t) > 0$ for $T-m \le t < T_1$. Consequently, $x(t) > 0$ for $t \ge t-m$. □

Proof of Theorem 3.8.1: Without loss of generality we assume that $x(t)$ is an eventually positive solution of Eq. (3.8.1) that every solution of Eq. (3.8.2) is oscillatory implies that its characteristic equation

$$f(\lambda) = \lambda - p_0\lambda e^{-\lambda\tau} + q_1 e^{-\lambda\sigma_1} - q_2 e^{-\lambda\sigma_2} = 0$$

has no real roots. Since $f(0) = q_1 - q_2$ and $f(\infty) = \infty$, we see that $q_1 > q_2$. Choose $\eta > 0$ so small that

$$q_1 - q_2 - 2\eta > 0 \tag{3.8.11}$$

and

$$1 - p_2 - M(q_2 + \eta)(\sigma_1 - \sigma_2) > 0. \tag{3.8.12}$$

Then by assumption (5) and (6) and (3.8.1), for sufficiently large t_1, when $t \geq t_1$, we have

$$\begin{aligned}&\big(x(t) - p(t)G(x(t-\tau))\big)' \\ &\quad + (q_1 - \eta)H_1\big(x(t-\sigma_1)\big) - (q_2 + \eta)H_2\big(x(t-\sigma_2)\big) \leq 0.\end{aligned} \tag{3.8.13}$$

Set

$$z(t) = x(t) - p(t)G\big(x(t-\tau)\big) - (q_2 + \eta)\int_{t-\sigma_1}^{t-\sigma_2} H_2\big(x(s)\big)ds. \tag{3.8.14}$$

From (3.8.11) and (3.8.13) and assumption (6)

$$\begin{aligned}z'(t) &= \big(x(t) - p(t)G(x(t-\tau))\big)' - (q_2 + \eta)\big(H_2(x(t-\sigma_2)) - H_2(x(t-\sigma_1))\big) \\ &\leq -(q_1 - \eta)H_1\big(x(t-\sigma_1)\big) + (q_2 + \eta)H_2\big(x(t-\sigma_1)\big) \\ &\leq -(q_1 - q_2 - 2\eta)H_2\big(x(t-\sigma_1)\big) \leq 0.\end{aligned} \tag{3.8.15}$$

Now we claim that $x(t)$ is bounded. Otherwise there exists a sequence $\{t_n\}$ such that $\lim_{n\to\infty} t_n = \infty$, $x(t_n) = \max_{s\leq t_n} x(s)$, and $\lim_{n\to\infty} x(t_n) = \infty$. Then by assumptions (3), (5), (6) and (3.8.12)

$$\begin{aligned}z(t_n) &= x(t_n) - p(t_n)G\big(x(t_n-\tau)\big) - (q_2 + \eta)\int_{t_n-\sigma_1}^{t_n-\sigma_2} H_2\big(x(s)\big)ds \\ &\geq x(t_n) - p(t_n)\,\frac{G\big(x(t_n-\tau)\big)}{x(t_n-\tau)}\,x(t_n-\tau) \\ &\qquad - (q_2 + \eta)\int_{t_n-\sigma_1}^{t_n-\sigma_2} \frac{H_2(x(s))}{x(s)}\,x(s)ds \\ &\geq x(t_n)[1 - p_2 - M(q_2 + \eta)(\sigma_1 - \sigma_2)] \to \infty, \quad \text{as} \quad n \to \infty,\end{aligned}$$

which contradicts (3.8.15). Therefore, $x(t)$ is bounded, and so $z(t)$ is bounded and $\lim_{t\to\infty} z(t) = \ell \in R$. Integrating (3.8.15) we have

$$\ell - z(t_1) \leq -(q_1 - q_2 - 2\eta)\int_{t_1}^{\infty} H_2\big(x(s-\sigma_1)\big)ds$$

which together with assumption (6) and the boundedness of $x(t)$, implies that $\liminf_{t\to\infty} x(t) = 0$. We claim that $\lim_{t\to\infty} x(t) = 0$. Otherwise, let $\mu = \limsup_{t\to\infty} x(t)$ and let $\{t_n\}$ and $\{t'_n\}$ be two sequences in the interval $[t_1, \infty)$ such that

$$\lim_{n\to\infty} t_n = \infty, \quad \lim_{n\to\infty} t'_n = \infty, \quad \lim_{t\to\infty} x(t_n) = 0 \quad \text{and} \quad \lim_{t\to\infty} x(t'_n) = \mu.$$

Since $z(t_n) \le x(t_n)$, then $\ell \le 0$. On the other hand, by choosing sufficiently small $\varepsilon' > 0$ we have

$$\begin{aligned} z(t'_n) &\ge x(t'_n) - p(t'_n)x(t'_n - \tau) - M(q_2 + \eta)\int_{t'_n - \sigma_1}^{t'_n - \sigma_2} x(s)ds \\ &\ge x(t'_n) - p(t'_n)(\mu + \varepsilon') - M(q_2 + \eta)(\sigma_1 - \sigma_2)(\mu + \varepsilon'). \end{aligned}$$

Letting $n \to \infty$ we obtain

$$\ell \ge \mu - p_2(\mu + \varepsilon') - M(q_2 + \eta)(\sigma_1 - \sigma_2)(\mu + \varepsilon').$$

Since ε' can be arbitrarily small, we have

$$\ell \ge \mu - p_2\mu - M(q_2 + \eta)(\sigma_1 - \sigma_2)\mu = [1 - p_2 - M(q_2 + \eta)(\sigma_1 - \sigma_2)]\mu$$

which, together with $\ell \le 0$ and (3.8.12), implies $\ell = \mu = 0$.

Hence $\lim_{t\to\infty} z(t) = 0$ and $\lim_{t\to\infty} x(t) = 0$. Then

$$-z(t) \le -(q_1 - q_2 - 2\eta)\int_t^\infty H_2\big(x(s - \sigma_1)\big)ds. \tag{3.8.16}$$

Now by using (3.8.14) and (3.8.16) we obtain

$$\begin{aligned} x(t) &\ge p(t)G\big(x(t-\tau)\big) + (q_2 + \eta)\int_{t-\sigma_1}^{t-\sigma_2} H_2\big(x(s)\big)ds \\ &\quad + (q_1 - q_2 - 2\eta)\int_{t-\sigma_1}^{\infty} H_2\big(x(s)\big)ds \\ &= p(t)\,\frac{G\big(x(t-\tau)\big)}{x(t-\tau)}\,x(t-\tau) + (q_2 + \eta)\int_{t-\sigma_1}^{t-\sigma_2} \frac{H_2\big(x(s)\big)}{x(s)}\,x(s)ds \\ &\quad + (q_1 - q_2 - 2\eta)\int_{t-\sigma_1}^{\infty} \frac{H_2\big(x(s)\big)}{x(s)}\,x(s)ds. \end{aligned}$$

Choose $0 < \varepsilon < \frac{1}{2} \min \{p_2, \frac{\eta}{q_2+\eta}\}$. Then by assumptions (3), (5), (6)

$$x(t) \geq (p_2 - \varepsilon)x(t-\tau) + (q_2+\eta)(1-\varepsilon)\int_{t-\sigma_1}^{t-\sigma_2} x(s)ds$$

$$+ (q_1 - q_2 - 2\eta)(1-\varepsilon)\int_{t-\sigma_1}^{\infty} x(s)ds, \qquad t \geq t_2$$

where t_2 is sufficiently large. By Lemma 3.8.2, the equation

$$v(t) = (p_2 - \varepsilon)v(t-\tau) + (q_2+\eta)(1-\varepsilon)\int_{t-\sigma_1}^{t-\sigma_2} v(s)ds$$

$$+ (q_1 - q_2 - 2\eta)(1-\varepsilon)\int_{t-\sigma_1}^{\infty} v(s)ds$$

has a continuous positive solution $v : [T-m, \infty) \to (0, \infty)$, where $T \geq t_2$ is sufficiently large. Clearly, $v(t)$ is a positive solution of the neutral equation

$$\big(v(t) - (p_2 - \varepsilon)v(t-\tau)\big)' + (q_1 - \varepsilon_1)v(t-\sigma_1) - (q_2+\varepsilon_2)v(t-\sigma_2) = 0$$

where $\varepsilon_1 = \eta + \varepsilon q_1 - \eta\varepsilon$ and $\varepsilon_2 = \eta - q_2\varepsilon - \eta\varepsilon$ are sufficiently small positive numbers. Hence, by Lemma 3.8.1., Eq. (3.8.2) has a positive solution. This contradiction proves the theorem. □

3.9. Equations with a Nonlinear Neutral Term

Consider the nonlinear neutral differential equations of the form

$$\big(x(t) - g(x(t-\tau))\big)' + h\big(t, x(t-\sigma)\big) = 0, \quad t \geq t_0 \tag{3.9.1}$$

where $\tau > 0,\ \sigma \geq 0,\ g \in C(R,R),\ h \in C([t_0,\infty) \times R, R)$.

Theorem 3.9.1. *Assume that*

(i) *g is nondecreasing, $xg(x) \geq 0$ and there exists $\alpha \in (0,1]$ such that either*

$$\limsup_{|x|\to\infty} \frac{|g(x)|}{|x|^{\alpha}} = M \in R_+,$$

whenever $\alpha \in (0,1)$, *or*

$$\limsup_{|x|\to\infty} \frac{g(x)}{x} = c \in [0,1), \quad \textit{if} \quad \alpha = 1;$$

(ii) *h is nondecreasing in* x, $\;h(t,x)x \geq 0$ *for* $x \neq 0$, *and*

$$\int_{t_0}^{\infty} |h(t,c)|dt = \infty \quad \text{for any} \quad c \neq 0;$$

(iii) *There exists a positive integer n such that the differential inequality*

$$y'(t) + h\big(t, y(t-\sigma) + g(y(t-\tau) \\ + g(y(t-\sigma-2\tau) + \cdots + g(y(t-\sigma-n\tau))\ldots))\big) \leq 0 \tag{3.9.2}$$

has no eventually positive solutions and

$$y'(t) + h\big(t, y(t-\sigma) + g(y(t-\sigma-\tau) \\ + g(y(t-\sigma-2\tau) + \cdots + g(y(t-\sigma-n\tau))\ldots))\big) \geq 0 \tag{3.9.3}$$

has no eventually negative solutions. Then every solution of Eq. (3.9.1) is oscillatory.

Proof: Assume the contrary, and let $x(t)$ be an eventually positive solution of Eq. (3.9.1), say $x(t) > 0, \quad t \geq T \geq t_0$. Set

$$z(t) = x(t) - g\big(x(t-\tau)\big). \tag{3.9.4}$$

Then $z'(t) \leq 0, \quad t \geq T + m, \quad m = \max\{\tau, \sigma\}$. If $z(t) < 0$ eventually, then there exists a $t_1 \geq T + m$ such that $z(t) \leq -\ell < 0$ for $t \geq t_1$. Thus, from (3.9.4), $g\big(x(t-\tau)\big) \geq \ell > 0$ for $t \geq t_1$. Integrating Eq. (3.9.1) and noting condition (ii) we have $\lim_{t\to\infty} z(t) = -\infty$. Consequently, $\lim_{t\to\infty} g\big(x(t-\tau)\big) = \infty$, which implies that $\lim_{t\to\infty} x(t) = \infty$. Thus there exists a sequence $\{\xi_n\}$ such that $\lim_{n\to\infty} \xi_n = \infty$ and

$x(\xi_n) = \max_{t_1 \le t \le \xi_n} \{x(t)\} \to \infty, \quad n \to \infty$. Then

$$\begin{aligned} z(\xi_n) &= x(\xi_n) - g\big(x(\xi_n - \tau)\big) \ge x(\xi_n) - g\big(x(\xi_n)\big) \\ &= x(\xi_n)[1 - g\big(x(\xi_n)\big)/x(\xi_n)] \to \infty, \quad \text{as} \quad n \to \infty, \end{aligned}$$

which contradicts the fact that $z(t)$ is eventually negative. Therefore $z(t) > 0$ for $t \ge T + m$. Hence

$$\begin{aligned} x(t) &= z(t) + g\big(x(t-\tau)\big) \\ &= z(t) + g\big(z(t-\tau) + g(x(t-2\tau))\big) \\ &\ge z(t) + g\big(z(t-\tau) + g(z(t-2\tau) + \cdots + g(z(t-n\tau))\ldots)\big) \end{aligned} \tag{3.9.5}$$

for $t \ge T + n\tau$. Substituting (3.9.5) into (3.9.1) we have

$$\begin{aligned} z'(t) + h\big(t, z(t-\sigma) &+ g(z(t-\sigma-\tau) \\ &+ g(z(t-\sigma-2\tau) + \cdots + g(z(t-\sigma-n\tau))\ldots))\big) \le 0 \end{aligned} \tag{3.9.6}$$

for $t \ge \sigma + T + n\tau$, which implies that (3.9.2) has an eventually positive solution. This is a contradiction.

A parallel argument is true if we assume that Eq. (3.9.1) has an eventually negative solution. □

Theorem 3.9.2. *Assume that*

(i) *g is nondecreasing, $xg(x) \ge 0$ and there exists a $d > 0$ such that $g(d) < d$,*

(ii) *h is nondecreasing in x, $h(t,x)x \ge 0$ and*

$$\int_{t_0}^{\infty} |h(t,c)|dt < \infty \quad \text{for any} \quad c \ne 0. \tag{3.9.7}$$

Then Eq. (3.9.1) has an eventually positive solution.

Proof: From (i) there exists a $\beta > 0$ such that $\beta + g(d) < d$. From (3.9.7) there exists a $T \geq t_0$ such that

$$\int_T^\infty h(t,d)dt \leq d - g(d) - \beta. \tag{3.9.8}$$

Define a sequence $\{w_k(t)\}$ on $[T-m,\infty)$ as follows: $w_1(t) = \beta, \quad t \geq T-m$, and for $k = 1, 2, \ldots$

$$w_{k+1}(t) = \begin{cases} \beta + g\big(w_k(t-\tau)\big) \\ \quad + \int_t^\infty h\big(s, w_k(s-\sigma)\big)ds, & t \geq T \\ w_{k+1}(T), & T-m \leq t \leq T. \end{cases} \tag{3.9.9}$$

We see that

$$\begin{aligned} \beta \leq w_2(t) &\leq \beta + g(\beta) + \int_t^\infty h(s,\beta)ds \\ &\leq \beta + g(d) + \int_t^\infty h(s,d)ds \leq d, \quad t \geq T. \end{aligned}$$

By induction we get

$$\beta = w_1(t) < w_2(t) \leq \cdots \leq w_k(t) \leq \cdots \leq d, \quad t \geq T.$$

Thus $\lim_{k\to\infty} w_k(t) = w(t)$ exists and $\beta \leq w(t) \leq d$. By the Lebesgue's convergence theorem, from (3.9.9) we have

$$w(t) = \beta + g\big(w(t-\tau)\big) + \int_t^\infty h\big(s, w(s-\sigma)\big)ds, \quad t \geq T,$$

which implies that $w(t)$ is a positive solution of Eq. (3.9.1). □

The linear equation

$$\big(x(t) - px(t-\tau)\big)' + q(t)x(t-\sigma) = 0 \tag{3.9.10}$$

is a special case of Eq. (3.9.1). From the above we have the following conclusions.

Corollary 3.9.1. *If $p \in (0,1)$, $\sigma \geq 0$, and $q \in C([t_0, \infty), R_+)$ and there exists a positive integer n such that every solution of*

$$y'(t) + q(t) \sum_{j=0}^{n} p^j y(t - \sigma - j\tau) = 0 \tag{3.9.11}$$

is oscillatory. Then every solution of Eq. (3.9.10) is oscillatory.

Corollary 3.9.2. *In Eq. (3.9.10), $p \in (0,1)$, $\sigma \geq 0$ and $q \in C([t_0, \infty), R_+)$, $\int_{t_0}^{\infty} q(t)dt < \infty$. Then Eq. (3.9.10) has an eventually positive solution.*

We now consider the nonlinear neutral differential equation

$$\left(x(t) - px^{\alpha}(t - \tau)\right)' + q(t)x^{\beta}(t - \sigma) = 0, \quad t \geq t_0, \tag{3.9.12}$$

where p, $\tau \in (0, \infty)$, $\sigma \in [0, \infty)$, α and β are quotients of odd positive integers, $q \in C([t_0, \infty), R_+)$.

Theorem 3.9.3. *If $\alpha \in (0,1)$, then every solution of Eq. (3.9.12) is oscillatory if and only if*

$$\int_{t_0}^{\infty} q(t)dt = \infty. \tag{3.9.13}$$

Proof: *Necessity.* Assume $\int_{t_0}^{\infty} q(t)dt < \infty$. Then it is not difficult to find that all conditions of Theorem 3.9.2 hold and hence Eq. (3.9.12) has an eventually positive solution.

Sufficiency. Assume (3.9.13) holds. We shall show that all assumptions of Theorem 3.9.1 hold. In fact, it is easy to see that (i) and (ii) in Theorem 3.9.1 are satisfied for Eq. (3.9.12). Now check (iii) in Theorem 3.9.1. For $\alpha \in (0,1)$ there exists a positive integer n such that $\alpha^n \beta < 1$. We show that the delay differential inequality

$$y'(t)+q(t)[y(t-\sigma)+p(y(t-\sigma-\tau)+p(\cdots+py^{\alpha}(t-\sigma-n\tau))^{\alpha}\ldots)^{\alpha}]^{\beta} \leq 0, \tag{3.9.14}$$

has no eventually positive solutions. Otherwise, let $y(t)$ be an eventually positive solution of (3.9.14). By the decreasing nature of y we have

$$y'(t) + q(t)p^{\frac{\beta(1-\alpha^n)}{1-\alpha}} y^{\alpha^n \beta}(t) \leq 0.$$

Thus

$$y'(t)y^{-\alpha^n \beta}(t) + q(t)p^{\frac{\beta(1-\alpha^n)}{1-\alpha}} \leq 0.$$

Integrating the above inequality from t to ∞ we have

$$\lim_{t\to\infty} y^{1-\alpha^n \beta}(t) = -\infty,$$

contradicting the positivity of y. Similarly, we can show that

$$y'(t) + q(t)[y(t-\sigma) + p(y(t-\sigma-\tau) + p(\cdots + py^{\alpha}(t-\sigma-n\tau))^{\alpha} \ldots)^{\alpha}]^{\beta} \geq 0$$

has no eventually negative solutions. □

Remark 3.9.1. If $\alpha = 1$ and $p = 1$, then (3.9.13) is sufficient but not necessary for every solution of Eq. (3.9.12) to be oscillatory. Thus condition $\alpha \in (0,1)$ is important for the validity of Theorem 3.9.3.

In a similar way, we can prove the following result.

Theorem 3.9.4. *If $\alpha = 1$ and $p,\ \beta \in (0,1)$, then every solution of Eq. (3.9.12) is oscillatory if and only if (3.9.13) holds.*

Remark 3.9.1. $\beta \in (0,1)$ is essential for the validity of Theorem 3.9.4. In fact, Theorem 3.9.4 may not be true if $\beta = 1$. We see this fact from the linear equation with the constant parameters of the form

$$\big(x(t) - p(t-\tau)\big)' + qx(t-\sigma) = 0 \tag{3.9.15}$$

where $p \in (0,1)$, $q, \tau \in (0,\infty)$, $\sigma \in (0,\infty)$. (3.9.13) holds for (3.9.15) but Eq. (3.9.15) may have positive solutions if p, q, r, σ satisfy some conditions, see Section 3.1.

Example 3.9.1. Consider

$$\big(x(t) - x^{1/3}(t-\pi)\big)' + (1 + 3\sin^2 t)x^{1/3}(t - \tfrac{\pi}{2}) = 0, \quad t \geq \pi. \tag{3.9.16}$$

It is easy to see that the conditions of Theorem 3.9.3 are satisfied. According to Theorem 3.9.3 every solution of Eq. (3.9.16) oscillates. In fact, $x(t) = \sin^3 t$ is such a solution.

Example 3.9.2. Consider

$$\big(x(t) - 3x^{1/3}(t-1)\big)' + \frac{(2t-1)(2t^2-2t+1)}{t(t-1)^5}\, x(t) = 0. \tag{3.9.17}$$

For this equation, (3.9.13) is not satisfied. According to Theorem 3.9.3, Eq. (3.9.17) has nonoscillatory solutions. In fact, $x(t) = (1 - \frac{1}{t})^3$ is such a solution of (3.9.17).

3.10. Forced Equations

We consider the forced nonlinear neutral equation

$$\left(x(t) - \sum_{i=1}^{m} p_i(t)x(h_i(t))\right)' + \sum_{j=1}^{n} f_j\big(t, x(g_j(t))\big) = Q(t), \quad t \geq t_0 \tag{3.10.1}$$

where $p_i, h_i, g_j,\ Q \in C([t_0,\infty), R)$, $\lim_{t\to\infty} h_i(t) = \infty$, $\lim_{t\to\infty} g_j(t) = \infty$ $i \in I_m = \{1,2,\ldots,m\}$, $j \in I_n = \{1,2,\ldots,n\}$.

In the following we will present some results for the existence of nonoscillatory solutions of (3.10.1).

Firstly we give a result for the case that p_i, $i = 1,\ldots,m$, do not change signs.

Theorem 3.10.1. *Assume that*

(i) $|f_j(t,x)| \leq |f_j(t,y)|$ *as* $|x| \leq |y|$, *and for each closed interval* $I = [c,\bar{c}]$, *where* $0 < c < \bar{c}$, *we have*

$$|f_j(t,x) - f_j(t,y)| \leq L_j(t)|x-y| \quad \text{when} \quad x, y \in I,$$

where $L_j(t) \in C([t_0,\infty), R_+)$ *are dependent on the interval* I *and* $\int_{t_0}^{\infty} L_j(t)dt < \infty$, $j \in I_n$;

(ii) $\int_{t_0}^{\infty} |Q(t)|dt < \infty$;

(iii) *there exists an* $r \in (0,1)$ *such that either*

(iii)$_1$ $p_i(t) \geq 0, \quad i \in I_m$, and $\sum_{i=1}^{m} p_i(t) \leq 1 - r, \quad t \geq t_0$ or

(iii)$_2$ $p_i(t) \leq 0, \quad i \in I_m$, and $-\sum_{i=1}^{m} p_i(t) \leq 1 - r, \quad t \geq t_0$;

(iv) $\sum_{j=1}^{n} \int_{t_0}^{\infty} |f_j(t,d)|dt < \infty$ *for some* $d \neq 0$.

Then Eq. (3.10.1) has a bounded nonoscillatory solution $x(t)$ with $\liminf_{t \to \infty} |x(t)| > 0$.

Proof: We denote by BC the space of all bounded continuous functions defined on $[t_0, \infty)$. Define a distance in BC by $\|x - y\| = \sup_{t \geq t_0} |x(t) - y(t)|$, for $x, y \in BC$. Let $\Omega = \{x \in BC : \; c \leq x(t) \leq |d|, \quad t \geq t_0\}$ where $0 < c < r|d|$. It is easy to see that Ω is a complete metric space. Define a mapping in Ω as follows:

$$(Tx)(t) = \begin{cases} c_1 + \sum_{i=1}^{m} p_i(t) x\big(h_i(t)\big) \\ \quad + \sum_{j=1}^{n} \int_t^{\infty} f_j\big(s, x(g_j(s))\big) ds - \int_t^{\infty} Q(s) ds, & t \geq T, \\ (Tx)(T), & t_0 \leq t < T, \end{cases} \tag{3.10.2}$$

where c_1 and $T \geq t_0$ satisfy the following conditions: when (iii)$_1$ holds, $c < c_1 < r|d|$ and T is sufficiently large such that $h_i(t) \geq t_0, \; i \in I_m, \; g_j(t) \geq t_0, \; j \in I_m$, for $t \geq T$,

$$\sum_{j=1}^{n} \int_T^{\infty} |f_j(s,d)| ds + \int_T^{\infty} |Q(s)| ds \leq \min\{c_1 - c, r|d| - c_1\}, \tag{3.10.3}$$

and

$$\sum_{j=1}^{n} \int_T^{\infty} L_j(s) ds \leq \tfrac{r}{2}\,; \tag{3.10.4}$$

when (iii)$_2$ holds, $c + (1-r)|d| < c_1 < \frac{1}{2}(c + (2-r)|d|)$, (3.10.4) holds, and

$$\sum_{j=1}^{n} \int_T^{\infty} |f_j(s,d)| ds + \int_T^{\infty} |Q(s)| ds \leq c_1 - c - (1-r)|d|. \tag{3.10.5}$$

It is easy to show that $c \le (\mathcal{T}x)(t) \le |d|$ for $t \ge t_0$, i.e., $\mathcal{T}\Omega \subset \Omega$. We shall show that $\mathcal{T}$ is a contraction on Ω. In fact, for $x, y \in \Omega$ and $t \ge T$

$$
\begin{aligned}
|(\mathcal{T}x)(t) - (\mathcal{T}y)(t)| &= \Bigg| \sum_{i=1}^{m} p_i(t)\big(x(h_i(t)) - y(h_i(t))\big) \\
&\quad + \sum_{j=1}^{n} \int_T^\infty \big(f_j(s, x(g_j(s)))\big) - f_j\big(s, y(g_j(s))\big) ds \Bigg| \\
&\le \left(\sum_{i=1}^{m} |p_i(t)| \right) \sup_{t \ge t_0} |x(t) - y(t)| \\
&\quad + \sum_{j=1}^{n} \int_T^\infty L_j(s) |x(g_j(s)) - y(g_j(s))| ds \\
&\le \left(\sum_{i=1}^{m} |p_i(t)| + \sum_{j=1}^{n} \int_T^\infty L_j(s) ds \right) \|x - y\| \\
&\le \left((1 - r) + \tfrac{r}{2}\right) \|x - y\| \\
&= \left(1 - \tfrac{r}{2}\right) \|x - y\|.
\end{aligned}
$$

Thus

$$
\|\mathcal{T}x - \mathcal{T}y\| \le \left(1 - \tfrac{r}{2}\right) \|x - y\|.
$$

By the contraction mapping theorem $\mathcal{T}$ has a fixed point x^* in Ω. Obviously, x^* is a bounded nonoscillatory solution of (3.10.1) and $\liminf_{t\to\infty} |x^*(t)| \ge c > 0$. □

The following result is for the case where p_i, $i = 1, \ldots, m$, are oscillatory functions.

Theorem 3.10.2. *Assume that (i), (ii), (iii)$_2$ in Theorem 3.10.1 hold and* $x f_j(t, x) \ge 0$ *for* $x \ne 0$, $j \in I_n$. *Then condition (iv) in Theorem 3.10.1 is a necessary and sufficient condition for Eq. (3.10.1) to have a bounded nonoscillatory solution $x(t)$ satisfying* $\liminf_{t\to\infty} |x(t)| > 0$.

Proof: Sufficiency follows from Theorem 3.10.1. We now prove the necessity.

Without loss of generality, assume that $x(t) \geq d > 0, \quad t \geq t_0$. Set

$$z(t) = x(t) - \sum_{i=1}^{m} p_i(t)x(h_i(t)).$$

If (iv) is not true, we have

$$z(t) - z(T) \leq \int_T^t Q(s)ds - \sum_{j=1}^{n} \int_T^t f_j(s,d)ds \to -\infty$$

as $t \to \infty$, contradicting the boundedness of x. □

Example 3.10.1. Consider the equation

$$\left(x(t) + \tfrac{1}{3}\sin x(h(t))\right)' + \frac{\cos t}{1+t^2}x\left(g(t)\right) = Q(t) \tag{3.10.6}$$

where $h, g, Q \in C([0,\infty), R)$ satisfying $\lim_{t\to\infty} h(t) = \infty$, $\lim_{t\to\infty} g(t) = \infty$ and $\int_{t_0}^{\infty} |Q(t)|dt < \infty$. By Theorem 3.10.2, (3.10.6) has a bounded nonoscillatory solution $x(t)$ satisfying $\liminf_{t\to\infty}|x(t)| > 0$.

3.11. Notes

Theorem 3.1.1 can be seen from Arino and Gyori [5], also from Gyori and Ladas [76]. The rest of Section 3.1 is from Erbe and Kong [40]. Theorem 3.2.1 is based on Jaros [87]. Theorem 3.2.2 is taken from Chuanxi, Ladas, Zhang and Zhao [25]. Theorem 3.2.3 is adopted from Zhang and Yu [211]. Theorem 3.2.4 is based on the results of Zhang and Gopalsamy [204]. Theorems 3.2.5 and 3.2.6 are taken from Zhang and Yu [214]. Theorem 3.2.9 is from Zhang, Yu and Liu. Theorem 3.2.10 is adopted from Gopalsamy, Lalli and Zhang [55], also see Yu [177]. Theorem 3.2.11 is taken from Yu, Wang and Qian [181], also see Wang [166], Kitamura and Kusano [94]. Theorem 3.2.12 is based on Ladas and Sticas [76]. Lemma 3.3.1 is from [25]. Lemma 3.3.2 is obtained by Gyori [75]. Theorem 3.3.1 to Theorem 3.3.3 are taken from Zhang and Yu [212]. Theorem 3.3.4 to 3.3.6 are adopted from Erbe and Kong [38]. Condition (3.3.54) is new. Theorem 3.3.7 and Theorem 3.3.8 are taken from Zhang and Gopalsamy [204]. The content of Section 3.4 is based on Lalli and Zhang [117] and Gopalsamy and Zhang [56]. Lemma 3.5.2 is from Gyori and Ladas [76]. Theorem 3.5.1 is adopted from Lalli and Zhang [116]. Theorem 3.5.4 is new. Theorem 3.5.5 is based on Erbe and Zhang [44]. Theorem 3.6.1 and 3.6.2 are from Jaros and Kusano [89]. Theorem 3.6.3 is adopted from Wang [165]. Theorem 3.7.1 is from Yu and Wang [179]. Theorem 3.7.2 is a modification of Yu [176]. Theorem 3.7.3 is by Gopalsamy and Zhang [203]. Theorem 3.8.1 is taken from Chuanxi and Ladas [22]. The content of Section 3.9 is taken from Zhang and Yu [211]. Theorem 3.10.1 is adopted from Lu Wudu [132].

4

Oscillation and Nonoscillation of Second Order Differential Equations with Deviating Arguments

4.0. Introduction

In Section 4.1 we introduce the linearized oscillation results for a certain class of second order nonlinear delay differential equations. In Section 4.2 we present results on the existence of oscillatory solutions of the delay differential equation. In Section 4.3 we introduce Sturm comparison theorems which are very useful for the oscillation theory as well as the boundary value problems. In Section 4.4 we establish some oscillation criteria for second order neutral differential equations. By way of these criteria the oscillation problem for neutral differential equations can be reduced to the same problem for the corresponding delay equations and sometimes to the corresponding ordinary differential equation. In Section 4.5 we introduce the classification of nonoscillatory solutions for second order nonlinear neutral differential equations, various existence results of nonoscillatory solutions of different types are given. In Section 4.6, we present several bounded oscillation criteria for the second order neutral differential equation of unstable type. Section 4.7 deals with the forced oscillation. In Section 4.8 equations with nonlinear neutral terms are investigated. In the last section, advanced type equations are investigated, also a relation between the oscillation of equation of the advanced type and the oscillation of the corresponding ordinary differential equation is established.

4.1. Linearized Oscillation

We consider the second order nonlinear delay differential equations

$$x''(t) + A(t)x'(t) + B(t)f\big(x(t-r)\big) = 0, \quad t \geq 0 \tag{4.1.1}$$

and

$$x''(t) - A(t)x'(t) + B(t)f\big(x(t-r)\big) = 0, \quad t \geq 0 \tag{4.1.2}$$

where

$$\begin{gathered} r > 0, \quad A, B \in C(R_+, (0,\infty)), \quad f \in C(R,R) \\ \lim_{t\to\infty} A(t) = a \in (0,\infty), \quad \lim_{t\to\infty} B(t) = b \in (0,\infty). \end{gathered} \tag{4.1.3}$$

The sunflowing equation

$$y''(t) + \frac{a}{r}\, y'(t) + \frac{b}{r}\, \sin y(t-r) = 0, \quad t \geq 0 \tag{4.1.4}$$

is a special case of (4.1.1).

Under some assumptions the following equations are called the linearized limiting equations of (4.1.1) and (4.1.2), respectively:

$$x''(t) + ax'(t) + bx(t-r) = 0, \quad t \geq 0 \tag{4.1.5}$$

and

$$x''(t) - ax'(t) + bx(t-r) = 0, \quad t \geq 0 \tag{4.1.6}$$

respectively.

We shall establish the relations between the oscillation of (4.1.1), (4.1.2) and that of their linearized limiting equations, (4.1.5), (4.1.6), respectively.

Theorem 4.1.1. *Assume that*

(i) $uf(u) > 0$ *for* $u \neq 0$, $|u| \leq H$, *where* $H \in (0, \infty)$ *and* $\lim_{u \to 0} \frac{f(u)}{u} = 1$,

(ii) *The characteristic equation of Eq. (4.1.5)*

$$f(\lambda) = \lambda^2 + a\lambda + be^{-\lambda r} = 0 \tag{4.1.7}$$

has no negative roots.

Then every solution of Eq. (4.1.1) whose graph lies eventually in the strip $R_+ \times [-H, H]$ *is oscillatory.*

Theorem 4.1.2. *Assume that*

(i) $uf(u) > 0$ *as* $u \neq 0$, $\lim_{|u| \to \infty} \frac{f(u)}{u} = 1$,

(ii) *The characteristic equation of (4.1.6)*

$$\lambda^2 - a\lambda + be^{-\lambda r} = 0 \tag{4.1.8}$$

has no positive roots.

Then every solution of Eq. (4.1.2) is oscillatory.

The methods of the proofs of Theorems 4.1.1 and 4.1.2 are similar. So we only show the proof of Theorem 4.1.1 in the following. We first prove some Lemmas which will be used in the proofs of Theorems 4.1.1 and 4.1.2.

Lemma 4.1.1. *Assume that* a, b, $r \in (0, \infty)$ *and every solution of Eq. (4.1.5) is oscillatory. Then there exists an* $\varepsilon \in (0, b)$ *such that every solution of the equation*

$$z''(t) + (a + \varepsilon)z'(t) + (b - \varepsilon)z(t - r) = 0 \tag{4.1.9}$$

is oscillatory also.

Proof: From Theorem 6.0.1 we will see that (4.1.5) is oscillatory is equivalent to the statement (4.1.7) has no real roots. It is easy to see that $\lambda \geq 0$ does not satisfy (4.1.7). Therefore, is oscillatory is equivalent to (4.1.7) having no negative roots. Since $f(0) = b > 0$, $f(-\infty) = \infty$, and (4.1.5) is oscillatory, we have

$$m = \min\{f(\lambda) : \lambda \leq 0\} > 0,$$

and hence

$$f(\lambda) = \lambda^2 + a\lambda + be^{-\lambda r} \geq m \quad \text{for} \quad \lambda \leq 0.$$

From the fact that

$$\lim_{\lambda \to -\infty} \left[\lambda^2 + (a + \frac{b}{2})\lambda + \frac{b}{2}\, e^{-\lambda r} \right] = \infty$$

we know that there exists a $\lambda_0 < 0$ such that for $\lambda \leq \lambda_0$

$$\lambda^2 + \left(a + \frac{b}{2} \right) \lambda + \frac{1}{2}\, be^{-\lambda r} \geq \frac{m}{2}.$$

Set

$$\varepsilon = \min \left\{ \frac{b}{2},\; me^{\lambda_0 r}/2(1 - \lambda_0 e^{\lambda_0 r}) \right\}.$$

Then for $\lambda \leq \lambda_0$

$$\lambda^2 + (a + \varepsilon)\lambda + (b - \varepsilon)e^{-\lambda r} \geq \lambda^2 + \left(a + \frac{b}{2} \right) \lambda + \frac{b}{2}\, be^{-\lambda r} \geq \frac{m}{2}\,.$$

For $\lambda \in (\lambda_0, 0]$

$$\begin{aligned} \lambda^2 + (a + \varepsilon)\lambda + (b - \varepsilon)e^{-\lambda r} &= f(\lambda) - \varepsilon(e^{-\lambda r} - \lambda) \\ &\geq m - \frac{me^{\lambda_0 r}}{2(1 - \lambda_0 e^{\lambda_0 r})}\,(e^{-\lambda_0 r} - \lambda_0) = \frac{m}{2}\,. \end{aligned}$$

Therefore,

$$\lambda^2 + (a + \varepsilon)\lambda + (b - \varepsilon)e^{-\lambda r} = 0$$

has no negative roots. Consequently every solution of (4.1.9) is oscillatory.

Lemma 4.1.2. *Assume that A, $B \in C(R_+, (0, \infty))$, $f \in C(R, R)$, $uf(u) > 0$ as $u \neq 0$ and $|u| \leq H$ where $H \in (0, \infty)$, and f is nondecreasing in $[-H, H]$. If*

$$y(t) \geq \int_t^\infty \int_T^s B(u) f\big(y(u - r)\big) \exp\Big(- \int_u^s A(v) dv \Big) du\, ds, \qquad t \geq T \quad (4.1.10)$$

has a positive solution $y(t) : [T-r,\infty) \to (0,H]$, then

$$z(t) = \int_t^{\infty} \int_T^{s} B(u)f(z(u-r))\exp\left(-\int_u^{s} A(v)dv \right) du\, ds \tag{4.1.11}$$

has a positive solution $z(t)$ on $[T-r,\infty)$ and

$$0 < z(t) \le y(t). \tag{4.1.12}$$

Proof: Let $C([T-r,\infty),R_+)$ be the space of all continuous and nonnegative functions on $[T-r,\infty)$. Define a set X by

$$X = \{w \in C([T-r,\infty),R_+) : \ 0 \le w(t) \le 1, \quad t \ge T-r\}. \tag{4.1.13}$$

Define an operator $\mathcal{T}$ on X by

$$(\mathcal{T}w)(t) = \begin{cases} \frac{1}{y(t)} \int_0^{\infty} \int_T^{s} B(u)f\big(y(u-r)w(u-r)\big) & \\ \qquad \times \exp\big(-\int_u^{s} A(v)dv\big) du\, ds, & t \ge T \\ \frac{t}{T}\,(\mathcal{T}w)(T) + (1-\frac{t}{T}), & T-r \le t \le T. \end{cases} \tag{4.1.14}$$

It is easy to see that $\mathcal{T}X \subset X$, $\mathcal{T}$ is continuous, and $w_1(t) \le w_2(t)$ implies that $(\mathcal{T}w_1)(t) \le (\mathcal{T}w_2)(t)$.

Define a sequence of functions $\{w_k(t)\}$, $k = 0,1,2,\ldots$ as follows:

$$\begin{aligned} &w_0(t) = 1, \\ &w_k(t) = (\mathcal{T}w_{k-1})(t), \quad k = 1,2,\ldots . \end{aligned} \tag{4.1.15}$$

By induction

$$0 \le w_k(t) \le w_{k-1}(t) \le \cdots \le w_0(t) = 1, \quad t \ge T-r.$$

Then the limit $w(t) = \lim_{k\to\infty} w_k(t)$, $t \ge T-r$ exists and $0 \le w(t) \le 1$, $t \ge$

$T-r$. By the Lebesgue's monotone convergence theorem

$$w(t)=\begin{cases}\frac{1}{y(t)}\int_t^\infty\int_T^s B(u)f\big(y(u-r)w(u-r)\big) \\ \qquad \exp\big(-\int_u^s A(v)dv\big)duds, & t\ge T \\ \frac{t}{T}\,w(T)+(1-\frac{t}{T}), & T-r\le t\le T.\end{cases} \tag{4.1.16}$$

$w(t)$ is continuous on $[T-r,\infty)$ and $w(t)>0$ for $t\in[T-r,T)$. Then $w(t)>0$ for $t\ge T-r$. Set

$$z(t)=y(t)w(t). \tag{4.1.17}$$

Then $z(t)$ is a positive solution of (4.1.11) and satisfies (4.1.12). □

Proof of Theorem 4.1.1: Assume the contrary. Let $x(t)$ be an eventually positive solution of (4.1.1) with $0<x(t)\le H$, $t\ge T$. Set

$$u(t)=x'(t)\exp\bigg(\int_0^t A(s)ds\bigg).$$

Then

$$u'(t)=-B(t)f\big(x(t-r)\big)\exp\bigg(\int_0^t A(s)ds\bigg)<0,\quad t\ge T_1=T+r. \tag{4.1.18}$$

There are two possibilities for u :

(i) $u(t)>0,\quad t\ge T_1$,
(ii) there exists $T_2\ge T_1$ such that $u(t)<0$ for $t\ge T_2$.

For (i), $x'(t)>0$ for $t\ge T_1$. Hence

$$x''(t)<-B(t)f\big(x(t-r)\big)\le -CB(t),\quad t\ge T_1+r$$

where $0<C=\min\{f(x):x(T_1)\le x\le H\}$, which implies that $\lim_{t\to\infty}x(t)=-\infty$, a contradiction.

For the case (ii), $x'(t)<0$ for $t\ge T_2$. Then $\lim_{t\to\infty}x(t)=\ell$ exists. We show that $\ell=0$. In fact, if $\ell>0$

$$B(t)f\big(x(t-r)\big)\to bf(\ell)>0,\quad \text{as}\quad t\to\infty,$$

and hence there exists a $T_3 \geq T_2$ such that

$$B(t)f\big(x(t-r)\big) > \frac{bf(\ell)}{2} \quad \text{and} \quad A(t) \leq a+1, \quad \text{for} \quad t \geq T_3.$$

From (4.1.1) we obtain

$$x''(t) + (a+1)x'(t) + \tfrac{b}{2}\, f(\ell) \leq 0, \quad t \geq T_3$$

which implies that

$$x'(t) + (a+1)x(t) \to -\infty, \quad \text{as} \quad t \to \infty,$$

and hence $x'(t) \to -\infty$, as $t \to \infty$. This is impossible. So $\lim_{t\to\infty} x(t) = 0$. It is easy to see that $\lim_{t\to\infty} x'(t) = 0$ also.

From condition (ii) and Lemma 4.1.1 there exists an $\varepsilon \in (0,b)$ such that

$$\lambda^2 + (a+\varepsilon)\lambda + (b-\varepsilon)e^{-\lambda r} > 0 \quad \text{for} \quad \lambda < 0. \tag{4.1.19}$$

From (4.1.3) and condition (i) there exists a T_4 such that

$$A(t) < a+\varepsilon, \quad \text{and} \quad B(t)\,\frac{f\big(x(t-r)\big)}{x(t-r)} > b-\varepsilon, \quad \text{for} \quad t \geq T_4.$$

Substituting them into (4.1.1) we obtain

$$x''(t) + (a+\varepsilon)x'(t) + (b-\varepsilon)x(t-r) \leq 0, \quad t \geq T_4,$$

and hence

$$[x'(t)e^{(a+\varepsilon)t}]' + (b-\varepsilon)e^{(a+\varepsilon)t}x(t-r) \leq 0, \quad t \geq T_4$$

which leads to the integral inequality

$$x(t) \geq (b-\varepsilon)\int_t^\infty \int_T^s x(u-r)e^{(a+\varepsilon)(u-s)}\,du\,ds, \quad t \geq T_4.$$

By Lemma 4.1.2, the integral equation

$$z(t) = (b-\varepsilon)\int_t^\infty \int_T^s z(u-r)e^{(a+\varepsilon)(u-s)}\,du\,ds, \quad t \geq T_4$$

has a positive solution and $0 < z(t) \leq x(t)$. It is easy to see that

$$z''(t) + (a+\varepsilon)z'(t) + (b-\varepsilon)z(t-r) = 0$$

which contradicts (4.1.9). Similarly we can prove that (4.1.1) has no eventually negative solution with $0 > x(t) \geq -H$. □

The following results are about the existence of nonoscillatory solutions, where condition (4.1.3) is no longer required.

Theorem 4.1.3. *Assume that*

(i) *there exist $a > 0$, $b > 0$ such that $A(t) \geq a$, $B(t) \leq b$, $t \geq 0$;*

(ii) *there exists an $H > 0$ such that $uf(u) > 0$, as $u \in (0, H]$, $f(u) \leq u$ for $u \in [0, H]$, and f is nondecreasing on $[0, H]$;*

(iii) *The characteristic equation (4.1.7) has a real root.*

Then Eq. (4.1.1) has an eventually positive solution $x(t)$ lying in the strip $R_+ \times (0, H]$ eventually.

Theorem 4.1.4. *Assume that*

(i) *$uf(u) > 0$ as $u \neq 0$;*

(ii) *there exists an $M > 0$ such that $f(u) \leq u$ for $u \geq M$*

(iii) *there exist $a > 0$, $b > 0$ such that $A(t) \geq a$, $B(t) \leq b$, $t \geq 0$; and f is nondecreasing on $[0, \infty)$;*

(iv) *(4.1.8) has a real root.*

Then (4.1.2) has an eventually positive solution.

Proof of Theorem 4.1.3: Let λ_0 be a real root of (4.1.7). Clearly, $\lambda_0 < 0$. Then $y(t) = e^{\lambda_0 t}$ is a solution of (4.1.5). Choose T so large that $y(t) \leq H$ for $t \geq T - r$. From (4.1.5)

$$[y'(t)e^{at}]' + be^{at}y(t-r) = 0, \quad t \geq T.$$

Integrating it twice we obtain

$$y(t) \geq b\int_t^{\infty} e^{-as}\int_T^s e^{au}y(u-r)duds, \quad t \geq T.$$

Then

$$y(t) \geq \int_t^\infty \int_T^s B(u) f(y(u-r)) \exp\Big(- \int_u^s A(v)dv \Big) du\, ds, \quad t \geq T.$$

By Lemma 4.1.2, the equation

$$z(t) = \int_t^\infty \int_T^s B(u) f(z(u-r)) \exp\Big(- \int_u^s A(v)dv \Big) du\, ds, \quad t \geq T$$

has a positive solution $z(t)$ and $0 < z(t) \leq y(t) \leq H$, $t \geq T - r$. Clearly $z(t)$ satisfies

$$z''(t) + A(t)z'(t) + B(t)f(z(t-r)) = 0, \quad t \geq T,$$

i.e., $z(t)$ is a positive solution of (4.1.1) on $[T, \infty)$ and $0 < z(t) \leq H$, $t \geq T$. □

The proof of Theorem 4.1.4 is similar to the above, we omit it here.

4.2. Existence of Oscillatory Solutions

In the oscillation theory of the differential equations with deviating arguments the problem of the existence of oscillatory solutions is a difficult problem.

4.2.1. We consider the second order delay differential equations of the form

$$x''(t) + a(t)x(t) + h(t, x(g_1(t)), \ldots, x(g_n(t))) = 0 \tag{4.2.1}$$

compared with the equation

$$y''(t) + a(t)y(t) = 0 \tag{4.2.2}$$

where $a \in C^1(R_+, R)$, $g_i \in C(R_+, R)$, $t - \tau \leq g_i(t) \leq t$, $t \geq 0$, $i = 1, 2, \ldots, n$, $\tau > 0$, and $h \in C(R_+ \times R^n, R)$.

Theorem 4.2.1. *Assume that*

(i) *there exist two constants A and B such that $0 < A \leq a(t) \leq B$, $t \in R_+$, $a'(t)$ is of a constant sign;*

(ii) $|h(t, u_1, u_2, \ldots, u_n)| \leq b(t)F(|u_1|, |u_2|, \ldots, |u_n|)$

where $b \in C(R_+, R_+)$, $F \in C(R_+^n, R_+)$, *and* $F > 0$ *when* $u_i > 0$, $1, 2, \ldots, n$;
(iii) *if* $0 < u_i \leq v_i$, $i = 1, 2, \ldots, n$, *then*

$$F(u_1, \ldots, u_n) \leq F(v_1, \ldots, v_n)$$

and

$$\int_0^\infty b(t)dt < \infty, \quad \int_0^\infty \frac{ds}{F(s, s, \ldots, s)} = \infty.$$

Then Eq. (4.2.1) has an oscillatory solution.

Proof: First we show some details of the behavior of the solutions of Eq. (4.2.2).

It is well known that condition (i) implies that every solution of (4.2.2) is oscillatory. Let $y_1(t), y_2(t)$ be solutions of (4.2.2) which satisfy the initial conditions $y_1(0) = 1$, $y_1'(0) = 0$, $y_2(0) = 0$, $y_2'(0) = 1$. Then

$$\begin{vmatrix} y_1(t) & y_2(t) \\ y_1'(t) & y_2'(t) \end{vmatrix} \equiv 1, \quad t \geq 0. \tag{4.2.3}$$

Suppose $y(t)$ is a solution of (4.2.2) and define

$$w_1(t, y, y') = y'^2 + a(t)y^2$$

$$w_2(t, y, y') = \frac{1}{a(t)}y'^2 + y^2.$$

Then along any solution of Eq. (4.2.2)

$$\begin{aligned} \frac{dw_1(t)}{dt} &= a'(t)y^2(t), \\ \frac{dw_2(t)}{dt} &= -\frac{a'(t)}{a^2(t)}[y'(t)]^2. \end{aligned} \tag{4.2.4}$$

If $a'(t) \leq 0$, then $\frac{dw_1}{dt} \leq 0$, $\frac{dw_2}{dt} \geq 0$, and so

$$\begin{aligned} [y_1'(t)]^2 + a(t)y_1^2(t) &\leq [y_1'(0)]^2 + a(0)y_1^2(0) = a(0) \\ [y_2'(t)]^2 + a(t)y_2^2(t) &\leq [y_2'(0)]^2 + a(0)y_2^2(0) = 1, \end{aligned} \tag{4.2.5}$$

and

$$\frac{1}{a(t)}\,[y_1'(t)]^2 + y_1^2(t) \geq \frac{1}{a(0)}[y_1'(0)]^2 + y_1^2(0) = 1,$$
$$\frac{1}{a(t)}\,[y_2'(t)]^2 + y_2^2(t) \geq \frac{1}{a(0)}[y_2'(0)]^2 + y_2^2(0) = \frac{1}{a(0)}. \tag{4.2.6}$$

If $a'(t) \geq 0$, then we have $\frac{dw_1}{dt} \geq 0, \quad \frac{dw_2}{dt} \leq 0$, and so

$$[y_1'(t)]^2 + a(t)y_1^2(t) \geq [y_1'(0)]^2 + a(0)y_1^2(0) = a(0)$$
$$[y_2'(t)]^2 + a(t)y_2^2(t) \geq [y_2'(0)]^2 + a(0)y_2^2(0) = 1; \tag{4.2.7}$$

and

$$\frac{1}{a(t)}[y_1'(t)]^2 + y_1^2(t) \leq \frac{1}{a(0)}[y_1'(0)]^2 + y_1^2(0) = 1,$$
$$\frac{1}{a(t)}[y_2'(t)]^2 + y_2^2(t) \leq \frac{1}{a(0)}[y_2'(0)]^2 + y_2^2(0) = \frac{1}{a(0)}. \tag{4.2.8}$$

From (4.2.5) and (4.2.8) we have

$$y_1^2(t) \leq \tfrac{a(0)}{a(t)} \leq \tfrac{a(0)}{A} \quad \text{or} \quad y_1^2(t) \leq 1,$$
$$[y_1'(t)]^2 \leq a(0) \quad \text{or} \quad [y_1^2(t)]^2 \leq a(t) \leq B,$$
$$y_2^2(t) \leq \tfrac{1}{a(t)} \leq \tfrac{1}{A} \quad \text{or} \quad y_2^2(t) \leq \tfrac{1}{a(0)},$$
$$[y_2'(t)]^2 \leq 1 \quad \text{or} \quad [y_2'(t)]^2 \leq \tfrac{a(t)}{a(0)} \leq \tfrac{B}{a(0)}.$$

Set

$$M_1 = \max\left\{\sqrt{\frac{a(0)}{A}}, 1, \sqrt{\frac{1}{A}}, \sqrt{\frac{1}{a(0)}}\right\},$$

and

$$M_1' = \max\left\{\sqrt{a(0)}, \sqrt{B}, 1, \sqrt{\frac{B}{a(0)}}\right\}.$$

Then

$$|y_1(t)| \le M_1, \qquad |y_2(t)| \le M_1, \tag{4.2.9}$$

$$|y_1'(t)| \le M_1', \qquad |y_2'(t)| \le M_1'. \tag{4.2.10}$$

Since y_1 is oscillatory, there exists a sequence $\{t_n\}$ such that $\lim_{n\to\infty} t_n = \infty$ $y_1'(t_i) = 0$, $i = 1, 2, \ldots$, and $y_1(t_{2m-1}) < 0$, $\quad y_1(t_{2m}) > 0$, $m = 1, 2, \ldots$. From (4.2.6) and (4.2.7) we have $y_1^2(t_i) \ge 1$ or $y_1^2(t_i) \ge \frac{a(0)}{a(t_i)} \ge \frac{a(0)}{B}$, $i = 1, 2, \ldots$. Let $M_2 = \min\{1, \sqrt{\frac{a(0)}{B}}\}$. Then

$$y_1(t_{2m-1}) \le -M_2 < 0, \quad y_1(t_{2m}) \ge M_2 > 0, \quad m = 1, 2, \ldots . \tag{4.2.11}$$

Now we turn to Eq. (4.2.1). Let

$$p = \frac{3M_1}{M_2}, \quad k = (1+p)M_1, \quad \Phi(u) = \int_k^u \frac{1}{F(s, \ldots, s)}\, ds. \tag{4.2.12}$$

Choose $t_0 \ge \tau$ so large that

$$\int_{t_0}^{\infty} b(t)dt < \min\left\{\frac{\Phi(k+1)}{2M_1^2}, \frac{1}{2M_1 F(k+1, \ldots k+1)}\right\}. \tag{4.2.13}$$

Define

$$x(t) = py_1(t) + y_2(t) - \int_{t_0}^{t} [y_1(s)y_2(t) - y_2(s)y_1(t)]h\big(s, x(g_1(s)), \ldots, x(g_n(s))\big)ds, \quad t \ge t_0 \tag{4.2.14}$$

and

$$x(t) = py_1(t_0) + y_2(t_0), \quad \text{for} \quad t_0 - \tau \le t \le t_0. \tag{4.2.15}$$

It is easy to show that $x(t)$ is a solution of Eq. (4.2.1) with the initial condition (4.2.15). Assume $x(t)$ exists on some interval $[t_0, T)$. Then we show that $x(t)$ and $x'(t)$ are bounded on $[t_0, T)$. In fact, in view of condition (ii) and (4.2.9) we have that for $t \in [t_0, T)$

$$|x(t)| \le p|y_1(t)| + |y_2(t)|$$

$$+\int_{t_0}^{t} |y_1(s)y_2(t) - y_2(s)y_1(t)|\,|h(s, x(g_1(s)), \ldots, x(g_n(s)))|ds$$

$$\le pM_1 + M_1 + 2M_1^2 \int_{t_0}^{t} b(s)F\big(|x(g_1(s))|, \ldots, |x(g_n(s))|\big)ds$$

$$= k + 2M_1^2 \int_{t_0}^{t} b(s)H(s)ds \tag{4.2.16}$$

where $H(t) = F\big(|x(g_1(t))|, \ldots, |x(g_n(t))|\big)$. By condition (iii) we have

$$H(t) \le F\Big(k + 2M_1^2 \int_{t_0}^{t} b(s)H(s)ds, \ldots, k + 2M_1^2 \int_{t_0}^{t} b(s)H(s)ds\Big). \tag{4.2.17}$$

From (4.2.17) and (4.2.12) we have

$$\frac{d}{dt}\,\Phi\Big(k + 2M_1^2 \int_{t_0}^{t} b(s)H(s)ds\Big)$$

$$\le \frac{2M_1^2 b(t)H(t)}{F(k + 2M_1^2 \int_{t_0}^{t} b(s)H(s)ds, \ldots, k + 2M_1^2 \int_{t_0}^{t} b(s)H(s)ds)}$$

$$\le 2M_1^2 b(t).$$

Integrating the above inequality from t_0 to t, and noting that $\Phi(k) = 0$ and (4.2.13) holds, we have

$$\Phi\Big(k + 2M_1^2 \int_{t_0}^{t} b(s)H(s)ds\Big) \le 2M_1^2 \int_{t_0}^{t} b(s)ds$$

$$\le 2M_1^2 \int_{t_0}^{\infty} b(s)ds \le \Phi(k+1).$$

Since $\Phi(u)$ is increasing we have

$$k + 2M_1^2 \int_{t_0}^{t} b(s)H(s)ds \le k + 1. \tag{4.2.18}$$

From (4.2.16) and (4.2.18) we obtain $|x(t)| \le k+1$ for $t_0 \le t < T$. On the other hand, for $t_0 - \tau \le t \le t_0$, $|x(t)| = |py_1(t_0) + y_2(t_0)| \le M_1 p + M_1 = k \le k+1$.

Then

$$|x(t)| \le k+1, \quad t_0 - \tau \le t < T, \quad \text{and}$$
$$|x(g_i(t))| \le k+1, \quad t_0 \le t < T. \tag{4.2.19}$$

In view of (4.2.10), (4.2.13), (4.2.14) and (4.2.19) we have

$$\begin{aligned}
|x'(t)| &\le p|y_1'(t)| + |y_2'(t)| \\
&\quad + \int_{t_0}^{t} |y_1(s)y_2'(t) - y_2(s)y'(t)|\, |h(s, x(g_1(s)), \ldots, x(g_n(s)))| ds \\
&\le PM_1' + M_1' + 2M_1M_1' \int_{t_0}^{t} b(s)F(|x(g_1(s))|, \ldots, |x(g_n(s))|) ds \\
&\le PM_1' + M_1' + 2M_1M_1' F(k+1, \ldots, k+1) \int_{t_0}^{t} b(s) ds \\
&\le PM_1' + M' + M_1' \\
&= (P+2)M_1', \quad t \in [t_0, T).
\end{aligned} \tag{4.2.20}$$

Now we claim that for the maximal existence interval of $x(t)$, $[t_0, T)$, we have $T = \infty$. Otherwise, $T < \infty$. Then rewrite (4.2.1) in the form

$$\begin{aligned}
x'(t) &= z(t) \\
z'(t) &= -a(t)x(t) - h(t, x(g_1(t)), \ldots, x(g_n(t))).
\end{aligned} \tag{4.2.21}$$

Define $\Omega = C([-\tau, 0], R^2)$ and $\|r\| = \sup_{-\tau \le \theta \le 0} |r(\theta)|$, for $r = (\varphi, \psi) \in \Omega$. Denote

$$f(t, r) := f(t, \varphi, \psi) = \begin{pmatrix} \psi(0) \\ -a(t)\varphi(0) - h(t, \varphi(g_1(t) - t)), \ldots, \varphi(g_n(t) - t) \end{pmatrix},$$

and

$$u(t) = \begin{pmatrix} x(t) \\ z(t) \end{pmatrix}, \quad u_t = u(t + \theta).$$

Then (4.2.21) becomes

$$u'(t) = f(t, u_t). \tag{4.2.22}$$

By (4.2.19) and (4.2.20) we have

$$\|u_t\| \le k + 1 + (P+2)M_1', \quad t \in [t_0, T) \tag{4.2.23}$$

and

$$\begin{aligned}&| - a(t)\varphi(0) - h\big(t, \varphi(g_1(t) - t), \dots, \varphi(g_n(t) - t)\big)| \\ &\qquad \le a(t)|\varphi(0)| + b(t)f\big(|\varphi(g_1(t) - t)|, \dots, |\varphi(g_n(t) - t)|\big).\end{aligned}$$

Since $a(t)$ and $b(t)$ are bounded on any finite interval, and $|\psi(0)|$, $|\varphi(0)|$, $|\varphi(g_i(t) - t)|$ are bounded if r is bounded, it is easy to see that f maps any bounded closed subset of $[t_0, \infty) \times \Omega$ into a bounded set. It follows that f is completely continuous. By the continuation theorem, for the bounded closed set

$$W = [t_0, T] \times \{r : \|r\| \le k + 1 + (P+2)M_1',\ r \in \Omega\} \subset [t_0, \infty) \times \Omega$$

there is a t_w, such that for $t_w \le t < T$, $\quad (t, u_t) \notin W$, which contradicts (4.2.23). It follows that $T = \infty$.

Finally, we conclude that $x(t)$ is an oscillatory solution. By (4.2.9), (4.2.11), (4.2.12), (4.2.13) and condition (ii) we have

$$\begin{aligned}x(t_{2m-1}) &= py_1(t_{2m-1}) + y_2(t_{2m-1}) \\ &\quad + \int_{t_0}^{t_{2m-1}} [y_1(s)y_2(t) - y_2(s)y_1(t)][-h(s, x(g_1(s)), \dots, x(g_n(s)))]ds \\ &\le -PM_2 + M_1 + 2M_1^2 \int_{t_0}^{t_{2m-1}} b(s)f\big(|x(g_1(s))|, \dots, (x(g_n(s))|\big)ds \\ &\le -3M_1 + M_1 + 2M_1^2 F(k+1, \dots, k+1) \int_{t_0}^{\infty} b(s)ds \\ &\le -3M_1 + M_1 + 2M_1^2 F(k+1, \dots, k+1) \frac{1}{2M_1 F(k+1, \dots, k+1)} \\ &\le -M_1 < 0\end{aligned}$$

$$x(t_{2m}) \ge PM_2 - M_1 - 2M_1^2 \int_{t_0}^{t_{2m}} b(s)F\big(|x(g_1(s))|, \dots, |x(g_n(s))|\big)ds$$

$$\geq 3M_1 - M_1 - M_1 = M_1 > 0.$$

Hence $x(t)$ is oscillatory. □

We consider the second order linear equations of the form

$$x''(t) + a(t)x(t) + \sum_{i=1}^{n} b_i(t)x(t - \tau_i) = f(t). \tag{4.2.24}$$

From Theorem 4.2.1 we have the following result.

Corollary 4.2.1. *Assume that* $0 < A \leq a(t) \leq B$, $a'(t)$ *is of a constant sign,* $\int_0^\infty |b_i(t)|dt < \infty$, $i = 1, 2$, *and* $\int_0^\infty |f(t)|dt < \infty$. *Then Eq. (4.2.24) has an oscillatory solution.*

Proof: Let $F(u_1, \ldots, u_n) = \sum_{i=1}^{n} u_i + 1$, $\quad b(t) = \max\{|b_1(t)|, \ldots, |b_n(t)|, |f(t)|\}$. Then $F(u_1, \ldots, u_n) > 0$ for $u_i > 0$, $\quad i = 1, 2, \ldots, n$, and

$$|h(t, u_1, \ldots, u_n)| = \left|\sum_{i=1}^{n} b_i(t)u_i - f(t)\right|$$

$$\leq b(t)\left(\sum_{i=1}^{n} |u_i| + 1\right) = |b(t)|\, F(|u_1|, \ldots, |u_m|).$$

Therefore, all assumptions of Theorem 4.2.1 are satisfied and hence Eq. (4.2.24) has an oscillatory solution.

Example 4.2.1. Consider

$$x''(t) + ax(t) + \frac{\sin t}{t^2 + 1}\, x(t - \tau) = 0 \tag{4.2.25}$$

where $a > 0$, $\tau > 0$. Since $\int_0^\infty |\frac{\sin t}{t^2+1}|dt < \infty$, by Corollary 3.2.1. Eq. (4.2.25) has an oscillatory solution.

4.2.2. We consider the initial value problem of the equation

$$y''(t) = p(t)y(g(t)) \tag{4.2.26}$$

where $p, g \in C([t_0, \infty), R_+)$, $g(t) \le t$, $\lim_{t\to\infty} g(t) = \infty$, and g is nondecreasing; with

$$\begin{aligned} y(s) = \varphi(s) \quad \text{for} \quad s \in E_{t_0}, \quad y(t_0) = y_0, \\ y'(t_0) = y_0', \quad \varphi \in C(E_{t_0}, R) \end{aligned} \tag{4.2.27}$$

where $E_{t_0} = \{t_0\} \cup \{g(t) : g(t) < t_0,\ t > t_0\}$. We denote the solution of (4.2.26) and (4.2.27) by $y(t, \sigma, y_0')$.

For a given $\varphi \in C(E_{t_0}, R)$, the solution $y(t, \varphi, y_0')$ is a one-parameter family. Define

$$S^{\infty} = \{y : \lim_{t\to\infty} y(t) = \infty\}$$

$$S^{-\infty} = \{y : \lim_{t\to\infty} y(t) = -\infty\}$$

$$S^{0} = \{y : \lim_{t\to\infty} y(t) = 0 \quad \text{monotonically}, \quad y(t) \not\equiv 0\}$$

$$S^{\sim} = \{y : y(t) \quad \text{is oscillatory}\}.$$

and

$$K^{\infty} = \{y_0' \in R :\ y(t, \varphi(t), y_0') \in S^{\infty}\}$$

$$K^{-\infty} = \{y_0' \in R :\ y(t, \varphi(t), y_0') \in S^{-\infty}\}$$

$$K^{0} = \{y_0' \in R :\ y(t, \varphi(t), y_0') \in S^{0}\}$$

$$K^{\sim} = \{y_0' \in R :\ y(t, \varphi(t), y_0') \in S^{\sim}\}$$

We state the following results without proof.

Theorem 4.2.2. *For any initial function $\varphi \in C(E_{t_0}, R)$, the sets $K^{\infty}, K^{-\infty}$ are nonempty and given by nonoverlapping half lines $(-\infty, \alpha)$ and (β, ∞) $(\alpha \le \beta)$, respectively. Denote $F = R - (K^{\infty} \cup K^{-\infty})$. If $\alpha < \beta$, then for every $y_0' \in F = [\alpha, \beta]$ the corresponding solution is unbounded and oscillatory.*

Theorem 4.2.3. *Let $\nu(t) = \sup\{s : g(s) < t\}$, $t \geq t_0$. Assume*

$$\limsup_{t\to\infty} \int_t^{\gamma(t)} (s-t)p(s)ds < \infty, \tag{4.2.28}$$

and

$$\limsup_{t\to\infty} \int_{g(t)}^{t} (s-g(t))p(s)ds > 1. \tag{4.2.29}$$

Then the initial value problem for Eq. (4.2.26) with $y(t) = \varphi(t)$ for $t \in E_{t_0}$ has a unique oscillatory solution.

In fact, (4.2.28) implies that $\alpha = \beta$. (4.2.29) guarantees that the corresponding solution is oscillatory.

Finding the conditions so that $\alpha < \beta$ is an open problem.

4.3. Sturm Comparison Theorems

We consider the second order linear delay equations of the form

$$\left(p(t)x'(t)\right)' + \sum_{i=1}^{n} q_i(t)x\left(\tau_i(t)\right) + \int_{\tau(a)}^{t} k(s,t)x(s)ds = 0, \quad t \in [a,b) \tag{4.3.1}$$

and

$$\left(p(t)y'(t)\right)' + \sum_{i=1}^{n} Q_i(t)y\left(\tau_i(t)\right) + \int_{\tau(a)}^{t} K(s,t)y(s)ds = 0, \quad t \in [a,b) \tag{4.3.2}$$

where $a < b \leq \infty$, $p \in C^1([a,b),(0,\infty))$, q_i, Q_i, $K(s,t)$, $k(s,t)$ are continuous functions over $[a,b)$ and $\{(s,t) : s \leq t, \ a \leq t < b\}$, respectively. Also, $\tau_i(t) \leq t$, where τ_i is continuous, $i = 1, \ldots, n$, and

$$\tau(t) = \min\{\tau_i(s), \ s \geq t, \ i = 1, \ldots, n\}. \tag{4.3.3}$$

For a given initial function $\varphi \in C[\tau(a), a]$, there exists a unique solution $x(t)$ to Eq. (4.3.1) in $[a,b)$ with

$$x(t) = \varphi(t) \quad t \in [\tau(a), a], \quad \text{and} \quad x'(a^+) = \varphi'(a^-). \tag{4.3.4}$$

Let $\psi(t) \in C[\tau(a), a]$ be an initial function for Eq. (4.3.2), and $y(t)$ be the corresponding solution to Eq. (4.3.2) with the initial condition given by $\psi(t)$.

For Eqs. (4.3.1) and (4.3.2) assume the following comparison conditions hold.

(A_1) $Q_i(t) \geq |q_i(t)|, \quad i = 1, \ldots, n,$

(A_2) $K(s,t) \geq |k(s,t)|, \quad s, t \in [a, b),$

(A_3) $\psi(t)/\psi(a) \geq |\varphi(t)/\varphi(a)|, \quad t \in [\tau(a), a].$

If we assume that all of the $q_i(t)$ and $k(s,t)$ are nonnegative, then we can relax (A_3) to get the following conditions:

(B_1) $Q_i(t) \geq q_i(t) \geq 0, \quad i = 1, \ldots, n,$

(B_2) $K(s,t) \geq k(s,t) \geq 0, \quad s, \ t \in [a, b),$

(B_3) $\psi(t)/\psi(a) \geq 0, \quad \psi(t)/\psi(a) \geq \varphi(t)/\varphi(a), \quad t \in [\tau(a), a].$

Likewise, if we assume that $\varphi(t)/\varphi(a) \geq 0$ we can relax (A_1) and (A_2) to get the following conditions:

(C_1) $Q_i(t) \geq 0, \quad Q_i(t) \geq q_i(t), \quad i = 1, \ldots, n,$

(C_2) $K(s,t) \geq 0, \quad K(s,t) \geq k(s,t), \quad s, t \in [a, b),$

(C_3) $\psi(t)/\psi(a) \geq \varphi(t)/\varphi(a) \geq 0, \quad t \in [\tau(a), a].$

In the following we will also use the conditions:

(D_1) $Q_i(t) \geq q_i(t) \geq 0, \quad i = 1, \ldots, n,$

(D_2) $K(s,t) \geq k(s,t) \geq 0, \quad s, t \in [a, b),$

(D_3) $\psi(a) \neq 0$ and $\psi(t)$ does not change sign in $[\tau(a), a]$, $\varphi(a) = 0$, $\varphi'(a) \neq 0$, and $\varphi(t)$ does not change sign in $[\tau(a), a)$.

From ((A_3/B_3/C_3), we obtain

$$\frac{\psi'(a^-)}{\psi(a)} \leq \frac{\varphi'(a^-)}{\varphi(a)} \,. \tag{4.3.5}$$

Conditions (D_1)-(D_3) imply conditions (B_1)-(B_3). In fact, from (D_3), we see that $x'(t)/x(t) \to \infty$ as $t \to a^+$. A new initial point a^- can be chosen so that with the shifting of the initial interval to $[\tau(a^-), a^-]$, the conditions (B_1)-(B_3) now hold.

Theorem 4.3.1. *Assume that one of the sets of comparison conditions $(A_1/B_1/C_1/D_1) - (A_3/B_3/C_3/D_3)$ holds, and that the solution $y(t)$ of Eq. (4.3.2) does not vanish in $[a, b)$. Then, for all $t \in (a, b)$*

$$\frac{y'(t)}{y(t)} \leq \frac{x'(t)}{x(t)} \tag{4.3.6}$$

and

$$\frac{y(t)}{y(a)} \leq \frac{x(t)}{x(a)} \,. \tag{4.3.7}$$

As a consequence, $x(t)$ does not vanish in (a, b).

Proof: Let $r(t) = -p(t)x'(t)/x(t)$, $\quad R(t) = -p(t)y'(t)/y(t)$, then, from (4.3.1) and (4.3.2), we obtain the following Riccati equations:

$$r'(t) = \frac{r^2(t)}{p(t)} + \sum_{i=1}^{n} q_i(t)\, \frac{x(\tau_i(t))}{x(t)} + \int_{\tau(a)}^{t} k(s,t)\, \frac{x(s)}{x(t)}\, ds \tag{4.3.8}$$

and

$$R'(t) = \frac{R^2(t)}{p(t)} + \sum_{i=1}^{n} Q_i(t) \frac{y(\tau_i(t))}{y(t)} + \int_{\tau(a)}^{t} K(s,t) \frac{y(s)}{y(t)}\, ds \tag{4.3.9}$$

According to the first two comparison conditions, $R(t)$ satisfies the differential inequality

$$R'(t) \geq \frac{R^2(t)}{p(t)} + \sum_{i=1}^{n} |q_i(t)| \frac{y(\tau_i(t))}{y(t)} + \int_{\tau(a)}^{t} |k(s,t)|\, \frac{y(s)}{y(t)} ds. \tag{4.3.10}$$

From (4.3.5)

$$R(a^-) \geq r(a^-). \tag{4.3.11}$$

To prove (4.3.6) it suffices to show that $R(t) \geq r(t)$, for all $t \in (a, b)$. To this end, we first assume that (4.3.10) and (4.3.11) hold strictly. Otherwise, we can first establish the required result with $y(t)$ replaced by $y_\varepsilon(t)$ ($\varepsilon > 0$), which satisfies the modified delay equation

$$(p(t)y_\varepsilon'(t))' + \sum_{i=1}^{n} Q_i(t)y_\varepsilon(\tau_i(t)) + \int_{\tau(a)}^{t} K(s,t)y_\varepsilon(s)ds + \varepsilon = 0 \qquad (4.3.12)$$

and the initial condition

$$R_\varepsilon(a^-) = \varepsilon + R(a^-) > r(a^-). \qquad (4.3.13)$$

From (4.3.12), we get the modified Riccati inequality

$$R_\varepsilon'(t) > \frac{R_\varepsilon^2(t)}{p(t)} + \sum_{i=1}^{n} |q_i(t)| \frac{y_\varepsilon(\tau_i(t))}{y_\varepsilon(t)} + \int_{\tau(a)}^{t} |k(s,t)| \frac{y_\varepsilon(s)}{y_\varepsilon(t)} ds. \qquad (4.3.14)$$

The general result then follows from the specialized result by letting $\varepsilon \to 0$.

We shall show that under the hypothesis of strict inequalities, the conclusion (4.3.6) actually holds with a strict inequality sign. Suppose the contrary, let $\sigma > a$ be the first point at which $R(\sigma) = r(\sigma)$. Then for all $t \in [a,\sigma)$, $R(t) > r(t)$. It follows that

$$\frac{y'(t)}{y(t)} < \frac{x'(t)}{x(t)}, \quad \text{for all} \quad t \in [a,\sigma). \qquad (4.3.15)$$

Integrating it over the interval $[s,t] \subset [a,\sigma]$, we obtain

$$\frac{y(s)}{y(t)} > \left|\frac{x(s)}{x(t)}\right| \qquad (4.3.16)$$

for all $s < t \in [a,\sigma]$. Combining this with the third comparison condition, we see that (4.3.16) actually holds for all $s < t \in [\tau(a),\sigma]$. Hence the right hand side of (4.3.10) is not smaller than that of (4.3.8) for all $t \in [a,\sigma]$. The various sign requirements in the comparison conditions are imposed to ensure that the product of the coefficient and the fraction in each of the terms of (4.3.8) and (4.3.9) do inherit the desirable comparison property. We have now concluded that $R'(t) > r'(t)$ for all $t \in [a,\sigma]$. In particular, $R'(\sigma) > r'(\sigma)$. This contradicts the assumption that $R(\sigma) = r(\sigma)$ and $R(t) > r(t)$ for $t < \sigma$ and completes the proof of the theorem. □

Associate (4.3.2) with the delay equation

$$\left(p(t)z'(t)\right)' + \sum_{i=1}^{n} Q_i(t)z(\overline{\tau}_i(t)) + \int_{\tau(a)}^{t} K(s,t)z(s)ds = 0 \tag{4.3.17}$$

where

$$\overline{\tau}_i(t) \leq \tau_i(t) \leq t, \quad i = 1, \ldots, n. \tag{4.3.18}$$

We assume that the initial condition

$$z'(a^+) = \psi'(a^-) \tag{4.3.19}$$

holds. Furthermore, assume that

$$Q_i(t), \quad K(s,t) \geq 0 \tag{4.3.20}$$

and

$$\psi'(t) < 0 \quad \text{or} \quad \geq 0 \quad \text{in} \quad [\tau(a), a]. \tag{4.3.21}$$

Theorem 4.3.2. *Let $y(t)$ and $z(t)$ be, respectively, positive solutions of (4.3.2) and (4.3.17) in $[a,b)$, with the same initial value given by $\psi(t)$. Suppose that (4.3.18) and (4.3.20) hold and that $\psi'(t) \leq 0$ in $[\tau(a), a]$. Then for all $t \in [a,b)$*

$$\frac{z'(t)}{z(t)} \leq \frac{y'(t)}{y(t)} \tag{4.3.22}$$

and

$$z(t) \leq y(t). \tag{4.3.23}$$

On the other hand, if $\psi'(t) \geq 0$ in $[\tau(a), a]$ and both $z'(t)$ and $y'(t)$ are nonnegative in $[a,b]$, then the reverse inequalities hold in (4.3.22) and (4.3.23).

Proof: The function $S(t) = -p(t)z'(t)/z(t)$ satisfies the Riccati equation

$$S'(t) = \frac{S^2(t)}{p(t)} + \sum_{i=1}^{n} Q_i(t)\,\frac{z(\overline{\tau}_i(t))}{z(t)} + \int_{\tau(a)}^{t} K(s,t)\frac{z(s)}{z(t)}ds. \tag{4.3.24}$$

Since $z(t)$ is decreasing (increasing), $z(\overline{\tau}_i(t)) \geq (\leq) z(\tau_i(t))$. Hence $S(t)$ satisfies the Riccati inequality

$$S'(t) \geq (\leq) \frac{S^2(t)}{p(t)} + \sum_{i=1}^{n} Q_i(t) \frac{z(\tau_i(t))}{z(t)} + \int_{\tau(a)}^{t} K(s,t) \frac{z(s)}{z(t)} ds.$$

The comparison between $S(t)$ and $R(t)$ can then be carried out as is done in the proof of Theorem 4.3.1 for $R(t)$ and $r(t)$.

Theorem 4.3.2 asserts that for a decreasing solution, a "shorter" memory slows down oscillation, whereas for an increasing solution, it speeds up oscillation (in the sense that the solution reaches its maximum or rebounces faster, and not that the solution becomes zero faster). We now apply Theorem 4.3.2 to a oscillation problem.

We consider delay equations of the form

$$\big(p(t)x'(t)\big)' + \sum_{i=1}^{n} q_i(t) x\big(\tau_i(t)\big) + \int_{\tau(t)}^{t} k(s,t) x(s) ds = 0, \quad t \geq a \tag{4.3.25}$$

with the assumptions that

$$q_i(t) \geq 0, \quad k(s,t) \geq 0 \tag{4.3.26}$$

and

$$\tau(t) = \min_{t \to \infty} \{\tau_i(t) : \ i = 0, 1, \ldots, n\} \to \infty, \quad t \to \infty. \tag{4.3.27}$$

We compare (4.3.25) with another delay equation

$$\big(p(t)y'(t)\big)' + \sum_{i=1}^{n} Q_i(t) y\big(\overline{\tau}_i(t)\big) + \int_{\overline{\tau}(t)}^{t} K(s,t) ds = 0, \quad t \geq a. \tag{4.3.28}$$

The following results can be easily deduced from Theorems 4.3.1 and 4.3.2.

Theorem 4.3.3. *Suppose that for sufficiently large s, t*

$$Q_i(t) \geq q_i(t) \geq 0, \quad i = 1, \ldots, n, \tag{4.3.29}$$

$$K(s,t) \geq k(s,t) \geq 0, \quad s < t \tag{4.3.30}$$

$$\overline{\tau}_i(t) \geq \tau_i(t), \quad i = 0, \ldots, n. \tag{4.3.31}$$

If Eq. (4.3.25) is oscillatory, so is Eq. (4.3.28).

Theorem 4.3.4. *Suppose that $p(t) \equiv 1$ and (4.3.26), (4.3.27) hold, and the ordinary differential equation*

$$y''(t) + \theta(t)y(t) = 0 \tag{4.3.32}$$

is oscillatory. Then so is the delay equation (4.3.25), where

$$\theta(t) = \frac{1}{t}\left[\sum_{i=1}^{n} q_i(t)\tau_i(t) + \int_{\tau(t)}^{t} sk(s,t)dt\right]. \tag{4.3.33}$$

Proof: Suppose the contrary, and (4.3.25) has an eventually positive solution $x(t)$. It is easy to see that $x'(t) > 0$ eventually.

Suppose that $\lim_{t\to\infty} x'(t) = \mu > 0$ and hence $x(t) \sim \mu t,\ t \to \infty$. From (4.3.25) $x''(t) \sim \mu t\theta(t)$. The convergence of $x'(t)$ implies the integrability of $x''(t)$. Hence

$$\int_a^{\infty} t\theta(t)dt < \infty. \tag{4.3.34}$$

But is is well known that (4.3.34) implies the nonoscillation of (4.3.32) which contradicts our hypothesis. Therefore, $\lim_{t\to\infty} x'(t) = 0$. Let us take a point on the solution curve, $(t_1, x(t_1))$. Denote by L the straight line joining this point and the origin $(0,0)$. Concavity implies that L may intersect the solution curve in at most two points. In the case that there are two points of intersection, t_1 and t_2, without loss of generality we may assume $t_1 < t_2$. The part of the straight line between these two points lies below the curve. Let $t_3 \geq t_2$ be so large that $\tau(t) > t_1$ for all $t > t_3$. For any $t > t_3$, the line joining $(0,0)$ and $(t, x(t))$ lies below L. Hence the part of this line between $\tau(t)$ and t lies below the solution curve. This implies that

$$x(\tau(t)) \geq \frac{\tau(t)}{t}\, x(t), \quad \text{for all} \quad t > t_3. \tag{4.3.35}$$

The case in which L is tangent to the solution curve at t_1 can be treated as the degenerate case with $t_1 = t_2$. Now suppose that $(t_1, x(t_1))$ is the only point of intersection. If L lies below the solution curve in the interval $[a, t_1]$, then (4.3.35) actually holds for all $t > a$. Finally, let us note that the remaining case is void since concavity and $\lim_{t\to\infty} x'(t) = 0$ dictate that the curve must meet L again.

We rewrite (4.3.25) in the form

$$\big(p(t)x'(t)\big)' + \left(\sum_{i=1}^{n} q_i(t)\,\frac{x\big(\tau_i(t)\big)}{x(t)} + \int_{\tau_0(t)}^{t} k(s,t)\,\frac{x(s)}{x(t)}\,ds\right)x(t) = 0 \qquad (4.3.36)$$

and regard it as a linear equation without delay. By (4.3.35) the coefficient is larger than $\theta(t)$.

Therefore, from the classical Sturmian theory (4.3.36) or equivalently (4.3.25) oscillates faster than (4.3.32), so we have a contradiction. □

4.4. Oscillation Criteria

4.4.3. Nonlinear case

In this section we deal with the oscillatory behavior of solutions of the neutral differential equation

$$\big(x(t) - px(t-\tau)\big)'' + q(t)f\big(x(t-\sigma(t)\big) = 0 \qquad (4.4.1)$$

under the assumption

p and τ are positive numbers;

$q,\ \sigma \in C(R_+, R_+),\ \lim_{t\to\infty}\big(t - \sigma(t)\big) = \infty, \quad \sigma(t) > \tau;$

$f \in C(R,R),\ f$ is increasing and $f(-x) = -f(x)$;

$f(xy) \geq f(x)f(y)$ as $xy > 0, \quad f(\infty) = \infty$, and

$\dfrac{f(y)}{y} \to \infty$ or 1 as $y \to 0$.

The purpose of this section is to establish a relation between the oscillation problems of (4.4.1) and a corresponding ordinary differential equation.

The following lemmas will be used to prove the main results.

Lemma 4.4.1. *Assume that $g \in C(R_+, R_+)$, $g(t) \leq t$ and $\lim_{t\to\infty} g(t) = \infty$, $y \in C^2([T,\infty), R)$ and*

$$y(t) > 0, \quad y'(t) > 0, \quad y''(t) \leq 0 \quad \text{on} \quad [T,\infty).$$

Then for each $k \in (0,1)$ *there is a* $T_k \geq T$ *such that*

$$y(g(t)) \geq k\,\frac{g(t)}{t}\,y(t), \quad t \geq T_k. \tag{4.4.2}$$

Proof: It suffices to consider only those t for which $g(t) < t$. Then we have for $t > g(t) \geq T$,

$$y(t) - y(g(t)) \leq y'(y(t))(y - g(t)),$$

by the mean value theorem and the monotone properties of y'. Hence

$$\frac{y(t)}{y(g(t))} \leq 1 + \frac{y'(g(t))}{y(g(t))}\,(t - g(t)), \quad t > g(t) \geq T.$$

Also,

$$y(g(t)) \geq y(T) + y'(g(t))(g(t) - T)$$

so that for any $0 < k < 1$ there is a $T_k \geq T$ with

$$\frac{y(g(t))}{y'(g(t))} \geq kg(t), \quad t \geq T_k.$$

Hence,

$$\frac{y(t)}{y(g(t))} \leq \frac{t + (k-1)g(t)}{kg(t)} \leq \frac{t}{kg(t)}, \quad t \geq T_k.$$

The proof is complete. □

We discuss Eq. (4.4.1) for the cases that $0 < p < 1$, $p = 1$, and $p > 1$, respectively.

Lemma 4.4.2. *Assume that* $p \in (0,1)$ *and (H) holds. If the equation*

$$z''(t) + q(t)f\left(\frac{\lambda(t - \sigma(t))}{t}\,z(t)\right) = 0 \tag{4.4.3}$$

is oscillatory for some $0 < \lambda < 1$, *then the nonoscillatory solutions of Eq. (4.4.1) tend to zero as* $t \to \infty$.

Proof: Without loss of generality let $x(t)$ be an eventually positive solution of (4.4.1) and define

$$z(t) = x(t) - px(t-\tau).$$

From Eq. (4.4.1) we have $z''(t) \leq 0$ for $t \geq T$. If $z'(t) < 0$ eventually, then $\lim_{t\to\infty} z(t) = -\infty$. But $z(t) < 0$ eventually implies that $\lim_{t\to\infty} x(t) = 0$ which contradicts the fact that $\lim_{t\to\infty} z(t) = -\infty$. Therefore, $z'(t) > 0$ for $t \geq T$. There are two possibilities for $z(t)$:

(a) $z(t) > 0$ for $t \geq T_1 \geq T$,
(b) $z(t) < 0$ for $t \geq T_1$.

For case (a), by Lemma 4.4.1, for each $k \in (0,1)$ there is a $T_k \geq T$ such that

$$z\big(t - \sigma(t)\big) \geq \frac{k\big(t - \sigma(t)\big)}{t} z(t), \quad t \geq T_k.$$

Since $0 < z(t) < x(t)$, from Eq. (4.4.1) we have

$$z''(t) + p(t)f\left(\frac{k(t - \sigma(t))}{t} z(t)\right) \leq 0.$$

Using Theorem 7.4 in [86] we see that (4.4.3) has an eventually positive solution. This contradicts the assumption. For the case (b), as mentioned before, we lead to that $\lim_{t\to\infty} x(t) = 0$. □

Theorem 4.4.1. *In addition to the conditions of Lemma 4.4.2, assume further that*

$$\limsup_{t\to\infty} \int_{t-\sigma(t)+\tau}^{t} (u-(t-\sigma(t)+\tau))q(u)du > \begin{cases} p, & \text{if } \lim_{y\to 0} \frac{f(y)}{y} = 1 \\ 0, & \text{if } \lim_{y\to 0} \frac{f(y)}{y} = \infty. \end{cases} \tag{4.4.4}$$

Then every solution of Eq. (4.4.1) is oscillatory.

Proof: As in the proof of Lemma 4.4.2, it is sufficient to show that $z(t) < 0$ for $t \geq T$ is impossible under the assumptions. Suppose that $x(t) > 0$, $z''(t) \leq 0$, $z'(t) > 0$, and $z(t) < 0$ eventually. Then

$$z\big(t - \sigma(t) + \tau\big) > -px\big(t - \sigma(t)\big).$$

Substituting this into Eq. (4.4.1) we have

$$z''(t) - \frac{q(t)}{p} f\big(z(t - \sigma(t) + \tau)\big) \leq 0. \tag{4.4.5}$$

Integrating (4.4.5) from s to t for $t > s$ we have

$$z'(t) - z'(s) - \frac{1}{p} \int_s^t q(u) f\big(z(u - \sigma(u) + \tau)\big) du \leq 0. \tag{4.4.6}$$

Integrating (4.4.6) in s from $t - \sigma(t) + \tau$ to t, we have

$$\begin{aligned} z'(t)\big(\sigma(t) - \tau\big) &- \int_{t-\sigma(t)+\tau}^t dz(s) \\ &- \frac{1}{p} \int_{t-\sigma(t)+\tau}^t [u - (t - \sigma(t) + \tau)] q(u) f\big(z(u - \sigma(u) + \tau)\big) du \leq 0. \end{aligned}$$

Hence for t sufficiently large

$$\begin{aligned} z\big(t - \sigma(t) + \tau\big) &- z(t) \\ &- \frac{1}{p} \int_{t-\sigma(t)+\tau}^t [u - \big(t - \sigma(t) + \tau\big)] q(u) f\big(z(u - \sigma(u) + \tau)\big) du \leq 0. \end{aligned}$$

Dividing the above inequality by $z\big(t - \sigma(t) + \tau\big)$ and noting the negativity of this term, we have

$$\begin{aligned} 1 &- \frac{z(t)}{z\big(t - \sigma(t) + \tau\big)} \\ &- \frac{1}{pz\big(t - \sigma(t) + \tau\big)} \int_{t-\sigma(t)+\tau}^t [u - \big(t - \sigma(t) + \tau\big)] q(u) f\big(z(u - \sigma(u) + \tau)\big) du \geq 0. \end{aligned} \tag{4.4.7}$$

We note that $z(t) < 0$ eventually implies that $\lim_{t \to \infty} z(t) = 0$. From (4.4.7) we have

$$1 > \frac{1}{p}\int_{t-\sigma(t)+\tau}^{t} [u - (t - \sigma(t) + \tau)]q(u)\,\frac{f(z(u - \sigma(u) + \tau))}{z(t - \sigma(t) + \tau)}du$$

which contradicts (4.4.4). □

Lemma 4.4.3. *Assume (H) holds and $p = 1$. Then the nonoscillatory solutions $x(t)$ of Eq. (4.4.1) are bounded provided every solution of the equation*

$$z''(t) + q(t)f\big(Q(t)z(t)\big) = 0 \tag{4.4.8}$$

is oscillatory, where $Q(t) = \frac{1}{3\tau t}\big(t - \sigma(t)\big)^2$.

Proof: Let $x(t)$ be an eventually positive solution of Eq. (4.4.1) and $z(t) = x(t) - x(t-\tau)$. Then $z''(t) \le 0$ for $t \ge t_0$. If $z'(t) < 0$ for $t \ge t_0$, then we have $\lim_{t\to\infty} z(t) = -\infty$. Thus

$$x(t) \le x(t-\tau) \quad \text{for all large} \quad t \tag{4.4.9}$$

which implies that $x(t)$ is bounded, a contradiction. Therefore $z'(t) > 0,\ t \ge t_1 \ge t_0$.

Assume $z(t) > 0,\ t \ge t_2 \ge t_1$. By Lemma 4.4.1, for any $k \in (0,1)$ and $i = 0, 1, 2, \ldots$ there exits $T_i \ge t_0 = T_0$ such that

$$z\big(t - \sigma(t) - i\tau\big) \ge \frac{k\big(t - \sigma(t) - i\tau\big)}{t}\,z(t), \quad t - \sigma(t) \ge T_i. \tag{4.4.10}$$

Since

$$\begin{aligned} x\big(t - \sigma(t)\big) &= \sum_{i=0}^{n-1} z\big(t - \sigma(t) - i\tau\big) + x\big(t - \sigma(t) - n\tau\big) \\ &\ge \sum_{i=0}^{n} z\big(t - \sigma(t) - i\tau\big) \end{aligned}$$

(here $\sum_{i=0}^{-1} = 0$), from Eq. (4.4.1) we have

$$z''(t) + q(t)f\Big(\sum_{i=0}^{n} z(t - \sigma(t) - i\tau)\Big) \le 0.$$

Using (4.4.10) we obtain

$$z''(t) + q(t)f\Big(\frac{k}{t}\sum_{i=0}^{n}(t - \sigma(t) - i\tau)z(t)\Big) \leq 0,$$

i.e.,

$$z''(t) + q(t)f\Big(\frac{k}{t}(n+1)(t - \sigma(t) - \frac{n}{2}\tau)z(t)\Big) \leq 0.$$

Since $n\tau \leq t - \sigma(t) - T_0 < (n+1)\tau$, we have

$$z''(t) + q(t)f\Big(\frac{k}{2\tau t}[(t - \sigma(t))^2 - T_0^2]z(t)\Big) \leq 0.$$

Choose $T \geq T_0$ large enough, then it follows that

$$z''(t) + q(t)f\Big(\frac{1}{3\tau t}(t - \sigma(t))^2 z(t)\Big) \leq 0, \quad t \geq T.$$

Noting that $z(t)$, $z(T)$ are upper and lower solutions of Eq. (4.4.8) respectively, and using a known result in [86], we see there is a solution $x(t)$ of Eq. (4.4.8) satisfying $z(T) \leq x(t) \leq z(t)$, contradicting the fact that Eq. (4.4.8) is oscillatory.

Assume $z(t) < 0$ for $t \geq t_2 \geq t_1$. Then $x(t) < x(t-\tau)$, $t \geq t_2$ which implies that $x(t)$ is bounded. □

Definition 4.4.1. Let E be a subset of R_+. Define

$$\rho_t(E) = \frac{\mu\{E \cap [0,t]\}}{t} \quad \text{and} \quad \rho(E) = \limsup_{t\to\infty} \rho_t(E) \tag{4.4.11}$$

where μ is the Lebesgue measure.

Lemma 4.4.4. *Assume (H) holds and $p > 1$. Then the nonoscillatory solutions $x(t)$ of Eq. (4.4.1) satisfy $x(t) < px(t-\tau)$ eventually provided the following conditions hold:*

i)

$$z''(t) + q(t)f\big(R(t,\lambda)z(t)\big) = 0 \tag{4.4.12}$$

is oscillatory for all $0 < \lambda < 1$, where $R(t,\lambda) = \frac{\lambda}{t}\, p^{\frac{t-\sigma(t)}{\tau}}$;

ii) $$\limsup_{\substack{t\to\infty \\ t\notin E}} p_1^{-t/\tau} \int_0^t (t-u)q(u)f(u-\sigma(u)+\tau)du > 0 \tag{4.4.13}$$

holds for some $p_1 > p$ *and any set* E *with* $\rho(E) = 0$.

Proof: First we claim that if a set $E \subset R_+$ and $\rho(E) = \rho > 0$, then for any $t_0 \in R_+$ and integer n, there exists a $T \in [t_0, t_0+\tau)$ such that the set $\{T+i\tau\}_{i=1}^{\infty}$ intersects E at least n times. If not, there exists a $t_0 \in R_+$ and an integer N, such that $\{T+i\tau\}_{i=1}^{\infty}$ intersects E at most N times for any $T \in [t_0, t_0+\tau)$. This implies that $\mu\{E\} < \infty$. But $\rho(E) = \rho > 0$ means there exists $t_n \to \infty$ such that $\rho_{t_n}(E) \geq \frac{\rho}{2} > 0$. Thus

$$\mu\{E \cap [0, t_n]\} \geq \frac{\rho}{2} t_n \to \infty \quad \text{as} \quad n \to \infty$$

and this is impossible.

Let $x(t)$ be an eventually positive solution of Eq. (4.3.1) and set $z(t) = x(t) - px(t-\tau)$. Then $z''(t) \leq 0$ eventually. There are three possibilities:

$$\text{(a) } z'(t) > 0,\ z(t) > 0, \quad \text{(b) } z'(t) < 0,\ z(t) < 0, \quad \text{(c) } z'(t) > 0,\ z(t) < 0$$

eventually.

(a) Assume $z'(t) > 0,\ z(t) > 0,\ t \geq t_0 \geq 0$. Then (4.4.10) holds and for any $t \in R_{T_0} = \{t;\ t+\sigma(t) \geq T_0\}$, there exists a positive integer n such that

$$T_0 \leq t - \sigma(t) - n\tau < T_0 + \tau.$$

Since

$$\begin{aligned} x(t-\sigma(t)) &= \sum_{i=0}^{n-1} p^i z(t-\sigma(t)-i\tau) + p^n x(t-\sigma(t)-n\tau) \\ &\geq \sum_{i=0}^{n} p^i z(t-\sigma(t)-i\tau), \end{aligned}$$

from Eq. (4.4.1) we have

$$z''(t) + q(t) f\left(\sum_{i=0}^{n} p^i z(t-\sigma(t)-i\tau)\right) \leq 0.$$

In view of (4.4.10) we have

$$z''(t) + q(t)f\left(\frac{k}{t}\sum_{i=0}^{n} p^i(t - \sigma(t) - i\tau)z(t)\right) \leq 0,$$

i.e.,

$$z''(t) + q(t)f\left[\left(\frac{k}{t}(t - \sigma(t))\frac{p^{n+1}}{p-1} - \frac{k\tau}{t}\sum_{i=1}^{n} p^i\right)z(t)\right] \leq 0. \tag{4.4.14}$$

Since

$$\sum_{i=1}^{n} ip^i = \frac{np^{n+2} - (n+1)p^{n+1} + p}{(p-1)^2}.$$

We have

$$\begin{aligned}
&\frac{k}{t}(t - \sigma(t))\frac{p^{n+1}-1}{p-1} - \frac{k\tau}{t}\sum_{i=1}^{n} ip^i \\
&= \frac{k}{(p-1)^2 t}\left[(t-\sigma(t))(p^{n+2} - p^{n+1} - p + 1) - \tau\left(np^{n+2} - (n+1)p^{n+1} + p\right)\right] \\
&= \frac{k}{(p-1)^2 t}[(t - \sigma(t) - n\tau)p^{n+2} - (t - \sigma(t) - (n+1)\tau)p^{n+1} \\
&\quad - (t - \sigma(t) + \tau)p + (t - \sigma(t))] \\
&\geq \frac{k}{(p-1)^2 t}[T_0 p^{n+2} - T_0 p^{n+1} - (t - \sigma(t) + \tau)p + (t - \sigma(t))] \\
&\geq \frac{1}{t}\, p^{n+2} \geq \frac{1}{t}\, p^{\frac{t-\sigma(t)-T_0+\tau}{\tau}} \geq \frac{\lambda}{t}\, p^{\frac{t-\sigma(t)}{\tau}}
\end{aligned} \tag{4.4.15}$$

for some $\lambda \in (0,1)$ if T_0 and t are sufficiently large. Substituting (4.4.15) into (4.4.14) we have

$$z''(t) + q(t)f\left(\frac{\lambda}{t}\, p^{\frac{t-\sigma(t)}{\tau}} z(t)\right) \leq 0.$$

Noting that $z(t)$, $z(T_0)$ are upper and lower solutions of (4.4.12) and by a known result in [86] we see there is a solution $y(t)$ of (4.4.12) satisfying $z(T_0) \leq y(t) \leq z(t)$, contradicting the fact that Eq. (4.4.12) is oscillatory for all $0 < \lambda < 1$.

(b) Assume $z'(t) < 0,\ z(t) < 0,\ t \geq t_0 \geq 0$. Then $z(t) \leq -\ell t,\ t \geq t_0$ for some $\ell > 0$.

We claim $z(t) \geq -p_1^{t/\tau}$ essentially, where $p_1 > p$ is arbitrary, i.e., if $E = \{t : z(t) < -p_1^{t/\tau}\}$. Then $\rho(E) = 0$.

Otherwise, $\rho(E) = \rho > 0$. As in the beginning of the proof, for any n, there exists a $T_1 \in [t_1, t_0 + \tau)$ such that the set $\{T_1 + i\tau\}_{i=1}^{\infty}$ intersects E at least n times. Assume $M = \max_{t_0 \leq t \leq t_0+\tau} x(t)$. Then if n is sufficiently large,

$$x(T_1 + n\tau) \leq p^n x(T_1) + z(T_1 + n\tau) \leq p^n M - p_1^{\frac{T_1+n\tau}{\tau}}$$

$$= p^n M - p_1^{n+\frac{T_1}{\tau}} < 0,$$

contradicting that $x(t) > 0$ eventually.

It is easy to see that condition ii) implies that

$$\int_0^{\infty} q(u) f(u - \sigma(u) + \tau) du = \infty. \tag{4.4.16}$$

Condition (ii) also implies that $z'(t) < -\mu$ for all $\mu > 0$ eventually. For otherwise, there exists a $\mu > 0$ such that $z'(t) \geq -\mu,\ t \geq T_2$. On the other hand, $x(t - \tau) \geq -\frac{1}{p}\, z(t)$, thus

$$z''(t) + q(t) f\Big(-\frac{1}{p} z(t - \sigma(t) + \tau)\Big) \leq 0. \tag{4.4.17}$$

Integrating it from T_2 to t we get

$$z'(t) + \int_{T_2}^{t} q(u) f\Big(-\frac{1}{p} z(u - \sigma(u) + \tau)\Big) du \leq 0.$$

Noting that $z(t - \sigma(t) + \tau) \leq -\ell(t - \sigma(t) + \tau)$ we have

$$z'(t) + \int_{T_2}^{t} q(u) f\Big(\frac{\ell}{p}(u - \sigma(u) + \tau)\Big) du \leq 0,$$

$$f\Big(\frac{\ell}{p}\Big) \int_{T_2}^{t} q(u) f(u - \sigma(u) + \tau) du \leq -z'(t) \leq \mu,$$

which contradicts (4.4.16). Therefore, $z'(t) < -\mu$ for all $\mu > 0$ eventually. From (4.4.17) there is a T_μ such that

$$z(t) + \int_{T_\mu}^{t} (t-u)q(u)f\Big(\frac{\mu}{p}(u-\sigma(u)+\tau)\Big)du \le 0, \quad \text{for} \quad t \ge T_\mu.$$

On $E^C \cap [T_\mu, \infty)$,

$$-p_1^{t/\tau} + \int_{T_\mu}^{t} (t-u)q(u)f\Big(\frac{\mu}{p}(u-\sigma(u)+\tau)\Big)du \le 0.$$

Thus

$$f(\frac{\mu}{p})p_1^{-t/\tau} \int_{T_\mu}^{t} (t-\mu)q(u)f\big(u-\sigma(u)+\tau\big)du \le 1,$$

and

$$p_1^{-t/\tau} \int_{T_\mu}^{t} (t-u)q(u)f\big(u-\sigma(u)+\tau\big)du \le \frac{1}{f(\frac{\mu}{p})}, \tag{4.4.18}$$

which contradicts (4.4.13) since $\mu > 0$ is arbitrary and $f(\infty) = \infty$.

(c) Assume $z'(t) > 0$, $z(t) < 0$, $t \ge t_0 \ge 0$. Then $x(t) < px(t-\tau)$ is obvious. □

Corollary 4.4.1. *In addition to the assumptions of Lemma 4.4.3, further assume that σ is a positive constant, and*

$$\sum_{i=0}^{\infty} \int_{T+i\tau}^{T+i\tau+\alpha} (u-T)q(u)du = \infty \tag{4.4.19}$$

holds for any $T \in R_+$, and $0 < \alpha \le \tau$, then all nonoscillatory solutions of Eq. (4.4.1) tend to zero as $t \to \infty$.

Proof: If not, there exists an eventually positive solution $x(t)$ satisfying $\limsup_{t\to\infty} x(t) > 0$, and this can only occur when $z''(t) \le 0$, $z'(t) > 0$ and $z(t) < 0$, $t \ge t_0 \ge 0$, hence $z'(t) \to 0$, $z(t) \to 0$ as $t \to \infty$. If $\liminf_{t\to\infty} x(t) > 0$, then $x(t) \ge a > 0$, $t \ge t_1 \ge t_0$. Integrating Eq. (4.4.1) twice we get

$$z(t) + \int_{t}^{\infty} (u-t)q(u)f(a)du < 0$$

which implies

$$\limsup_{t\to\infty} \int_t^\infty (u-t)q(u)du \le 0,$$

contradicting (4.4.19). Thus

$$\limsup_{t\to\infty} x(t) > 0 \quad \text{and} \quad \liminf_{t\to\infty} x(t) = 0.$$

Then we can choose $t_2 > t_1 \ge t_0$ such that $x(t_2-\sigma) > x(t_1-\sigma)$. We claim that

$$\liminf_{n\to\infty} x(t_2-\sigma+n\tau) > 0. \tag{4.4.20}$$

In fact

$$x(t_j-\sigma+n\tau) = \sum_{i=1}^n z(t_j-\sigma+i\tau) + x(t_j-\sigma), \quad j=1,2.$$

Since $z(t_2-\sigma+i\tau) \ge z(t_1-\sigma+i\tau)$ for $i=1,2,\dots,n$, and

$$\liminf_{n\to\infty} x(t_1-\sigma+n\tau) \ge 0.$$

We have

$$\liminf_{n\to\infty} x(t_2-\sigma+n\tau) \ge x(t_2-\sigma) - x(t_1-\sigma) > 0.$$

Now, choose $t_0 \le t_1 < t_2 < t_3$ such that for any $T \in [t_2, t_3]$,

$$x(t_1-\sigma) < x(t_2-\sigma) \le x(T-\sigma).$$

From the above discussion, we see that (4.4.20) holds, i.e., there exists a $\mu > 0$ such that $x(t_2-\sigma+n\tau) \ge \mu$ for all n. It is easy to see that for $T \in [t_2, t_3]$,

$$\begin{aligned} x(T-\sigma+n\tau) &= \sum_{i=1}^n z(T-\sigma+i\tau) + x(T-\sigma) \\ &\ge \sum_{i=1}^n z(t_2-\sigma+i\tau) + x(t_2-\sigma) \\ &= x(t_2-\sigma+n\tau) \ge \mu. \end{aligned}$$

From (4.4.1) we have

$$-z'(s) + \int_s^t q(u)f\big(x(u-\sigma)\big)du \le 0, \quad t_0 \le s \le t,$$

$$z(t_0) + \int_{t_0}^t (u-t_0)q(u)f\big(x(u-\sigma)\big)du \le 0, \quad t \ge t_0. \tag{4.4.21}$$

Hence

$$z(t_0) + f(\mu)\sum_{i=0}^{n} \int_{t_2+i\tau}^{t_3+i\tau} (u-t_0)q(u)du \le 0,$$

and then

$$z(t_0) + f(\mu)\sum_{i=0}^{n} \int_{t_2+i\tau}^{t_3+i\tau} (u-t_2)q(u)du \le 0$$

contradicting (4.4.19). □

Corollary 4.4.2. *In addition to the assumptions of Lemma 4.4.4, assume further that σ is a positive constant,*

$$\int_t^\infty (u-t)q(u)du = \infty \tag{4.4.22}$$

and

$$\sum_{i=0}^{\infty} f(p^i) \int_{T+i\tau}^{T+i\tau+\alpha} (u-T)q(u)du = \infty \tag{4.4.23}$$

hold for any $T \in R_+$ and $0 < \alpha \le \tau$. Then all nonoscillatory solutions of Eq. (4.4.1) tend to zero as $t \to \infty$.

Proof: If not, similar to the proof of Corollary 4.4.1 we see that there exists an eventually positive solution $x(t)$ satisfying

$$\limsup_{t\to\infty} x(t) > 0 \quad \text{and} \quad \liminf_{t\to\infty} x(t) = 0.$$

From the proof of Lemma 4.4.4 this can only occur when $z''(t) \le 0$, $z'(t) > 0$, and $z(t) < 0$, $t \ge t_0$. Choose $t_2 > t_1 \ge t_0$ such that $x(t_2 - \sigma) > x(t_1 - \sigma)$.

Since

$$x(t_2 - \sigma + n\tau) = \sum_{i=1}^{n} p^{n-i} z(t_2 - \sigma + i\tau) + p^n x(t_2 - \sigma),$$

$$x(t_1 - \sigma + n\tau) = \sum_{i=1}^{n} p^{n-i} z(t_1 - \sigma + i\tau) + p^n x(t_1 - \sigma),$$

$$z(t_2 - \sigma + i\tau) \geq z(t_1 - \sigma + i\tau), \quad i = 1, 2, \ldots, n,$$

and $x(t_1 - \sigma + n\tau) > 0, \ n = 0, 1, \ldots,$ we see

$$x(t_2 - \sigma + n\tau) \geq p^n \big(x(t_2 - \sigma) - x(t_1 - \sigma)\big) = Ap^n.$$

Similar to the proof of Corollary 4.4.1, we can show that there is an interval $[t_2, t_3]$ such that

$$x(T - \sigma + nt) \geq Ap^n$$

for $T \in [t_2, t_3]$ and all n. From (4.4.21) we get

$$z(t_0) + f(A) \sum_{i=0}^{n} \int_{t_2+i\tau}^{t_3+i\tau} (u - t_2) q(u) f(p^n) du \leq 0$$

which contradicts (4.4.23).

We are now ready to state an oscillation criterion.

Theorem 4.4.2. *Assume (H) holds and $p \geq 1$. In addition to the conditions of Lemma 4.4.3 and Lemma 4.4.4 for the cases $p = 1$ and $p > 1$, respectively, we assume (4.4.4) holds. Then Eq. (4.4.1) is oscillatory.*

Proof: If not, let $x(t)$ be an eventually positive solution of Eq. (4.4.1). Then we have $z'(t) > 0, \ z(t) < 0$ eventually, where

$$z(t) = x(t) - px(t - \tau).$$

The following is similar to the proof of Theorem 4.4.1. We omit it here. □

4.4.4. Further discussion for the linear case

Now we pay attention to the following linear equation

$$\big(x(t) + px(t-\tau)\big)'' + q(t)x(t-\sigma) = 0. \tag{4.4.24}$$

Lemma 4.4.5. *Assume that $p \geq 0$ and $q \in C(R_+, R_+)$. Let $x(t)$ be an eventually positive solution of (4.4.24). Set*

$$y(t) = x(t) + px(t-\tau),$$

and

$$z(t) = y(t) + py(t-\tau). \tag{4.4.25}$$

Then $z(t) > 0$, $z'(t) > 0$ and $z''(t) \leq 0$ and

$$z''(t) + \frac{q^*(t)}{1+p}\, z(t-\sigma) \leq 0 \tag{4.4.26}$$

eventually, where $q^*(t) = \min\{q(t), q(t-\tau)\}$.

Proof: Suppose $x(t) > 0$ for $t \geq t_0$. Then $y(t) > 0$ for $t \geq t_0 + \tau$, and $y''(t) < 0$ for $t \geq t_1 = t_0 + \max\{\sigma, \tau\}$. Therefore $y'(t) > 0$ for $t \geq t_1$. Then

$$\begin{aligned} z''(t) &= y''(t) + py''(t-\tau) \\ &\leq -q^*(t)[x(t-\sigma) + px(t-\tau-\sigma)] \\ &= -q^*(t)y(t-\sigma). \end{aligned} \tag{4.4.27}$$

Similar to the above we have $z(t) > 0$, $z'(t) > 0$ and $z''(t) \leq 0$ for $t \geq t_2 \geq t_1$ and

$$\begin{aligned} z''(t) &\leq -q^*(t)y(t-\sigma) \\ &\leq -\frac{q^*(t)}{1+p}\,[y(t-\sigma) + py(t-\sigma-z)] \\ &= -\frac{q^*(t)}{1+p}\, z(t-\sigma). \end{aligned}$$

The proof is complete. □

Theorem 4.4.3. *Let $p > 0$, $q \in C(R_+, R_+)$. Assume the second order ODE*

$$y''(t) + \lambda q(t)\,\frac{t-\sigma}{t}\,y(t) = 0 \tag{4.4.28}$$

is oscillatory for some $\lambda \in (0,1)$. Then every solution of Eq. (4.4.24) is oscillatory.

Proof: Otherwise, let $x(t)$ be an eventually positive solution of (4.4.24) and $z(t)$ is defined by (4.4.25). Then $z(t)$ satisfies all conditions of Lemma 4.4.1. So for every $k \in (0,1)$, there exists a $t_k \geq 0$ such that

$$z(t-\sigma) \geq k\,\frac{t-\sigma}{t}\,z(t), \quad \text{for} \quad t \geq t_k. \tag{4.4.29}$$

By Lemma 4.4.5, (4.4.26) is true. Combining (4.4.26) and (4.4.29) we have

$$z''(t) + \frac{k(t-\sigma)q^*(t)}{(1+p)t}\,z(t) \leq 0, \quad \text{for} \quad t \geq t_k, \tag{4.4.30}$$

which implies that

$$z''(t) + \frac{k(t-\sigma)q^*(t)}{(1+p)t}\,z(t) = 0 \tag{4.4.31}$$

has a nonoscillatory solution, contradicting the assumption. □

From this theorem every oscillation criterion for the second order ODE (4.4.28) becomes an oscillation criterion for the second order NDDF (4.4.24).

For example, we have the following Corollary.

Corollary 4.4.3. *Let $p > 0$, $q \in C(R_+, R_+)$. Then every solution of Eq. (4.4.24) is oscillatory if for some $\alpha \in (0,1)$*

$$\int_0^\infty t^\alpha q(t)dt = \infty. \tag{4.4.32}$$

Then we turn to the equation with variable p as follows:

$$\left(x(t)+p(t)x(t-\tau)\right)''+q(t)x(t-\sigma)=0, \quad t\geq t_0. \tag{4.4.33}$$

Theorem 4.4.4. *Assume that*

(i) $\tau>0, \quad \sigma>0$;

(ii) $q\in C(R_+,R_+)$ *and* $q(t)\geq q_0>0$;

(iii) $p\in C(R_+,R)$ *and there exist constants* p_1 *and* p_2 *such that* $p_1\leq p(t)\leq p_2$, *and* $p(t)$ *is not eventually negative.*

Then every solution of Eq. (4.4.33) is oscillatory.

Proof: If not, let $x(t)$ be an eventually positive solution. Set

$$z(t)=x(t)+p(t)x(t-\tau). \tag{4.4.34}$$

It is not difficult to show that $z(t)<0$ eventually, which contradicts (iii). □

Example 4.4.1. Consider the equation

$$\left(x(t)+\left(\frac{1}{2}+\sin t\right)x(t-2\pi)\right)''+\left(\frac{3}{2}+\sin t\right)x(t-4\pi)=0, \quad t\geq 0. \tag{4.4.35}$$

It is easy to see that the assumptions of Theorem 4.4.4 are satisfied here. Therefore, every solution of Eq. (4.4.35) oscillates. For example, $x(t)=\frac{\sin t}{\frac{3}{2}+\sin t}$ is such a solution.

The following example shows that, if the hypothesis (ii) of Theorem 4.4.4 is violated, the result may be wrong.

Example 4.4.2. Consider the equation

$$\left(x(t)+(t-1)^{-1/2}x(t-1)\right)''+\frac{1}{4}\,t^{-3/2}(t-2)^{-1/2}x(t-2)=0, \quad t\geq 2. \tag{4.4.36}$$

All assumptions of Theorem 4.4.4, except (ii) are satisfied. Note, however, that $x(t)=t^{1/2}$ is a nonoscillatory solution.

4.5. Classification of Nonoscillatory Solutions

We consider the second order nonlinear neutral differential equations of the form

$$\left(x(t)-\sum_{i=1}^{m}p_i(t)x(t-\tau_i)\right)'' + \sum_{j=1}^{n} f_j(t, x(g_{j1}(t)), \ldots, x(g_{j\ell}(t))) = 0, \quad t \ge t_0. \tag{4.5.1}$$

In this section we assume that

(i) $\tau_i > 0$, $p_i \in C([t_0,\infty), R_+)$, $i = 1,\ldots,m$ and there exists $\delta \in (0,1]$ such that $\sum_{i=1}^{m} p_i(t) \le 1-\delta$, $t \ge t_0$;
(ii) $g_{js} \in C([t_0,\infty), R)$, $\lim_{t\to\infty} g_{js}(t) = \infty$, $j = 1,\ldots,n$, $s = 1,\ldots,\ell$;
(iii) $f_j \in C([t_0,\infty)\times R^\ell, R)$, $y_1 f_j(t,y_1,\ldots,y_\ell) > 0$ for $y_1 y_i > 0$, $i = 1,\ldots,\ell$, $j = 1,\ldots,n$. Moreover

$$|f_j(t,x_1,\ldots,x_\ell)| \ge |f_j(t,y_1,\ldots,y_\ell)|$$

whenever $|y_i| \le |x_i|$ and $x_i y_i > 0$, $i = 1,\ldots,\ell$, $j = 1,\ldots,n$.
Set

$$y(t) = x(t) - \sum_{i=1}^{m} p_i(t)x(t-\tau_i). \tag{4.5.2}$$

We shall give the classification of nonoscillatory solutions of Eq. (4.5.1).

First we show some lemmas which will be useful for the main results.

Lemma 4.5.1. *Let $x(t)$ be an eventually positive (or negative) solution of Eq. (4.5.1). If $\lim_{t\to\infty} x(t) = 0$, then $y(t)$ is eventually negative (or positive) and $\lim_{t\to\infty} y(t) = 0$. Otherwise, $y(t)$ is eventually positive (or negative).*

Proof: Let $x(t)$ be an eventually positive solution of Eq. (4.5.1). From (4.5.1), $y''(t) < 0$ eventually. Thus $y'(t)$ is decreasing and $y'(t) > 0$ or $y'(t) < 0$ eventually. Also $y(t) > 0$ or $y(t) < 0$ eventually. If $\lim_{t\to\infty} x(t) = 0$, from (4.5.2) we have $\lim_{t\to\infty} y(t) = 0$. Since $y(t)$ is monotonic, so $\lim_{t\to\infty} y'(t) = 0$ which implies that $y'(t) > 0$. Therefore, $y(t) < 0$ eventually. If $\lim_{t\to\infty} x(t) = 0$ fail, then

$\limsup_{t\to\infty} x(t) > 0$. We show that $y(t) > 0$ eventually. If not, then $y(t) < 0$ eventually. If $x(t)$ is unbounded, then there exists a sequence $\{t_n\}$ such that $\lim_{n\to\infty} t_n = \infty$, $x(t_n) = \max_{t_0 \le t \le t_n} x(t)$, and $\lim_{n\to\infty} x(t_n) = \infty$. From (4.5.2), we have

$$y(t_n) = x(t_n) - \sum_{i=1}^{m} p_i(t_n)x(t_n - \tau_i) \ge x(t_n)\Big(1 - \sum_{i=1}^{m} p_i(t_n)\Big). \tag{4.5.3}$$

Thus $\lim_{n\to\infty} y(t_n) = \infty$, a contradiction. If $x(t)$ is bounded, then there exists a sequence $\{t_n\}$ such that $\lim_{n\to\infty} t_n = \infty$, and $\lim_{n\to\infty} x(t_n) = \limsup_{t\to\infty} x(t)$. Since sequences $\{p_i(t_n)\}$ and $\{x(t_n - \tau_i)\}$ are bounded, there exist convergent subsequences. Without loss of generality we may assume that $\lim_{n\to\infty} x(t_n - \tau_i)$ and $\lim_{n\to\infty} p_i(t_n)$, $i = 1, \ldots, m$, exist. Hence

$$\begin{aligned} 0 \ge \lim_{n\to\infty} y(t_n) &= \lim_{n\to\infty} \Big(x(t_n) - \sum_{i=1}^{m} p_i(t_n)x(t_n - \tau_i)\Big) \\ &\ge \limsup_{t\to\infty} x(t)\Big(1 - \lim_{n\to\infty} \sum_{i=1}^{m} p_i(t_n)\Big) > 0, \end{aligned}$$

a contradiction again. Therefore, $y(t) > 0$ eventually. A similar proof can be given if $x(t) < 0$ eventually. □

Lemma 4.5.2. *Assume that* $\lim_{t\to\infty} \sum_{i=1}^{m} p_i(t) = p \in [0, 1)$, *and* $x(t)$ *is an eventually positive (or negative) solution of Eq. (4.5.1). If* $\lim_{t\to\infty} y(t) = a \in R$, *then* $\lim_{t\to\infty} x(t) = \frac{a}{1-p}$. *If* $\lim_{t\to\infty} y(t) = \infty$ *(or* $-\infty$*), then* $\lim_{t\to\infty} x(t) = \infty$ *(or* $-\infty$*).*

Proof: Let $x(t)$ be an eventually positive solution of Eq. (4.5.1), then $x(t) \ge y(t)$ eventually. If $\lim_{t\to\infty} y(t) = \infty$, then $\lim_{t\to\infty} x(t) = \infty$. Now we consider the case that $\lim_{t\to\infty} y(t) = a \in R$. Thus $y(t)$ is bounded which implies that $x(t)$ is bounded (see (4.5.3)). Therefore, there exists a sequence $\{t_n\}$ such that $\lim_{n\to\infty} t_n = \infty$ and $\lim_{n\to\infty} x(t_n) = \limsup_{t\to\infty} x(t)$. As before, without loss of generality, we may assume that $\lim_{n\to\infty} p_i(t_n)$ and $\lim_{n\to\infty} x(t_n - \tau_i)$, $i = 1, \dots, n$, exist. Hence

$$\begin{aligned} a &= \lim_{n\to\infty} y(t_n) \\ &= \lim_{n\to\infty} x(t_n) - \sum_{i=1}^{m} \lim_{n\to\infty} p_i(t_n) \lim_{n\to\infty} x(t_n - \tau_i) \\ &\geq \limsup_{t\to\infty} x(t)(1-p) \end{aligned}$$

i.e.

$$\frac{a}{1-p} \geq \limsup_{t\to\infty} x(t). \tag{4.5.4}$$

On the other hand, there exists $\{t'_n\}$ such that $\lim_{n\to\infty} x(t'_n) = \liminf_{t\to\infty} x(t)$. Without loss of generality, we assume that $\lim_{n\to\infty} p_i(t'_n)$ and $\lim_{n\to\infty} x(t'_n - \tau_i)$, $i = 1, \dots, m$, exist. Hence

$$\begin{aligned} a &= \lim_{n\to\infty} y(t'_n) \\ &= \lim_{n\to\infty} x(t'_n) - \sum_{i=1}^{m} \lim_{n\to\infty} p_i(t'_n) \lim_{n\to\infty} x(t'_n - t_i) \\ &\leq \liminf_{t\to\infty} x(t)(1-p) \end{aligned}$$

or

$$\frac{a}{1-p} \leq \liminf_{t\to\infty} x(t). \tag{4.5.5}$$

Combining (4.5.4) and (4.5.5) we obtain $\lim_{t\to\infty} x(t) = \frac{a}{1-p}$. A similar proof can be given if $x(t) < 0$. □

We are now ready to prove the following results.

Theorem 4.5.1. *Assume that $\lim_{t\to\infty} \sum_{i=1}^{m} p_i(t) = p \in [0,1)$. Let $x(t)$ be a nonoscillatory solution of Eq. (4.5.1). Let S denote the set of all nonoscillatory solutions of Eq. (4.5.1), and define*

$$S(0,0,0) = \{x \in S : \lim_{t\to\infty} x(t) = 0, \ \lim_{t\to\infty} y(t) = 0, \ \lim_{t\to\infty} y'(t) = 0\},$$

$$S(b,a,0) = \{x \in S: \lim_{t\to\infty} x(t) = b := \frac{a}{1-p}, \lim_{y\to\infty} y(t) = a, \lim_{t\to\infty} y'(t) = 0\},$$

$$S(\infty,\infty,0) = \{x \in S: \lim_{t\to\infty} x(t) = \infty, \lim_{t\to\infty} y(t) = \infty, \lim_{t\to\infty} y'(t) = 0\},$$

$$S(\infty,\infty,d) = \{x \in S: \lim_{t\to\infty} x(t) = \infty, \lim_{t\to\infty} y(t) = \infty, \lim_{t\to\infty} y'(t) = d \neq 0\}.$$

Then

$$S = S(0,0,0) \cup S(b,a,0) \cup S(\infty,\infty,0) \cup S(\infty,\infty,d).$$

Proof: Without loss of generality, let $x(t)$ be an eventually positive solution. If $\lim_{t\to\infty} x(t) = 0$, by Lemma 4.5.1, $\lim_{t\to\infty} y(t) = 0$ and $\lim_{t\to\infty} y'(t) = 0$, i.e., $x \in S(0,0,0)$. If $\lim_{t\to\infty} x(t) = 0$, fails, by Lemma 4.5.1, $y(t) > 0$ eventually, it is easy to see that $y'(t) > 0$, $y''(t) < 0$ eventually. If $\lim_{t\to\infty} y(t) = a > 0$ exists. Then $\lim_{t\to\infty} y'(t) = 0$, by Lemma 4.5.2, we have $\lim_{t\to\infty} x(t) = \frac{a}{1-p} = b$, i.e., $x \in S(b,a,0)$. If $\lim_{t\to\infty} y(t) = \infty$, by Lemma 4.5.2, $\lim_{t\to\infty} x(t) = \infty$. Since $y''(t) < 0$ and $y'(t) > 0$, we have $\lim_{t\to\infty} y'(t) = d$, where $d = 0$ or $d > 0$. Then either $s \in S(\infty,\infty,0)$, or $x \in S(\infty,\infty,d)$. □

In the following we shall show some existence results for each kind of nonoscillatory solutions of Eq. (4.5.1).

Theorem 4.5.2. *Assume there exist two constants $k_1 > k_2 > 0$ such that*

$$|p_i(t_2) - p_i(t_1)| \le k_1 |t_2 - t_1|, \quad i = 1, \ldots, m,$$

$$\sum_{i=1}^{m} p_i(t)\exp(k_1\tau_i) > 1 \ge \sum_{i=1}^{m} p_i(t)\exp(k_2\tau_i), \tag{4.5.6}$$

and

$$\Big(\sum_{i=1}^{m} p_i(t)\exp(k_1\tau_i) - 1\Big)\exp(-k_1 t) \ge \int_t^\infty (u-t)\sum_{j=1}^{n} f_j(u, \exp(-k_2 g_{j1}(u)), \ldots, \exp(-k_2 g_{j\ell}(u)))du \tag{4.5.7}$$

eventually. Then Eq. (4.5.1) has a solution $x \in S(0,0,0)$.

Proof: Let BC be the space of all bounded continuous functions on $[t_0, \infty)$ with the sup norm $\|x\| = \sup_{t \geq t_0} |x(t)|$. Define

$$\Omega = \{x \in BC : \exp(-k_1 t) \leq x(t) \leq \exp(-k_2 t) \text{ and } |x(t_2) - x(t_1)| \leq L|t_2 - t_1|, \quad t_1, t_2 \geq t_0\},$$

where $L \geq k_1$. Then Ω is a nonempty, closed convex bounded subset of BC.

For the sake of convenience, denote

$$\begin{aligned} f(u, x(g(u))) &= \sum_{j=1}^{n} f_j(u, x(g_{j1}(u)), \ldots, x(g_{j\ell}(u))), \\ f(u, \exp(-k_2 g(u))) &= \sum_{j=1}^{n} f_j(u, \exp(-k_2 g_{j1}(u)), \ldots, \exp(-k_2 g_{j\ell}(u))). \end{aligned} \tag{4.5.8}$$

Define a mapping $\mathcal{T}$ on Ω as follows:

$$(\mathcal{T}x)(t) = \begin{cases} \sum_{i=1}^{m} p_i(t) x(t - \tau_i) - \int_t^{\infty} (u - t) f(u, x(g(u))) du, & t \geq T \\ \exp(-K(x)t), & t_0 \leq t < T, \end{cases} \tag{4.5.9}$$

where $K(x) = -\frac{\ln(\mathcal{T}x)(T)}{T}$, T is a sufficiently large number so that $t - \tau_i \geq t_0$, $g_{js}(t) \geq t_0$, $i = 1, \ldots, m$, $j = 1, \ldots, n$, $s = 1, \ldots, \ell$, for $t \geq T$.

Condition (4.5.7) implies that

$$\int_T^{\infty} f(u, \exp(-k_2 g(u))) du < \infty.$$

Assumption (i) at the beginning of this section implies that for given $\alpha \in (1-\delta, 1)$

$$\left(\alpha - \sum_{i=1}^{m} p_i(t)\right) L \geq [\alpha - (1 - \delta)] L > 0. \tag{4.5.10}$$

Therefore, T can be chosen so large that for $t \geq T$

$$\int_T^{\infty} f(u, \exp(-k_2 g(u))) du \leq \left(\alpha - \sum_{i=1}^{m} p_i(t)\right) L, \tag{4.5.11}$$

and

$$\alpha + \sum_{i=1}^{m} \exp\big(- k_2(t - \tau_i)\big) \le 1.$$

Hence by (4.5.6) and (4.5.7)

$$\begin{aligned}
(\mathcal{T}x)(t) &\le \sum_{i=1}^{m} p_i(t)x(t - \tau_i) \\
&\le \sum_{i=1}^{m} p_i(t)\exp\big(- k_2(t - \tau_i)\big) \\
&= \Big(\sum_{i=1}^{m} p_i(t)\exp(k_2\tau_1) \Big) \exp(-k_2 t) \\
&\le \exp(-k_2 t), \quad \text{for} \quad t \ge T,
\end{aligned}$$

and

$$\begin{aligned}
(\mathcal{T}x)(t) &\ge \sum_{i=1}^{m} p_i(t)\exp\big(- k_1(t - \tau_i)\big) - \int_t^\infty (u - t)f\big(u, \exp(-k_2 g(u))\big)du \\
&= \exp(-k_1 t) + \Big(\sum_{i=1}^{m} p_i(t)\exp(k_1\tau_i) - 1 \Big) \exp(-k_1 t) \\
&\quad - \int_t^\infty (u - t)f\big(u, \exp(-k_2 g(u))\big)du \\
&\ge \exp(-k_1 t) \quad \text{for} \quad t \ge T,
\end{aligned}$$

i.e.,

$$\exp(-k_1 t) \le (\mathcal{T}x)(t) \le \exp(-k_2 t), \quad t \ge T.$$

By the definition of $K(x)$, and $\exp(-k_1 T) \le (\mathcal{T}x)(T) \le \exp(-k_2 T)$, we know that $k_2 \le K(x) \le k_1$. Hence

$$\exp(-k_1 t) \le (\mathcal{T}x)(t) \le \exp(-k_2 t), \quad t_0 \le t < T.$$

Now we show that

$$|(\mathcal{T}x)(t_2) - (\mathcal{T}x)(t_1)| \le L|t_2 - t_1| \quad \text{for} \quad t_1, t_2 \in [t_0, \infty). \tag{4.5.12}$$

Without loss of generality assume $t_2 \ge t_1 \ge t_0$. For $t_2 \ge t_1 \ge T$, using (4.5.11) and (4.5.12), we have

$$\begin{aligned}
&|(\mathcal{T}x)(t_1) - (\mathcal{T}x)(t_2)| \\
&\quad \le \sum_{i=1}^{m} |p_i(t_1)x(t_1 - \tau_i) - p_i(t_2)x(t_2 - \tau_i)| \\
&\qquad + \left| \int_{t_1}^{\infty} (u - t_1) f(u, x(g(u)))\,du - \int_{t_2}^{\infty} (u - t_2) f(u, x(g(u)))\,du \right| \\
&\le \sum_{i=1}^{m} \big(p_i(t_2)|x(t_2 - \tau_i) - x(t_1 - \tau_i)| + |p_i(t_2) - p_i(t_1)|x(t_1 - \tau_i)\big) \\
&\qquad + \left| \int_{t_1}^{t_2} (u - t_1) f(u, x(g(u)))\,du + \int_{t_2}^{\infty} (t_2 - t_1) f(u, x(g(u)))\,du \right| \\
&\le \left[\sum_{i=1}^{m} \big(p_i(t_2) + \exp(-k_2(t - \tau_i))\big) L \right. \\
&\qquad \left. + \int_{t_1}^{\infty} f(u, \exp(-k_2 g(u)))\,du \right] |t_2 - t_1| \\
&\le \left\{ \left[\sum_{i=1}^{m} p_i(t_2) + \sum_{i=1}^{m} \exp(-k_2(t_1 - \tau_i)) \right] + \left(\alpha - \sum_{i=1}^{m} p_i(t_2) \right) \right\} L|t_2 - t_1| \\
&= \left[\sum_{i=1}^{m} \exp(-k_2(t_1 - \tau_i)) + \alpha \right] L|t_2 - t_1| \\
&\le L|t_2 - t_1|.
\end{aligned}$$

For $t_0 \le t_1 \le t_2 \le T$, we have

$$\begin{aligned}
|(\mathcal{T}x)(t_2) - (\mathcal{T}x)(t_1)| &= \big|\exp(-K(x)(t_2)) - \exp(-K(x)(t_1))\big| \\
&\le |K(x)(t_2) - K(x)(t_1)| \\
&\le L|t_2 - t_1|.
\end{aligned}$$

For $t_0 < t_1 \le T \le t_2$, we have

$$\begin{aligned}|(\mathcal{T}x)(t_2) - (\mathcal{T}x)(t_1)| &\le |(\mathcal{T}x)(t_2) - (\mathcal{T}x)(T)| + |(\mathcal{T}x)(T) - (\mathcal{T}x)(t_2)| \\ &\le L|t_2 - T| + L|T - t_1| \\ &= L|t_2 - t_1|.\end{aligned}$$

We have proved that (4.5.12) holds for all $t_0 \le t_1 \le t_2$. Therefore, $\mathcal{T}\Omega \subseteq \Omega$. Clearly, $\mathcal{T}$ is continuous. Since $\mathcal{T}\Omega \subseteq \Omega$, $\mathcal{T}\Omega$ is uniformly bounded.

Set $x \in \Omega$, we see that

$$|(\mathcal{T}x)(t)| \le \exp(-k_2 t)$$

$$|(\mathcal{T}x)(t_2) - (\mathcal{T}x)(t_1)| \le L|t_2 - t_1|, \quad t_0 \le t_1 \le t_2.$$

For given $\varepsilon > 0$ there exists a $T' > t_0$ such that $\exp(-k_2 t) < \frac{\varepsilon}{2}$ for $t \ge T'$. Then

$$|(\mathcal{T}x)(t_2) - (\mathcal{T}x)(t_1)| \le \exp(-k_2 t_2) + \exp(-k_2 t_1) < \varepsilon \tag{4.5.13}$$

for $t_2 \ge t_1 \ge T'$.

For $t_0 \le t_1 \le t_2 \le T'$, we have

$$|(\mathcal{T}x)(t_2) - (\mathcal{T}x)(t_1)| \le L|t_2 - t_1| < \varepsilon, \quad \text{whenever} \quad |t_2 - t_1| < \frac{\varepsilon}{L}. \tag{4.5.14}$$

(4.5.13) and (4.5.14) implies that $\mathcal{T}\Omega$ is equicontinuous. Therefore, $\mathcal{T}\Omega$ is relatively compact. By Schauder's fixed point theorem there is an $x^* \in \Omega$ such that $x^* = \mathcal{T}x^*$. Then x^* is a positive solution of Eq. (4.5.1), and $x^* \in S(0,0,0)$. □

Theorem 4.5.3. *Assume that* $\lim_{t\to\infty} \sum_{i=1}^{m} p_i(t) = p \in [0,1)$. *Then Eq. (4.5.1) has a nonoscillatory solution* $x \in S(b,a,0)$ $(b \ne 0, a \ne 0)$ *if and only if*

$$\int_{t_0}^{\infty} u \left| \sum_{j=1}^{n} f_j(u, b_1, \ldots, b_1) \right| du < \infty \quad \textit{for some} \quad b_1 \ne 0. \tag{4.5.15}$$

Proof: Necessity. Without loss of generality, let $x(t) \in S(b,a,0)$ be an eventually positive solution of Eq. (4.5.1). By Theorem 4.5.1 we know that $b > 0$, $a > 0$.

Using the notation (4.5.8), from (4.5.1) and (4.5.2) we have

$$y''(t) = -f\big(t, x(g(t))\big).$$

Integrating it from s to ∞ for $s \geq t_0$ we have

$$y'(s) = \int_s^\infty f\big(u, x(g(u))\big)du.$$

Integrating it from T to t for T sufficiently large, we have

$$y(t) = y(T) + \int_T^t (u-T) f\big(u, x(g(u))\big)du + \int_t^\infty (t-T) f\big(u, x(g(u))\big)du. \tag{4.5.16}$$

Since $\lim_{u\to\infty} x\big(g_{jh}(u)\big) = b > 0, \quad j = 1, \ldots, n, \ h = 1, \ldots, \ell$, there exists a $T \geq t_0$ such that $x\big(g_{jh}(u)\big) \geq \frac{b}{2}$ for $u \geq T$. Hence from (4.5.16) we have

$$\int_T^t (u-T)\left|\sum_{j=1}^n f_j\left(u, \frac{b}{2}, \ldots, \frac{b}{2}\right)\right| du < y(t) - y(T)$$

which implies that (4.5.15) holds.

Sufficiency. Set $b_1 > 0$ and $A > 0$ so that $A < (1-p)b_1$. From (4.5.15) there exists a sufficiently large T so that for $t \geq T$ we have $t - \tau_i \geq t_0, \ i = 1, \ldots, m$, and $g_{jh}(t) \geq t_0, \ j = 1, \ldots, n, \ h = 1, \ldots, \ell$, and

$$\frac{A}{b_1} + \sum_{i=1}^m p_i(t) + \frac{1}{b_1}\int_T^\infty u \sum_{j=1}^n f_j(u, b_1, \ldots, b_1)du \leq 1. \tag{4.5.17}$$

Define Ω to be the set of all continuous functions $x(t)$ on $[t_0, \infty)$ such that $0 \leq x(t) \leq b_1$, $t \geq t_0$, and define a mapping $\mathcal{T}$ in Ω as follows:

$$(\mathcal{T}x)(t) = \begin{cases} A + \sum_{i=1}^m p_i(t)x(t-\tau_i) + \int_T^t u f\big(u, x(g(u))\big)du \\ \quad + \int_t^\infty t f\big(u, x(g(u))\big)du, & t \geq T \\ (\mathcal{T}x)(T), & t_0 \leq t < T. \end{cases} \tag{4.5.18}$$

Set

$$x_0(t) = 0, \quad t \geq t_0$$
$$x_k(t) = (\mathcal{T} x_{k-1})(t), \quad t \geq t_0, \quad k = 1, 2, \ldots . \tag{4.5.19}$$

It is easy to see that $x_0(t) < x_1(t) = A \leq b_1$, $t \geq t_0$. By induction, we have

$$A \leq x_k(t) \leq x_{k+1}(t) \leq b_1, \quad t \geq t_0, \quad k = 1, 2, \ldots .$$

Thus $\lim_{k\to\infty} x_k(t) = x(t)$ exists and $A \leq x(t) \leq b_1$, $t \leq [t_0, \infty)$. By the Lebesgue's convergence theorem, from (4.5.19) we obtain

$$x(t) = \begin{cases} A + \sum_{i=1}^{m} p_i(t) x(t-\tau_i) + \int_T^t u f\big(u, x(g(u))\big) du \\ \quad + \int_t^\infty t f\big(u, x(g(u))\big) du, & t \geq T \\ x(T), & t_0 \leq t < T. \end{cases}$$

Hence, $x(t)$ is a positive solution of Eq. (4.5.1). Since $0 < A \leq x(t) \leq b_1$, from Theorem 4.5.1, $x \in S(b, a, 0)$. □

Similarly, we can prove the following results.

Theorem 4.5.4. *Assume that* $\lim_{t\to\infty} \sum_{i=1}^{m} p_i(t) = p \in [0, 1)$. *Then Eq. (4.5.1) has a nonoscillatory solution* $x \in S(\infty, \infty, d)$ $(d \neq 0)$ *if and only if*

$$\int_{t_0}^{\infty} \left| \sum_{j=1}^{n} f_j\big(u, d_1 g_{j1}(u), \ldots, d_1 g_{j\ell}(u)\big) \right| du < \infty, \quad \textit{for some} \quad d_1 \neq 0. \tag{4.5.20}$$

Theorem 4.5.5. *Assume that* $\lim_{t\to\infty} \sum_{i=1}^{m} p_i(t) = p \in [0, 1)$. *Further assume that*

$$\int_{t_0}^{\infty} \left| \sum_{j=1}^{n} f_j\big(u, d_1 g_{j1}(u), \ldots, d_1 g_{j\ell}(u)\big) \right| du < \infty \quad \textit{for some} \quad d_1 \neq 0 \tag{4.5.21}$$

and

$$\int_{t_0}^{\infty} u \left| \sum_{j=1}^{n} f_j(u, b_1, \ldots, b_1) \right| du = \infty \quad \text{for some} \quad b_1 \neq 0, \tag{4.5.22}$$

where $b_1 d_1 > 0$. Then Eq. (4.5.1) has a nonoscillatory solution $x \in S(\infty, \infty, 0)$.

Example 4.5.1. Consider

$$\left(x(t) - \tfrac{1}{2}x(t-1)\right)'' + \frac{2(t-1)^3 - t^3}{(t-1)^6}\, x^3(t) = 0 \tag{4.5.23}$$

where $q(t) = \frac{2(t-1)^3 - t^3}{(t-1)^6}$. It is easy to see that (4.5.15) holds. Therefore, (4.5.23) has a nonoscillatory solution $x \in S(b, a, 0)$, $b \neq 0$, $a \neq 0$. In fact, $x(t) = 1 - \frac{1}{t}$ is such a solution, where $a = \frac{1}{2}$, $b = 1$.

Example 4.5.2. Consider

$$\left(x(t) - \tfrac{1}{2}x(t-t)\right)'' + q(t)x^{1/3}(t) = 0 \tag{4.5.24}$$

where

$$q(t) = \tfrac{1}{4}\left(t^{-3/2} - \tfrac{1}{2}(t-1)^{-3/2}\right)t^{-1/6}.$$

For large t, $q(t) \sim kt^{-5/3}$. It is easy to see that (4.5.21) and (4.5.22) are satisfied. According to Theorem 4.5.5, Eq. (4.5.28) has a solution $x \in S(\infty, \infty, 0)$. In fact, $x(t) = \sqrt{t}$ is such a solution of (4.5.24).

Remark 4.5.1. The above arguments can be applied to the equation

$$\left(x(t) - \sum_{i=1}^{m} p_i(t)x(t-\tau_i)\right)'' = \sum_{j=1}^{n} f_i\big(t, x(g_{j1}(t)), \ldots, x(g_{j\ell}(t))\big), \quad t \geq t_0. \tag{4.5.25}$$

For instance, under the assumptions of Theorem 4.5.1, we have

$$S = S(0,0,0) \cup S(b,a,0) \cup S(\infty,\infty,d) \cup S(\infty,\infty,\infty).$$

Theorems 4.5.3 and 4.5.4 hold also for Eq. (4.5.25). Furthermore, Eq. (4.5.25) has a nonoscillatory solution $x(t) \in S(\infty, \infty, \infty)$ if

$$\int_{t_0}^{\infty} \left| \sum_{j=1}^{n} f_j(t, d_1 g_{j1}(t), \ldots, d_1 g_{j\ell}(t)) \right| dt < \infty \quad \text{for some} \quad d_1 \neq 0. \tag{4.5.26}$$

We now present another result for the linear equation

$$\big(x(t) - px(t-\tau)\big)'' + q(t)x\big(g(t)\big) = 0, \quad t \geq t_0, \tag{4.5.27}$$

where the condition $p \in [0, 1)$ is not required.

Theorem 4.5.6. *Assume that*

(i) $p,\ \tau > 0,\quad q \in C([t_0, \infty), R_+),\quad g \in C([t_0, \infty), R),\quad g(t+\tau) < t,\quad t \geq t_0,$ *and* $\lim_{t\to\infty} g(t) = \infty$;

(ii) there exists a constant $\alpha > 0$ *such that for sufficiently large* t

$$\frac{1}{p}e^{-\alpha\tau} + \frac{1}{p}\int_{t+\tau}^{\infty} (s - t - \tau)q(s)\exp[\alpha(t - g(s))]ds \leq 1. \tag{4.5.28}$$

Then Eq. (4.5.27) has a positive solution $x(t)$ *satisfying that* $x(t) \to 0$ *as* $t \to \infty$.

Proof: If the equality in (4.5.28) holds eventually, then we can verify that $x(t) = e^{-\alpha t}$ is an expected solution. Otherwise, we assume that there exists $T > t_0$, such that $t + \tau \geq 0,\quad g(t+\tau) \geq t_0$ for $t \geq T$, and

$$\beta := \frac{1}{p}e^{-\alpha\tau} + \frac{1}{p}\int_{T+\tau}^{\infty} (s - T - \tau)q(s)\exp[\alpha\big(T - g(s)\big)]ds < 1, \tag{4.5.29}$$

and (4.5.28) holds for $t \geq T$.

Let BC denote the Banach space of all continuously bounded functions defined on $[t_0, \infty)$ endowed with the sup norm. Let Ω be the subset of BC defined by

$$\Omega = \{y \in BC : \ 0 \leq y(t) \leq 1 \quad \text{for} \quad t \geq t_0\}.$$

Define a map $\mathcal{T} : \Omega \to BC$ as follows:

$$(\mathcal{T}y)(t) = (\mathcal{T}_1 y)(t) + (\mathcal{T}_2 y)(t)$$

where

$$(\mathcal{T}_1 y)(t) = \begin{cases} \frac{1}{p} e^{-\alpha\tau} y(t+\tau), & t \geq T \\ (\mathcal{T}_1 y)(T) + \exp[\varepsilon(T-t)] - 1, & t_0 \leq t \leq T \end{cases}$$

and

$$(\mathcal{T}_2 y)(t) = \begin{cases} \frac{1}{p} \int_{t+\tau}^{\infty} (s-t-\tau) q(s) \exp[\alpha(t-g(s))] y(g(s)) ds, & t \geq T \\ (\mathcal{T}_2 y)(T), & t_0 \leq t \leq T \end{cases}$$

where $\varepsilon = \ln(2-\beta)/(T-t_0)$.

We can show that the map $\mathcal{T}$ satisfies all the assumptions of Krasnoselskii's fixed point theorem, and so $\mathcal{T}$ has a fixed point y in Ω. Clearly, $y(t) > 0$ for $t \geq t_0$. It is easy to check that $x(t) = y(t)e^{-\alpha t}$ is a solution of Eq. (4.5.27). □

Corollary 4.5.1. *Assume that* $0 < p < 1$, $\tau > 0$, *and there exist constants* q^*, $r > \tau$ *such that eventually*

$$0 \leq q(t) \leq q^*, \quad g(t) \geq t - r.$$

If the "majorant" equation

$$\left(x(t) - px(t-\tau)\right)'' + q^* x(t-r) = 0, \quad t \geq t_0 \tag{4.5.30}$$

has a positive solution, then Eq. (4.5.27) also has a positive solution.

Proof: (4.5.30) has a positive solution if and only if

$$\lambda^2(1 - pe^{-\lambda r}) + q^* e^{-\lambda r} = 0$$

has a real root α. Clearly α must be negative. Let $\mu = -\alpha > 0$, then

$$1 - pe^{\mu\tau} + \frac{q^*}{\mu^2} e^{\mu r} = 0$$

or

$$\frac{1}{p} e^{-\mu\tau} + \frac{q^*}{p\mu^2} e^{\mu(r-\tau)} = 1.$$

Hence

$$\frac{1}{p}e^{-\mu\tau} + \frac{1}{p}\int_{t+\tau}^{\infty}(s-t-\tau)q(s)\exp[\mu(t-g(s))]ds$$
$$\le \frac{1}{p}e^{-\mu\tau} + \frac{q^*}{p\mu^2}e^{\mu(r-\tau)} = 1,$$

i.e., (4.5.28) holds. Then, by Theorem 4.5.5, Eq. (4.5.27) has a positive solution.

4.6. Unstable Type Equations

4.6.5. Equations with constant p

Consider the second order linear neutral differential equation of the form

$$\big(x(t) - px(t-\tau)\big)'' = q(t)x\big(g(t)\big), \quad t \ge t_0, \tag{4.6.1}$$

where $p \in R, q \in C([t_0,\infty), R_+)$, $g \in C([t_0,\infty), R)$, $\lim_{t\to\infty} g(t) = \infty$, $\tau > 0$.

We will show later that Eq. (4.6.1) always has an unbounded, nonoscillatory solution. Therefore, we only need to find conditions for all bounded solutions of Eq. (4.6.1) to be oscillatory.

Theorem 4.6.1. *Assume that*

(i) $0 < p < 1$, $\tau > 0$ *are constants;*

(ii) $g(t) \le t$ *and* g *is nondecreasing for* $t \ge t_0$;

(iii)
$$\limsup_{t\to\infty} \int_{g(t)}^{t} \big(s - g(t)\big)q(s)ds > 1. \tag{4.6.2}$$

Then every bounded solution of Eq. (4.6.1) is oscillatory.

Proof: Assume the contrary, and let $x(t)$ be an eventually positive bounded solution of Eq. (4.6.1). Define

$$z(t) = x(t) - px(t-\tau). \tag{4.6.3}$$

We have $z''(t) > 0$ for $t \ge T \ge t_0$. If $z'(t) > 0$ for $t \ge T_1 > T$, then $\lim_{t\to\infty} z(t) = \infty$, which contradicts the boundedness of x. Therefore, $z'(t) \le 0$ for $t \ge T$.

There are two possibilities for $z(t)$:

(a) $z(t) > 0$ for $t \geq T$; (b) $z(t) < 0$ for $t \geq T_2 \geq T$.

In case (a), integrating (4.6.1) from s to t we have

$$z'(t) - z'(s) = \int_s^t q(u)x(g(u))du. \tag{4.6.4}$$

Integrating (4.6.4) in s from $g(t)$ to t, we have

$$\begin{aligned} z'(t)(t-g(t)) - z(t) + z(g(t)) &= \int_{g(t)}^t \int_s^t q(u)x(g(u))du\, ds \\ &= \int_{g(t)}^t (s-g(t))q(s)x(g(s))ds \\ &> \int_{g(t)}^t (s-g(t))q(s)z(g(s))ds \\ &\geq z(g(t)) \int_{g(t)}^t (s-g(t))q(s)ds. \end{aligned}$$

Hence for $t \geq T$

$$z(t) + z(g(t))\Big(\int_{g(t)}^t (s-g(t))q(s)ds - 1\Big) \leq 0$$

which contradicts the positivity of $z(t)$ and (4.6.2).

In case (b), we have

$$x(t) < px(t-\tau) < p^2x(t-2\tau) < \cdots < p^n x(t-n\tau)$$

for $t \geq T_2 + n\tau$, which implies that $\lim_{t\to\infty} x(t) = 0$. Consequently, $\lim_{t\to\infty} z(t) = 0$, a contradiction. □

Remark 4.6.1. Theorem 4.6.1 is also true for $p = 0$.

Theorem 4.6.2. *Assume that*

(i) $p < 0$ and $q(t) > 0$, $t \geq t_0$;
(ii) $g(t) = t - \sigma$, where σ is a constant, $\sigma > \tau$;

(iii) *There exists* $\alpha > 0$ *such that*

$$\limsup_{t\to\infty} q(t)/q(t-\tau) = \alpha \tag{4.6.5}$$

and

$$\limsup_{t\to\infty} \int_{t-(\sigma-\tau)}^{t} (s-(t-(\sigma-\tau)))q(s)ds > 1-\alpha p. \tag{4.6.6}$$

Then every bounded solution of (4.6.1) is oscillatory.

Proof: Assume the contrary, and let $x(t)$ be a bounded, eventually positive solution of Eq. (4.6.1) and $z(t)$ be defined as in (4.6.3). As shown before, $z''(t) > 0$, $z'(t) < 0$, and $z(t) > 0$ eventually, where $z(t)$ is defined by (4.6.3).

From (4.6.5) and (4.6.6), there exists a constant $k > 1$ such that

$$\limsup_{t\to\infty} \int_{t-(\sigma-\tau)}^{t} (s-(t-(\sigma-\tau)))q(s)ds > 1-k\alpha p, \tag{4.6.7}$$

$$q(t)/q(t-\tau) < k\alpha, \quad t \geq t_1 \tag{4.6.8}$$

where t_1 is a sufficiently large number. We rewrite Eq. (4.6.1) in the form

$$z''(t) - p\,\frac{q(t)}{q(t-\tau)} z''(t-\tau) = q(t)z(t-\sigma). \tag{4.6.9}$$

Substituting (4.6.8) into (4.6.9) we have

$$z''(t) - k\alpha p z''(t-\tau) \geq q(t)z(t-\sigma), \quad t \geq t_1. \tag{4.6.10}$$

Set

$$w(t) = z(t) - k\alpha p z(t-\tau). \tag{4.6.11}$$

Then

$$w''(t) \geq q(t)z(t-\sigma) > 0, \quad t \geq t_1. \tag{4.6.12}$$

By the boundedness of $x(t)$, it is easy to see that $w(t) > 0$, $w'(t) \leq 0$, $t \geq t_2 \geq t_1$. Since $z(t)$ is decreasing,

$$w(t) = z(t) - k\alpha p z(t-\tau) \leq \big(1-k\alpha p\big)z(t-\tau), \quad t \geq t_2. \tag{4.6.13}$$

Combining (4.6.12) and (4.6.13) we have

$$w''(t) \geq \frac{1}{1-k\alpha p} q(t) w\big(t-(\sigma-\tau)\big). \tag{4.6.14}$$

Integrating (4.6.14) from s to t for $t \geq s \geq t_2$ we have

$$w'(t) - w'(s) \geq \frac{1}{1-k\alpha p} \int_s^t q(u) w\big(u-(\sigma-\tau)\big) du. \tag{4.6.15}$$

Integrating (4.6.15) in s from $t-(\sigma-\tau)$ to t we have

$$\begin{aligned}
& w'(t)(\sigma-\tau) - w(t) + w\big(t-(\sigma-\tau)\big) \\
& \geq \frac{1}{1-k\alpha p} \int_{t-(\sigma-\tau)}^t \int_s^t q(u) w\big(u-(\sigma-\tau)\big) du ds \\
& = \frac{1}{1-k\alpha p} \int_{t-(\sigma-\tau)}^t \big(u-(t-(\sigma-\tau))\big) q(u) w\big(u-(\sigma-\tau)\big) du \\
& > \frac{w\big(t-(\sigma-\tau)\big)}{1-k\alpha p} \int_{t-(\sigma-\tau)}^t \big(u-(t-(\sigma-\tau))\big) q(u) du, \quad t \geq t_2.
\end{aligned}$$

Thus

$$w(t) + w\big(t-(\sigma-\tau)\big) \left[\frac{1}{1-k\alpha p} \int_{t-(\sigma-\tau)}^t \big(u-(t-(\sigma-\tau))\big) q(u) du - 1\right] \leq 0$$

which contradicts (4.6.7). □

Theorem 4.6.3. *Assume that*

(i) $p = 1, \quad \tau > 0$;

(ii) $g(t) \leq t$, *and* g *is nondecreasing for* $t \geq t_0$;

(iii) *either*

$$\int_{t_0}^{\infty} t q(t) dt = \infty \tag{4.6.16}$$

or

$$\lim_{t \to \infty} t \int_t^{\infty} q(s) ds = \infty. \tag{4.6.17}$$

Then every bounded solution of Eq. (4.6.1) is oscillatory.

Proof: Assume the contrary and let $x(t)$ be a bounded eventually positive solution of Eq. (4.6.1) and $z(t)$ be defined as in (4.6.3). There are two possibilities:

(a) $z''(t) \geq 0, \quad z'(t) \leq 0, \quad z(t) < 0$ for $t \geq t_1 \geq t_0$,
(b) $z''(t) \geq 0, \quad z'(t) \leq 0$ and $z(t) > 0, \quad t \geq t_1 \geq t_0$.

In case (a), there exists a finite number $\alpha > 0$ such that $\lim_{t\to\infty} z(t) = -\alpha$. Thus there exists $t_2 \geq t_1$ such that $-\alpha < z(t) < -\frac{\alpha}{2}, \quad t \geq t_2$, i.e.,

$$-\alpha < x(t) - x(t-\tau) < -\frac{\alpha}{2}, \quad t \geq t_2.$$

Hence $x(t-\tau) > \frac{\alpha}{2}$, $t \geq t_2$, then there exists $t_3 \geq t_2$ such that $x\big(g(t)\big) > \frac{\alpha}{2}$, $t \geq t_3$. From Eq. (4.6.1), we have

$$z''(t) \geq \frac{\alpha}{2}\, q(t), \quad t \geq t_3. \tag{4.6.18}$$

In case (b), we have

$$x(t) > x(t-\tau), \quad t \geq t_1$$

then there exists $M > 0$ such that $x(t) \geq M, \quad t \geq t_1$. Hence

$$z''(t) \geq Mq(t), \quad t \geq t_3. \tag{4.6.19}$$

In both cases we are lead to the same inequality (4.6.19). Integrating (4.6.19) from t to T for $T > t \geq t_3$ we have

$$z'(T) - z'(t) \geq M \int_t^T q(s)ds, \quad t_3 \leq t < T.$$

Hence

$$-z'(t) \geq M \int_t^T q(s)ds, \quad t_3 \leq t < T,$$

which implies that $\int_{t_0}^{\infty} q(s)ds < \infty$ and so

$$-z'(t) \geq M \int_t^{\infty} q(s)ds.$$

Integrating the above inequality from t to T for $T > t$ we have

$$z(t) \geq z(T) + M \int_t^T \int_s^\infty q(u)duds$$

$$= z(T) + M\left[\int_t^T (u-t)q(u)du + (T-t)\int_T^\infty q(u)du\right], \quad t \geq t_2$$

which leads to a contradiction to the boundedness of z in either case of (4.6.16) and (4.6.17). □

Example 4.5.1. Consider

$$\big(x(t) - x(t-2\pi)\big)'' = \frac{2\pi}{t-\pi}x(t-\pi). \tag{4.6.20}$$

It is easy to see that all assumptions of Theorem 4.6.3 are satisfied. Therefore, every bounded solution of Eq. (4.6.20) is oscillatory.

Eq. (4.6.1) may have unbounded oscillatory solutions. For example, (4.6.20) has a solution $x(t) = t\sin t$.

Theorem 4.6.4. *Assume that* $p > 1$. *Then Eq. (4.6.1) has a bounded positive solution if and only if*

$$\int_{t_0}^\infty tq(t)dt < \infty. \tag{4.6.21}$$

Proof: *Necessity.* Let $x(t)$ be a bounded positive solution of Eq. (4.6.1) and $z(t)$ be defined by (4.6.3). Then Eq. (4.6.1) becomes

$$z''(t) = q(t)x\big(g(t)\big).$$

There are only two possibilities for $z(t)$:

(a) $z''(t) \geq 0, \quad z'(t) \leq 0, \quad z(t) < 0, \quad t \geq t_1 \geq t_0;$
(b) $z''(t) \geq 0, \quad z'(t) \leq 0, \quad z(t) > 0, \quad t \geq t_1 \geq t_0.$

As in the proof of Theorem 4.6.3, in either case we are lead to the inequality

$$z''(t) \geq Mq(t), \quad \text{for} \quad t \geq t_2,$$

where $M > 0$ is a constant and t_2 is a sufficiently large number. Integrating it twice we have

$$\begin{aligned} z(t) - z(T) &\geq M \int_t^T \int_s^T q(u)duds \\ &= M \int_t^T (u-t)q(u)du, \quad T > t \geq t_2. \end{aligned}$$

Letting $T \to \infty$, we see that (4.6.21) holds.

The sufficiency part of Theorem 4.6.4 follows from the following more general result. □

Theorem 4.6.5. *Assume $p > 1$ and $q \in C([t_0, \infty), R)$ such that*

$$\int_{t_0}^{\infty} s|q(s)|ds < \infty. \tag{4.6.22}$$

Then Eq. (4.6.1) has a bounded positive solution.

Proof: Let $T \geq t_0$ be sufficiently large so that $t + \tau \geq t_0$, $g(t+\tau) \geq t_0$ for $t \geq T$, and

$$\int_{t+\tau}^{\infty} s|q(s)|ds \leq \frac{p-1}{4}, \quad t \geq T. \tag{4.6.23}$$

Consider the Banach space BC of all continuous bounded functions defined on $[t_0, \infty)$ with the sup norm. Set

$$\Omega = \left\{ x \in BC : \frac{p}{2} \leq x(t) \leq 2p, \ t \geq t_0 \right\}.$$

Clearly, Ω is a bounded closed convex subset of BC. Define a map $\mathcal{T} : \Omega \to BC$ as follows:

$$(\mathcal{T}x)(t) = \begin{cases} p - 1 + \frac{1}{p}x(t+\tau) & \\ \quad - \frac{1}{p}\int_{t+\tau}^{\infty}(s-t-\tau)q(s)x(g(s))ds, & t \geq T \\ (\mathcal{T}x)(T), & t_0 \leq t \leq T. \end{cases} \tag{4.6.24}$$

For any $x \in \Omega$, from (4.6.23) we have

$$(\mathcal{T}x)(t) \leq p + 1 + \frac{1}{p}\int_{t+\tau}^{\infty}(s-t-\tau)|q(s)|\,|x(g(s))|ds < 2p, \quad \text{for} \quad t \geq T,$$

and

$$(\mathcal{T}x)(t) \geq p - \frac{1}{2} - \frac{1}{p}\int_{t+\tau}^{\infty}(s-t-\tau)|q(s)|\,|x(g(s))|ds \geq \frac{p}{2}, \quad \text{for} \quad t \geq T.$$

Therefore, $\mathcal{T}\Omega \subseteq \Omega$.

We shall show that $\mathcal{T}$ is a contraction on Ω. In fact, for any $x_1,\ x_2 \in \Omega$, we have

$$\begin{aligned}
|(\mathcal{T}x_1)(t) - (\mathcal{T}x_2)(t)| &\leq \frac{1}{p}\,|x_1(t+\tau) - x_2(t+\tau)| \\
&\quad + \frac{1}{p}\int_{t+\tau}^{\infty}(s-t-\tau)|q(s)|\,|x_1(g(x)) - x_2(g(s))|ds \\
&\leq \frac{1}{p}\|x_1 - x_2\| + \frac{1}{p}\|x_1 - x_2\|\int_{t+\tau}^{\infty}(s-t-\tau)|q(s)|ds \\
&\leq \|x_1 - x_2\|\,\frac{1}{p}\left(1 + \frac{p-1}{4}\right) \\
&= \frac{1}{4}\left(1 + \frac{3}{p}\right)\|x_1 - x_2\|, \quad t \geq T
\end{aligned}$$

which implies that

$$\begin{aligned}
\|\mathcal{T}x_1 - \mathcal{T}x_2\| &= \sup_{t \geq t_0}|(\mathcal{T}x_1)(t) - (\mathcal{T}x_2)(t)| \\
&= \sup_{t \geq T}|(\mathcal{T}x_1)(t) - (\mathcal{T}x_2)(t)| \leq \frac{1}{4}\left(1 + \frac{3}{p}\right)\|x_1 - x_2\|.
\end{aligned}$$

Since $\frac{1}{4}(1 + \frac{3}{p}) < 1$, it follows that $\mathcal{T}$ is a contraction. Hence there exists a fixed point $x \in \Omega$. Then

$$x(t) = p - 1 + \frac{1}{p}x(t+\tau) - \frac{1}{p}\int_{t+\tau}^{\infty}(s-t-\tau)q(s)x(g(s))ds, \quad t \geq T.$$

Differentiating it twice, we have

$$\big(x(t+\tau)-x(t)\big)''=q(t+\tau)x\big(g(t+\tau)\big),\quad t\geq T,$$

i.e., $x(t)$ is a bounded positive solution of Eq. (4.6.1). □

Remark 4.6.2. By a similar proof we can show that Theorem 4.6.5 is also true for
$p\in(0,1)$.

The following result is about the existence of asymptotically decaying positive solutions of Eq. (4.6.1).

Theorem 4.6.6. *Assume that* $0<p<1$ *and there exists a constant* $\alpha>0$ *such that eventually*

$$pe^{\alpha\tau}+\int_t^\infty(s-t)q(s)\exp\left[\alpha(t-g(s))\right]ds\leq 1. \tag{4.6.25}$$

Then Eq. (4.6.1) has a positive solution $x(t)$ *satisfying* $x(t)\to 0$ *as* $t\to\infty$.

Proof: It is easy to see that if the equality in (4.6.25) holds eventually, then Eq. (4.6.17) has a positive solution $x(t)=e^{-\alpha t}$. In the rest of the proof we may assume that there exists a number $T>t_0$ such that $t-\tau\geq t_0$, $g(t)\geq t_0$ for $t\geq T$ and

$$\beta:=pe^{\alpha\tau}+\int_T^\infty(s-T)q(s)\exp[\alpha(T-g(s))]ds<1, \tag{4.6.26}$$

and condition (4.6.25) holds for $t\geq T$.

Let BC denote the Banach space of all continuous bounded functions defined on $[t_0,\infty)$ with the sup norm. Let Ω be the subset of BC defined by

$$\Omega=\{y\in BC:\ 0\leq y(t)\leq 1,\ t\geq t_0\}.$$

Define a map $\mathcal{T}:\Omega\to BC$ as follows:

$$(\mathcal{T}y)(t)=(\mathcal{T}_1y)(t)+(\mathcal{T}_2y)(t)$$

where

$$(\mathcal{T}_1 y)(t) = \begin{cases} pe^{\alpha\tau} y(t-\tau), & t \geq T \\ (\mathcal{T}_1 y)(T) + \exp[\varepsilon(T-t)] - 1, & t_0 \leq t \leq T \end{cases} \tag{4.6.27}$$

and

$$(\mathcal{T}_2 y)(t) = \begin{cases} \int_t^{\infty} (s-t) q(s) \exp[x(t-g(s))] y(g(s))\, ds, & t \geq T \\ (\mathcal{T}_2 y)(T), & t_0 \leq t \leq T, \end{cases} \tag{4.6.28}$$

where $\varepsilon = \ln(2-\beta)/(T-t_0)$.

It is easy to see that the integral in (4.6.28) is defined whenever $y \in \Omega$.

Clearly, the set Ω is closed, bounded and convex in BC. We shall show that for every pair $x,\ y \in \Omega$

$$\mathcal{T}_1 x + \mathcal{T}_2 y \in \Omega. \tag{4.6.29}$$

In fact, for any $x, y \in \Omega$, we have

$$\begin{aligned} (\mathcal{T}_1 x)(t) + (\mathcal{T}_2 y)(t) &= pe^{\alpha\tau} x(t-\tau) + \int_t^{\infty} (s-t) q(s) \exp[\alpha(t-g(s))] y(g(s))\, ds \\ &\leq pe^{\alpha\tau} + \int_t^{\infty} (s-t) q(s) \exp[\alpha(t-g(s))]\, ds \\ &\leq 1, \quad \text{for} \quad t \geq T, \end{aligned}$$

and

$$\begin{aligned} (\mathcal{T}_1 x)(t) + (\mathcal{T}_2 y)(t) &= (\mathcal{T}_1 x)(T) + (\mathcal{T}_2 y)(T) + \exp[\varepsilon(T-t)] - 1 \\ &= \beta + \exp[\varepsilon(T-t)] - 1 \\ &\leq \beta + \exp[\varepsilon(T-t_0)] - 1 = 1 \quad \text{for} \quad t_0 \leq t \leq T. \end{aligned}$$

Obviously, $(\mathcal{T}_1 x)(t) + (\mathcal{T}_2 y)(t) \geq 0$ for $t \geq t_0$. Thus, (4.6.29) is true.

From (4.6.26), we know that $pe^{\alpha\tau} < 1$, which implies that $\mathcal{T}_1$ is a contraction.

We now shall show that $\mathcal{T}_2$ is completely continuous. In fact, from (4.6.25), there exists a positive constant M such that

$$\int_t^{\infty} q(s) \exp[\alpha(t-g(s))]\, ds \leq M, \quad \text{for} \quad t \geq T.$$

Thus we have

$$\left|\frac{d}{dt}(T_2y)(t)\right| = \left|\int_t^\infty q(s)\exp[\alpha(t-g(s))]y(g(s))\,ds + \alpha\int_t^\infty (s-t)q(s)\exp[\alpha(t-g(s))]y(g(s))\,ds\right| \le M+\alpha, \quad \text{for} \quad t > T,$$

and

$$\frac{d}{dt}(T_2y)(t) = 0, \quad \text{for} \quad t_0 \le t < T,$$

which shows the equicontinuity of the family $T_2\Omega$. On the other hand, it is easy to see that T_2 is continuous and the family of $T_2\Omega$ is uniformly bounded. Therefore T_2 is completely continuous.

By Krasnoselskii's fixed point theorem, T has a fixed point $y \in \Omega$. That is,

$$y(t) = \begin{cases} pe^{\alpha\tau}y(t-\tau) \\ \quad + \int_t^\infty (s-t)q(s)\exp[\alpha(t-g(s))]y(g(s))\,ds, & t \ge T \\ y(T) + \exp[\varepsilon(T-t)] - 1, & t_0 \le t \le T. \end{cases} \tag{4.6.30}$$

Since $y(t) \ge \exp[\varepsilon(T-t)] - 1 > 0$ for $t_0 \le t < T$, it follows that $y(t) > 0$ for $t \ge t_0$. Set $x(t) = y(t)e^{-\alpha t}$. Then (4.6.30) becomes

$$x(t) = px(t-\tau) + \int_t^\infty (s-t)q(s)x(g(s))\,ds, \quad t \ge T. \tag{4.6.31}$$

Thus $x(t)$ is a positive solution of Eq. (4.6.1) and $x(t) \to 0$ as $t \to \infty$. □

Remark 4.6.3. For the case $p = 0$, if $g(t) < t$, then the conclusion of Theorem 4.6.6 still holds.

Corollary 4.6.1. *Assume that $0 < p < 1$, and there exist constants $q^* > 0$, $\sigma > 0$ such that*

$$0 \le q(t) \le q^*, \quad g(t) \ge t - \sigma. \tag{4.6.32}$$

If the "majorant" equation

$$\big(x(t) - px(t-\tau)\big)'' = q^* x(t-\sigma), \quad t \geq t_0 \tag{4.6.33}$$

has a bounded positive solution, then Eq. (4.6.1) also has a positive solution $x(t)$ satisfying that $x(t) \to 0$ as $t \to \infty$.

Proof: Eq. (4.6.33) has a bounded positive solution if and only if its characteristic equation

$$\lambda^2(-1 + pe^{\lambda\tau}) + q^* e^{\lambda\sigma} = 0 \tag{4.6.34}$$

has a real root $\alpha \in (0, \infty)$. Thus, we have

$$pe^{\alpha\tau} + \frac{1}{\alpha^2} q^* e^{\alpha\sigma} = 1. \tag{4.6.35}$$

Combining (4.6.32) and (4.6.35) we have that for sufficiently large t

$$\begin{aligned} pe^{\alpha\tau} &+ \int_t^\infty (s-t) q(s) \exp[\alpha(t-\sigma)] ds \\ &\leq pe^{\alpha\tau} + \frac{1}{\alpha^2} q^* e^{\alpha\sigma} = 1. \end{aligned}$$

By Theorem 4.6.6, Eq. (4.6.1) has a positive solution $x(t)$ which tends to zero as $t \to \infty$. □

Example 4.6.1. Consider

$$\left(x(t) - \frac{1}{2e}\, x(t-2)\right)'' = \left(\frac{1}{8e} - \frac{1}{t}\right) x(t-2). \tag{4.6.36}$$

In our notation, $p = \frac{1}{2e}$, $q^* = \frac{1}{8e}$, $\tau = 2$ and $\sigma = 2$. The "majorant" equation is

$$\left(x(t) - \frac{1}{2e}\, x(t-2)\right)'' = \frac{1}{8e}\, x(t-2) \tag{4.6.37}$$

and (4.6.34) becomes

$$\lambda^2\left(-1 + \frac{1}{2e}\, e^{2\lambda}\right) + \frac{1}{8e}\, e^{2\lambda} = 0. \tag{4.6.38}$$

It is obvious that $\lambda = \frac{1}{2}$ is a real root of (4.6.38), and hence (4.6.37) has a bounded positive solution. By Corollary 4.6.1, Eq. (4.6.36) has a positive solution $x(t)$ satisfying $x(t) \to 0$ as $t \to \infty$.

4.6.6. Equations with variable p

We now consider the second order neutral differential equation

$$\big(x(t) - p(t)x(t-\tau)\big)'' = q(t)x(t-\sigma), \quad t \geq t_0 \tag{4.6.39}$$

where $\tau,\ \sigma \in (0,\infty)$, $\quad p,\ q \in C([t_0,\infty), R)$.

Theorem 4.6.7. *Assume that*

(i) $0 \leq p(t) \leq 1, \quad t \geq t_0;$
(ii) $0 < k_1 \leq q(t) \leq k_2, \quad t \geq t_0;$
(iii) *for any* $\lambda > 0$

$$\liminf_{t\to\infty}\left\{p(t-\sigma)\,\frac{q(t)}{q(t-\tau)}e^{\lambda\tau} + \frac{1}{\lambda^2}\,q(t)e^{\lambda\sigma}\right\} > 1. \tag{4.6.40}$$

Then every bounded solution of Eq. (4.6.39) is oscillatory.

Proof: Let $x(t)$ be a bounded, eventually positive solution, say $x(t-\tau) > 0$, $x(t-\sigma) > 0$ for $t \geq t_1 \geq t_0$. Set

$$z(t) = x(t) - p(t)x(t-\tau). \tag{4.6.41}$$

It is not difficult to show that $z''(t) > 0$, $\quad z'(t) < 0$, $\quad z(t) > 0$ for $t \geq t_1$, if t_1 is large enough and $\lim_{t\to\infty} z(t) = \lim_{t\to\infty} z'(t) = 0$. Then from (ii)

$$z''(t) \geq k_1 x(t-\sigma), \quad t \geq t_1 \tag{4.6.42}$$

and

$$z''(t) \leq k_2 x(t-\sigma), \quad t \geq t_1. \tag{4.6.43}$$

Define a set Λ as follows:

$$\Lambda = \{\lambda > 0 : \quad z''(t) > \lambda^2 z(t) \quad \text{eventually}\}. \tag{4.6.44}$$

It is easy to see that $\sqrt{k_1} \in \Lambda$, i.e., Λ is nonempty. We shall show that Λ is bounded above. In fact, (4.6.42) implies that

$$z''(t) \geq k_1 z(t-\sigma), \quad t \geq t_1 + \sigma.$$

Integrating from t to $t + \frac{\sigma}{2}$ we have

$$\begin{aligned} z'\big(t+\frac{\sigma}{2}\big) - z'(t) &\geq k_1 \int_t^{t+\frac{\sigma}{2}} z(s-\sigma)ds \\ &> k_1 \frac{\sigma}{2} z\big(t-\frac{\sigma}{2}\big), \quad t \geq t_1 + \sigma, \end{aligned}$$

and then

$$\begin{aligned} z(t) - z\big(t+\frac{\sigma}{4}\big) &> k_1 \frac{\sigma}{2} \int_t^{t+\frac{\sigma}{4}} z\big(s-\frac{\sigma}{2}\big)ds \\ &> k_1 \big(\frac{\sigma^2}{8}\big) z\big(t-\frac{\sigma}{4}\big). \end{aligned}$$

This implies that

$$z(t) > \alpha z\big(t-\frac{\sigma}{4}\big), \quad t \geq t_1 + \sigma, \tag{4.6.45}$$

where $\alpha = k_1 \left(\frac{\sigma^2}{8}\right)$. Applying (4.6.45) four times we obtain that

$$z(t) > \alpha^4 z(t-\sigma), \quad t \geq t_1 + 2\sigma. \tag{4.6.46}$$

In view of the boundedness of $x(t)$ it is not difficult to see that $\liminf_{t\to\infty} x(t) = 0$. Choose a sequence $\{s_n\}$ such that $s_n \geq t_1 + 2\sigma, \quad n = 1, 2, \ldots, \lim_{n\to\infty} s_n = \infty$, and

$$x(s_n - \sigma) = \min\{x(s) : \ t_1 \leq s \leq s_n - \sigma\}, \quad n = 1, 2, \ldots .$$

Integrating (4.6.42) twice we have

$$z(t-\sigma) > k_1 \int_{t-\sigma}^{t} \int_s^t x(u-\sigma)duds, \quad t \geq t_1 + \sigma$$

and hence

$$z(s_n - \sigma) > k_1 \int_{s_n-\sigma}^{s_n} \int_{s}^{s_n} x(u-\sigma)duds$$
$$\geq \frac{\sigma^2}{2} k_1 x(s_n - \sigma), \quad n = 1, 2, \ldots$$

i.e.,

$$x(s_n - \sigma) < \beta z(s_n - \sigma), \quad n = 1, 2, \ldots$$

where $\beta = 2/k_1\sigma^2$. Then from (4.6.43), (4.6.46) we obtain

$$z''(s_n) \leq k_2 x(s_n - \sigma) < \beta k_2 z(s_n - \sigma) < \alpha^{-4}\beta k_2 z(s_n), \quad n = 1, 2, \ldots$$

which implies that $\sqrt{\alpha^{-4}\beta k_2} \,\overline{\in}\, \Lambda$, i.e., Λ is bounded above.

Set $\lambda_0 = \sup \Lambda$. Then $\lambda_0 \in (0, \sqrt{\alpha^{-4}\beta k_2}]$. For any $\alpha \in (0, 1)$ we have that for sufficiently large t

$$z''(t) \geq (\alpha\lambda_0)^2 z(t). \tag{4.6.47}$$

Set $\overline{z}(t) = z'(t) + \alpha\lambda_0 z(t)$. Then

$$\overline{z}'(t) - \alpha\lambda_0 \overline{z}(t) = z''(t) - (\alpha\lambda_0)^2 z(t) \geq 0$$

eventually. It implies that $\overline{z}(t)e^{-\alpha\lambda_0 t}$ is nondecreasing. Since $z(t) \to 0$, $z'(t) \to 0$ as $t \to \infty$ so $\overline{z}(t) \to 0$ as $t \to \infty$. Thus $\overline{z}(t) < 0$, i.e., $z'(t) + \alpha\lambda_0 z(t) \leq 0$ eventually. Set $w(t) = z(t)e^{\alpha\lambda_0 t}$. Then

$$w'(t) = [z'(t) + \alpha\lambda_0 z(t)]e^{\alpha\lambda_0 t} \leq 0.$$

We can rewrite (4.6.39) in the form

$$z''(t) = p(t-\sigma)\,\frac{q(t)}{q(t-\tau)} z''(t-\tau) + q(t)z(t-\sigma). \tag{4.6.48}$$

Then by (4.6.47) we have

$$z''(t) \geq (\alpha\lambda_0)^2 p(t-\sigma)\,\frac{q(t)}{q(t-\tau)}\, z(t-\tau) + q(t)z(t-\sigma)$$

$$= (\alpha\lambda_0)^2 p(t-\sigma)\,\frac{q(t)}{q(t-\tau)} w(t-\tau)e^{-\alpha\lambda_0(t-\tau)} + q(t)w(t-\sigma)e^{-\alpha\lambda_0(t-\sigma)}$$

$$\geq \left[(\alpha\lambda_0)^2 p(t-\sigma)\,\frac{q(t)}{q(t-\tau)} e^{\alpha\lambda_0\tau} + q(t)e^{\alpha\lambda_0\sigma}\right] z(t)$$

which implies that

$$\inf_{t\geq T}\left\{(\alpha\lambda_0)^2 p(t-\sigma)\,\frac{q(t)}{q(t-\tau)}\, e^{\alpha\lambda_0\tau} + q(t)e^{\alpha\lambda_0\sigma}\right\} \leq \lambda_0^2.$$

Letting $\alpha \to 1$ we have

$$\inf_{t\geq T}\left\{\lambda_0^2 p(t-\sigma)\,\frac{q(t)}{q(t-\tau)} e^{\lambda_0\tau} + q(t)e^{\lambda_0\sigma}\right\} \leq \lambda_0^2$$

which contradicts (4.6.40). □

Remark 4.6.1. Noting that $e^x \geq 1$, $e^x \geq \frac{e^2}{4}x^2$, for $x \geq 0$, (4.6.40) can be replaced by

$$\liminf_{t\to\infty}\left\{p(t-\sigma)\,\frac{q(t)}{q(t-\tau)} + \frac{e^2}{4}\sigma^2 q(t)\right\} > 1.$$

Remark 4.6.2. In the case $p(t) \equiv p$, $q(t) \equiv q$ are constants (4.6.40) is also a necessary condition for the bounded oscillation of Eq. (4.6.39).

Theorem 4.6.8. *Assume that* $p(t) \leq 0$, $q(t) > 0$, $\sigma > \tau$,

$$\limsup_{t\to\infty}\left\{-p(t-\sigma)\,\frac{q(t)}{q(t-\tau)}\right\} = \alpha \in (0,\infty) \tag{4.6.49}$$

and

$$\limsup_{t\to\infty}\int_{t-(\sigma-\tau)}^{t}\big(s-t+(\sigma-\tau)\big)q(s)ds > 1-\alpha. \tag{4.6.50}$$

Then every bounded solution of Eq. (4.6.39) is oscillatory.

Proof: Let $x(t)$ be a bounded, eventually positive solution with $x(t-r) \geq 0$ for $t \geq t_1$. Then $z''(t) \geq 0$, $z'(t) < 0$, $z(t) > 0$, $t \geq t_1$. From (4.6.50) there exists a constant $k > 1$ such that

$$\limsup_{t\to\infty} \int_{t-(\sigma-\tau)}^{t} (s - t + \sigma - \tau) q(s) ds > 1 - k\alpha \tag{4.6.51}$$

and there is a $t_2 > t_1$ such that

$$-p(t-r)\,\frac{q(t)}{q(t-\tau)} \le k\alpha, \quad t \ge t_2. \tag{4.6.52}$$

From (4.6.48) we have

$$z''(t) - k\alpha z''(t-\sigma) \ge q(t) z(t-\sigma), \quad t \ge t_2.$$

Set $w(t) = z(t) - k\alpha z(t-\tau)$. Then

$$w''(t) \ge q(t) z(t-\sigma), \quad t \ge t_2, \tag{4.6.53}$$

it is easy to see that $w(t) > 0, \quad w'(t) \le 0, \quad t \ge t_2$. By the monotone property of $z(t)$ we have

$$w(t) = z(t) - k\alpha z(t-\tau) \le (1 - k\alpha) z(t-\tau), \quad t \ge t_2,$$

or

$$z(t) \ge \frac{1}{1-k\alpha} w(t+\tau), \quad t \ge t_2.$$

Substituting this into (4.6.53) we obtain

$$w''(t) \ge \frac{1}{1-k\alpha} q(t) w\big(t - (\sigma-\tau)\big), \quad t \ge t_2 + \sigma.$$

Integrating from s to t for $t \ge s$ we have

$$w'(t) - w'(s) \ge \frac{1}{1-k\alpha} \int_s^t q(u) w\big(u - (\sigma-\tau)\big) du, \quad s \ge t_2.$$

Integrating this inequality in s from $t - (\sigma - \tau)$ to t we have

$$w'(t)(\sigma-\tau) - w(t) + w\big(t - (\sigma-\tau)\big)$$

$$\geq \frac{1}{1-k\alpha}\int_{t-\sigma+\tau}^{t}(u-t+\sigma-\tau)q(u)w(u-\sigma+\tau)du$$

$$\geq \frac{w(t-\sigma+\tau)}{1-k\alpha}\int_{t-\sigma+\tau}^{t}(u-t+\sigma-\tau)q(u)du, \quad t\geq t_2.$$

Thus

$$w(t)+w\big(t-(\sigma-\tau)\big)\Big[-1+\frac{1}{1-k\alpha}\int_{t-\sigma+\tau}^{t}(u-t+\sigma-\tau)q(u)du\Big]\leq 0$$

which implies that

$$\int_{t-(\sigma-\tau)}^{t}(u-t+\sigma-\tau)q(u)du<1-k\alpha, \quad t\geq t_2.$$

We reach a contradiction. □

4.7. Forced Oscillation

We consider the second order neutral differential equation with a forcing term of the form

$$\big(x(t)+px(t-\tau)\big)''+f\big(t,x(t-\sigma)\big)=R(t), \quad t\geq t_0 \tag{4.7.1}$$

under the assumptions:

(H_1) $p,\tau>0$, and $\sigma\geq 0$;

(H_2) $f\in C([t_0,\infty)\times R,R)$, $xf(t,x)>0$ as $x\neq 0$;

(H_3) there exists a function $r\in C^2([t_0,\infty),R)$ such that $R(t)=r''(t)$ and r changes sign on $[T,\infty)$ for any $T\geq t_0$.

Set

$$r_*(t)=\min\Big\{\frac{\alpha(t-\tau)}{2},\ \frac{\alpha(t)}{2p}\Big\},$$
$$r_*^+(t)=\max\big\{r_*(t),0\big\}, \quad r_*^-(t)=\max\big\{-r_*(t),0\big\}.$$

Lemma 4.7.1. *Assume $y\in C([t_0,\infty),R)$, $\beta\in C([t_0,\infty),R_+)$, and $y(t)+py(t-\tau)\geq\beta(t)\geq 0$, $t\geq t_0$, where p, $\tau>0$. Then for each $t^*\geq t_0+\tau$, there* *exists a set*

$$A = \{t|\ t^* \le t \le t^* + 2\tau,\ y(t-\tau) \ge \beta_*(t)\}$$

with the measure mes $(A) \ge \tau$, *where* $\beta_*(t) = \min\ \{\frac{\beta(t-\tau)}{2}, \frac{\beta(t)}{2p}\}$.

Proof: For any fixed $t^* \ge t_0 + \tau$ we define a set

$$B = \left\{t|t \in [t^*, t^* + \tau],\ y(t) > \frac{\beta(t)}{2}\right\}.$$

If $B = \emptyset$(empty set), then $py(t-\tau) \ge \frac{\beta(t)}{2}$, for $t \in [t^*, t^* + \tau]$ i.e., $A = [t^*, t^* + \tau]$. Now we consider the case that $B \ne \emptyset$, then mes $(B) = \alpha \in (0, \tau)$. Let $\overline{B}$ denote the closure of B. In view of the continuity of y, we have $y(t) \ge \frac{\beta(t)}{2}$, $t \in \overline{B}$. Define a set $\overline{B} + \tau = \{t, t - \tau \in \overline{B}\}$. Then $y(t-\tau) \ge \frac{\beta(t-\tau)}{2}$ for $t \in (\overline{B} + \tau)$. Set

$$A = \{[t^*, t^* + \tau] \backslash B\} \cup (\overline{B} + \tau)$$

then mes $A = \tau$ and $y(t-\tau) \ge \beta_*(\tau)$ on A. □

Theorem 4.7.1. *Assume that* (H_1)*-*(H_3) *hold. Further assume that* f *is nondecreasing in* y *and*

$$\int_E f(t, r_*^+(t+\tau-\sigma))dt = \infty, \quad \int_E f(t, -r_*^-(t+\tau-\sigma))dt = -\infty \tag{4.7.2}$$

for every closed set E *whose intersection with every segment of the form* $[t-\tau, t+\tau]$, $t \ge t_0 + \tau$, *has a measure not smaller than* τ. *Then every solution of Eq. (4.7.1) is oscillatory.*

Proof: If not, without loss of generality, let $x(t)$ be an eventually positive solution, say $x(t) > 0$ for $t \ge t_0$. Set

$$z(t) = x(t) + px(t-\tau),$$

Then $\big(z(t) + r(t)\big)'' < 0$ for $t \ge t_0 + \tau$. It is easy to see that $\big(z(t) - r(t)\big)' > 0$

eventually, which implies that

$$\int_{t_0}^{\infty} f\big(t, x(t-\sigma)\big)dt < \infty. \tag{4.7.3}$$

On the other hand, it is easy to see that $\big(z(t) - r(t)\big) > 0$ eventually. Then we have

$$z(t) = x(t) + px(t-\tau) \geq r^+(t), \quad t \geq t_0 + \tau.$$

By Lemma 4.7.1, for every $t^* \geq t_0 + 2\tau$ there exists a set $A = \{t|\ t^* \leq t \leq t^* + 2\tau,\ x(t-\tau) \geq r_*^+(t)\}$ with mes $(A) \geq \tau$. Let us consider the set $A - (\tau - \sigma) = \{t|\ t + (\tau - \sigma) \in A\}$. It is obvious that mes $\big(A - (\tau - \sigma)\big) \geq \tau$ and $x(t-\sigma) \geq r_*^+\big(t + (\tau - \sigma)\big), \quad t \in (A - (\tau - \sigma))$. From (4.7.3) we have

$$\infty > \int_E f\big(t, x(t-\sigma)\big)dt \geq \int_E f\big(t, r_*^+(t + (\tau - \sigma))\big)dt$$

which contradicts the assumption (4.7.2). □

Example 4.7.1. Consider the equation

$$\big(x(t) + px(t-\tau)\big)'' + t^{\alpha}|x(t)|^{\nu} \text{ sign } x(t) = t^{\delta} \sin\ t, \quad t \geq t_0 \tag{4.7.4}$$

where $p, \tau, \nu > 0,\ \alpha, \delta \in R$. If $\alpha + \nu\delta > -1$, then condition (4.7.2) holds. From Theorem 4.7.1 every solution of (4.7.4) oscillates. For instance, every solution of the equation

$$\big(x(t) + x(t-\pi)\big)'' + x(t) = \sin t \tag{4.7.5}$$

is oscillatory. In fact, $x(t) = \sin t$ is such a solution.

The following result is for equations of the more general form

$$\big(x(t) + p(t)x(t-\tau)\big)'' + f\big(t, x(g(t)), x'(\sigma(t))\big) = R(t). \tag{4.7.6}$$

Theorem 4.7.2. *Assume that*

(i) $p,\ g,\ \sigma$ *and* R *are continuous on* $t \geq t_0$;

(ii) $p(t) \geq 0$, g *is nondecreasing and* $\lim_{t\to\infty} g(t) = \infty$;

(iii) $f \in C([t_0, \infty) \times R^2, R)$ *and* $f(t,u,v)u > 0$ *as* $u \neq 0$;

(iv) *for any* $T \geq t_0$

$$\begin{aligned} \liminf_{t\to\infty} \textstyle\int_T^t R(s)ds &= -\infty, & \limsup_{t\to\infty} \textstyle\int_T^t R(s)ds &= \infty, \\ \liminf_{t\to\infty} \textstyle\int_T^t \int_T^s R(u)duds &= -\infty, & \limsup_{t\to\infty} \textstyle\int_T^t \int_T^s R(u)duds &= \infty. \end{aligned} \tag{4.7.7}$$

Then every solution of Eq. (4.7.6) is oscillatory.

Proof: Assume the contrary. Then without loss of generality we assume there is an eventually positive solution $x(t)$. Set $z(t) = x(t) + p(t)x(t-\tau)$. Then $z(t) > 0$, $t \geq T \geq t_0$. From (4.7.6) we have $z''(t) < R(t)$. Thus

$$z'(t) - z'(T) < \int_T^t R(s)ds. \tag{4.7.8}$$

By (iv), there exists a sufficiently large $T^* \geq t_0$ such that $z'(T^*) < 0$. Replacing T by T^* in (4.7.8) we have

$$z'(t) < \int_{T^*}^t R(s)ds$$

and

$$z(t) - z(T^*) < \int_{T^*}^t \int_{T^*}^s R(u)duds.$$

Therefore, $\liminf_{t\to\infty} z(t) = -\infty$, which contradicts the positivity of $z(t)$. □

Example 4.7.2. Consider

$$\big(x(t) + x(t-\pi)\big)'' + tx(t-2\pi) = t\sin t. \tag{4.7.9}$$

It is easy to see that all assumptions of Theorem 4.7.2 are satisfied. Therefore, every solution of (4.7.9) is oscillatory. In fact, $x(t) = \sin t$ is such a solution.

As an application we give sufficient conditions for oscillation of the nonlinear hyperbolic equations of neutral type

$$\begin{cases} u_{tt}(x,t) + pu_{tt}(x,t-\tau) \\ \quad -\Delta u(x,t) + q(s,t,u) = f(x,t), \quad (x,t) \in G \\ \frac{\partial u}{\partial n}(x,t) = g(x,t), \qquad\qquad\qquad (x,t) \in \sigma \end{cases} \tag{4.7.10}$$

where p and τ are positive constants, Δ is the Laplacian in R^n. D is a bounded domain in R^n with the smooth boundary ∂D, the cylindrical domain
$G = D \times (0,\infty)$, $\sigma = \partial D \times (0,\infty)$, n is the vector of the external normal to σ,
$G_\alpha = D \times (\alpha,\infty)$, $\alpha \in R$.

Assume that

(A_1) $q(x,t,u) \in C(G \times R, R)$, $f(x,t) \in C(G,R)$, and $g(x,t) \in C(\sigma, R)$;

(A_2) $q(x,t,\xi) \geq \beta(t)\varphi(\xi)$ for $(x,t,\xi) \in G\times(0,\infty)$ where β, $\varphi \in C((0,\infty),(0,\infty))$ and φ is a convex function on $(0,\infty)$;
(A_3) $q(x,t,-\xi) = -q(x,t,\xi)$ for $(x,t,\xi) \in G \times (0,\infty)$.

Let $u(x,t)$ be the solution of problem (4.7.10). Define

$$u(t) = \frac{1}{|D|}\int_D u(x,t)dx, \quad t \in [-\tau,\infty),$$

where $|D| = \int_D dx$. Introduce the notation that

$$G(t) = \frac{1}{|D|}\int_{\partial D} g(x,t)d\sigma \quad \text{and} \quad F(t) = \frac{1}{|D|}\int_D f(x,t)ds.$$

Definition 4.7.1. The solution $u(x,t)$ of problem (4.7.10) is said to be oscillatory in G if u has a zero in G_α for any $\alpha \geq 0$.

Lemma 4.7.2. *Let conditions (A_1)-(A_3) hold. Assume the differential inequalities*

$$u''(t) + pu''(t-\tau) + \beta(t)\varphi(u) \leq G(t) + F(t) \tag{4.7.11}$$

and

$$u''(t) + pu''(t-\tau) + \beta(t)\varphi(u) \le -\big(G(t) + F(t)\big) \tag{4.7.12}$$

have no eventually positive solutions. Then every solution of (4.7.10) is oscillatory in G.

Similar to Theorem 4.7.1, we can prove the following lemma.

Lemma 4.7.3. *In addition to (A_1) and (A_2) assume that φ is nondecreasing for $u > 0$, and there exists $r \in C^2([t_0, \infty), R)$ such that $r''(t) = G(t) + F(t)$ and r is oscillatory and*

$$\int_E p(t)\varphi\big(r_*^+(t+\tau)\big)\,dt = \infty \tag{4.7.13}$$

where set E and $r_*^+(t)$ are defined as in Theorem 4.7.1. Then Eq. (4.7.11) is oscillatory.

We are now ready to state the main result for oscillation of the problem (4.7.10).

Theorem 4.7.3. *Assume that (A_1)-(A_3) hold. Further assume that $\varphi(\xi)$ is nondecreasing for $\xi > 0$ and there exists an oscillatory function $r \in C^2(R_+, R)$ such that $r''(t) = G(t) + F(t)$,*

$$\int_E p(t)\varphi\big(r_*^+(t+\tau)\big)\,dt = \infty,$$

and (4.7.14)

$$\int_E p(t)\varphi\big(r_*^-(t+\tau)\big)\,dt = \infty.$$

Then every solution of problem (4.7.10) is oscillatory in G.

Example 4.7.3. Consider the problem

$$\begin{aligned} u_{tt}(x,t) + u_{tt}(x,t-\pi) - u_{xx}(x,t) + u(x,t) \\ = 2e^t \cos x(\sin t + \cos t - e^{-\pi} \cos t) \end{aligned} \tag{4.7.15}$$

where $(x,t) \in G = (0, \frac{\pi}{2}) \times (0, \infty)$ with the conditions

$$u_x(0,t) = 0, \quad u_x(\frac{\pi}{2}, t) = -e^t \sin t, \quad t \in (0, \infty).$$

It is obvious that the problem (4.7.15) satisfies (A_1)-(A_3) and

$$F(t) + G(t) = \frac{2}{\pi} e^t \sin t + \frac{4}{\pi}(1 - e^{-\pi}) e^t \cos t.$$

Thus

$$r(t) = -\frac{1}{\pi} e^t \sin t + \frac{2}{\pi}(1 - e^{-\pi}) e^t \sin t.$$

It is easy to see that

$$\int_E r_*^+(t+\pi)dt = \infty, \quad \int_E r_*^-(t+\pi)dt = \infty.$$

Therefore, by Theorem 4.7.3, every solution of the problem (4.7.15) oscillates in G. In fact, $u(x,t) = e^t \sin t \cos x$ is such a solution.

4.8. Equations with a Nonlinear Neutral Term

We consider the nonlinear neutral differential equations of the form

$$\left(x(t) - p x^\alpha(t-\tau)\right)'' + q(t) x^\beta(t-\sigma) = 0, \quad t \geq t_0 \tag{4.8.1}$$

where $p, \tau, \sigma \in R$, α and β are quotients of odd positive integers, and $q \in C([t_0, \infty), R)$.

Theorem 4.8.1. *Assume that*

(i) $p > 0$, $\sigma > \tau > 0$, α, $\beta \in (0,1]$, $q(t) \geq 0$, $t \geq t_0$;

(ii) *either*

$$\limsup_{t\to\infty} \int_{t-(\sigma-\tau)}^{t} [u - (\sigma - \tau)] q(u) du > 0 \quad \textit{for} \quad \beta < \alpha, \tag{4.8.2}$$

or

$$\limsup_{t\to\infty}\int_{t-(\sigma-\tau)}^{t}[u-(t-(\sigma-\tau))]q(u)du > p \quad \text{for} \quad \beta=\alpha, \tag{4.8.3}$$

where $p\in(0,1)$ *for* $\alpha=1$, $p\in(0,\infty)$ *for* $\alpha\in(0,1)$;

(iii) *every solution of the second order ordinary differential equation*

$$z''(t)+\lambda q(t)\left(\frac{t-\sigma}{t}\right)^{\beta}z^{\beta}(t)=0 \tag{4.8.4}$$

is oscillatory, where $0<\lambda<1$ *is a constant.*

Then every solution of Eq. (4.8.1) is oscillatory.

Proof: Without loss of generality, let $x(t)$ be an eventually positive solution of Eq. (4.8.1) and define

$$z(t)=x(t)-px^{\alpha}(t-\tau). \tag{4.8.5}$$

From (4.8.1), we know that $z''(t)\le 0$. If $z'(t)<0$ eventually, then $\lim_{t\to\infty}z(t)=-\infty$. Thus $\lim_{t\to\infty}x(t)=\infty$ and there exists a sequence $\{\xi_n\}$ such that $\lim_{n\to\infty}\xi_n=\infty$ and $x(\xi_n)=\max_{t_0\le t\le\xi_n}x(t)\to\infty$ as $n\to\infty$. Hence

$$\begin{aligned}z(\xi_n)=x(\xi_n)-px^{\alpha}(\xi_n-\tau)&\ge x(\xi_n)-px^{\alpha}(\xi_n)\\&=x(\xi_n)[1-px^{\alpha-1}(\xi_n)]\to\infty, \quad \text{as} \quad n\to\infty,\end{aligned}$$

a contradiction. Therefore, $z'(t)>0$. If $z(t)<0$, then $z(t)>-px^{\alpha}(t-\tau)$. Then

$$x(t-\tau)>\left(-\frac{z(t)}{p}\right)^{1/\alpha}. \tag{4.8.6}$$

Substituting (4.8.6) into (4.8.1) we have

$$z''(t)-q(t)\left(\frac{z(t-(\sigma-\tau))}{p}\right)^{\beta/\alpha}\le 0. \tag{4.8.7}$$

As in the proof of Theorem 4.3.1, (4.8.7) has no negative solution under the assumptions. This contradiction shows that $z(t)>0$. By Lemma 4.3.1, for each

$k \in (0,1)$, there is a $t_k \geq t_0$ such that

$$z(t-\sigma) \geq k \, \frac{t-\sigma}{t} \, z(t), \quad \text{for} \quad t \geq t_k. \tag{4.8.8}$$

Substituting (4.8.8) into (4.8.1) we have

$$z''(t) + k^\beta \left(\frac{t-\sigma}{t} \right)^\beta q(t) z^\beta(t) \leq 0 \tag{4.8.9}$$

which implies that Eq. (4.8.4) has a nonoscillatory solution, contradicting the assumptions (iii). □

Example 4.8.1. We consider

$$\left(x(t) - 6x^{1/3}(t-\pi) \right)'' + 9 \sin^2 t + x^{1/3}(t-2\pi) = 0, \quad t \geq 2\pi. \tag{4.8.10}$$

We see that

$$9 \int_{t-\pi}^{t} \left(u - (t-\pi) \right) \sin^2 u du = \frac{9\pi^2}{4} > 6.$$

That is, (4.8.3) holds. It is known (see [110]) that every solution of

$$z''(t) + 9\lambda \sin^2 t \left(\frac{t-2\pi}{t} \right)^{1/3} z^{1/3}(t) = 0 \tag{4.8.11}$$

is oscillatory for $0 < \lambda < 1$. Then, by Theorem 4.8.1, every solution of Eq. (4.8.10) oscillates. In fact, $x(t) = \sin^3 t$ is such a solution.

Example 4.8.2. We consider

$$\left(x(t) - \frac{1}{6} x^{1/3}(t-\pi) \right)'' + 9 \sin^{12/5} t \, x^{1/5}(t-2\pi) = 0, \quad t \geq 2\pi. \tag{4.8.12}$$

It is easy to see that all assumptions of Theorem 4.8.1 hold. Therefore, by Theorem 4.8.1, every solution of (4.8.12) oscillates. In fact, $x(t) = \sin^3 t$ is such a solution.

We now consider (4.8.1) of the unstable type. For the sake of convenience, we put $Q(t) \equiv -q(t) \geq 0$, $t \geq t_0$.

Theorem 4.8.2. *Assume that*

(i) $p,\ \tau,\ \sigma > 0$, and $\beta \in (0,1]$, $Q(t) \geq 0,\ t \geq t_0$;

(ii)
$$\limsup_{t\to\infty} \int_{t-\sigma}^{t} \big(s-(t-\sigma)\big)Q(s)ds > 1. \tag{4.8.13}$$

Then every bounded solution of (4.8.1) is oscillatory.

Proof: Assume the contrary, and let $x(t)$ be a bounded eventually positive solution of (4.8.1). Then $z''(t) \geq 0$. By the boundedness of z, we have $z'(t) < 0$ eventually. If $z(t) > 0$ eventually, integrating (4.8.1) twice we have

$$\begin{aligned} z'(t)\sigma - z(t) + z(t-\sigma) &= \int_{t-\sigma}^{t}\int_{s}^{t} Q(u)x^{\beta}(u-\sigma)duds \\ &= \int_{t-\sigma}^{t} [u-(t-\sigma)]Q(u)x^{\beta}(u-\sigma)du \\ &\geq z^{\beta}(t-\sigma)\int_{t-\sigma}^{t} [u-(t-\sigma)]Q(u)du. \end{aligned} \tag{4.8.14}$$

It is easy to see that $\lim_{t\to\infty} z(t) = 0$. Hence there exists a $T \geq t_0$ such that $z(t-\sigma) < 1$, for $t \geq T$. Thus (4.8.14) leads to

$$z(t) + z(t-\sigma)\left[\int_{t-\sigma}^{t} [u-(t-\sigma)]Q(u)du - 1\right] \leq 0$$

which contradicts (4.8.13).

If $z(t) < 0$, then $z(t) \leq -d < 0$ for some $d > 0$. Hence $-px^{\alpha}(t-\tau) \leq -d$, or $x^{\alpha}(t-\tau) \geq \frac{d}{p} > 0$. From (4.8.1) we get

$$z''(t) \geq \left(\frac{d}{p}\right)^{\beta/\alpha} Q(t). \tag{4.8.15}$$

We note from (4.8.13) that

$$\int_{T}^{\infty} tQ(t)dt = \infty. \tag{4.8.16}$$

Hence (4.8.15) implies that $\lim_{t\to\infty} z(t) = \infty$, we reach a contradiction. □

Example 4.8.3. We consider

$$\left(x(t) + \frac{4}{3}x^3(t-\pi)\right)'' = 9\sin^{8/3} t x^{1/3}(t-\pi). \tag{4.8.17}$$

It is easy to see that the conditions of Theorem 4.8.2 hold for (4.8.17). Hence every bounded solution of (4.8.17) oscillates. In fact, $x(t) = \sin t$ is such a solution.

Theorem 4.8.3. *Assume that*

(i) $p,\ \tau > 0,\ \sigma \geq 0,\ \alpha \geq 1,\ \beta > 0,\quad Q(t) \geq 0,\ t \geq t_0$;

(ii) *There exists a constant* $\lambda > 0$ *such that*

$$\alpha p \exp\{\lambda\alpha\tau + \lambda t(1-\alpha)\} \leq L < 1 \tag{4.8.18}$$

and

$$p\exp\{\lambda\alpha\tau + \lambda t(1-\alpha)\} + \int_t^\infty (s-t)Q(t)\exp\{\lambda(t-\beta(s-\sigma))\}ds \leq 1 \tag{4.8.19}$$

hold eventually.

Then Eq.(4.8.1) has a positive solution $x(t)$ which tends to zero as $t \to \infty$.

Proof: If the equality in (4.8.19) holds eventually. Then $x(t) = \exp(-\lambda t)$ is a positive solution of Eq. (4.8.1). Therefore, we may assume that there exists a $T > t_0$ such that $t - \tau \geq t_0,\ t - \sigma \geq t_0$ for $t \geq T$,

$$r := p\exp\{\lambda\alpha\tau + \lambda T(1-\alpha)\} + \int_T^\infty (s-T)Q(s)\exp\{\lambda(T-\beta(s-\sigma))\}ds < 1,$$

and (4.8.19) holds for all $t \geq T$.

As before, let BC denote the Banach space of all bounded continuous functions defined on $[t_0, \infty)$ with the sup norm. Let Ω be the subset of BC defined by

$$\Omega = \{y \in BC : 0 \leq y(t) \leq 1, \quad \text{for} \quad t \geq t_0\}.$$

It is easy to see that Ω is closed, convex and bounded in BC. Define a map $\mathcal{T} : \Omega \to BC$ by follows

$$(\mathcal{T}y)(t) = (\mathcal{T}_1 y)(t) + (\mathcal{T}_2 y)(t)$$

where

$$(\mathcal{T}_1 y)(t) = \begin{cases} p\exp\{\lambda\alpha\tau + \lambda t(1-\alpha)\}y^\alpha(t-\tau), & t \geq T \\ (\mathcal{T}_1 y)(T) + \exp\{\varepsilon(T-t)\} - 1, & t_0 \leq t \leq T, \end{cases}$$

$$(\mathcal{T}_2 y)(t) = \begin{cases} \int_t^\infty (s-t)Q(s)\exp\{\lambda(t-\beta(s-\sigma))\}y^\beta(s-\sigma)ds, & t \geq T \\ (\mathcal{T}_2 y)(T), & t_0 \leq t \leq T \end{cases}$$

and $\varepsilon = \ln\, 2/(T - t_0)$.

It is easy to see that for every pair $x, y \in \Omega$, $\mathcal{T}_1 x + \mathcal{T}_2 y \in \Omega$. And $\mathcal{T}_1$ is a contraction and $\mathcal{T}_2$ is completely continuous. By Krasnoselskii's fixed point theorem $\mathcal{T}$ has a fixed point $y \in \Omega$. That is,

$$y(t) = \begin{cases} p\exp\{\lambda\alpha\tau + \lambda t(1-\alpha)\}y^\alpha(t-\tau) & \\ \quad + \int_t^\infty (s-t)Q(s)\exp\{\lambda(t-\beta(s-\sigma))\}y^\beta(s-\sigma)ds, & t \geq T \\ y(T) + \exp\{\varepsilon(T-t)\} - 1, & t_0 \leq t \leq T. \end{cases}$$

Since $y(t) > 0$ for $t_0 \leq t < T$, it follows that $y(t) > 0$ for $t \geq t_0$. Set $x(t) = y(t)\exp(-\lambda t)$. Then

$$x(t) = px^\alpha(t-\tau) + \int_t^\infty (s-t)Q(s)x^\beta(s-\sigma)ds, \quad t \geq T,$$

which implies that $x(t)$ is a positive solution of (4.8.1). It is obvious that $x(t) \to 0$ as $t \to \infty$. □

Example 4.8.4. We consider

$$\left(x(t) - \frac{1}{2e}x^3(t-1)\right)'' = Q(t)x(t), \quad t > 10 \tag{4.8.20}$$

where

$$Q(t) = 1 - \frac{2}{t} - \frac{1}{t}e^{2t}\,(9t^3 - 45t^2 - 3t - 33). \tag{4.8.21}$$

It is not difficult to see that all assumptions of Theorem 4.8.3 are satisfied. Therefore, (4.8.20) has a bounded solution $x(t)$ which tends to zero as $t \to \infty$. In fact, $x(t) = te^{-t}$ is such a solution of (4.8.20).

4.9. Advanced Type Equations

We consider the second order advanced type differential equations of the form

$$\big(r(t)x'(t)\big)' + q(t)x\big(h(t)\big) = 0 \tag{4.9.1}$$

where $r \in C([t_0, \infty), (0, \infty))$, $\int_0^\infty \frac{dt}{r(t)} = \infty$, $q \in C([t_0, \infty), R_+)$, $q \not\equiv 0$, $h \in C([t_0, \infty), (0, \infty))$ and $h(t) \geq t$.

Theorem 4.9.1. *Assume that*

$$\int_{t_0}^\infty q(t)dt = \infty. \tag{4.9.2}$$

Then every solution of Eq. (4.9.1) is oscillatory.

Proof: Assume the contrary, and let $x(t)$ be an eventually positive solution. It is easy to see that $r(t)x'(t) > 0$ for $t \geq T \geq t_0$. Then

$$\int_T^\infty q(t)x\big(h(t)\big)dt < \infty \tag{4.9.3}$$

which contradicts (4.9.2). □

In the following we want to derive some oscillation criteria for (4.9.1) when

$$\int_{t_0}^\infty q(t)dt < \infty. \tag{4.9.4}$$

Lemma 4.9.1. *Let $x(t) > 0$, $t \geq t_1$, be a solution of Eq. (4.9.1). Set*

$$w(t) = r(t)x'(t)/x(t). \tag{4.9.5}$$

Then $w(t) > 0, \quad \lim_{t\to\infty} w(t) = 0,$

$$\int_{t_1}^{\infty} \frac{w^2(t)}{r(t)}\,dt < \infty, \tag{4.9.6}$$

and

$$w(t) = \int_t^{\infty} \frac{w^2(s)}{r(s)}\,ds + \int_r^{\infty} q(s)\exp\Big(\int_s^{h(s)} \frac{w(u)}{r(u)}du\Big)ds, \quad t \geq t_2 \geq t_1. \tag{4.9.7}$$

Proof: From (4.9.1) we have

$$\big(x(t)w(t)\big)' + q(t)x\big(h(t)\big) = 0,$$

since

$$\frac{x\big(h(t)\big)}{x(t)} = \exp\int_t^{h(t)} \frac{w(s)}{r(s)}\,ds,$$

$$w'(t) + \frac{w^2(t)}{r(t)} + q(t)\exp\int_t^{h(t)} \frac{w(s)}{r(s)}\,ds = 0. \tag{4.9.8}$$

Integrating it from t to T for $T \geq t \geq t_1$, we have

$$w(T) - w(t) + \int_t^T \frac{w^2(s)}{r(s)}ds + \int_t^T q(s)\exp\bigg(\int_s^{h(s)} \frac{w(u)}{r(u)}du\bigg)ds = 0. \tag{4.9.9}$$

Because $r(t)x'(t) > 0$, so $w(t) > 0$. We shall show that $\lim_{t\to\infty} w(t) = 0$. In fact, if $\lim_{t\to\infty} r(t)x'(t) = c > 0$, then there exists a $t_2 \geq t_1$ such that for $t \geq t_2$

$$x(t) \geq \bigg[x(t_2) + \int_{t_2}^{t} \frac{c}{2r(s)}\,ds\bigg] \to \infty, \quad \text{as} \quad t \to \infty,$$

and hence $\lim_{t\to\infty} w(t) = 0$. If $\lim_{t\to\infty} r(t)x'(t) = 0$, then $\lim_{t\to\infty} w(t) = 0$ also. Letting $T \to \infty$ in (4.9.9) we obtain (4.9.7). □

Lemma 4.9.2. *Eq. (4.9.1) has a nonoscillatory solution if and only if there* *exists a positive differentiable function* $\varphi(t)$ *such that*

$$\varphi'(t) + \frac{\varphi^2(t)}{r(t)} \leq -q(t)\exp\biggl(\int_t^{h(t)} \frac{\varphi(s)}{r(s)}\, ds\biggr), \quad \text{for} \quad t \geq t_2. \tag{4.9.10}$$

Proof: The necessity follows from Lemma 4.8.1. Now we assume that (4.9.10) holds. Then $\varphi'(t) < 0$ and hence $\lim_{t\to\infty} \varphi(t) = d \geq 0$. If $d > 0$, (4.9.10) leads to $\lim_{t\to\infty} \varphi(t) = -\infty$, a contradiction. Therefore, $\lim_{t\to\infty} \varphi(t) = 0$. Integrating (4.9.10) from t to ∞ we obtain

$$\int_t^\infty \frac{\varphi^2(s)}{r(s)} ds + \int_t^\infty q(s)\exp\biggl(\int_s^{h(s)} \frac{\varphi(u)}{r(u)}\, du\biggr) ds \leq \varphi(t), \quad t \geq t_2 \tag{4.9.11}$$

which implies that

$$\int_t^\infty \frac{\varphi^2(s)}{r(s)}\, ds < \infty \quad \text{and} \quad \int_t^\infty q(s)\exp\biggl(\int_s^{h(s)} \frac{\varphi(u)}{r(u)}\, du\biggr) ds < \infty.$$

For all functions $y(t)$ satisfying $0 \leq y(t) \leq \varphi(t)$, $t \geq t_2$, define a mapping $\mathcal{T}$ by

$$(\mathcal{T}y)(t) = \int_t^\infty \frac{y^2(s)}{r(s)} ds + \int_t^\infty q(s)\exp\biggl(\int_s^{h(s)} \frac{y(u)}{r(u)} du\biggr) ds, \quad t \geq t_2.$$

It is easy to see that $0 \leq y_1(t) \leq y_2(t)$, $t \geq t_2$, implies $(\mathcal{T}y_1)(t) \leq (\mathcal{T}y_2)(t)$, $t \geq t_2$.

Define $x_0(t) \equiv 0$ and $x_n(t) = \mathcal{T}x_{n-1}(t)$, $n = 1, 2, \ldots$. Then $x_{n-1}(t) \leq x_n(t) \leq \varphi(t)$, $n = 1, 2, \ldots$, and $\lim_{n\to\infty} x_n(t) = w(t) \leq \varphi(t)$. By the Lebesgue's dominated convergence theorem we have

$$w(t) = \int_t^\infty \frac{w^2(s)}{r(s)} ds + \int_t^\infty q(s)\exp\biggl(\int_s^{h(s)} \frac{w(u)}{r(u)} du\biggr) ds, \quad t \geq t_2.$$

Set

$$x(t) = \exp\biggl(\int_{t_2}^t \frac{w(u)}{r(u)}\, du\biggr), \quad t \geq t_2.$$

Then $w(t) = r(t)x'(t)/x(t)$ and

$$\bigl(r(t)x'(t)\bigr)' + q(t)x\bigl(h(t)\bigr) = 0, \quad t \geq t_2,$$

i.e., $x(t)$ is a nonoscillatory solution of (4.9.1). □

Theorem 4.9.2. *If Eq. (4.9.1) has a nonoscillatory solution, then the second order linear differential equation*

$$\left(r(t)x'(t)\right)' + q(t)x(t) = 0 \tag{4.9.12}$$

is nonoscillatory. Conversely, if Eq. (4.9.12) is oscillatory, then every solution of Eq. (4.9.1) is oscillatory.

Proof: Assume Eq. (4.9.1) has a nonoscillatory solution. By Lemma 4.9.2 there exists a positive differentiable function$\varphi(t)$ such that

$$\varphi'(t) + \frac{\varphi^2(t)}{r(t)} \leq -q(t)\exp\left(\int_t^{h(t)} \frac{\varphi(u)}{r(u)}du\right), \quad t \geq t_2 \tag{4.9.13}$$

which implies that

$$\varphi'(t) + \frac{\varphi^2(t)}{r(t)} \leq -q(t). \tag{4.9.14}$$

By Lemma 4.9.2 for the case that $h(t) \equiv t$, (4.9.12) is nonoscillatory.

Then the second part of the theorem is immediately obtained. □

4.10. Notes

Section 4.1 is extracted from Yu [175]. Section 4.2.1 is taken from Chen [17]. Section 4.2.2 is taken from [110]. The existence of oscillatory solutions of first order unstable delay differential equations can be seen from Györi [72]. Theorem 4.4.1 and Theorem 4.4.2 are adopted from Erbe and Kong [39] also see Erbe and Zhang [43]. Theorem 4.4.3 is based on [43]. Theorem 4.4.4 is adopted from Grammatikopoulos, Ladas and Meimaridon [63]. Theorem 4.5.1 to Theorem 4.5.5 are from Lu [129]. Theorem 4.5.6 is adopted from Zhang and Yu [213]. The material in Section 4.6 is taken from [213], Yu [279], Yu and Zhang [185], and Zhang [196]. Theorem 4.7.1 and Theorem 4.7.3 are from Zhang [197]. Theorem 4.7.2 is new. The oscillation properties of the hyperbolic equations with neutral type was studied first by Mishev and Bainov [137], also see Yu and Cui [188]. Section 4.8 is from Zhang [190]. Section 4.9 is taken from Lu [131].

5

Oscillation of Higher Order Neutral Differential Equations

5.0. Introduction

In this chapter we investigate the neutral differential equations of odd order and of even order separately.

In Section 5.1 we present some comparison results for oscillation of odd order equations. In Section 5.2 we give some criteria for oscillation and for the existence of positive solutions of odd order equations. In Section 5.3 we obtain some oscillation results for even order equations. Section 5.4 deals with the classification of nonoscillatory solutions of higher order equations, the existence results for solutions of various asymptotic behavior are presented. Section 5.5 is concerned with the existence of oscillatory solutions of higher order equations. In Section 5.6 we study the oscillation and nonoscillation of equations with nonlinear neutral terms. And finally, in Section 5.7, the behavior of solutions of even order neutral differential equations of unstable type are investigated in detail.

5.1. Comparison Theorems for Odd Order Equations

We consider the odd order neutral delay differential equations of the form

$$(x(t) - p(t)x(t-\tau))^{(n)} + q(t)x(t-\sigma(t)) = 0, \quad t \geq t_0 \tag{5.1.1}$$

where

$$\begin{gathered} n \geq 1 \text{ is an odd integer, } \tau \in (0,\infty),\ p,\, q,\, \sigma \in C([t_0,\infty), R_+), \\ q(t) \not\equiv 0, \quad t - \sigma(t) \to \infty \text{ as } t \to \infty. \end{gathered} \tag{5.1.2}$$

For higher order equations the following lemmas are useful. We only state the results and only prove Lemmas 5.1.4 and 5.1.5.

Lemma 5.1.1. *Let $y(t)$ be an n times differentiable function on R_+ of constant sign, $y^{(n)}(t) \not\equiv 0$ on $[t_1,\infty)$ which satisfies $y^{(n)}(t)y(t) \leq 0$. Then*

(i) There exists $t_2 \geq t_1$ such that the functions $y^{(j)}(t)$, $j = 1, 2, \ldots, n-1$, are of constant sign on $[t_2,\infty)$.

(ii) There exists a number $k \in \{1,3,5,\ldots,n-1\}$ when n is even, or $k \in \{0,2,4,\ldots,n-1\}$ when n is odd, such that

$$\begin{aligned} y(t)y^{(j)}(t) > 0 \quad & \text{for } j = 0,1,\ldots,k,\ t \geq t_2 \\ (-1)^{n+j+1}y(t)y^{(j)}(t) > 0 \quad & \text{for } j = k+1,\ldots,n,\ t \geq t_2. \end{aligned} \tag{5.1.3}$$

Lemma 5.1.2. *Assume that $y, y', \ldots, y^{(n-1)}$ are absolutely continuous and of constant sign on the interval (t_0,∞), and assume $y^{(n)}(t)y(t) \geq 0$. Then either $y^{(k)}(t)y(t) \geq 0$, $k = 0,1,\ldots,n$ or there exists an integer ℓ, $0 \leq \ell \leq n-2$, which is even when n is even and odd when n is odd, such that*

$$\begin{aligned} y^{(k)}(t)y(t) \geq 0, \quad & k = 0,1,\ldots,\ell \\ (-1)^{n+k}y^{(k)}(t)y(t) \geq 0, \quad & k = \ell+1,\ldots,n. \end{aligned}$$

Lemma 5.1.3. *Assume that the hypotheses of Lemma 5.1.1 hold and $y^{(n)}(t)y^{(n-1)}(t) \leq 0$ for $t \geq t_0$. Then for every $\lambda \in (0,1)$, there exists an $M > 0$ such that $|y(\lambda t)| \geq Mt^{n-1}|y^{(n-1)}(t)|$ for all large t.*

Lemma 5.1.4. *Assume that (5.1.2) holds and that $0 \le p(t) \le 1$. Let $x(t)$ be an eventually positive solution of the inequality*

$$(x(t) - p(t)x(t-\tau))^{(n)} + q(t)x(t-\sigma(t)) \le 0 \tag{5.1.4}$$

and set

$$y(t) = x(t) - p(t)x(t-\tau). \tag{5.1.5}$$

Then $y(t) > 0$ eventually.

Proof: From (5.1.4) and (5.1.5), $y^{(n)}(t) \le -q(t)x(t-\sigma(t)) \le 0$. Since $q(t) \not\equiv 0$, $y^{(n-1)}(t)$ is eventually positive or eventually negative. Hence $y^{(i)}(t)$ is strictly monotonic for $i = 0, 1, \ldots, n-2$. This implies that $y(t)$ is eventually positive or eventually negative. If $y(t)$ is eventually negative, i.e., $y^{(n)}(t)y(t) > 0$ eventually, by Lemma 5.1.2, $y'(t) < 0$ eventually. Therefore, there exist $\ell > 0$ and $t_1 \ge t_0$ such that $y(t) \le -\ell, \quad t \ge t_1$. Thus

$$x(t) \le -\ell + p(t)x(t-\tau) \le -\ell + x(t-\tau), \quad t \ge t_1.$$

In particular,

$$x(t_1 + i\tau) \le -(i+1)\ell + x(t_1 - \tau) \to -\infty \quad \text{as } i \to \infty$$

which is impossible. Therefore, $y(t)$ must be eventually positive. □

Lemma 5.1.5. *Assume that (5.1.2) holds, and either*

$$p(t) + q(t)\sigma(t) > 0 \quad \text{for } t \ge T, \tag{5.1.6}$$

or

$$\sigma(t) > 0 \quad \text{and} \quad q(s) \not\equiv 0 \quad \text{for } s \in [t, T^*] \quad \text{and} \quad t \ge T \tag{5.1.7}$$

where T^* satisfies that $T^* - \sigma(T^*) = t$.

Let $b = \max\{\tau,\ T - \min_{t \ge T}\{t - \sigma(t)\}\}$, and assume that the integral inequality

$$z(t) \ge p(t)z(t-\tau) + \frac{1}{(n-1)!}\int_t^\infty (s-t)^{n-1} q(s) z(s-\sigma(s))\, ds, \quad t \ge T \tag{5.1.8}$$

has a continuous positive solution $y : [T-b,\infty) \to (0,\infty)$. *Then the corresponding integral equation*

$$x(t) = p(t)x(t-\tau) + \frac{1}{(n-1)!}\int_t^\infty (s-t)^{n-1}q(s)x(s-\sigma(s))\,ds, \quad t \geq T \quad (5.1.9)$$

has a continuous positive solution $x : [T-b,\infty) \to (0,\infty)$.

Proof: Define a set of functions Ω and a mapping $\mathcal{T}$ as follows:

$$\Omega = \{w \in C([T-b,\infty), R_+) : 0 \leq w(t) \leq 1,\ t \geq T-b\},$$

$$(\mathcal{T}w)(t) = \begin{cases} \dfrac{1}{z(t)}\Bigg[p(t)w(t-\tau)z(t-\tau) + \dfrac{1}{(n-1)!} \\ \qquad \displaystyle\int_t^\infty (s-t)^{n-1}q(s)w(s-\sigma(s))z(s-\sigma(s))\,ds\Bigg], & t \geq T \\ \dfrac{t-T+b}{b}(\mathcal{T}w)(T) + 1 - \dfrac{t-T+b}{b}, & T-b \leq t \leq T. \end{cases} \quad (5.1.10)$$

It is easy to see from (5.1.8) that $\mathcal{T}\Omega \subset \Omega$ and $(\mathcal{T}w)(t) > 0$ on $[T-b,T)$ for any $w \in \Omega$.

Define a sequence $\{w_k(t)\}$ in Ω as follows: $w_0(t) = 1$, and $w_{k+1}(t) = (\mathcal{T}w_k)(t)$, $k = 0,1,2,\ldots$, $t \geq T-b$. From (5.1.8), by induction, we have

$$0 \leq w_{k+1}(t) \leq w_k(t) \leq 1, \quad t \geq T-b,\ k = 0,1,2,\ldots.$$

Then $w(t) = \lim_{k\to\infty} w_k(t)$, $\quad t \geq T-b$, exists, and

$$w(t) = \frac{1}{z(t)}\Bigg[p(t)w(t-\tau)z(t-\tau) + \frac{1}{(n-1)!}\int_t^\infty (s-t)^{n-1}q(s)w(s-\sigma(s))z(s-\sigma(s))\,ds\Bigg], \quad t \geq T,$$

and

$$w(t) = \frac{t-T+b}{b}w(T) + 1 - \frac{t-T+b}{b} > 0 \quad \text{for } T-b \leq t < T.$$

Set $x(t) = w(t)z(t)$. Then $x(t)$ satisfies (5.1.9) and $x(t) > 0$ for $t \in [T-b,T)$. Clearly, $x(t)$ is continuous on $[T-b,T]$. Then, in view of (5.1.9), we see that $x(t)$ is continuous on $[T-b,\infty)$.

Finally, it remains to show that $x(t) > 0$ for $t \geq T - b$. Assume that there exists $t^* \geq T$ such that $x(t) > 0$ for $T - b \leq t < t^*$ and $x(t^*) = 0$. Thus, by (5.1.9), we have

$$0 = x(t^*) = p(t^*)x(t^* - \tau) + \frac{1}{(n-1)!}\int_{t^*}^{\infty} (s - t^*)^{n-1} q(s)x(s - \sigma(s))\, ds$$

which implies that $p(t^*) = 0$ and $q(s)x(s - \sigma(s)) = 0$ for all $s \geq t^*$ which contradict (5.1.6) and (5.1.7). Therefore, $x(t) > 0$ on $[T - b, \infty)$. □

Lemma 5.1.6. *Assume (5.1.2) holds, $k \in \{1, 2, \ldots, n-1\}$, $c > 0$, and the integral inequality*

$$z(t) \geq c + p(t)z(t - \tau) + \int_T^t (t - u)^{k-1} \times \int_u^{\infty} (s - u)^{n-k-1} q(s)z(s - \sigma(s))\, ds, \quad t \geq T, \tag{5.1.11}$$

has a continuous positive solution $z : [T - b, \infty) \to (0, \infty)$ where b is defined as in Lemma 5.1.5. Then the corresponding integral equation

$$x(t) = c + p(t)x(t - \tau) + \int_T^t (t - u)^{k-1} \int_u^{\infty} (s - u)^{n-k-1} q(s)x(s - \sigma(s))\, ds, \quad t \geq T, \tag{5.1.12}$$

also has a continuous positive solution $x : [T - b, \infty) \to (0, \infty)$.

The proof of Lemma 5.1.6 is similar to the proof of Lemma 5.1.5 and hence we omit it here.

Theorem 5.1.1. *Let $0 \leq p(t) \leq 1$. Assume that (5.1.2) and the assumptions of Lemma 5.1.5 hold. Then every solution of Eq. (5.1.1) is oscillatory if and only if the corresponding differential inequality (5.1.4) has no eventually positive solutions.*

Proof: The sufficiency is obvious. To prove the necessity, we assume that (5.1.4) has an eventually positive solution $x(t)$. Set $y(t) = x(t) - p(t)x(t - \tau)$.

By Lemma 5.1.5, $y(t) > 0$ eventually. According to Lemma 5.1.1, there exists an even number k such that $0 \le k \le n-1$, and

$$\begin{aligned} (-1)^i y^{(i)}(t) > 0 \quad \text{for } i = k+1, \dots, n-1, \\ y^{(i)}(t) > 0 \quad \text{for } i = 0, 1, \dots, k. \end{aligned} \tag{5.1.13}$$

If $k = 0$, integrating (5.1.4) from t to ∞ we have

$$y(t) \ge \frac{1}{(n-1)!} \int_t^\infty (s-t)^{n-1} q(s) x(s - \sigma(s))\, ds.$$

That is,

$$x(t) \ge p(t)x(t-\tau) + \frac{1}{(n-1)!} \int_t^\infty (s-t)^{n-1} q(s) x(s - \sigma(s))\, ds.$$

By Lemma 5.1.5, the corresponding integral equation

$$z(t) = p(t)z(t-\tau) + \frac{1}{(n-1)!} \int_t^\infty (s-t)^{n-1} q(s) z(s - \sigma(s))\, ds$$

also has a positive solution $z(t)$. Clearly, $z(t)$ is an eventually positive solution of (5.1.1), contradicting the assumption.

If $2 \le k \le n-1$, then integrating (5.1.4) from t to ∞ $n-k$ times, we have

$$y^{(k)}(t) \ge \frac{1}{(n-k-1)!} \int_t^\infty (s-t)^{n-k-1} q(s) x(s - \sigma(s))\, ds. \tag{5.1.14}$$

Then integrating (5.1.4) from T to t k times we have

$$\begin{aligned} y(t) \ge y(T) + \frac{1}{(k-1)!\,(n-k-1)!} \int_T^t (t-u)^{k-1} \\ \times \int_u^\infty (s-u)^{n-k-1} q(s) x(s-\sigma(s))\, ds\, du, \quad t \ge T, \end{aligned}$$

where T is sufficiently large such that $y(T) > 0$. Thus we have, for $t \ge T$,

$$\begin{aligned} x(t) &\ge y(T) + p(t)x(t-\tau) \\ &+ \frac{1}{(k-1)!\,(n-k-1)!} \int_T^t (t-u)^{k-1} \int_u^\infty (s-u)^{n-k-1} q(s) x(s-\sigma(s))\, ds\, du. \end{aligned}$$

By Lemma 5.1.6, the corresponding integral equation

$$z(t) = y(T) + p(t)z(t-\tau) + \frac{1}{(k-1)!\,(n-k-1)!}\int_T^t (t-u)^{k-1}$$
$$\times \int_u^\infty (s-u)^{n-k-1} q(s) z(s-\sigma(s))\,ds\,du, \quad t \ge T,$$

has a positive solution $z(t)$. Clearly, $z(t)$ is a positive solution of Eq. (5.1.1). This contradiction completes the proof. □

We now show some applications of Theorem 5.1.1. We compare Eq. (5.1.1) with the equation

$$\left(x(t) - p^*(t)x(t-\tau)\right)^{(n)} + q^*(t)x(t-\sigma(t)) = 0 \tag{5.1.15}$$

where $p^*,\ q^* \in C([t_0,\infty), R_+)$.

Theorem 5.1.2. *Assume that the assumptions of Theorem 5.1.1 hold and*

$$p(t) \le p^*(t) \le 1, \qquad q(t) \le q^*(t), \quad \text{for } t \ge t_0. \tag{5.1.16}$$

Then every solution of Eq. (5.1.1) is oscillatory implies the same for Eq. (5.1.15).

Proof: Assume the contrary, and let $x(t)$ be an eventually positive solution of (5.1.15). Let $y(t) = x(t) - p^*(t)x(t-\tau)$. As in the proof of Theorem 5.1.1 we see that $y(t) > 0,\ y'(t) < 0$, for large t and (5.1.13) holds.

If $k = 0$, integrating (5.1.15) from t to ∞ n times we have

$$y(t) = y(\infty) + \frac{1}{(n-1)!}\int_t^\infty (s-t)^{n-1} q^*(s) x(s-\sigma(s))\,ds,$$

and so

$$x(t) = y(\infty) + p^*(t)x(t-\tau) + \frac{1}{(n-1)!}\int_t^\infty (s-t)^{n-1} q^*(s)x(s-\sigma(s))\,ds$$
$$\ge p(t)x(t-\tau) + \frac{1}{(n-1)!}\int_t^\infty (s-t)^{n-1} q(s)x(s-\sigma(s))\,ds$$

where $y(\infty) = \lim_{t\to\infty} y(t) \geq 0$. Using a similar method as in the proof of Theorem 5.1.1, one can see that Eq. (5.1.1) also has an eventually positive solution which contradicts the assumption.

If $k > 0$, as shown in the proof of Theorem 5.1.1, we obtain that for $t \geq T$

$$y(t) \geq y(T) + \frac{1}{(k-1)!\,(n-k-1)!} \int_T^t (t-u)^{k-1} \times \int_u^\infty (s-u)^{n-k-1} q^*(s) x(s-\sigma(s))\, ds\, du.$$

Hence for $t \geq T$

$$\begin{aligned} x(t) &\geq y(T) + p^*(t)x(t-\tau) + \frac{1}{(k-1)!\,(n-k-1)!} \\ &\quad \times \int_T^t (t-u)^{k-1} \int_u^\infty (s-u)^{n-k-1} q^*(s) x(s-\sigma(s))\, ds\, du \\ &\geq y(T) + p(t)x(t-\tau) + \frac{1}{(k-1)!\,(n-k-1)!} \\ &\quad \times \int_T^t (t-u)^{k-1} \int_u^\infty (s-u)^{n-k-1} q(s) x(s-\sigma(s))\, ds\, du. \end{aligned}$$

The rest of the proof is the same as that of Theorem 5.1.1. □

We now state a similar result for equations of the more general form

$$\big(x(t) - p(t)x(t-\tau)\big)^{(n)} + q(t)x(t-\sigma(t)) + F(t, x(t), x(\sigma_1(t)), \ldots, x(\sigma_k(t))) = 0. \tag{5.1.17}$$

Theorem 5.1.3. *Assume that*

(i) *the assumptions of Theorem 5.1.1 hold;*

(ii) $F \in C([t_0,\infty) \times R^{k+1}, R)$, $F(t, y_0, y_1, \ldots, y_k)y_0 > 0$ *as* $y_0 y_i > 0$, $i = 1, 2, \ldots, k$;

(iii) $\sigma_i \in C([t_0,\infty), R)$, $\lim_{t\to\infty} \sigma_i(t) = \infty$, $i = 1, 2, \ldots, k$.

Then every solution of Eq. (5.1.1.) is oscillatory implies the same for Eq. (5.1.17).

Remark 5.1.1. Every solution of Eq. (5.1.1) with $p(t) \equiv 0$ is oscillatory implies the same for Eq. (5.1.1) with $0 \leq p(t) \leq 1$. In fact, if (5.1.1) with $0 \leq p(t) \leq 1$

has an eventually positive solution $x(t)$, by Lemma 5.1.4, $y(t) > 0$ and hence

$$y^{(n)}(t) + q(t)y(t - \sigma(t)) \leq 0,$$

which implies that

$$x^{(n)}(t) + q(t)x(t - \sigma(t)) = 0$$

has a nonoscillatory solution.

But the inverse proposition is not true in general. To see this, we show the following example.

Example 5.1.1. Every solution of

$$\left(x(t) - px(t-1)\right)' + x(t - \tfrac{1}{e}) = 0, \quad p \in (0,1), \tag{5.1.18}$$

oscillates. But, for $p = 0$, (5.1.18) has a nonoscillatory solution.

From the above discussion we find that the presence of the neutral term may create oscillation of solutions.

In the sequel, for the sake of convenience, we define

$$\prod_{j=1}^{0} p(s - \sigma - (j-1)\tau) \equiv 1. \tag{5.1.19}$$

Theorem 5.1.4. *Let $0 \leq p(t) \leq 1$. Assume (5.1.2) holds with $\sigma(t) \equiv \sigma > 0$. Suppose that there exist two nonnegative integers m and N with $m \leq N$, and there exists an $i_0 \in \{m, m+1, \ldots, N\}$ such that for all large t*

$$q(s) \prod_{j=1}^{i_0} p(s - \sigma - (j-1)\tau) \not\equiv 0 \quad \text{for} \quad s \in [t, t + \sigma + i_0\tau]. \tag{5.1.20}$$

Then every solution of the delay differential equation

$$u^{(n)}(t) + \sum_{i=m}^{N} q(t) \left(\prod_{j=1}^{i} p(t - \sigma - (j-1)\tau) \right) u(t - \sigma - i\tau) = 0 \tag{5.1.21}$$

is oscillatory implies the same for Eq. (5.1.1).

Proof: Assume the contrary, and let $x(t)$ be an eventually positive solution of Eq. (5.1.1). Then $y(t) > 0$ eventually and (5.1.13) holds. From (5.1.1)

$$y^{(n)}(t) + q(t)y(t-\sigma) + q(t)p(t-\sigma)x(t-\sigma-\tau) = 0.$$

By induction

$$y^{(n)}(t) + \sum_{i=0}^{N} q(t)\Big(\prod_{j=1}^{i} p(t-\sigma-(j-1)\tau)\Big)y(t-\sigma-i\tau)$$
$$+ q(t)\Big(\prod_{j=1}^{N+1} p(t-\sigma-(j-1)\tau)\Big)x(t-\sigma-(N+1)\tau) = 0.$$

Hence for sufficiently large t

$$y^{(n)}(t) + \sum_{i=m}^{N} q(t)\Big(\prod_{j=1}^{i} p(t-\sigma-(j-1)\tau)\Big)y(t-\sigma-i\tau) \le 0. \qquad (5.1.22)$$

Consider the k defined by (5.1.3). If $k = 0$, integrating (5.1.22) n times, we have

$$y(t) \ge \frac{1}{(n-1)!}\sum_{i=m}^{N}\int_t^\infty (s-t)^{n-1}q(s)\Big(\prod_{j=1}^{i} p(s-\sigma-(j-1)\tau)\Big)y(s-\sigma-i\tau)\,ds.$$

By a slight modification of the proof of Lemma 5.1.2, we have

$$z(t) = \frac{1}{(n-1)!}\sum_{i=m}^{N}\int_t^\infty (s-t)^{n-1}q(s)\Big(\prod_{j=1}^{i} p(s-\sigma-(j-1)\tau)\Big)z(s-\sigma-i\tau)\,ds$$

has a continuous positive solution $z(t)\colon [T-b,\infty) \to (0,\infty)$ for some $T \ge t_0$, where $b = \max\{T,\sigma\}$. It is easy to see that $z(t)$ is a positive solution of (5.1.21).

If $k > 0$, similar to the proof of Lemma 5.1.3 we derive that the integral equation

$$y(t) = y(T) + \frac{1}{(k-1)(n-k-1)!}\sum_{i=m}^{N}\int_T^t (t-u)^{k-1}\int_u^\infty (s-u)^{n-k-1}q(s)$$
$$\times\Big(\prod_{j=1}^{i} p(s-\sigma-(j-1)\tau)\Big)y(s-\sigma-i\tau)\,ds\,du$$

has a continuous positive solution $y(t)$ on $[T-b,\infty)$. Clearly, $y(t)$ is a solution of (5.1.21). □

Lemma 5.1.7. *In addition to (5.1.2) assume that* $0 \le p(t) \le p < 1$ *and*

$$\int_{t_0}^{\infty} q(t)(t-\sigma(t))^{n-2}\,dt = \infty \quad \text{for } n \ge 3, \tag{5.1.23}$$

or

$$\int_{t_0}^{\infty} q(t)\,dt = \infty, \quad \text{for } n = 1. \tag{5.1.24}$$

Then every nonoscillatory solution $x(t)$ *of Eq. (5.1.1) tends to zero as* $t \to \infty$, *and* $(-1)^i y^{(i)}(t) > 0$ *eventually. Moreover,* $\lim_{t\to\infty} y^{(i)}(t) = 0$, $i = 0,1,2,\ldots,n-1$, *where* y *is defined by (5.1.5).*

Proof: Without loss of generality assume $x(t)$ is an eventually positive solution of (5.1.1). By Lemmas 5.1.1 and 5.1.4, $y(t) > 0$ eventually and (5.1.13) holds. We claim that $k = 0$. Otherwise $k \ge 2$ and hence $y^{(i)}(t) > 0$, $i = 0,\ldots,k$, which implies that there exist $d > 0$ and $T \ge t_0$ such that $y(t) \ge dt^{k-1}$, for $t \ge T$, and hence $x(t) \ge y(t) \ge dt^{k-1}$ for $t \ge T$. Substituting this into Eq. (5.1.1) we have

$$y^{(n)}(t) + dq(t)(t-\sigma(t))^{k-1} \le 0, \quad t \ge T. \tag{5.1.25}$$

Hence

$$t^{n-k-1}y^{(n)}(t) \le -dq(t)(t-\sigma(t))^{n-2}, \quad t \ge T.$$

In view of (5.1.23), we get

$$\int_T^t s^{n-k-1}y^{(n)}(s)\,ds \longrightarrow -\infty \quad \text{as} \quad t \to \infty. \tag{5.1.26}$$

On the other hand,

$$\int_T^t s^{n-k-1}y^{(n)}(s)\,ds = F(t) - F(T) \tag{5.1.27}$$

where

$$F(t) = t^{n-k-1}y^{(n-1)}(t) + \sum_{i=0}^{n-k-2} (-1)^i(i+1)\cdots(n-k-1)t^i y^{(k+i)}(t) > 0. \quad (5.1.28)$$

This contradicts (5.1.26) and (5.1.13). Therefore, $k = 0$ and $\lim_{t\to\infty} y^{(i)}(t) = 0$, $i = 1, 2, \ldots, n-1$. From (5.1.13) $y'(t) < 0$. Thus $\lim_{t\to\infty} y(t) = \ell \geq 0$. If $\ell > 0$, then $x(t) \geq \ell > 0$, $t \geq T \geq t_0$, and hence from (5.1.1) we have $y^{(n)}(t) + \ell q(t) \leq 0$, $t \geq T$. From (5.1.23), we have

$$\int_T^t s^{n-2} y^{(n)}(s)\, ds \longrightarrow -\infty \quad \text{as} \quad t \to \infty.$$

It contradicts (5.1.13). Therefore, $\ell = 0$. We show that $x(t)$ is bounded. Otherwise, there exists a sequence $\{t_n\}$ such that $\lim_{n\to\infty} t_n = \infty$, $x(t_n) = \max_{s\leq t_n} x(s)$, and $\lim_{n\to\infty} x(t_n) = \infty$. Then

$$y(t_n) = x(t_n) - p(t_n)x(t_n - \tau) \geq (1-p)x(t_n) \to \infty \quad \text{as } n \to \infty$$

which is impossible. Let $\xi_n \to \infty$ be such that

$$\alpha = \limsup_{t\to\infty} x(t) = \lim_{n\to\infty} x(\xi_n).$$

Since

$$p(\xi_n)x(\xi_n) \geq p(\xi_n)x(\xi_n - \tau) = x(\xi_n) - y(\xi_n),$$

we get $0 \leq \alpha \leq p\alpha$. This implies that $\alpha = 0$ and hence $x(t) \to 0$ as $t \to \infty$. □

Using Lemma 5.1.7 we can compare the odd order delay differential equation

$$x^{(n)}(t) + p(t)x(g(t)) = 0 \quad (5.1.29)$$

with the equation

$$x^{(n)}(t) + q(t)x(h(t)) = 0. \quad (5.1.30)$$

Theorem 5.1.5. *Assume that* $p(t) \le q(t)$, $h(t) \le g(t) < t$,

$$\int_{t_0}^{\infty} q(t)h(t)^{n-2}\,dt = \infty, \quad \text{for } n \ge 3, \tag{5.1.31}$$

or

$$\int_{t_0}^{\infty} q(t)\,dt = \infty, \quad \text{for } n = 1. \tag{5.1.32}$$

Then every solution of Eq. (5.1.29) is oscillatory implies the same for Eq. (5.1.30).

Proof: Assume the contrary, and let $x(t)$ be an eventually positive solution of (5.1.30). By Lemma 5.1.7, $x'(t) < 0$ and $\lim_{t\to\infty} x(t) = 0$. We see that

$$0 = x^{(n)}(t) + q(t)x(h(t)) \ge x^{(n)}(t) + p(t)x(g(t)).$$

By Theorem 5.1.1, (5.1.29) has a positive solution. □

Remark 5.1.2. If (5.1.23) is replaced by

$$\int_{t_0}^{\infty} q(t)(t - \sigma(t))^{n-1}\,dt = \infty, \tag{5.1.33}$$

then Lemma 5.1.7 is true for the bounded nonoscillatory solutions of (5.1.1).

Remark 5.1.3. The above results can be extended to equations with several delays of the form

$$\big(x(t) - p(t)x(t-\tau)\big)^{(n)} + \sum_{i=1}^{n} q_i(t)x(t - \sigma_i(t)) = 0. \tag{5.1.34}$$

We now discuss the nonlinear neutral differential equation

$$\big(x(t) - p(t)x(t-\tau)\big)^{(n)} + \sum_{i=0}^{k} q_i(t)f_i(x(t - \sigma_i(t))) = 0, \quad t \ge t_0, \tag{5.1.35}$$

and present the following comparison theorem which is parallel to Theorem 5.1.4 for the linear case.

Theorem 5.1.6. *Assume that*

(i) $\tau > 0$, p, q_i, $\sigma_i \in C([t_0, \infty), R_+)$, $\lim_{t\to\infty}(t - \sigma_i(t)) = \infty$, $i = 0, 1, \ldots, k$, *and there exists* $i_0 \in \{0, 1, \ldots, k\}$ *such that* $q_{i_0}(t) > 0$, $\sigma_{i_0}(t) > 0$ *and (5.1.23) or (5.1.24) holds;*

(ii) $f_i(x) \in C(R, R)$, $xf_i(x) > 0$ *as* $x \neq 0$, $f_i(x)$ *is nondecreasing, and* $\liminf_{\substack{t\to\infty \\ x\to 0}} \frac{f_i(x)}{x} > r_i > 0$, $i = 0, 1, \ldots, k$;

(iii) *there exist* α *and* β *such that* $0 \le \alpha \le p(t) \le \beta \le 1$;

(iv) *there exists a positive integer* N *such that every bounded solution of*

$$u^{(n)}(t) + \sum_{i=0}^{k} r_i q_i(t) \sum_{j=0}^{N} \alpha^j u(t - j\tau - \sigma_i(t)) = 0 \tag{5.1.36}$$

is oscillatory.

Then every solution of Eq. (5.1.35) is oscillatory.

Proof: Assume the contrary, and let $x(t)$ be an eventually positive solution of (5.1.35). Set

$$z(t) = x(t) - p(t)x(t - \tau). \tag{5.1.37}$$

By Lemma 5.1.7, we have $\lim_{t\to\infty} z^{(i)}(t) = 0$, $z^{(i)}(t)z^{(i+1)}(t) < 0$, $i = 0, 1, 2, \ldots, n-1$. Hence $z(t) > 0$, $z(t - N\tau - \sigma_i(t)) > 0$, $i = 0, 1, \ldots, k$, for all large t. From (5.1.37), if $\alpha \neq 0$,

$$x(t) \ge \sum_{j=0}^{N} \alpha^j z(t - j\tau) + \alpha^{N+1} x(t - (j+1)\tau) \ge \sum_{j=0}^{N} \alpha^j z(t - j\tau). \tag{5.1.38}$$

Denote $0^0 = 1$. Then (5.1.38) holds also for $\alpha = 0$. Substituting (5.1.38) into (5.1.35) we have

$$z^{(n)}(t) + \sum_{i=0}^{k} q_i(t) f_i\left(\sum_{j=0}^{N} \alpha^j z(t - j\tau - \sigma_i(t))\right) \le 0. \tag{5.1.39}$$

Since $\lim_{t\to\infty} z(t) = 0$, from condition (ii) and (5.1.39) we get that

$$z^{(n)}(t) + \sum_{i=0}^{k} r_i q_i(t) \sum_{j=0}^{N} \alpha^j z(t - j\tau - \sigma_i(t)) \le 0. \tag{5.1.40}$$

By Theorem 5.1.1, (5.1.36) has a positive solution $u(t)$ satisfying $u(t) \leq z(t)$ which contradicts (iv). □

The following result can be similarly proved.

Theorem 5.1.7. *Let $1 \leq \alpha \leq p(t) \leq \beta$. In addition to the conditions (i) and (ii) of Theorem 5.1.6 where r_i are replaced by μ_i, $i = 0, 1, \ldots, k$, further assume that there exists a positive integer N such that every unbounded solution of the advanced equation*

$$u^{(n)}(t) - \sum_{i=0}^{k} \mu_i q_i(t+\tau) \sum_{j=0}^{N} \left(\frac{1}{\beta}\right)^{j+1} u(t + (j+1)\tau - \sigma_i(t+\tau)) = 0$$

is oscillatory. Then every solution of Eq. (5.1.35) is oscillatory.

5.2. Oscillation and Nonoscillation of Odd Order Equations

We consider the odd order linear neutral equation

$$\left(x(t) - px(t-\tau)\right)^{(n)} + q(t)x(g(t)) = 0, \quad t \geq t_0 \tag{5.2.1}$$

where $n \geq 1$ is an odd integer, $p \geq 0$, $\tau \in (0, \infty)$, $q, g \in C([t_0, \infty), R_+)$, and $\lim_{t\to\infty} g(t) = \infty$.

Theorem 5.2.1. *Assume that*

(i) *$p \in [0, 1)$, g is nondecreasing, $g(t) \leq t$, and (5.1.23) or (5.1.24) holds;*

(ii) *either*

$$\limsup_{t\to\infty} \left\{ \int_{g(t)}^{t} (g(s) - g(t))^{n-1} q(s)\, ds + p \int_{g(t)-\tau}^{g(t)} (g(s) - g(t))^{n-1} q(s)\, ds \right\} > (n-1)!\,(1-p), \tag{5.2.2}$$

or

$$\limsup_{t\to\infty}\left\{\int_{g(t)}^{t}(s-g(t))^{n-1}q(s)\,ds+p\int_{g(t)-\tau}^{g(t)}(s-g(t)+\tau)^{n-1}q(s)\,ds\right\}$$
$$>(n-1)!\,(1-p). \tag{5.2.3}$$

Then every solution of Eq. (5.2.1) is oscillatory.

Proof: Assume the contrary, and let $x(t)$ be an eventually positive solution of (5.2.1). Set $y(t)=x(t)-px(t-\tau)$. By Lemma 5.1.4, we know that

$$(-1)^i y^{(i)}(t)>0 \quad \text{and} \quad \lim_{t\to\infty} y^{(i)}(t)=0, \quad i=0,1,2,\ldots,n-1. \tag{5.2.4}$$

For any positive integer N we have

$$x(t)=\sum_{i=0}^{N+1}p^i y(t-i\tau)+p^{N+2}x(t-(N+2)\tau) \tag{5.2.5}$$

and

$$\sum_{i=0}^{N+1}p^i y(t-i\tau)\geq y(t)+\sum_{i=1}^{N+1}p^i y(t-\tau), \quad t\geq\tau\geq t_0. \tag{5.2.6}$$

Consider the case that (5.2.2) holds. By expanding $(1-p)^{-1}$ we see that there exists a sufficiently large N such that

$$\limsup_{t\to\infty}\left\{\sum_{i=0}^{N}p^i\int_{g(t)}^{t}(g(s)-g(t))^{n-1}q(s)\,ds\right.$$
$$\left.+\sum_{i=1}^{N+1}p^i\int_{g(t)-\tau}^{g(t)}(g(s)-g(t))^{n-1}q(s)\,ds\right\}>(n-1)!. \tag{5.2.7}$$

Substituting (5.2.5) into (5.2.1) we have

$$y^{(n)}(t)+q(t)\left[y(g(t))+\sum_{i=1}^{N+1}p^i y(g(t)-\tau)\right]\leq 0. \tag{5.2.8}$$

Using Taylor's formula and noting (5.2.4), we have that for s, t large and $s \le t$

$$y(s) > \frac{y^{(n-1)}(t)}{(n-1)!}(s-t)^{n-1}. \tag{5.2.9}$$

And hence

$$y(g(s)) > \frac{y^{(n-1)}(g(t))}{(n-1)!}(g(s)-g(t))^{n-1} \tag{5.2.10}$$

and

$$y(g(s)-\tau) > \frac{y^{(n-1)}(g(t)-\tau)}{(n-1)!}(g(s)-g(t))^{n-1}. \tag{5.2.11}$$

Substituting (5.2.10) and (5.2.11) into (5.2.8) we have, for s, t large and $s \le t$

$$y^{(n)}(s) + \frac{1}{(n-1)!}(g(s)-g(t))^{n-1}q(s)\left[y^{(n-1)}(g(t)) + \sum_{i=1}^{N+1} p^i y^{(n-1)}(g(t)-\tau)\right] \le 0. \tag{5.2.12}$$

Integrating (5.2.12) in s from $g(t)$ to t we have

$$y^{(n-1)}(t) - y^{(n-1)}(g(t)) + \frac{1}{(n-1)!}\int_{g(t)}^{t}(g(s)-g(t))^{n-1}q(s)\,ds$$
$$\times\left[y^{(n-1)}(g(t)) + \sum_{i=1}^{N+1} p^i y^{(n-1)}(g(t)-\tau)\right] \le 0. \tag{5.2.13}$$

Integrating (5.2.12) in s from $g(t)-\tau$ to t we have

$$y^{(n-1)}(t) - y^{(n-1)}(g(t)-\tau) + \frac{1}{(n-1)!}\int_{g(t)-\tau}^{t}(g(s)-g(t))^{n-1}\,q(s)\,ds$$
$$\times\left[y^{(n-1)}(g(t)) + \sum_{i=1}^{N+1} p^i y^{(n-1)}(g(t)-\tau)\right] \le 0. \tag{5.2.14}$$

Combining (5.2.13) and (5.2.14) we have

$$\sum_{i=0}^{N+1} p^i \int_{g(t)}^{t}(g(s)-g(t))^{n-1}q(s)\,ds + \sum_{i=1}^{N+1} p^i \int_{g(t)-\tau}^{g(t)}(g(s)-g(t))^{n-1}q(s)\,ds < (n-1)!$$

which contradicts (5.2.7). The proof corresponding to (5.2.3) is similar, so we omit it here. □

The following is a nonoscillation result for Eq. (5.2.1).

Theorem 5.2.2. *Assume that* $0 \le p < 1$, $q \in C([t_0, \infty), R_+)$, *and there exists* $\alpha > 0$ *such that eventually*

$$pe^{\alpha\tau} + \frac{1}{(n-1)!}\int_t^\infty (s-t)^{n-1} q(s) \exp[\alpha(t - g(s))]\, ds \le 1. \tag{5.2.15}$$

Then Eq. (5.2.1) has an eventually positive solution which tends to zero as $t \to \infty$.

Proof: If the equality in (5.2.15) holds eventually, then $x(t) = \exp(-\alpha t)$ is a positive solution of Eq. (5.2.1). Now we may assume that there exists $T > t_0$ such that $t - \tau \ge t_0$ and $g(t) \ge t_0$ for $t \ge T$,

$$\beta = pe^{\alpha\tau} + \frac{1}{(n-1)!}\int_T^\infty (s-T)^{n-1} q(s) \exp[\alpha(T - g(s))]\, ds < 1,$$

and (5.2.15) holds for all $t \ge T$.

Let Y be the set of all continuous functions y defined on $[t_0, \infty)$ satisfying $0 \le y(t) \le 1$ for $t \ge t_0$. Let Y be endowed with the usual pointwise ordering: $y_1 \le y_2 \iff y_1(t) \le y_2(t)$ for all $t \ge t_0$. It is easy to see that for any subset A of Y there exist $\inf A$ and $\sup A$. Define a mapping $\mathcal{T}$ on Y as follows:

$$(\mathcal{T}y)(t) = \begin{cases} pe^{\alpha\tau} y(t-\tau) + \dfrac{1}{(n-1)!} \\ \qquad \times \displaystyle\int_t^\infty (s-t)^{n-1} q(s) \exp[\alpha(t - g(s))] y(g(s))\, ds, & t \ge T \\ (\mathcal{T}y)(T) + \exp[\varepsilon(T-t)] - 1, & t_0 \le t \le T, \end{cases}$$

where $q = \ln(2-\beta)/(T - t_0)$.

For any $y \in Y$, $(\mathcal{T}y)(t) \ge 0$ for $t \ge t_0$, and

$$(\mathcal{T}y)(t) \le pe^{\alpha\tau} + \frac{1}{(n-1)!}\int_t^\infty (s-t)^{n-1} q(s) \exp[\alpha(t - g(s))]\, ds \le 1, \quad t \ge T,$$

$$(\mathcal{T}y)(t) = (\mathcal{T}y)(T) + \exp[\varepsilon(T-t)] - 1 \le \beta + \exp(T - t_0) - 1 = 1, \quad t_0 \le t \le T.$$

This means that $\mathcal{T}Y \subset Y$. Moreover, $\mathcal{T}$ is a nondecreasing mapping. By Knaster's fixed point theorem there exists $y \in Y$ such that $\mathcal{T}y = y$. That is,

$$y(t) = \begin{cases} pe^{\alpha\tau}y(t-\tau) + \dfrac{1}{(n-1)!} \\ \qquad \times \displaystyle\int_t^\infty (s-t)^{n-1} q(s) \exp[\alpha(t-g(s))] y(g(s))\, ds, & t \geq T \\ y(T) + \exp[\varepsilon(T-t)] - 1, & t_0 \leq t \leq T. \end{cases} \tag{5.2.16}$$

Since $y(t) > 0$ for $t_0 \leq t < T$, it follows that $y(t) > 0$ for all $t \geq t_0$. Set $x(t) = y(t)\exp(-\alpha t)$. Then we get

$$\left(x(t) - px(t-\tau)\right)^{(n)} + q(t)x(g(t)) = 0, \quad t \geq T,$$

i.e., $x(t)$ is a positive solution of Eq. (5.2.1) and $\lim_{t\to\infty} x(t) = 0$. □

Corollary 5.2.1. *Assume there exist constants $q \in (0,1)$ and $\sigma \geq 0$ such that*

$$0 \leq q(t) \leq q, \qquad g(t) \geq t - \sigma. \tag{5.2.17}$$

If the "majorant" equation

$$\left(x(t) - px(t-\tau)\right)^{(n)} + qx(t-\sigma) = 0, \quad t \geq t_0 \tag{5.2.18}$$

has a positive solution, then Eq. (5.2.1) has a positive solution which tends to zero as $t \to \infty$.

Proof: It is easy to see that (5.2.18) has a positive solution implies that the characteristic equation has a negative real root. Hence there exists a $\mu > 0$ such that $1 = pe^{\mu t} + qe^{\mu\sigma}/\mu^n$. Then we can show that (5.2.15) is satisfied for $\alpha = \mu$. The conclusion of the corollary follows from Theorem 5.2.2. The following result depicts the asymptotic behavior of decaying nonoscillatory solutions of Eq. (5.2.1), where a nonoscillatory solution x is said to be decaying if $x \to 0$ as $t \to \infty$.

Theorem 5.2.3. *Assume the conditions of Theorem 5.2.2 hold. Then the decaying nonoscillatory solutions of Eq. (5.2.1) are of the following two types:*

(i) *Solution x satisfies*

$$x(t) = \omega(t)p^{\frac{t}{\tau}} + o(p^{\frac{t}{\tau}}) \quad \text{as } t \to \infty \tag{5.2.19}$$

where ω is a τ-periodic continuous function such that

$$x(t)\omega(t) > 0 \quad \text{for all large } t. \tag{5.2.20}$$

(ii) *Solution x satisfies*

$$\frac{|x(t)|}{p^{\frac{t}{\tau}}} \longrightarrow \infty \quad \text{as } t \to \infty. \tag{5.2.21}$$

Furthermore, for any continuous positive and τ-periodic function ω, Eq. (5.2.1) has a decaying nonoscillatory solution x with the asymptotic property (5.2.19) if and only if

$$\int_{t_0}^{\infty} q(s)p^{(g(s)-s)/\tau}\, ds < \infty. \tag{5.2.22}$$

We omit the proof here.

For the existence of bounded nonoscillatory solutions of Eq. (5.2.1) we first present a general result where q changes sign, then we employ it to obtain a necessary and sufficient condition for the case that q is nonnegative.

Theorem 5.2.4. *Assume that $p > 0$, $p \neq 1$, $q \in C([t_0, \infty), R)$ and*

$$\int_{t_0}^{\infty} s^{n-1}|q(s)|\, ds < \infty. \tag{5.2.23}$$

Then Eq. (5.2.1) has a bounded positive solution.

Proof: We only prove it for the case $p > 1$. The proof for the case $0 < p < 1$ is in a similar way. Choose $T \geq t_0$ such that $g(t + \tau) \geq t_0$ for $t \geq T$ and

$$\frac{1}{(n+1)!}\int_{t+\tau}^{\infty} s^{n-1}|q(s)|\, ds \leq \frac{p-1}{4}, \quad t \geq T. \tag{5.2.24}$$

Consider the Banach space BC of all continuously bounded functions defined on $[t_0, \infty)$ with the sup norm $\|x\| = \sup\{|x(t)|: t \geq t_0\}$. Set

$$\Omega = \{x \in BC: \tfrac{1}{2}p \leq x(t) \leq 2p \text{ for } t \geq t_0\}.$$

Then Ω is a bounded closed convex subset of BC. Let the mapping $\mathcal{T}: \Omega \to BC$ be defined as follows

$$(\mathcal{T}x)(t) = \begin{cases} p - 1 + \dfrac{1}{p}x(t+\tau) \\ \qquad - \dfrac{1}{p(n-1)!}\displaystyle\int_{t+\tau}^{\infty}(s-t-\tau)^{n-1}q(s)x(g(s))\,ds, & t \geq T \\ (\mathcal{T}x)(T), \quad t_0 \leq t \leq T. \end{cases}$$

Since for any $x \in \Omega$

$$\begin{aligned} (\mathcal{T}x)(t) &\leq p - 1 + 2 + \frac{2}{(n-1)!}\int_{t+\tau}^{\infty}(s-t-\tau)^{n-1}|q(s)|\,ds \\ &\leq p + 1 + \frac{p-1}{2} < 2p \quad \text{for } t \geq T, \\ (\mathcal{T}x)(t) &\geq p - 1 + \frac{1}{2} - \frac{2}{(n-1)!}\int_{t+\tau}^{\infty}(s-t-\tau)^{n-1}|q(s)|\,ds \\ &\geq p - \frac{1}{2} - \frac{p-1}{2} = \frac{p}{2} \quad \text{for } t \geq T, \end{aligned}$$

and

$$\frac{p}{2} \leq (\mathcal{T}x)(t) = (\mathcal{T}x)(T) \leq 2p, \quad t_0 \leq t \leq T.$$

It follows that $\mathcal{T}\Omega \subset \Omega$. We now prove that $\mathcal{T}$ is a contraction mapping. In fact, for any $x_1, x_2 \in \Omega$ we have that for $t \geq T$

$$\begin{aligned} &|(\mathcal{T}x_1)(t) - (\mathcal{T}x_2)(t)| \\ &\quad \leq \frac{1}{p}|x_1(t+\tau) - x_2(t+\tau)| \\ &\quad + \frac{1}{p(n-1)!}\int_{t+\tau}^{\infty}(s-t-\tau)^{n-1}|q(s)|\,|x_1(g(s)) - x_2(g(s))|\,ds \end{aligned}$$

$$\leq \frac{1}{p}\|x_1 - x_2\| + \frac{1}{p}\|x_1 - x_2\|\frac{p-1}{4}$$

$$= \frac{1}{4}\left(1 + \frac{3}{p}\right)\|x_1 - x_2\|$$

which implies that

$$\|\mathcal{T}x_1 - \mathcal{T}x_2\| = \sup_{t \geq t_0} |(\mathcal{T}x_1)(t) - (\mathcal{T}x_2)(t)| = \sup_{t \geq T} |(\mathcal{T}x_1)(t) - (\mathcal{T}x_2)(t)|$$

$$\leq \frac{1}{4}\left(1 + \frac{3}{p}\right)\|x_1 - x_2\|.$$

Since $\frac{1}{4}(1 + \frac{3}{p}) < 1$, $\mathcal{T}$ is a contraction on Ω. Hence there is an $x \in \Omega$ such that $\mathcal{T}x = x$. It is easy to see that $x(t)$ is a bounded positive solution of Eq. (5.2.1). □

Theorem 5.2.5. *Assume that $p > 1$, and $q \in C([t_0, \infty), R_+)$. Then Eq. (5.2.1) has a bounded positive solution if and only if*

$$\int_{t_0}^{\infty} s^{n-1} q(s)\, ds < \infty. \tag{5.2.25}$$

Proof: Sufficiency follows from Theorem 5.2.4. To prove the necessity we assume that $x(t)$ is an eventually bounded positive solution of Eq. (5.2.1). Set $z(t) = x(t) - px(t-\tau)$. Then $z^{(n)}(t) \leq 0$ eventually. Since $z(t)$ is bounded, there are only two possibilities for $z(t)$:

(i) $(-1)^i z^{(i)}(t) > 0$, $i = 1, 2, \ldots, n$, and $z(t) > 0$;
(ii) $(-1)^i z^{(i)}(t) > 0$, $i = 1, 2, \ldots, n$, and $z(t) < 0$.

If (i) holds, since $p > 1$ and $x(t) > px(t-\tau)$, we conclude that $x(t) \to \infty$ as $t \to \infty$, which contradicts the boundedness of $x(t)$.

If (ii) holds, then there is a $d > 0$ such that $\lim_{t\to\infty} z(t) = -d$. Hence $-d \leq z(t) \leq -\frac{d}{2}$ eventually, and $px(g(t)) \geq \frac{d}{2}$ eventually. Substituting this into Eq. (5.2.1) we have $pz^{(n)}(t) \leq -\frac{d}{2}q(t)$. Integrating it from t to T n times for $T > t$ sufficiently large and noting the signs of $z^{(i)}(T)$, $i = 0, 1, \ldots, n-1$, we obtain

$$-pz(T) \geq \frac{d}{2(n-1)!}\int_t^T (s-t)^{n-1} q(s)\, ds.$$

Letting $T \to \infty$ we get (5.2.25). □

We now consider the equation

$$(x(t) - x(t-\tau))^{(n)} + \sum_{i=1}^{m} q_i(t)x(t-\sigma_i) = 0 \tag{5.2.26}$$

where $n \geq 1$ is an odd integer, $\tau > 0$, $\sigma_i \geq 0$, $q_i \in C([t_0, \infty), R_+)$, $i = 1, 2, \ldots, m$.

Theorem 5.2.6. *Assume that for some $T \geq t_0$*

$$\int_T^{\infty} t^{n-1} \sum_{i=1}^{m} q_i(t)\, dt = \infty. \tag{5.2.27}$$

Then every solution of Eq. (5.2.26) is oscillatory.

Proof: If not, let $x(t)$ be an eventually positive solution of (5.2.26). By Lemma 5.1.4

$$y(t) = x(t) - x(t-\tau) > 0 \tag{5.2.28}$$

eventually. Then according to Lemma 5.1.1 there exists a $k \in \{0, 2, \ldots, n-1\}$ such that

$$\begin{aligned} y^{(i)}(t) &> 0, \quad i = 0, \ldots, k, \\ (-1)^{i+1} y^{(i)}(t) &> 0, \quad i = k+1, \ldots, n-1. \end{aligned} \tag{5.2.29}$$

If $k = 0$, then there exist $M > 0$ and $T \geq t_0$ such that $x(t) \geq M$, $t \geq T$. From (5.2.26) and (5.2.29) we have

$$y^{(n)}(t) \leq -M \sum_{i=1}^{m} q_i(t). \tag{5.2.30}$$

In view of (5.2.27), we have

$$\int_T^t s^{n-1} y^{(n)}(s)\, ds \to -\infty \quad \text{as } t \to \infty. \tag{5.2.31}$$

On the other hand from (5.2.29)

$$\int_T^t s^{n-1}y^{(n)}(s)\,ds = F(t) - F(T) \tag{5.2.32}$$

where

$$F(t) = t^{n-1}y^{(n-1)}(t) + \sum_{i=0}^{n-1}(-1)^i(i+1)\cdots(n-1)t^i y^i(t) > 0. \tag{5.2.33}$$

It is easy to see that (5.2.32) and (5.2.33) contradict (5.2.31).

If $k > 0$, from (5.2.29), there exist $M_1 > 0$ and $T_1 \geq t_0$ such that

$$y(t) \geq M_1 t^{k-1} \quad \text{for } t \geq T_1. \tag{5.2.34}$$

We claim that there exists $M_2 > 0$ such that

$$x(t) \geq M_2 t^k \quad \text{for } t \geq T_1. \tag{5.2.35}$$

In fact, from (5.2.34)

$$x(t) \geq x(t-\tau) + M_1 t^{k-1}, \quad t \geq T_1. \tag{5.2.36}$$

Let $L = \min\{x(t) \colon T_1 \leq t \leq T_1 + \tau\}$. Choose α_1 such that $0 < \alpha_1 \leq \min\{L/(T_1 + \tau)^k,\ M_1/2\tau k\}$ and

$$G(t) = (M_1 - k\alpha_1\tau)t^{k-1} + \alpha_1 \sum_{i=1}^{k}(-1)^i C_k^i \tau^i t^{k-i} \geq 0, \quad t \geq T_1 \tag{5.2.37}$$

provided T_1 is sufficiently large.

Since $x(t) \geq L$ for $T_1 \leq t \leq T_1 + \tau$, it follows that $x(t) \geq \alpha_1 t^k$, $T \leq t \leq T_1 + \tau$. From (5.2.36) and (5.2.37), for $T_1 + \tau \leq t \leq T_1 + 2\tau$

$$x(t) \geq \alpha_1(t-\tau)^k + M_1 t^{k-1} = \alpha_1 t^k + G(t) \geq \alpha_1 t^k.$$

By induction we have

$$x(t) \geq \alpha_1 t^k \quad \text{for } T_1 + i\tau \leq t \leq T_1 + (i+1)\tau,\ i = 0, 1, 2, \ldots.$$

Therefore, (5.2.35) is true for $M_2 = \alpha_1$. Substituting (5.2.35) into (5.2.26) we have

$$y^{(n)}(t) + M_3 t^k \sum_{i=1}^{m} q_i(t) \leq 0 \tag{5.2.38}$$

where $M_3 > 0$ is a constant. (5.2.38) and (5.2.27) lead to

$$\int_T^t s^{n-k-1} y^{(n)}(s)\, ds \longrightarrow -\infty \quad \text{as } t \to \infty. \tag{5.2.39}$$

On the other hand, from (5.2.29)

$$\int_T^t s^{n-k-1} y^{(n)}(s)\, ds = F_{n-k}(t) - F_{n-k}(T) \tag{5.2.40}$$

where

$$\begin{aligned} F_{n-k}(t) = t^{n-k-1} y^{(n-1)}(t) - (n-k-1) t^{n-k-2} y^{(n-2)}(t) \\ + \cdots + (n-k-1)! y^{(k)}(t) > 0. \end{aligned} \tag{5.2.41}$$

(5.2.40) with (5.2.41) contradicts (5.2.39). The proof is complete. □

Theorem 5.2.7. *Eq. (5.2.26) has a bounded positive solution if and only if*

$$\sum_{i=0}^{\infty} \int_{t_0+i\tau}^{\infty} t^{n-1} \sum_{i=1}^{m} q_i(t)\, dt < \infty. \tag{5.2.42}$$

Proof: *Sufficiency.* Choose $T \geq t_0$ such that

$$\sum_{i=0}^{\infty} \int_{T+i\tau}^{\infty} t^{n-1} \sum_{i=1}^{m} q_i(t)\, dt \leq 1. \tag{5.2.43}$$

Set

$$K(t) = \begin{cases} \displaystyle\int_t^{\infty} s^{n-1} \sum_{i=1}^{m} q_i(s)\, ds, & t \geq T \\ (t - T + \tau) K(T)/\tau, & T - \tau \leq t \leq T \\ 0, & t \leq T - \tau \end{cases} \tag{5.2.44}$$

and $y(t) = \sum_{i=0}^{\infty} K(t-i\tau)$, $t \geq T$. Clearly, $K \in C(R, R_+)$, $y \in C([T,\infty),(0,1])$ and $y(t) = y(t-\tau) + K(t)$. Define a set of functions as follows:

$$X = \{x \in C([T,\infty), R) : 0 \leq x(t) \leq y(t),\ t \geq T\}. \tag{5.2.45}$$

In X, $x_1 \leq x_2$ means that $x_1(t) \leq x_2(t)$ for all $t \geq T$. It is easy to see that for any subset A of X, there exist $\inf A$ and $\sup A$. Define a mapping $\mathcal{T}$ on X as follows:

$$(\mathcal{T}x)(t) = \begin{cases} x(t-\tau) + \displaystyle\int_t^\infty \frac{(s-t)^{n-1}}{(n-1)!} \sum_{i=1}^m q_i(s) x(s-\sigma_i)\, ds, & t \geq T + r \\ (\mathcal{T}x)(T+r) \dfrac{t y(t)}{(T+r) y(T+r)} + y(t)\left(1 - \dfrac{t}{T+r}\right), & t \in [T, T+r] \end{cases} \tag{5.2.46}$$

where $r = \max_{1 \leq i \leq m}\{\tau, \sigma_i\}$.

From (5.2.43) and noting that $y(t) \leq 1$ we have

$$0 \leq (\mathcal{T}x)(t) \leq y(t-\tau) + K(t) = y(t), \quad t \geq T + r$$

and

$$0 \leq (\mathcal{T}x)(t) \leq y(t), \quad t \in [T, T+r].$$

Therefore, $\mathcal{T}X \subset X$. Clearly, $\mathcal{T}$ is nondecreasing. By Knaster's fixed point theorem, there is an $x \in X$ such that $\mathcal{T}x = x$. That is,

$$x(t) = \begin{cases} x(t-\tau) + \displaystyle\int_t^\infty \frac{(s-t)^{n-1}}{(n-1)!} \sum_{i=1}^m q_i(s) x(s-\sigma_i)\, ds, & t \geq T + r \\ x(T+r) \dfrac{t y(t)}{(T+r) y(T+r)} + y(t)\left(1 - \dfrac{t}{T+r}\right), & t \in [T, T+r]. \end{cases} \tag{5.2.47}$$

It is easy to see that $x(t) > 0$ for $t \in [T, T+r)$ and hence $x(t) > 0$ for all $t \geq T$. Therefore, x is a bounded positive solution of Eq. (5.2.26).

Necessity. Let $x(t)$ be a bounded positive solution of (5.2.26). Set $z(t) = x(t) - x(t-\tau)$. Then $z^{(n)}(t) \leq 0$, $(-1)^i z^{(i)}(t) > 0$, $i = 0, 1, 2, \ldots, n-1$, eventually. Hence there exist $\alpha > 0$, $t_1 \geq t_0 + r$, such that $x(t) \geq \alpha$, $t \geq t_1 - r$. Integrating

Eq. (5.2.26) from t to ∞ for $t \geq t_1$ n times we have

$$z(t) \geq \int_t^\infty \frac{(s-t)^{n-1}}{(n-1)!} \sum_{i=1}^m q_i(s)x(s-\sigma_i)\, ds, \tag{5.2.48}$$

and hence for $\ell = 0, 1, 2, \ldots,$

$$x(t_1 + \ell\tau) \geq x(t_1) + \alpha \sum_{i=1}^{\ell} \int_{t_1+i\tau}^\infty \frac{(s-(t_1+i\tau))^{n-1}}{(n-1)!} \sum_{i=1}^m q_i(s)\, ds.$$

From the boundedness of x the above inequality implies that

$$\sum_{i=1}^{\infty} \int_{t_1+i\tau}^\infty \frac{(s-(t_1+i\tau))^{n-1}}{(n-1)!} \sum_{i=1}^m q_i(s)\, ds < \infty.$$

Then

$$\sum_{i=1}^{\infty} \int_{t_1+i\tau}^\infty s^{n-1} \sum_{i=1}^m q_i(s)\, ds < \infty.$$

□

From Theorem 3.2.5 we obtain the following result.

Theorem 5.2.8. *Eq. (5.2.26) has a bounded positive solution if and only if*

$$\int_T^\infty t^n \sum_{i=1}^m q_i(t)\, dt < \infty. \tag{5.2.49}$$

Example 5.2.1. Consider

$$(x(t) - x(t-1))''' + \frac{6(2t-1)(2t^2-2t+1)}{t^4(t-1)^3(t-2)} x(t-1) = 0. \tag{5.2.50}$$

Clearly, (5.2.49) is satisfied and so (4.2.50) has a bounded solution. In fact, $x(t) = 1 - \frac{1}{t}$ is such a solution.

Example 5.2.2. Consider

$$(x(t) - x(t-2\pi))' + \frac{4\pi}{2t-\pi} x(t - \tfrac{\pi}{2}) = 0, \quad t \geq \pi. \tag{5.2.51}$$

We see that

$$\int_T^\infty q(t)\,dt = \int_T^\infty \frac{4\pi}{2t-\pi}\,dt = \infty. \tag{5.2.52}$$

(5.2.27) is satisfied. Therefore, every solution of (5.2.51) oscillates. In fact, $x(t) = t(\sin t + \cos t)$ is such a solution.

In the following we will investigate equations of a mixed type. Before doing this, we state a result for the odd order delay differential equation

$$z^{(n)}(t) + \sum_{i=1}^{L} q_i^n z(t - n\delta_i) = 0 \tag{5.2.53}$$

where n is an odd integer, q_i and δ_i are positive constants, $i = 1, 2, \ldots, L$.

Lemma 5.2.1. *If*

$$\sum_{i=1}^{L} (q_i \delta_i e)^n > 1, \tag{5.2.54}$$

then every solution of (5.2.53) is oscillatory.

We now consider the mixed type equation

$$\big(x(t) - px(t-\tau)\big)^{(n)} + \sum_{i=1}^{m} q_i x(t - \sigma_i) = 0 \tag{5.2.55}$$

where τ, σ_i and q_i are constant, $p \in (0,1)$, $\tau, q_i \in (0,\infty)$, and $\sigma_i \in R$, $i = 1, \ldots, m$. Let m_i be the least nonnegative integer such that $\overline{\tau}_i = m_i\tau + \sigma_i > 0$, $i = 1, 2, \ldots, m$. Let ℓ be a positive integer such that $\ell \geq \max\{m_i,\ i = 1, 2, \ldots, m\}$. Set

$$q_i^* = q_1 p^{m_1 + (i-1)}, \quad i = 1, \ldots, \ell - m_1 + 1,$$

$$q_{\ell - m_1 + 1 + i}^* = q_2 p^{m_2 + (i-1)}, \quad i = 1, \ldots, \ell - m_2 + 1,$$

$$\cdots$$

$$q^*_{(m-1)(\ell+1)-(\sum_{j=1}^{m-1} m_j)+i} = q_m p^{m_m+(i-1)}, \quad i = 1, \ldots, \ell - m_m + 1;$$

and

$$n\delta_i = \overline{\tau}_1 + (i-1)\tau, \quad i = 1, \ldots, \ell - m_1 + 1,$$

$$n\delta_{\ell+m_1-1+i} = \overline{\tau}_2 + (i-1)\tau, \quad i = 1, \ldots, \ell - m_2 + 1,$$

$$\cdots$$

$$n\delta_{(m-1)(\ell-1)-(\sum_{j=1}^{m-1} m_j)+i} = \overline{\tau}_m + (i-1)\tau, \quad i = 1, \ldots, \ell - m_m + 1;$$

$$L = m(\ell+1) - \sum_{i=1}^{m} m_i.$$

Theorem 5.2.9. *If*

$$\sum_{i=1}^{L} q_i^*(\delta_i e)^n > 1, \tag{5.2.56}$$

then every solution of Eq. (5.2.55) is oscillatory.

Proof: Let $x(t)$ be an eventually positive solution of Eq. (5.2.55). Set $z(t) = x(t) - px(t-\tau)$. Then $(-1)^i z^{(i)}(t) > 0$ eventually for $i = 0, 1, \ldots, n-1$ by Lemma 5.1.7. Thus

$$x(t) = \sum_{j=0}^{\ell} p^j z(t - j\tau) + p^{\ell+1} x(t - (\ell+1)\tau)$$

$$> \sum_{j=0}^{\ell} p^j z(t - j\tau)$$

and

$$x(t - \sigma_i) > \sum_{j=0}^{\ell} p^j z(t - \sigma_i - j\tau) = \sum_{j=0}^{\ell} p^j z(t - \overline{\tau}_i - (j - m_i)\tau)$$

$$\geq \sum_{j=m_i}^{\ell} p^j z(t - \overline{\tau}_i - (j - m_i)\tau)$$

$$= p^{m_i} \sum_{j=0}^{\ell - m_i} p^j z(t - \overline{\tau}_i - j\tau), \quad i = 1, 2, \ldots, m. \tag{5.2.57}$$

Substituting (5.2.56) into (5.2.55) we have

$$z^{(n)}(t) + \sum_{i=1}^{m} q_i p^{m_i} \sum_{j=0}^{\ell - m_i} p^j z(t - \overline{\tau}_i - j\tau) \leq 0,$$

i.e.,

$$z^{(n)}(t) + \sum_{i=1}^{L} q_i^* z(t - n\delta_i) \leq 0. \tag{5.2.58}$$

By Theorem 5.4.1, (5.2.58) implies that the equation

$$z^{(n)}(t) + \sum_{i=1}^{L} q_i^* z(t - n\delta_i) = 0 \tag{5.2.59}$$

has a nonoscillatory solution. By Lemma 5.2.1 the condition (5.2.56) is a sufficient condition for all solutions of (5.2.59) to be oscillatory. This contradiction proves the theorem. □

5.3. Oscillation of Even Order Equations

Consider the nth order neutral differential equation

$$(x(t) + p(t)x(t - \tau))^{(n)} + q(t)x(\sigma(t)) = 0, \quad t \geq t_0 \tag{5.3.1}$$

where n is an even integer, p, q, $\sigma \in C([t_0, \infty), R)$.

Lemma 5.3.1. *Assume that*

(i) $0 \leq p(t) \leq 1$, $q(t) \geq 0$ *and* $q(t) \not\equiv 0$ *for* $t \geq t_0$;

(ii) $0 < \sigma(t) < t$, $0 < \sigma'(t) \leq 1$, $t \geq t_0$ *and* $\lim_{t\to\infty} \sigma(t) = \infty$.

Let $x(t)$ be an eventually positive solution of Eq. (5.3.1) and $y(t) = x(t) + p(t)x(t-\tau)$. Then $y(t)$ eventually satisfies the following differential inequality

$$y^{(n)}(t) + q(t)\big(1 - p(\sigma(t))\big)y(\sigma(t)) \le 0. \tag{5.3.2}$$

Proof: From (5.3.1) $y(t) > 0$, $y^{(n)}(t) \le 0$ eventually. By Lemma 5.1.1 $y^{(n-1)}(t) > 0$ and $y'(t) > 0$ eventually. Rewrite (5.3.1) as

$$y^{(n)}(t) + q(t)y(\sigma(t)) - p(\sigma(t))q(t)x(\sigma(t) - \tau) = 0.$$

Since for large t $y(t) \ge x(t) > 0$ and so

$$y^{(n)}(t) + q(t)y(\sigma(t)) - p(\sigma(t))q(t)y(\sigma(t) - \tau) \le 0.$$

In view of $y'(t) > 0$ we have

$$y^{(n)}(t) + q(t)[1 - p(\sigma(t))]y(\sigma(t)) \le 0$$

eventually. □

Theorem 5.3.1. *In addition to the assumptions of Lemma 5.3.1 assume that there exists a function $\rho \in C^1([t_0, \infty), (0, \infty))$ such that*

$$\lim_{t\to\infty} \frac{1}{t^{m-1}} \int_{t_0}^{t} \frac{(t-s)^{m-3}(\rho'(s)(t-s) - (m-1)\rho(s))^2}{\sigma'(s)\sigma^{n-2}(s)\rho(s)}\, ds < \infty \tag{5.3.3}$$

and

$$\lim_{t\to\infty} \frac{1}{t^{m-1}} \int_{t_0}^{t} (t-s)^{m-1}\rho(s)q(s)(1 - p(\sigma(s)))\, ds = \infty \tag{5.3.4}$$

where $m \ge 3$ is an integer. Then every solution of Eq. (5.3.1) is oscillatory.

Proof: Assume the contrary, and let $x(t)$ be an eventually positive solution of (5.3.1), and let $y(t)$ be defined as in Lemma 5.3.1. Then there exists a $T \ge t_0$ such that $y(t) > 0$, $y'(t) > 0$, $y^{(n-1)}(t) > 0$, and $y^{(n)}(t) \le 0$ for $t \ge T$. By Lemma 5.3.1

$$y^{(n)}(t) + q(t)\big(1 - p(\sigma(t))\big)y(\sigma(t)) \le 0, \quad t \ge T \tag{5.3.5}$$

if T is sufficiently large. Choose $T_1 \geq T$ so large that $\sigma(t) \geq 2T$ for $t \geq T_1$. Applying Lemma 5.1.3 for y' with $\lambda = \frac{1}{2}$ we know that there exist $M > 0$ and $T_2 \geq T_1$ such that

$$y'(\tfrac{1}{2}\sigma(t)) \geq M\sigma^{n-2}(t)y^{(n-1)}(t) \quad \text{for } t \geq T_2. \tag{5.3.6}$$

Set $w(t) = y^{(n-1)}(t)/y(\frac{1}{2}\sigma(t))$. From (5.3.5) we have

$$w'(t) \leq -q(t)\big(1 - p(\sigma(t))\big) - M_1\sigma'(t)\sigma^{n-2}(t)w^2(t).$$

Then

$$\rho(t)q(t)\big(1 - p(\sigma(t))\big) \leq -\rho(t)w'(t) - M_1\sigma'(t)\sigma^{n-2}(t)\rho(t)w^2(t) \tag{5.3.7}$$

where $M_1 > 0$ is a constant. Multiplying (5.3.7) by $(t-s)^{m-1}$ and integrating it from T_2 to t we have that

$$\begin{aligned}
&\int_{T_2}^{t} (t-s)^{m-1}\rho(s)q(s)(1 - p(\sigma(s)))\,ds \\
&\leq -\int_{T_2}^{t} (t-s)^{m-1}\rho(s)w'(s)\,ds - M_1\int_{T_2}^{t} (t-s)^{m-1}\sigma'(s)\sigma^{n-2}(s)\rho(s)w^2(s)\,ds \\
&= (t-T_2)^{m-1}\rho(T_2)w(T_2) + \int_{T_2}^{t} (t-s)^{m-2}[(t-s)\rho'(s) - (m-1)\rho(s)]w(s)\,ds \\
&\qquad - M_1\int_{T_2}^{t} (t-s)^{m-1}\sigma'(s)\sigma^{n-2}(s)\rho(s)w^2(s)\,ds \\
&= (t-T_2)^{m-1}\rho(T_2)w(T_2) - \int_{T_2}^{t} \Big[\sqrt{M_1}(t-s)^{\frac{m-1}{2}}\sqrt{\sigma'(s)\sigma^{n-2}(s)\rho(s)}\,w(s) \\
&\qquad - \frac{1}{2\sqrt{M_1}}\frac{(t-s)^{\frac{m-3}{2}}}{\sqrt{\sigma'(s)\sigma^{n-2}(s)\rho(s)}}(\rho'(s)(t-s) - (m-1)\rho(s))\Big]^2 ds \\
&\qquad + \frac{1}{4M_1}\int_{T_2}^{t} \frac{(t-s)^{m-3}}{\sigma'(s)\sigma^{n-2}(s)\rho(s)}(\rho'(s)(t-s) - (m-1)\rho(s))^2\,ds \\
&\leq (t-T_2)^{m-1}\rho(T_2)w(T_2) \\
&\qquad + \frac{1}{4M_1}\int_{T_2}^{t} \frac{(t-s)^{m-3}}{\sigma'(s)\sigma^{n-2}(s)\rho(s)}(\rho'(s)(t-s) - (m-1)\rho(s))^2\,ds.
\end{aligned}$$

Since for every $t \geq T_2$

$$\int_{t_0}^{t} (t-s)^{m-1}\rho(s)q(s)(1-p(\sigma(s))\,ds - \int_{T_2}^{t} (t-s)^{m-1}\rho(s)q(s)(1-p(\sigma(s)))\,ds$$

$$= \int_{t_0}^{T_2} (t-s)^{m-1}\rho(s)q(s)(1-p(\sigma(s)))\,ds$$

$$\leq (t-t_0)^{m-1}\int_{t_0}^{T_2} \rho(s)q(s)(1-p(\sigma(s)))\,ds;$$

for $t \geq T_2$

$$\frac{1}{t^{m-1}}\int_{t_0}^{t} (t-s)^{m-1}\rho(s)q(s)(1-p(\sigma(s)))\,ds$$

$$\leq \left(1-\frac{t_0}{t}\right)^{m-1}\int_{t_0}^{T_2} \rho(s)q(s)(1-p(\sigma(s)))\,ds + \left(1-\frac{T_2}{t}\right)^{m-1}\rho(T_2)w(T_2)$$

$$+\frac{1}{4M_1 t^{m-1}}\int_{t_0}^{t} \frac{(t-s)^{m-3}}{\sigma'(s)\sigma^{n-2}(s)\rho(s)}(\rho'(s)(t-s)-(m-1)\rho(s))^2\,ds.$$

Hence there exists a $C_1 > 0$ such that

$$\lim_{t\to\infty}\frac{1}{t^{m-1}}\int_{t_0}^{t} (t-s)^{m-1}\rho(s)q(s)(1-p(\sigma(s)))\,ds$$

$$\leq C_1 + \lim_{t\to\infty}\frac{1}{4M_1 t^{m-1}}\int_{t_0}^{t} \frac{(t-s)^{m-3}}{\sigma'(s)\sigma^{n-2}(s)\rho(s)}(\rho'(s)(t-s)-(m-1)\rho(s))^2\,ds$$

which contradicts (5.3.3) and (5.3.4). □

Corollary 5.3.1. *In addition to the assumptions of Lemma 5.3.1 assume that there exist an* $\alpha \in [0, n-1)$ *and an integer* $m \geq 3$ *such that*

$$\lim_{t\to\infty}\frac{1}{t^{m-1}}\int_{t_0}^{t} (t-s)^{m-1}\sigma^{\alpha}(s)q(s)(1-p(\sigma(s)))\,ds = \infty. \tag{5.3.8}$$

Then every solution of (5.3.1) is oscillatory.

In particular, if

$$\lim_{t\to\infty}\int_{t_0}^{t} q(s)(1-p(\sigma(s)))\,ds = \infty, \tag{5.3.9}$$

then every solution of (5.3.1) is oscillatory.

We now consider the case that

$$\int_{t_0}^{\infty} q(s)(1 - p(\sigma(s)))\, ds < \infty. \tag{5.3.10}$$

In this case we define a sequence

$$Q_0(t) = \int_t^{\infty} q(s)(1 - p(\sigma(s)))\, ds, \tag{5.3.11}$$

$$Q_k(t) = \int_t^{\infty} h(s)Q_{k-1}^2(s)\, ds + Q_0(s), \quad k = 1, 2, \dots, \; t \geq t_0, \tag{5.3.12}$$

if the right hand side of (5.3.12) exists, where $h(t) = M_1\sigma'(t)\sigma^{n-2}(t)$. Set $H(t) = \int_{t_0}^t h(s)\, ds$. Then $\lim_{t\to\infty} H(t) = \infty$.

Theorem 5.3.2. *Assume that the conditions of Lemma 5.3.1 are satisfied. Then every solution of Eq. (5.3.1) is oscillatory if either one of the following holds:*

(a) *There exists an integer $\ell > 0$ such that functions $Q_k(t)$, $k = 1, 2, \dots, \ell - 1$, are defined and*

$$\lim_{t\to\infty} \int_{t_0}^t h(s)Q_{\ell-1}^2(s)ds = \infty; \tag{5.3.13}$$

(b) *$Q_k(t)$ is defined for $k = 1, 2, \dots,$ but there is a $t^* \geq t_0$ such that $\lim_{k\to\infty} Q_k(t^*) = \infty$.*

(c) *There exists a k such that $Q_k(t)$ is defined and $\limsup_{t\to\infty} Q_k(t)H(t) > 1$.*

Proof: Let $x(t)$ be an eventually positive solution. Set $w(t) = y^{(n-1)}(t)/y(\frac{1}{2}\sigma(t))$. Then

$$w'(t) + q(t)(1 - p(\sigma(t))) + h(t)w^2(t) \leq 0. \tag{5.3.14}$$

Hence

$$w'(t) + h(t)w^2(t) \leq 0$$

and

$$w(t) \leq \frac{1}{H(t) - c} \tag{5.3.15}$$

where $c \geq 0$ is a constant. Since $\lim_{t\to\infty} H(t) = \infty$, we have $\lim_{t\to\infty} w(t) = 0$. Integrating (5.3.14) from t to ∞ we have

$$\begin{aligned} w(t) &\geq \int_t^\infty q(s)(1 - p(\sigma(s)))\, ds + \int_t^\infty h(s)w^2(s)\, ds \\ &= Q_0(t) + \int_t^\infty h(s)w^2(s)\, ds \end{aligned} \tag{5.3.16}$$

which implies that

$$\int_t^\infty h(s)w^2(s)\, ds < \infty. \tag{5.3.17}$$

For case (a), from (5.3.15), we have $w(t) \geq Q_0(t)$, and so

$$\int_t^\infty h(s)Q_0^2(s)\, ds < \infty.$$

Again from (5.3.15)

$$Q_1(t) = \int_t^\infty h(s)Q_0^2(s)\, ds + Q_0(t) \leq \int_t^\infty h(s)w^2(s)\, ds + Q_0(t) \leq w(t),$$

and so

$$\int_t^\infty h(s)Q_1^2(s)\, ds < \infty.$$

By induction we have $Q_n(t) \leq w(t), \quad n = 1, 2, \ldots, \ell - 1$, and

$$\int_t^\infty h(s)Q_{\ell-1}^2(s)\, ds \leq \int_t^\infty h(s)w^2(s)\, ds < \infty$$

which contradicts (a).

For case (b), as before we have $Q_n(t) \leq w(t)$. Then $\lim_{n\to\infty} Q_n(t) \leq w(t) < \infty$, which contradicts (b).

For case (c), in view of (5.3.15) we have

$$H(t)Q_n(t) \le H(t)w(t) \le H(t)/(H(t) - c).$$

Hence $\limsup_{t\to\infty} H(t)Q_n(t) \le 1$, which contradicts (c). □

Corollary 5.3.2. *In addition to the assumptions of Lemma 5.3.1 assume either one of the following conditions holds:*

(i) $Q_0(t) \ge K/H(t)$;

(ii) $q(t)(1 - p(\sigma(t))) \ge Kh(t)/H^2(t)$;

(iii) $\int_t^\infty h(s)Q_0^2(s)\,ds \ge KQ_0(t)$

where $K > \frac{1}{4}$. Then every solution of Eq. (5.3.1) is oscillatory.

Proof: For case (i) we have

$$Q_1(t) = \int_t^\infty h(s)Q_0^2(s)\,ds + Q_0(t) \ge K^2 \int_t^\infty \frac{dH(s)}{H^2(s)} + \frac{K}{H(t)} = \frac{K_1}{H(t)}$$

where $K_1 = K^2 + K$.

$$Q_2(t) = \int_t^\infty h(s)Q_1^2(s)\,ds + Q_0(t) \ge \frac{K_1^2}{H(t)} + \frac{K}{H(t)} = \frac{K_2}{H(t)}$$

where $K_2 = K_1^2 + K$. In general we have $Q_n(t) \ge K_n/H(t)$ where $K_n = K_{n-1}^2 + K$. By induction we see that $\{K_n\}$ is an increasing sequence. Therefore, $\lim_{n\to\infty} K_n = K^* \in (0, \infty]$. If $K^* < \infty$, then K^* satisfies the equation $K^* = K^{*2} + K$. Since $K > \frac{1}{4}$, it has no real roots. Therefore, $\lim_{n\to\infty} K_n = \infty$. Hence for fixed $t^* \ge t_0$

$$\lim_{n\to\infty} Q_n(t^*) \ge \lim_{n\to\infty} K_n/H(t^*) = \infty.$$

By Theorem 5.3.2 (b), every solution of (5.3.1) is oscillatory.

For case (ii) we have that

$$Q_0(t) = \int_t^\infty q(s)(1 - p(\sigma(s)))\,ds \ge K \int_t^\infty \frac{dH(s)}{H^2(s)} = \frac{K}{H(t)},$$

i.e., (ii) implies (i).

For case (iii) we have that

$$\int_t^\infty h(s)Q_0^2(s)\,ds \geq KQ_0(t),$$

$$Q_1(t) = \int_t^\infty h(s)Q_0^2(s)\,ds + Q_0(t) \geq (K+1)Q_0(t) = K_1Q_0(t),$$

$$Q_2(t) = \int_t^\infty h(s)Q_1^2(s)\,ds + Q_0(t) \geq K_1^2 \int_t^\infty h(s)Q_0^2(s)\,ds + Q_0(t)$$

$$\geq (K_1^2K + 1)Q_0(t) = K_2Q_0(t).$$

In general, $Q_n(t) \geq K_nQ_0(t)$ where $K_n = K_{n-1}^2K+1$. Then $\{K_n\}$ is an increasing sequence and $\lim_{n\to\infty} K_n = \infty$. Since $Q_0(t) \not\equiv 0$, there exists a $t^* \geq t_0$ such that $Q_0(t^*) > 0$. Then

$$Q_n(t^*) \geq K_nQ_0(t^*) \longrightarrow \infty \quad \text{as } n \to \infty.$$

By Theorem 5.3.2 (b) every solution of (5.3.1) is oscillatory.

5.4. Classification of Nonoscillatory Solutions

Consider the higher order nonlinear neutral delay differential equation

$$\big(x(t) + px(t-\tau)\big)^{(n)} + f\big(t, x(t-\sigma_1(t)), \ldots, x(t-\sigma_m(t))\big) = 0 \tag{5.4.1}$$

where $n \geq 2$, $p \in R$, $\tau > 0$; $\sigma_i \in C([t_0,\infty), R_+)$, $0 \leq \sigma_i(t) \leq h < \infty$, $i = 1,\ldots,m$; $f(t,x_1,\ldots,x_m) \in C([t_0,\infty)\times R^m, R)$ and $x_1f(t,x_1,\ldots,x_m) > 0$ if $t \geq t_0$, $x_1x_i > 0$, $i = 1,2,\ldots,m$.

If there are continuous functions $q_i(t) \geq 0$, $i = 1,\ldots,m$, such that $\sum_{i=1}^m q_i(t) > 0$ and

$$\frac{f(t,y_1,\ldots,y_m)}{\sum_{i=1}^m q_i(t)y_i} \geq (\text{or} \leq)\ \frac{f(t,x_1,\ldots,x_m)}{\sum_{i=1}^m q_i(t)x_i} \tag{5.4.2}$$

for $y_i \geq x_i > 0$ (or $y_i \leq x_i < 0$), $i = 1,2,\ldots,m$, then f is said to be superlinear (or sublinear).

From the definitions of superlinearity and sublinearity it is easy to obtain the following lemma.

Lemma 5.4.1. *Suppose that $0 < a \le x_i \le b$, $i = 1, 2, \ldots, m$. Then for $t \ge t_0$, we have*

$$f(t, a, \ldots, a) \le f(t, x_1, \ldots, x_m) \le f(t, b, \ldots, b) \tag{5.4.3}$$

if f is superlinear, and

$$\frac{a}{b} f(t, b, \ldots, b) \le f(t, x_1, \ldots, x_m) \le \frac{b}{a} f(t, a, \ldots, a) \tag{5.4.4}$$

if f is sublinear.

Remark 5.4.1. Similar results to Lemma 5.4.1 hold for the case that $-b \le x_i \le -a < 0$, $i = 1, \ldots, m$.

Lemma 5.4.2. *Assume that $p \ge 0$, $p \ne 1$, $x(t)/t^i$ is bounded, nonoscillatory on $[t_0, \infty)$, where i is a nonnegative integer. Let $z(t) = x(t) + px(t - \tau)$. If $\lim_{t\to\infty} z(t)/t^i = b$, then*

$$\lim_{t\to\infty} \frac{x(t)}{t^i} = \frac{b}{1+p}. \tag{5.4.5}$$

Proof: Without loss of generality we assume that $x(t)/t^i > 0$, $t \ge t_0$. Denote

$$\limsup_{t\to\infty} \frac{x(t)}{t^i} = Q \quad \text{and} \quad \liminf_{t\to\infty} \frac{x(t)}{t^i} = q.$$

Then there exist sequences $\{t_n\}$ and $\{s_n\}$ such that $\lim_{n\to\infty} t_n = \infty$, $\lim_{n\to\infty} s_n = \infty$, $\lim_{n\to\infty} x(t_n)/t_n^i = Q$ and $\lim_{n\to\infty} x(s_n)/s_n^i = q$. If $p \in [0, 1)$, then

$$b = \lim_{n\to\infty} \frac{z(t_n)}{t_n^i} = \lim_{n\to\infty} \frac{x(t_n) + px(t_n - \tau)}{t_n^i} \ge Q + pq$$

and

$$b = \lim_{n\to\infty} \frac{z(s_n)}{s_n^i} = \lim_{n\to\infty} \frac{x(s_n) + px(s_n - \tau)}{s_n^i} \le q + pQ.$$

Hence $q + pQ \ge Q + pq$ or $(1 - p)q \ge (1 - p)Q$, i.e., $q \le Q$, which implies that $q = Q$.

Similarly, if $p > 1$, we have

$$b = \lim_{n\to\infty} \frac{z(t_n + \tau)}{(t_n + \tau)^i} \geq q + pQ, \qquad b = \lim_{n\to\infty} \frac{z(s_n + \tau)}{(s_n + \tau)^i} \leq Q + pq.$$

Then $Q + pq \geq q + pQ$ or $(p-1)q \geq (p-1)Q$, i.e., $q \geq Q$, which implies that $q = Q$. Therefore,

$$b = \lim_{t\to\infty} \frac{z(t)}{t^i} = \lim_{t\to\infty} \frac{x(t) + px(t-\tau)}{t^i} = q + pq,$$

or $q = b/(1+p)$. Hence $\lim_{t\to\infty} x(t)/t^i = b/(1+p)$.

Theorem 5.4.1. *Assume that $p \geq 0$, $p \neq 1$, $x(t)$ is a nonoscillatory solution of Eq. (5.4.1).*

(a) *If n is even, then $x(t)$ belongs to one of the following types:*

$$A_{2j-1}(\infty, a)\text{-type}: \quad \lim_{t\to\infty} \frac{x(t)}{t^{2j-2}} = \infty, \qquad \lim_{t\to\infty} \frac{x(t)}{t^{2j-1}} = a \neq 0,$$

$$A_{2j-1}(\infty, 0)\text{-type}: \quad \limsup_{t\to\infty} \frac{x(t)}{t^{2j-2}} = \infty, \qquad \lim_{t\to\infty} \frac{x(t)}{t^{2j-1}} = 0,$$

$$A_{2j-1}(a, 0)\text{-type}: \quad \lim_{t\to\infty} \frac{x(t)}{t^{2j-2}} = a \neq 0, \quad \lim_{t\to\infty} \frac{x(t)}{t^{2j-1}} = 0$$

where $j = \{1, 2, \ldots, \frac{n}{2}\}$ and a is a constant.

(b) *If n is odd, then $x(t)$ belongs to one of the following types:*

$$A_{2j}(\infty, a)\text{-type}: \quad \lim_{t\to\infty} \frac{x(t)}{t^{2j-1}} = \infty, \qquad \lim_{t\to\infty} \frac{x(t)}{t^{2j}} = a \neq 0,$$

$$A_{2j}(\infty, 0)\text{-type}: \quad \limsup_{t\to\infty} \frac{x(t)}{t^{2j-1}} = \infty, \qquad \lim_{t\to\infty} \frac{x(t)}{t^{2j}} = 0,$$

$$A_{2j}(a, 0)\text{-type}: \quad \lim_{t\to\infty} \frac{x(t)}{t^{2j-1}} = a \neq 0, \quad \lim_{t\to\infty} \frac{x(t)}{t^{2j}} = 0,$$

$$A_0(a)\text{-type}: \quad \lim_{t\to\infty} x(t) = a \neq 0,$$

$$A_0(0)\text{-type}: \quad \lim_{t\to\infty} x(t) = 0$$

where $j \in \{1, 2, \ldots, \frac{n-1}{2}\}$ and a is a constant.

Proof: Without loss of generality we assume that $x(t)$ is an eventually positive solution of (5.4.1). Then $z(t) > 0$ and $z^{(n)}(t) < 0$ for $t \geq T_1 \geq t_0$.

If n is even, by Lemma 5.1.1, there exist $T_2 \geq T_1$ and $\ell = 2j - 1$, where $j \in \{1, 2, \ldots, \frac{n}{2}\}$, such that

$$z^{(k)}(t) > 0 \quad \text{for} \quad t \geq T_2, \ k = 0, 1, \ldots, \ell - 1$$

and

$$(-1)^{k+1} z^{(k)}(t) > 0 \quad \text{for} \quad t \geq T_2, \ k = \ell, \ell + 1, \ldots, n - 1.$$

Especially, $z^{(2j-2)}(t) > 0, \quad z^{(2j-1)}(t) > 0, \quad z^{(2j)}(t) < 0, \quad t \geq T_2$. Denote

$$\lim_{t \to \infty} z^{(2j-1)}(t) = \ell_{2j-1} \quad \text{and} \quad \lim_{t \to \infty} z^{(2j-2)}(t) = \ell_{2j-2}.$$

Then $0 \leq \ell_{2j-1} < \infty$, $0 < \ell_{2j-2} \leq \infty$. If $\ell_{2j-1} > 0$, by l'Hospital's rule, we have

$$\lim_{t \to \infty} \frac{z(t)}{t^{2j-1}} = \lim_{t \to \infty} \frac{z'(t)}{(2j-1)t^{2j-2}} = \cdots = \lim_{t \to \infty} \frac{z^{(2j-1)}(t)}{(2j-1)!} = \frac{\ell_{2j-1}}{(2j-1)!}.$$

Since $0 \leq x(t)/t^{2j-1} \leq z(t)/t^{2j-1}$, $x(t)/t^{2j-1}$ is bounded on $[t_0, \infty)$. By Lemma 5.5.1, we have

$$\lim_{t \to \infty} \frac{x(t)}{t^{2j-1}} = \frac{\ell_{2j-1}}{(2j-1)!\,(1+p)} \neq 0.$$

It follows that $\lim_{t \to \infty} x(t)/t^{2j-2} = \infty$. Hence $x(t)$ is an $A_{2j-1}(\infty, a)$-type solution.

If $\ell_{2j-1} = 0$ and $\ell_{2j-2} = \infty$, then by l'Hospital's rule, it is easy to see that

$$\lim_{t \to \infty} \frac{z(t)}{t^{2j-1}} = 0 \quad \text{and} \quad \lim_{t \to \infty} \frac{z(t)}{t^{2j-2}} = \infty.$$

By Lemma 5.4.2 we have

$$\lim_{t \to \infty} \frac{x(t)}{t^{2j-1}} = 0 \quad \text{and} \quad \limsup_{t \to \infty} \frac{x(t)}{t^{2j-2}} = \infty.$$

Hence, $x(t)$ is an $A_{2j-1}(\infty, 0)$-type solution. If $\ell_{2j-1} = 0$ and $0 < \ell_{2j-2} < \infty$, then, by l'Hospital's rule, we have

$$\lim_{t \to \infty} \frac{z(t)}{t^{2j-2}} = \frac{\ell_{2j-2}}{(2j-2)!} \neq 0.$$

By Lemma 5.4.1 we have

$$\lim_{t\to\infty}\frac{x(t)}{t^{2j-2}}=\frac{\ell_{2j-2}}{(2j-2)!(1+p)}\neq 0.$$

It follows that $\lim_{t\to\infty} x(t)/t^{2j-1}=0$. Hence $x(t)$ is an $A_{2j-1}(a,0)$-type solution.

For the case that n is odd, the proof is similar. □

Theorem 5.4.2. *Suppose that n is even, $p\geq 0$, $p\neq 1$, and f is superlinear or sublinear. Then Eq. (5.4.1) has an $A_{2j-1}(\infty,a)$-type nonoscillatory solution if and only if there is a $k\neq 0$ such that*

$$\int_{t_0}^{\infty} s^{n-2j}\,|f(s,k(s-\sigma_1(s))^{2j-1},\ldots,k(s-\sigma_m(s))^{2j-1})|\,ds<\infty \tag{5.4.6}$$

where $j\in\{1,2,\ldots,\frac{n}{2}\}$.

Proof: *Necessity.* Let $x(t)$ be an eventually positive $A_{2j-1}(\infty,a)$-type solution. Then $z(t)>0$ for $t\geq T_1\geq t_0$. Since $z^{(n)}(t)<0$, $z^{(i)}(t)$ is eventually monotonic, $i=1,2,\ldots,n-1$. Since $\lim_{t\to\infty} x(t)/t^{2j-1}=a>0$, there exists $T_2\geq T_1$ such that

$$\frac{1}{2}at^{2j-1}\leq x(t)\leq\frac{3}{2}at^{2j-1},\quad t\geq T_2. \tag{5.4.7}$$

We show that

$$\lim_{t\to\infty} z^{(2j-1)}(t)=(2j-1)!\,(1+p)a. \tag{5.4.8}$$

In fact, by l'Hospital's rule

$$\lim_{t\to\infty}\frac{z(t)}{t^{2j-1}}=\cdots=\lim_{t\to\infty}\frac{z^{(2j-1)}(t)}{(2j-1)!}=(1+p)a. \tag{5.4.9}$$

For the case that $j<\frac{n}{2}$, since $z^{(i)}(t)$ is eventually monotonic for $i=2j$, $2j+1,\ldots,n-1$, from (5.4.8) we have

$$\lim_{t\to\infty} z^{(2j)}(t)=\lim_{t\to\infty} z^{(2j+1)}(t)=\cdots=\lim_{t\to\infty} z^{(n-1)}(t)=0. \tag{5.4.10}$$

Integrating (5.4.1) $n-2j$ times and using (5.4.10), we have

$$z^{(2j)}(t) = (-1)^{n-2j} \int_t^\infty \int_{s_{n-2j-1}}^\infty \cdots \int_{s_1}^\infty G(s) ds\, ds_1 \cdots ds_{n-2j-1} \tag{5.4.11}$$

where $G(s) = f(s, x(s-\sigma_1(s)), \ldots, x(s-\sigma_m(s)))$.

Integrating (5.4.11) from T_2 to ∞ and noting (5.4.8) we have

$$(2j-1)!(1+p)a - z^{(2j-1)}(T_2) = (-1)^{n-2j} \int_{T_2}^\infty \int_{s_{n-2j}}^\infty \cdots \int_{s_1}^\infty G(s)\, ds\, ds_1 \cdots ds_{n-2j}.$$

This together with (5.4.3), (5.4.4) and (5.4.7) implies that

$$\int_{T_2}^\infty \int_{s_{n-2j}}^\infty \cdots \int_{s_1}^\infty F_{j,k}(s)\, ds\, ds_1 \cdots ds_{n-2j} < \infty \tag{5.4.12}$$

where $F_{j,k}(s) = f\big(s, k(s-\sigma_1(s))^{2j-1}, \ldots, k(s-\sigma_m(s))^{2j-1}\big)$, $k = \frac{a}{2}$ if f is superlinear, $k = \frac{3a}{2}$ if f is sublinear. This shows that (5.4.6) holds.

Sufficiency. Assume that (5.4.6) holds. We may assume that $k > 0$. Let $d = k/2$ if f is superlinear and $d = k$ if f is sublinear. Set $R(t) = t^{2j-1}$.

If $p \in (0,1)$, we have

$$\lim_{t\to\infty} p \frac{R(t)}{R(t-\tau-h)} = p \qquad \text{and} \qquad \lim_{t\to\infty} \frac{R(t-\tau)}{R(t)} = 1 > 1 - \frac{1-p}{4p}.$$

Choose $p_1 \in (p,1)$ and $T \geq t_0 + \tau + h$ such that

$$p\frac{R(t)}{R(t-\tau-h)} < p_1 \qquad \text{and} \qquad \frac{R(t-\tau)}{R(t)} > 1 - \frac{1-p}{4p} \quad \text{for } t \geq T, \tag{5.4.13}$$

and

$$\int_T^\infty \int_{s_{n-2j}}^\infty \cdots \int_{s_1}^\infty F_{j,k}(s)\, ds\, ds_1 \cdots ds_{n-2j} < \frac{(1-p)d}{8} \tag{5.4.14}$$

where $F_{j,k}(s) = f(s, k(s-\sigma_1(s))^{2j-1}, \ldots, k(s-\sigma_m(s))^{2j-1})$. Let $T^* = T-\tau-h$. Then $pR(T)/R(T^*) < p_1$. Introduce the Banach space:

$$C_R[T^*,\infty) = \left\{x : x \in C([T^*,\infty), R),\ \sup_{t\geq T^*} \frac{|x(t)|}{R^2(t)} < \infty\right\}$$

with the norm $\|x\|_R = \sup_{t \geq T^*} |x(t)|/R^2(t)$. Define a subset Ω of $C_R[T^*, \infty)$ as follows:

$$\Omega = \{x\colon x \in C_R[T^*, \infty),\ dR(t) \leq x(t) \leq 2dR(t)\}.$$

Then Ω is a bounded, convex and closed subset of $C_R[T^*, \infty)$. Define two maps $\mathcal{T}_1$ and $\mathcal{T}_2 : \Omega \to C_R[T^*, \infty)$ as follows:

$$(\mathcal{T}_1 x)(t) = \begin{cases} \dfrac{3d(1+p)}{2} R(t) - \dfrac{px(T-\tau)}{R(T)} R(t), & T^* \leq t < T, \\ \dfrac{3d(1+p)}{2} R(t) - px(t-\tau), & t \geq T, \end{cases}$$

and

$$(\mathcal{T}_2 x)(t) = \begin{cases} 0, \quad T^* \leq t < T, \\ \displaystyle\int_T^t \int_T^{s_{n-1}} \cdots \int_T^{s_{n-2j+2}} \int_{s_{n-2j+1}}^{\infty} \cdots \\ \displaystyle\qquad \cdots \int_{s_1}^{\infty} G(s)\, ds\, ds_1 \cdots ds_{n-1}, \quad t \geq T, \end{cases}$$

where $G(s) = f(s, x(s - \sigma_1(s)), \ldots, x(s - \sigma_m(s)))$. It is not difficult to see that for any $x,\ y \in \Omega,\ \mathcal{T}_1 x + \mathcal{T}_2 y \in \Omega$.

We show that $\mathcal{T}_1$ is a contraction. Let $x,\ y \in \Omega$. Then, for $T^* \leq t < T$

$$\begin{aligned} \frac{|(\mathcal{T}_1 x)(t) - (\mathcal{T}_1 y)(t)|}{R^2(t)} &= p\, \frac{|x(T-\tau) - y(T-\tau)|}{R(t)R(T)} \\ &= p\, \frac{|x(T-\tau) - y(T-\tau)|}{R^2(T-\tau)} \, \frac{R^2(T-\tau)}{R(t)R(T)} \\ &\leq p\, \frac{|x(T-\tau) - y(T-\tau)|}{R^2(T-\tau)} \, \frac{R(T)}{R(T^*)} \leq p_1 \sup_{t \geq T^*} \frac{|x(t) - y(t)|}{R^2(t)}, \end{aligned}$$

and for $t \geq T$

$$\begin{aligned} \frac{|(\mathcal{T}_1 x)(t) - (\mathcal{T}_1 y)(t)|}{R^2(t)} &= p\, \frac{|x(T-\tau) - y(T-\tau)|}{R^2(t)} \leq p\, \frac{|x(t-\tau) - y(t-\tau)|}{R^2(t-\tau)} \\ &\leq p_1 \sup_{t \geq T^*} \frac{|x(t) - y(t)|}{R^2(t)}. \end{aligned}$$

Then $\|\mathcal{T}_1 x - \mathcal{T}_1 y\| \leq p_1 \|x - y\|$. Since $p_1 \in (0,1)$, $\mathcal{T}_1$ is a contraction.

We now show that $\mathcal{T}_2$ is completely continuous. Let $x_k \in \Omega$ be such that $x_k \to x$ as $k \to \infty$. Since Ω is closed, $x \in \Omega$. For $t \geq T$ we have

$$|(\mathcal{T}_2 x_k)(t) - (\mathcal{T}_2 x)(t)| \leq \int_T^t \int_T^{s_{n-1}} \cdots \int_T^{s_{n-2j+2}} \int_{s_{n-2j+1}}^{\infty} \cdots \cdots \int_{s_1}^{\infty} |H_k(s)|\, ds ds_1 \cdots ds_{n-1}$$

where

$$\begin{aligned}
|H_k(s)| &= |f(s, x_k(s - \tau_1(s)), \ldots, x_k(s - \tau_m(s))) \\
&\qquad - f(s, x(s - \tau_1(s)), \ldots, x(s - \tau_m(s)))| \\
&\leq f(s, x_k(s - \tau_1(s)), \ldots, x_k(s - \tau_m(s))) \\
&\qquad + f(s, x(s - \tau_1(s)), \ldots, x(s - \tau_m(s))) \\
&\leq 4f(s, k(s - \tau_1(s))^{2j-1}, \ldots, k(s - \tau_m(s))^{2j-1}) = 4F_j(s).
\end{aligned}$$

Hence, for $t \geq T$, we obtain

$$\begin{aligned}
|(\mathcal{T}_2 x_k)(t) - (\mathcal{T}_2 x)(t)| &\leq \left(\int_T^{\infty} \int_{s_{n-2j}}^{\infty} \cdots \int_{s_1}^{\infty} |H_k(s)|\, ds ds_1 \cdots ds_{n-2j} \right) \\
&\qquad \times \int_T^t \int_T^{s_{n-1}} \cdots \int_T^{s_{n-2j+2}} ds_{n-2j+1} \cdots ds_{n-1} \\
&\leq \frac{(t-T)^{2j-1}}{(2j-1)!} \int_T^{\infty} \int_{s_{n-2j}}^{\infty} \cdots \int_{s_1}^{\infty} |H_k(s)| ds\, ds_1 \cdots ds_{n-2j} \\
&\leq R(t) \int_T^{\infty} \int_{s_{n-2j}}^{\infty} \cdots \int_{s_1}^{\infty} |H_k(s)|\, ds\, ds_1 \cdots ds_{n-2j}.
\end{aligned}$$

For $T^* \leq t < T$, $|(\mathcal{T}_2 x_k)(t) - (\mathcal{T}_2 x)(t)| = 0$. Then

$$\|\mathcal{T}_2 x_k - \mathcal{T}_2 x\| \leq \sup_{t \geq T^*} R^{-1}(t) \int_T^{\infty} \int_{s_{n-2j}}^{\infty} \cdots \int_{s_1}^{\infty} |H_k(s)|\, ds\, ds_1 \cdots ds_{n-2j}.$$

By the Lebesgue's dominated convergence theorem, we get

$$\lim_{k \to \infty} \|\mathcal{T}_2 x_k - \mathcal{T}_2 x\| = 0.$$

It follows that $\mathcal{T}_2$ is continuous. Next, we show that Ω is relatively compact. It suffices to show that the family of functions $\{R^{-2}\mathcal{T}_2 x\colon x \in \Omega\}$ is uniformly bounded and equicontinuous on $[T^*, \infty)$. The uniform boundedness is obvious. For the equicontinuity, according to Levitan's result, we only need to show that, for any given $\varepsilon > 0$, $[T^*, \infty)$ can be decomposed into finite subintervals in such a way that on each subinterval all functions of the family have change of amplitude less than ε. For any $\varepsilon > 0$, take $T_1 \geq T$ large enough so that

$$\frac{1}{R(t)} < \frac{\varepsilon}{(1-p)d} \quad \text{for } t \geq T_1.$$

Then for $x \in \Omega,\ t_2 > t_1 \geq T_1$

$$\begin{aligned}
|(R^{-2}\mathcal{T}_2 x)(t_2) - (R^{-2}\mathcal{T}_2 x)(t_1)| &\leq (R^{-2}\mathcal{T}_2 x)(t_2) + (R^{-2}\mathcal{T}_2 x)(t_1) \\
&\leq \sum_{i=1}^{2} \frac{1}{R^2(t_i)} \int_T^{t_i} \int_T^{s_{n-1}} \cdots \int_T^{s_{n-2j+2}} \int_{s_{n-2j+1}}^{\infty} \cdots \int_{s_1}^{\infty} 2F_{j,k}(s)\, ds\, ds_1 \cdots ds_{n-1} \\
&\leq \sum_{i=1}^{2} \left[\frac{1}{R^2(t_i)} \frac{(1-p)d}{4} R(t_i)\right] = \frac{(1-p)d}{4} \sum_{i=1}^{2} \frac{1}{R(t_i)} < \frac{\varepsilon}{2} < \varepsilon.
\end{aligned}$$

For $x \in \Omega$ and $T \leq t_1 < t_2 \leq T_1$

$$\begin{aligned}
&|(R^{-2}\mathcal{T}_2 x)(t_2) - (R^{-2}\mathcal{T}_2 x)(t)| \\
&\quad \leq \frac{1}{R^2(t_1)} \int_{t_1}^{t_2} \int_T^{s_{n-1}} \cdots \int_T^{s_{n-2j+2}} \frac{(1-p)d}{4}\, ds_{n-2j+1} \cdots ds_{n-1} \\
&\qquad + \left|\frac{1}{R^2(t_2)} - \frac{1}{R^2(t_2)}\right| \int_T^{t_2} \int_T^{s_{n-1}} \cdots \int_T^{s_{n-2j+2}} \frac{(1-p)d}{4}\, ds_{n-2j+1} \cdots ds_{n-1}.
\end{aligned}$$

Then there exists an $\delta > 0$ such that

$$|(R^{-2}\mathcal{T}_2 x)(t_2) - (R^{-2}\mathcal{T}_2 x)(t_1)| < \varepsilon \quad \text{if } 0 < t_2 - t_1 < \delta.$$

For any $x \in \Omega,\ T^* \leq t_1 < t_2 \leq T$, it is easy to see that

$$|(R^{-2}\mathcal{T}_2 x)(t_2) - (R^{-2}\mathcal{T}_2 x)(t_1)| = 0 < \varepsilon.$$

Therefore $\{R^{-2}\mathcal{T}_2 x\colon x \in \Omega\}$ is uniformly bounded and equicontinuous on $[T^*, \infty)$, and hence $\mathcal{T}\Omega$ is relatively compact. By Krasnoselskii's fixed point theorem, there is a $x_0 \in \Omega$ such that $(\mathcal{T}_1 + \mathcal{T}_2)x_0 = x_0$. That is, for $t \geq T$

$$x_0(t) = \frac{3(1+p)}{2} dR(t) - px_0(t-\tau) + \int_T^t \int_T^{s_{n-1}} \cdots \int_T^{s_{n-2j+2}} \int_{s_{n-2j+1}}^{\infty} \cdots \int_{s_1}^{\infty} G_0(s)\, ds\, ds_1 \cdots ds_{n-1}$$

where $G_0(s) = f\big(s, x_0(s-\tau_1(s)), \ldots, x_0(s-\tau_m(s))\big)$. It is easy to see that $x_0(t)$ is a nonoscillatory solution of (5.5.1). Let $z_0(t) = x_0(t) + px_0(t-\tau)$. Then for $t \geq T$

$$z_0(t) = \frac{3(1+p)}{2} dR(t) + \int_T^t \int_T^{s_{n-1}} \cdots \int_T^{s_{n-2j+2}} \int_{s_{n-2j+1}}^{\infty} \cdots \int_{s_1}^{\infty} G_0(s)\, ds\, ds_1 \cdots ds_{n-1} \tag{5.4.15}$$

where $0 \leq G_0(s) \leq 2F_j(s)$.

By applying l'Hospital's rule $2j-1$ times we obtain

$$\begin{aligned} &\lim_{t\to\infty} \frac{1}{t^{2j-1}} \int_T^t \int_T^{s_{n-1}} \cdots \int_T^{s_{n-2j+2}} \int_{s_{n-2j+1}}^{\infty} \cdots \int_{s_1}^{\infty} F_j(s)\, ds\, ds_1 \cdots ds_{n-1} \\ &= \lim_{t\to\infty} \frac{1}{(2j-1)!} \int_t^{\infty} \int_{s_{n-2j}}^{\infty} \cdots \int_{s_1}^{\infty} F_j(s) ds\, ds_1 \cdots ds_{n-2j} = 0. \end{aligned} \tag{5.4.16}$$

In view of (5.4.15) and (5.4.16) we have

$$\lim_{t\to\infty} \frac{z_0(t)}{t^{2j-1}} = \frac{3(1+p)d}{2}.$$

By Lemma 5.5.1, $\lim_{t\to\infty} x_0(t)/t^{2j-1} = 3d/2$. Consequently, $\lim_{t\to\infty} x_0(t)/t^{2j-2} = \infty$. Then $x_0(t)$ is an $A_{2j-1}(\infty, a)$-type solution.

If $p = 0$, let T be so large that $T \geq t_0 + \tau + h$ and

$$\int_T^{\infty} \int_{s_{n-2j}}^{\infty} \cdots \int_{s_1}^{\infty} F_j(s)\, ds\, ds_1 \cdots ds_{n-2j} < \frac{d}{8}$$

where $F_j(s) = f\big(s, k(s-\sigma_1(s))^{2j-1}, \ldots, k(s-\sigma_m(s))^{2j-1}\big)$.

Let $T^* = T - \tau - h$. Introduce the Banach space $C_R[T^*, \infty)$ as before. Define the operator $\mathcal{T}_1$ by $(\mathcal{T}_1 x)(t) = \frac{3}{2} dR(t)$ and the operator $\mathcal{T}_2$ as before, where

$R(t) = t^{2j-1}$. By a similar argument as before we can show that Eq. (5.4.1) has an $A_{2j-1}(\infty, a)$-type solution.

Let us consider the case that $p > 1$. Choose p_1 such that $0 < \frac{1}{p} < p_1 < 1$. Let T be so large that $T \geq t_0 + \tau + h$,

$$\frac{1}{p}\frac{R^2(t+\tau)}{R^2(t-\tau-h)} < p_1 < 1, \qquad \frac{R(t+\tau)}{R(t)} < 1 + \frac{p-1}{4}, \quad t \geq T \tag{5.4.17}$$

and

$$\int_T^\infty \int_{s_{n-2j}}^\infty \cdots \int_{s_1}^\infty F_j(s)\,ds ds_1 \cdots ds_{n-2j} < \frac{(p-1)d}{8}.$$

Denote $T^* = T - \tau - h$, and define $C_R[T^*, \infty)$ and Ω as before with the norm (as in the proof of Theorem 5.4.2), and let the operators $\mathcal{T}_1$ and $\mathcal{T}_2 \colon \Omega \to C_R[T^*, \infty)$ be given by

$$(\mathcal{T}_1 x)(t) = \begin{cases} \dfrac{3(1+p)}{2p}\, dR(t) - \dfrac{1}{p}\dfrac{x(T+\tau)}{R(T)} R(t), & T^* \leq t < T, \\[2ex] \dfrac{3(1+p)}{2p}\, dR(t) - \dfrac{1}{p} x(t+\tau), & t \geq T \end{cases}$$

and

$$(\mathcal{T}_2 x)(t) = \begin{cases} 0, \quad T^* \leq t < T, \\ \dfrac{1}{p}\displaystyle\int_{T+\tau}^{t+\tau} \int_T^{s_{n-1}} \cdots \int_T^{S_{n-2j+2}} \int_{s_{n-2j+1}}^\infty \cdots \\ \qquad \cdots \displaystyle\int_{s_1}^\infty G(s)ds\,ds_1 \cdots ds_{n-1}, \quad t \geq T \end{cases}$$

where $G(s) = f\big(s, x(s - \tau_1(s), \ldots, x(s - \tau_m(s))\big)$.

Similar to the proof for the case that $p \in (0,1)$, we can show that the conditions of the Krasnoselskii's fixed point theorem are satisfied. Hence, there is $x_0 \in \Omega$ such that $(\mathcal{T}_1 + \mathcal{T}_2)x_0 = x_0$. That is,

$$x_0(t) = \frac{3(1+p)}{2p} dR(t) - \frac{1}{p} x_0(t+\tau) + \frac{1}{p}\int_{T+\tau}^{t+\tau} \int_T^{s_{n-1}}$$

$$\cdots \int_T^{S_{n-2j+2}} \int_{s_{n-2j+1}}^\infty \cdots \int_{s_1}^\infty G_0(s)ds\,ds_1 \cdots ds_{n-1}, \quad t \geq T$$

where $G_0(s) = f\big(s, x_0(s-\sigma_1(s)), \ldots, x_0(s-\sigma_m(s))\big)$. It is not difficult to show that $x_0(t)$ is an $A_{2j-1}(\infty, a)$-type solution. □

Theorem 5.4.3. *Suppose that n is even, $p \geq 0$, $p \neq 1$, f is superlinear or sublinear. Then Eq. (5.4.1) has an $A_{2j-1}(a,0)$-type nonoscillatory solution if and only if there is a $k \neq 0$ such that*

$$\int_{t_0}^{\infty} s^{n-2j+1} \left| f(s, k(s-\sigma_1(s))^{2j-2}, \ldots, k(s-\sigma_m(s)))^{2j-2} \right| ds < \infty \qquad (5.4.18)$$

where $j \in \{1, 2, \ldots, \frac{n}{2}\}$.

Proof: *Necessity.* The proof is similar to that of Theorem 5.4.2 and hence is omitted.

Sufficiency. Without loss of generality we may assume that $k > 0$. Let $R^*(t) = t^{2j-2}$. By an argument similar to the proof of Theorem 5.4.2 we can show that for $p \in (0,1)$ there exist $T \geq t_0$ and $x^*(t)$ such that

$$dR^*(t) \leq x^*(t) \leq 2dR^*(t), \quad t \geq T^* = T - \tau - h$$

and

$$x^*(t) = \frac{3(1+p)}{2} dR^*(t) - px^*(t-\tau)$$
$$+ \int_T^t \int_T^{s_{n-1}} \cdots \int_T^{S_{n-2j+2}} \int_{s_{n-2j+1}}^{\infty} \cdots \int_{s_1}^{\infty} G^*(s) ds\, ds_1 \cdots ds_{n-1}, \quad t \geq T$$

where $G^*(s) = f\big(s, x^*(s-\sigma_1(s)), \ldots, x^*(s-\sigma_m(s))\big)$. It is not difficult to prove that

$$\lim_{t\to\infty} \frac{x^*(t)}{t^{2j-2}} = \frac{3}{2}d + \frac{\alpha}{1+p}$$

where α is a constant and

$$0 < \alpha \leq \int_T^{\infty} \int_{s_{n-2j+1}}^{\infty} \cdots \int_{s_1}^{\infty} 2F_j^*(s)\, ds\, ds_1 \cdots ds_{n-1}.$$

It follows that $\lim_{t\to\infty} x^*(t)/t^{2j-1} = 0$. Then $x^*(t)$ is an $A_{2j-1}(a,0)$-type solution of Eq. (5.4.1). Similarly, for the cases $p = 0$ and $p > 1$, we can prove that Eq. (5.4.1) has an $A_{2j-1}(a,0)$-type solution. □

Theorem 5.4.4. *Suppose that n is odd, $p \geq 0$, $p \neq 1$, f is superlinear or sublinear. Then Eq. (5.4.1) has an $A_{2j}(\infty, a)$-type nonoscillatory solution if and only if there is a $k \neq 0$ such that*

$$\int_{t_0}^{\infty} s^{n-2j-1} |f(s, k(s - \sigma_1(s))^{2j}, \ldots, k(s - \sigma_m(s))^{2j-1})| \, ds < \infty$$

where $j \in \{1, 2, \ldots, \frac{n-1}{2}\}$.

Proof: The proof is similar to that of Theorem 5.4.2. We need only note that $R(t)$ and $\mathcal{T}_2$ there are replaced by $R(t) = t^{2j}$ and

$$(\mathcal{T}_2 x)(t) = \begin{cases} 0, & T^* \leq t < T \\ \int_T^t \int_T^{s_{n-1}} \cdots \int_T^{S_{n-2j+1}} \int_{s_{n-2j}}^{\infty} \cdots \int_{s_1}^{\infty} G(s) ds \, ds_1 \cdots ds_{n-1}, & t \geq T, \end{cases}$$

if $p \in [0, 1)$, or

$$(\mathcal{T}_2 x)(t) = \begin{cases} 0, & T^* \leq t < T \\ \frac{1}{c} \int_{T+\tau}^{t+\tau} \int_T^{s_{n-1}} \cdots \int_T^{S_{n-2j+1}} \int_{s_{n-2j}}^{\infty} \cdots \\ \qquad \cdots \int_{s_1}^{\infty} G(s) ds \, ds_1 \cdots ds_{n-1}, & t \geq T, \end{cases}$$

if $p > 1$, respectively. □

Theorem 5.4.5. *Suppose that n is odd, $p \geq 0$, $p \neq 1$, f is superlinear or sublinear. Then Eq. (5.4.1) has an $A_{2j}(a, 0)$-type nonoscillatory solution if and only if there is a $k \neq 0$ such that*

$$\int_{t_0}^{\infty} s^{n-2j} |f(s, k(s - \sigma_1(s))^{2j-1}, \ldots, k(s - \sigma_m))^{2j-1})| \, ds < \infty$$

where $j \in \{1, 2, \ldots, \frac{n-1}{2}\}$.

Proof: The proof is similar to that of Theorems 5.4.3 and 5.4.4. □

Theorem 5.4.6. *Suppose that n is odd, $p \geq 0$, $p \neq 1$, f is superlinear or sublinear. Then Eq. (5.4.1) has an $A_0(a)$-type nonoscillatory solution if and only if there is a $k \neq 0$ such that*

$$\int_{t_0}^{\infty} s^{n-1}|f(s,k,\ldots,k)|\,ds < \infty.$$

Proof: The proof is similar to that of of Theorem 5.4.2. We need only note that $R(t)$ and $\mathcal{T}_2$ there are replaced by $R(t) = 1$ and

$$(\mathcal{T}_2 x)(t) = \begin{cases} \displaystyle\int_T^{\infty}\int_{s_{n-1}}^{\infty}\cdots\int_{s_1}^{\infty} G(s)\,ds\,ds_1\cdots ds_{n-1}, & T^* \leq t < T \\ \displaystyle\int_t^{\infty}\int_{s_{n-1}}^{\infty}\cdots\int_{s_1}^{\infty} G(s)\,ds\,ds_1\cdots ds_{n-1}, & t \geq T, \end{cases}$$

if $p \in [0,1)$, or

$$(\mathcal{T}_2 x)(t) = \begin{cases} \displaystyle\frac{1}{p}\int_{T+\tau}^{\infty}\int_{s_{n-1}}^{\infty}\cdots\int_{s_1}^{\infty} G(s)\,ds\,ds_1\cdots ds_{n-1}, & T^* \leq t < T \\ \displaystyle\frac{1}{p}\int_{t+d}^{\infty}\int_{s_{n-1}}^{\infty}\cdots\int_{s_1}^{\infty} G(s)\,ds\,ds_1\cdots ds_{n-1}, & t \geq T, \end{cases}$$

if $p > 1$, respectively. □

Then we consider the linear neutral differential equation

$$\big(x(t) - p(t)x(\tau(t))\big)^{(n)} + \sum_{i=1}^{N} q_i(t)x(g_i(t)) = 0, \quad t \geq t_0. \tag{5.4.19}$$

The following result is about the existence of positive solutions of Eq. (5.4.19).

Theorem 5.4.7. *Let n be odd and assume that*

(i) $p \in C([t_0,\infty), R_+)$;

(ii) τ *is continuous on* $[t_0,\infty)$, $\tau(t) < t$ *for* $t \geq t_0$, *and* $\lim_{t\to\infty} \tau(t) = \infty$;

(iii) q_i, $g_i \in C([t_0,\infty), R)$, $q_i(t) \geq 0$ and $\lim_{t\to\infty} g_i(t) = \infty$, $i = 1, 2, \ldots, N$;

(iv) *there exist* $\alpha > 0$ *and* $\ell \in (0,1)$ *such that* $0 \le p(t)e^{\alpha(t-\tau(t))} \le \ell$ *and*

$$p(t)e^{\alpha(t-\tau(t))} + \frac{e^{\alpha t}}{(n-1)!}\int_t^\infty (s-t)^{n-1}\sum_{i=1}^N q_i(s)e^{-\alpha g_i(s)}\,ds \le 1 \quad (5.4.20)$$

hold eventually.

Then Eq. (5.4.19) has an eventually positive solution $x(t)$ *which approaches zero exponentially.*

Proof: Let $t_1 \ge t_0$ be such that

$$\min\{\inf_{t\ge t_1}\tau(t),\ \inf_{t\ge t_1} g_1(t),\ldots,\inf_{t\ge t_1} g_N(t)\} = T \ge t_0.$$

Let BC be the Banach space of all bounded and continuous functions defined on $[t_0,\infty)$ endowed with the sup norm. and Ω be the subset of BC defined as follows:

$$\Omega = \{y \in BC\colon 0 \le y(t) \le 1,\ t \ge T\}.$$

Now we define operators T_1 and T_2 on Ω as follows:

$$(T_1y)(t) = \begin{cases} p(t)y(\tau(t))e^{\alpha(t-\tau(t))}, & t \ge t_1 \\ \dfrac{t}{t_1}(T_1y)(t_1) + \left(1 - \dfrac{t}{t_1}\right), & T \le t \le t_1, \end{cases}$$

$$(T_2y)(t) = \begin{cases} \dfrac{e^{\alpha t}}{(n-1)!}\displaystyle\int_t^\infty \sum_{i=1}^N q_i(s)y(g_i(s))e^{-\alpha g_i(s)}\,ds, & t \ge t_1 \\ \dfrac{t}{t_1}(T_2y)(t_1), \quad T \le t \le t_1. \end{cases}$$

In view of (iv), for sufficiently large T and every pair $x,\ y \in \Omega$, we have that $T_1x + T_2x \in \Omega$ and T_1 is a contraction. We also see that $\left|\frac{d}{dt}(T_2y)(t)\right| \le M$ for some $M > 0$, and hence T_2 is completely continuous on Ω. By Krasnoselskii's fixed point theorem there exists a $y \in \Omega$ such that $T_1 y + T_2 y = y$. That is,

$$y(t) = \begin{cases} p(t)y(\tau(t))e^{\alpha(t-\tau(t))} + \dfrac{e^{\alpha t}}{(n-1)!}\displaystyle\int_t^\infty (s-t)^{n-1} \\ \qquad \times \displaystyle\sum_{i=1}^N q_i(s)y(g_i(s))e^{-\alpha g_i(s)}ds, \quad t \geq t_1, \\ \dfrac{t}{t_1}y(t_1) + \left(1 - \dfrac{t}{t_1}\right), \quad T \leq t \leq t_1. \end{cases} \tag{5.4.21}$$

Since $y(t) > 0$ on $[T, t_1)$, it follows that $y(t) > 0$ for all $t \geq T$. Set $x(t) = y(t)e^{-\alpha t}$. Then (5.4.21) implies that

$$x(t) = p(t)x(\tau(t)) + \int_t^\infty \frac{(s-t)^{n-1}}{(n-1)!} \sum_{i=1}^N q_i(s)x(g_i(s))\, ds, \quad t \geq t_1.$$

Consequently,

$$\big(x(t) - p(t)x(\tau(t))\big)^{(n)} + \sum_{i=1}^n q_i(s)x(g_i(s)) = 0, \quad t \geq t_1.$$

Thus $x(t)$ is a solution of Eq. (5.4.19), which tends to zero exponentially as $t \to \infty$. □

Consider the special case of (5.4.19)

$$\big(x(t) - px(t-\tau)\big)^{(n)} + \sum_{i=1}^N q_i(t)x(t-\sigma_i) = 0 \tag{5.4.22}$$

where $\tau,\ \sigma_i > 0,\ i = 1, \ldots, N,\ 0 < p < 1,\ 0 \leq q_i(t) \leq q_i^*,\ i = 1, \ldots, N$. (5.4.20) becomes

$$pe^{\alpha\tau} + \sum_{i=1}^N q_i^* \frac{e^{\alpha\sigma_i}}{\alpha^n} \leq 1. \tag{5.4.23}$$

Condition (5.4.23) provides not only a sufficient condition but also a necessary condition for the existence of a nonoscillatory solution of Eq. (5.4.19). In fact, the characteristic equation of (5.4.22) is

$$\lambda^n - p\lambda^n e^{-\lambda\tau} + \sum_{i=1}^{n} q_i^* e^{-\lambda\sigma_i} = 0. \tag{5.4.24}$$

Eq. (5.4.22) has a nonoscillatory solution if and only if (5.4.24) has a negative real root, say λ^*. Put $\alpha = -\lambda^*$. Then (5.4.24) becomes

$$p\alpha^n e^{\alpha\tau} + \sum_{i=1}^{N} q_i^* e^{\alpha\sigma_i} = \alpha^n,$$

which implies that (5.4.23) holds.

Example 5.4.1. Consider the third order neutral differential equation

$$\left(x(t) - \frac{1}{2e}x(t-1)\right)''' + \left(\frac{1}{2} - \frac{1}{t}\right)x(t) = 0, \quad t \geq 3 \tag{5.4.25}$$

where condition (5.4.23) holds with $\alpha = 1$. By Theorem 5.4.7 there exists a positive solution of (5.4.25). In fact, $x(t) = te^{-t}$ is such a solution.

5.5. Existence of Oscillatory Solutions

We consider the neutral differential equations

$$\left(x(t) + px(t-\tau)\right)^{(n)} + f\left(t, x(g_1(t)), \ldots, x(g_N(t))\right) = 0 \tag{5.5.1}$$

and

$$\left(x(t) - px(t-\tau)\right)^{(n)} + f\left(t, x(g_1(t)), \ldots, x(g_N(t))\right) = 0 \tag{5.5.2}$$

where $n \geq 1$, p and τ are positive constants, $f \in C([t_0,\infty) \times R^N, R)$ and $g_i \in C([t_0,\infty), R)$, $\lim_{t\to\infty} g_i(t) = \infty$, $i \in I_N$. We further assume that there exists a continuous function
$F(t, v_1, \ldots, v_N)$ on $[t_0,\infty) \times R_+^N$ which is nondecreasing in each v_i, $i = 1, \ldots, N$, and

$$|f(t, u_1, \ldots, u_N)| \leq F(t, |u_1|, \ldots, |u_N|), \quad (t, u_1, \ldots, u_N) \in [t_0,\infty) \times R^N. \tag{5.5.3}$$

We shall establish conditions for the existence of oscillatory solutions of (5.5.1) and (5.5.2).

Theorem 5.5.1. *Suppose that $0 < p \leq 1$ and there exist constants $\mu \in (0,p)$ and $a > 0$ such that*

$$\int_{t_0}^{\infty} t^{n-1}\mu^{-t/\tau} F(t, ap^{g_1(t)/\tau}, \dots, ap^{g_N(t)/\tau})\, dt < \infty. \tag{5.5.4}$$

Then

(i) *for any continuous periodic oscillatory function $\omega_-(t)$ with period τ, Eq. (5.5.2) has a bounded oscillatory solution $x_-(t)$ such that*

$$x_-(t) = k_1 p^{t/\tau}\omega_-(t) + o(p^{t/\tau}) \quad \text{as } t \to \infty; \tag{5.5.5}$$

(ii) *for any continuous oscillatory function $\omega_+(t)$ such that $\omega_+(t+\tau) = -\omega_+(t)$ for all t, Eq. (5.5.1) has a bounded oscillatory solution $x_+(t)$ such that*

$$x_+(t) = k_2 p^{t/\tau}\omega_+(t) + o(p^{t/\tau}) \quad \text{as } t \to \infty \tag{5.5.6}$$

where k_1 and k_2 are constants.

Proof: Let $\omega_\pm(t)$ be given and denote $\omega_\pm^* = \max_{t\in[t_0,t_0+\tau)} |\omega_\pm(t)|$. Choose $c > 0$ and $T \geq t_0$ such that

$$pc(1+\omega_\pm^*)/(p-\mu) \leq a,$$

$$T_0 = \min\{T-\tau,\ \inf_{t\geq T} g_1(t), \dots, \inf_{t\geq T} g_N(t)\} \geq \max\{t_0, \tau\}, \tag{5.5.7}$$

and

$$\int_{T}^{\infty} t^{n-1}\mu^{-t/\tau} F(t, ap^{g_1(t)/\tau}, \dots, ap^{g_N(t)/\tau})\, dt \leq c. \tag{5.5.8}$$

Let X be the set of functions defined by

$$X = \{x \in C[T_0,\infty) \colon |x(t)| \leq c\mu^{t/\tau} \text{ for } t \geq T_0\}$$

which is a closed convex subset of Frechet space $C[T_0,\infty)$ of continuous functions on $[T_0,\infty)$ with the usual metric topology. With each $x \in X$ we associate the functions $\widehat{x_\pm}(t) : [T_0, \infty) \to R$ defined by

$$\widehat{x_\pm}(t) = \begin{cases} \dfrac{pc}{p-\mu}\, p^{t/\tau}\omega_\pm(t) - \displaystyle\sum_{i=1}^{\infty}(\mp 1)^i p^{-i} x(t+i\tau), & t \geq T-\tau, \\ \widehat{x_\pm}(t)(T-\tau), \quad T_0 \leq t \leq T-\tau. \end{cases} \tag{5.5.9}$$

It is easy to see that, for each $x \in X$, $\widehat{x_\pm}(t)$ are well defined and continuous on $[T_0, \infty)$ and satisfy

$$\widehat{x_\pm}(t) \pm p\widehat{x_\pm}(t-\tau) = x(t), \quad t \geq T. \tag{5.5.10}$$

$x \in X$ implies that

$$\sum_{i=1}^{\infty} p^{-i}|x(t+i\tau)| \leq \frac{\mu c}{p-\mu}\, \mu^{t/\tau}, \quad t \geq T-\tau. \tag{5.5.11}$$

Substituting (5.5.11) into (5.5.9) we obtain

$$|\widehat{x_\pm}(t)| \leq \frac{pc\omega_\pm^*}{p-\mu}\, p^{t/\tau} + \frac{\mu c}{p-\mu}\, \mu^{t/\tau} \leq \frac{pc(\omega_\pm^*+1)}{p-\mu} p^{t/\tau}, \quad t \geq T-\tau,$$

and hence we have

$$|\widehat{x_\pm}(g_i(t))| \leq \frac{pc(\omega_\pm^*+1)}{p-\mu}\, p^{g_i(t)/\tau} \leq a p^{g_i(t)/\tau}, \quad t \geq T,\ i = 1, \ldots, N. \tag{5.5.12}$$

Now we define the mappings $\mathcal{T}_\pm \colon X \to C[T_0, \infty)$ as follows:

$$\mathcal{T}_\pm x(t) = \begin{cases} (-1)^{n-1} \displaystyle\int_t^{\infty} \frac{(s-t)^{n-1}}{(n-1)!} f(s, \widehat{x_\pm}(g(s)))\, ds, & t \geq T \\ \mathcal{T}_\pm x(T), \quad T_0 \leq t \leq T \end{cases} \tag{5.5.13}$$

where $f(s, \widehat{x_\pm}(g(s)) \equiv f(s, \widehat{x_\pm}(g_1(s)), \ldots, \widehat{x_\pm}(g_N(s)))$. It needs to show that $\mathcal{T}_\pm$ are continuous and map X into compact subsets of X, so that Schauder-Tychonoff fixed point theorem is applicable to $\mathcal{T}_\pm$. In fact, if $x \in X$, from (5.5.13), (5.5.3), (5.5.12) and (5.5.8) we obtain

$$|\mathcal{T}_\pm x(t)| \leq \int_t^{\infty} \frac{(s-t)^{n-1}}{(n-1)!}\, |f(s, \widehat{x_\pm}(g(s)))|\, ds$$

$$\leq \mu^{t/\tau} \int_t^\infty s^{n-1} \mu^{-s/\tau} F\big(s, |\widehat{x_\pm}(g_1(s))|, \ldots, |\widehat{x_\pm}(g_N(s))|\big)\, ds$$

$$\leq \mu^{t/\tau} \int_t^\infty s^{n-1} \mu^{-s/\tau} F\big(s, ap^{g_1(s)/\tau}, \ldots, ap^{g_N(s)/\tau}\big)\, ds$$

$$\leq c\mu^{t/\tau}, \quad t \geq T,$$

which implies that $\mathcal{T}_\pm x \in X$. Thus $\mathcal{T}_\pm(X) \subset X$. Let $\{x_k\}$ be a sequence in X converging to an $x \in X$ in the topology of $C[T_0, \infty)$. Since

$$\sum_{i=1}^\infty (\pm 1)^i p^{-i} x_k(t + i\tau) \longrightarrow \sum_{i=1}^\infty (\pm 1)^i p^{-i} x(t + i\tau) \quad \text{as } k \to \infty$$

uniformly on every compact subinterval of $[T_0, \infty)$, the Lebesgue dominated convergence theorem shows that $\mathcal{T}_\pm x_k(t) \to \mathcal{T}_\pm x(t)$ uniformly on compact subintervals of $[T_0, \infty)$ as $k \to \infty$. This shows the continuity of $\mathcal{T}_\pm$. Finally, since the inequalities

$$|(\mathcal{T}_\pm x)'(t)| \leq F(t, ap^{g_1(t)/\tau}, \ldots, ap^{g_N(t)/\tau}), \quad t \geq T, \text{ for } n = 1$$

and

$$|(\mathcal{T}_\pm x)'(t)| \leq \int_T^\infty \frac{(s-t)^{n-2}}{(n-2)!} F(s, ap^{g_i(s)/\tau}, \ldots, ap^{g_N(s)/\tau})\, ds, \quad t \geq T, \text{ for } n \geq 2$$

hold for all $x \in W$, we conclude that the sets $\mathcal{T}_\pm(X)$ have compact closure in $C[T_0, \infty)$ by the Ascoli-Arzela Theorem. Therefore, applying the Schauder-Tychonoff Theorem we see that there exist $\xi_\pm \in X$ satisfying

$$\xi_\pm(t) = (-1)^{n-1} \int_t^\infty \frac{(s-t)^{n-1}}{(n-1)!} f(s, \hat{\xi}_\pm(g(s)))\, ds, \quad t \geq T.$$

From (5.5.10) we have

$$\hat{\xi}_\pm(t) \pm p\hat{\xi}_\pm(t-\tau) = (-1)^{n-1} \int_t^\infty \frac{(s-t)^{n-1}}{(n-1)!} f(s, \hat{\xi}_\pm(g(s)))\, ds, \quad t \geq T. \quad (5.5.14)$$

This concludes that $\hat{\xi}_\pm(t)$ are solutions of Eq. (5.5.1) and Eq. (5.5.2) on $[T_0, \infty)$, respectively.

From (5.5.9) and (5.5.11) we have

$$\left|\hat{\xi}_{\pm}(t) - \frac{pc}{p-\mu} p^{t/\tau} \omega_{\pm}(t)\right| \leq \frac{\mu p}{p-\mu} \mu^{t/\tau}, \quad t \geq T - \tau \tag{5.5.15}$$

which shows that $\hat{\xi}_{\pm}(t)$ satisfy the relations (5.5.6) and (5.5.5), respectively. It is obvious that $\hat{\xi}_{\pm}(t)$ are oscillatory. □

Theorem 5.5.2. *Suppose that $p > 1$ and there exists a constant $a > 0$ such that*

$$\int_{t_0}^{\infty} t^{n-1} F(t, ap^{g_1(t)/\tau}, \ldots, ap^{g_N(t)/\tau})\, dt < \infty. \tag{5.5.16}$$

Then

(i) *for any continuous periodic oscillatory function $\omega_-(t)$ with period τ, Eq. (5.5.2) has an unbounded oscillatory solution $x_-(t)$ such that*

$$x_-(t) = k_1 p^{t/\tau} \omega_-(t) + o(1) \quad \text{as } t \to \infty; \tag{5.5.17}$$

(ii) *for any continuous oscillatory function $\omega_+(t)$ such that $\omega_+(t+\tau) = -\omega_+(t)$ for all t, Eq. (5.5.1) has an unbounded oscillatory solution $x_+(t)$ such that*

$$x_+(t) = k_2 p^{t/\tau} \omega_+(t) + o(1) \quad \text{as } t \to \infty \tag{5.5.18}$$

where k_1 and k_2 are constants.

Proof: The proof is similar to that of Theorem 5.5.1. Choose $c > 0$ and $T \geq t_0$ such that, $pc(1 + \omega_{\pm}^*)/(p-1) \leq a$, (5.5.7) holds, and

$$\int_{T}^{\infty} t^{n-1} F(t, ap^{g_1(t)/\tau}, \ldots, ap^{g_N(t)/\tau})\, dt \leq c. \tag{5.5.19}$$

Define a set $Y \subset C[T_0, \infty)$ and mappings $\mathcal{T}_{\pm}\colon Y \to C[T_0, \infty)$ as follows:

$$Y = \{y \in C[T_0, \infty)\colon |y(t)| \leq c, \; t \geq T_0\}$$

$$\mathcal{T}_{\pm} y(t) = \begin{cases} (-1)^{n-1} \displaystyle\int_t^{\infty} \frac{(s-t)^{n-1}}{(n-1)!} f(s, \widehat{y}_{\pm}(g(s)))\, ds, & t \geq T \\ \mathcal{T}_{\pm} y(T), \quad T_0 \leq t \leq T, & \end{cases} \tag{5.5.20}$$

where $\widehat{y}_{\pm}\colon [T_0, \infty) \to R$ are defined by

$$\widehat{y}_{\pm}(t) = \frac{pc}{p-1} p^{t/\tau} \omega_{\pm}(t) - \sum_{i=1}^{\infty} (\mp 1)^i p^{-i} y(t + i\tau), \quad t \geq T - \tau,$$

$$\widehat{y}_{\pm}(t) = \frac{pc}{p-1} p^{t/\tau} \omega_{\pm}(t) - \sum_{i=1}^{\infty} (\mp 1)^i p^{-i} y(T + (i-1)\tau), \quad T_0 \leq t \leq T - \tau.$$

It is easy to see that $y \in Y$ implies that $\widehat{y}_{\pm} \in C[T_0, \infty)$ and

$$\widehat{y}_{\pm}(t) \pm p\widehat{y}_{\pm}(t - \tau) = y(t), \quad t \geq T.$$

Furthermore, since

$$\sum_{i=1}^{\infty} p^{-i} |y_{\pm}(t + i\tau)| \leq \frac{c}{p-1}, \quad t \geq T - \tau,$$

$$|\widehat{y}_{\pm}(t)| \leq \frac{pc\omega_{\pm}^*}{p-1} p^{t/\tau} + \frac{c}{p-1} \leq \frac{pc(\omega_{\pm}^* + 1)}{p-1} p^{t/\tau}, \quad t \geq T_0.$$

And hence

$$|\widehat{y}_{\pm}(g_i(t))| \leq a p^{g_i(t)/\tau}, \quad t \geq T, \ i \in I_N = \{1, \ldots, N\}.$$

Proceeding as in Theorem 5.5.1, by the Schauder-Tychonoff fixed point theorem, $\mathcal{T}_{\pm}$ have fixed points $\eta_{\pm} \in Y$ and $\widehat{\eta}_{\pm}(t)$ satisfy Eq. (5.5.1) and (5.5.2), respectively. Since $\eta_{\pm}(t) \to 0$ as $t \to \infty$, the solutions $\widehat{\eta}_{\pm}(t)$ satisfy (5.5.17) and (5.5.18), respectively. □

With a slightly different discussion we can show the following result.

Theorem 5.5.3. *Suppose that $p > 1$, and there exist constants $\mu \in (1, p)$ and $a > 0$ such that*

$$\int_{t_0}^{\infty} \mu^{-t/\tau} F(t, a p^{g_1^*(t)/\tau}, \ldots, a p^{g_N^*(t)/\tau}) \, dt < \infty; \tag{5.5.21}$$

where $g_i^*(t) = \max\{g_i(t), t\}$, $i \in I_N$. Then

(i) *for any continuous periodic oscillatory function* $\omega_-(t)$ *with period* τ, *Eq. (5.5.2) has an unbounded oscillatory solution* $x_-(t)$ *such that*

$$x_-(t) = k_1 p^{t/\tau}\omega_-(t) + o(p^{t/\tau}) \quad \text{as } t \to \infty; \tag{5.5.22}$$

(ii) *for any continuous oscillatory function* $\omega_+(t)$ *such that* $\omega_+(t+\tau) = -\omega_+(t)$ *for all* t, *Eq. (5.5.1) has an unbounded oscillatory solution* $x_+(t)$ *such that*

$$x_+(t) = k_2 p^{t/\tau}\omega_+(t) + o(p^{t/\tau}) \quad \text{as } t \to \infty$$

where k_1 *and* k_2 *are constants.*

Remark 5.5.1. For $m = 1, 2, \ldots$ the functions

$$a\cos(2m\pi t/\tau) + b\sin(2m\pi t/\tau) \tag{5.5.23}$$

and

$$a\cos((2m-1)\pi t/\tau) + b\sin((2m-1)\pi t/\tau) \tag{5.5.24}$$

where a and b are constants with $a^2 + b^2 > 0$, can be used as $\omega_-(t)$ and $\omega_+(t)$, respectively, in the above theorems.

Example 5.5.1. Consider the equation

$$\left(x(t) \pm e^{-1}x(t-1)\right)^{(n)} + (1 \pm e^{-2})e^{(1-\nu\theta)t}|x(\theta t)|^{\nu}\operatorname{sgn} x(\theta t) = 0 \qquad (5.5.25)_{\pm}$$

where ν and θ are positive constants with $\nu\theta > 1$. Here $F(t, v) = (1+e^{-2})e^{(1-\nu\theta)t}v^{\nu}$. Since

$$\begin{aligned}\int_{t_0}^{\infty} t^{n-1}\mu^{-t/\tau}F(t, ap^{g(t)/\tau})\,dt &= a^{\nu}(1+e^{-2})\int_{t_0}^{\infty} t^{n-1}\mu^{-t}e^{(1-2\nu\theta)t}\,dt \\ &= a^{\nu}(1+e^{-2})\int_{t_0}^{\infty} t^{n-1}(e^{1-2\nu\theta}/\mu)^t\,dt < \infty\end{aligned}$$

for any $\mu \in (e^{1-2\nu\theta}, e^{-1})$. Therefore, the condition (5.5.4) holds. By Theorem 5.5.1, given any continuous periodic oscillatory function $\omega_-(t)$ of period 1 and any continuous oscillatory function $\omega_+(t)$ with $\omega_+(t+1) = -\omega_+(t)$ for all t, there exist bounded oscillatory solutions $x_\pm(t)$ of Eq. $(5.5.25)_\pm$ satisfying

$$x_\pm(t) = e^{-t}\omega_\pm(t) + o(e^{-t}) \quad \text{as } t \to \infty.$$

In particular, $(5.5.25)_\pm$ have infinitely many bounded oscillatory solutions

$$\begin{aligned} x_m^-(t) &= \text{const} \cdot e^{-t}\cos(2m\pi t) + o(e^{-t}) \quad \text{as } t \to \infty \\ x_m^+(t) &= \text{const} \cdot e^{-t}\cos((2m-1)\pi t) + o(e^{-t}) \quad \text{as } t \to \infty, \end{aligned} \qquad m = 1, 2, \ldots.$$

5.6. Equations with Nonlinear Neutral Terms

Consider the nth order nonlinear neutral delay differential equation

$$\big(x(t) - p(t)g(x(t-\tau))\big)^{(n)} + q(t)h(x(t-\sigma)) = 0, \quad t \geq t_0 \tag{5.6.1}$$

where $n \geq 1$ is an odd integer.

The following lemma will be used to establish a linearized oscillation result.

Lemma 5.6.1. *Assume $p \in (0,1]$, q, $\tau \in (0,\infty)$, $\sigma \in [0,\infty)$ and every solution of the equation*

$$\big(x(t) - px(t-\tau)\big)^{(n)} + qx(t-\sigma) = 0, \quad t \geq t_0 \tag{5.6.2}$$

is oscillatory. Then there exists an $\varepsilon_0 > 0$ such that for each $\varepsilon \in [0, \varepsilon_0]$ every solution of the equation

$$\big(y(t) - (p-\varepsilon)y(t-\tau)\big)^{(n)} + (q-\varepsilon)y(t-\sigma) = 0, \quad t \geq t_0 \tag{5.6.3}$$

is oscillatory.

Proof: If not, for any $\varepsilon_0 > 0$ there exists $\varepsilon \in [0, \varepsilon_0]$ such that (5.6.3) has a positive solution. Then the characteristic equation of (5.6.3)

$$f(\lambda) = \lambda^n - (p-\varepsilon)\lambda^n e^{-\lambda\tau} + (q-\varepsilon)e^{-\lambda\sigma} = 0 \tag{5.6.4}$$

has a negative real root λ^*. On the other hand, by the assumption the characteristic equation of (5.6.2)

$$F(\lambda) = \lambda^n - p\lambda^n e^{-\lambda\tau} + qe^{-\lambda\sigma} = 0$$

has no real roots. $F(\infty) = \infty$ and $F(-\infty) = \infty$ implies that $F(\lambda) \geq m > 0$. Set

$$G(\mu) = \mu^n - \frac{p}{2}\mu^n e^{-\mu\tau} + \frac{q}{2}e^{-\mu\sigma} \quad \text{for } \mu < 0.$$

Since $G(-\infty) = \infty$, there exists $\alpha > 0$ such that $G(\mu) > 0$ for $\mu < -\alpha$. Choose $\varepsilon_0 \in (0, \frac{1}{2})$ such that $p - \varepsilon_0 \geq p/2$, $q - \varepsilon_0 \geq q/2$, and $\varepsilon_0(\alpha^n e^{\alpha\tau} + e^{\alpha\sigma}) \leq m/2$. For $\varepsilon \leq \varepsilon_0$, if $\mu < -\alpha$, then

$$f(\mu) \geq \mu^n - \frac{p}{2}\mu^n e^{-\mu\tau} + \frac{q}{2}e^{-\mu\sigma} = G(\mu) > 0;$$

if $0 > \mu \geq -\alpha$, then

$$\begin{aligned} f(\mu) &= \mu^n - p\mu^n e^{-\mu\tau} + qe^{-\mu\sigma} + \varepsilon\mu^n e^{-\mu\tau} - \varepsilon e^{-\mu\sigma} \\ &\geq F(\mu) - \varepsilon\alpha^n e^{\alpha\tau} - \varepsilon e^{\alpha\sigma} \geq m - \frac{m}{2} = \frac{m}{2} > 0. \end{aligned}$$

That is, there is an $\varepsilon_0 > 0$ such that for every $\varepsilon \in [0, \varepsilon_0]$ (5.6.4) has no real roots. We reach a contradiction. □

Theorem 5.6.1. *Assume that*

(i) $\tau > 0$, $\sigma \geq 0$, p, $q \in C([t_0, \infty), R_+)$, g, $h \in C(R, R)$;

(ii) $g(u)u \geq 0$ *as* $u \neq 0$, $\lim_{u\to 0} g(u)/u = 1$, *and* $|g(u)| \leq L|u|$ *for* $L > 0$;

(iii) $uh(u) > 0$ *for* $u \neq 0$, $\lim_{u\to 0} h(u)/u = 1$, *and* $|h(u)| \geq h_0 > 0$ *for* $|u|$ *sufficiently large;*

(iv) $0 < p_1 \leq p(t) \leq p_2 < 1$ *and* $Lp_2 < 1$;

(v) $\liminf_{t\to\infty} q(t) = q_0 > 0$, $\liminf_{t\to\infty} q(t)/q(t-\tau) = M > 0$, *and* $p_1 M < 1$;

(vi) *every solution of the linear equation*

$$\big(y(t) - p_1 M y(t-\tau)\big)^{(n)} + q_0 y(t-\sigma) = 0, \quad t \geq t_0 \tag{5.6.5}$$

is oscillatory.

Then every solution of (5.6.1) is oscillatory.

Proof: Assume the contrary, and let $x(t)$ be an eventually positive solution of (5.6.1). Set $z(t) = x(t) - p(t)g(x(t-\tau))$. Then $z^{(n)}(t) \le 0$. If $\lim_{t\to\infty} z^{(n-1)}(t) = -\infty$, then

$$\lim_{t\to\infty} z^{(i)}(t) = -\infty, \quad i = 0, 1, 2, \ldots, n-1. \tag{5.6.6}$$

Thus $x(t)$ is an unbounded solution and there exists a sequence $\{t_k\}$ such that $t_k \to \infty$ and $\lim_{k\to\infty} x(t_k) = \infty$ where $x(t_k) = \max_{t_0 \le s \le t_k} x(s)$ for $k = 1, 2, \ldots$. Hence

$$\begin{aligned} z(t_k) &= x(t_k) - p(t_k)g(x(t_k - \tau)) \\ &= x(t_k) - p(t_k)\frac{g(x(t_k-\tau))}{x(t_k-\tau)}x(t_k - \tau) \\ &\ge x(t_k)(1 - Lp_2) \to \infty \quad \text{as } k \to \infty \end{aligned}$$

which contradicts (5.6.6). Therefore, $\lim_{t\to\infty} z^{(n-1)}(t) = \ell \in R$. Then integrating (5.6.1) from t_1 to ∞ we find that

$$\int_{t_1}^{\infty} q(s)h(x(s-\sigma))\, ds < \infty \tag{5.6.7}$$

which implies that $\liminf_{t\to\infty} x(t) = 0$. Let $\{t_k\}$ be a sequence such that $t_k \to \infty$ and $\lim_{k\to\infty} x(t_k) = \liminf_{t\to\infty} x(t) = 0$. Then $\lim_{k\to\infty} z(t_k) \le 0$. On the other hand,

$$z(t_k + \tau) \ge -p(t_k + \tau)\frac{g(x(t_k))}{x(t_k)}x(t_k) \longrightarrow 0 \quad \text{as } k \to \infty.$$

Since $z(t)$ is monotonic, we have $\lim_{t\to\infty} z(t) = 0$. Then it is easy to see that

$$\lim_{t\to\infty} z^{(i)}(t) = 0, \quad i = 0, 1, 2, \ldots, n-1, \tag{5.6.8}$$

and $\lim_{t\to\infty} x(t) = 0$. We rewrite Eq. (5.6.1) in the form

$$\left(x(t) - P(t)x(t-\tau)\right)^{(n)} + Q(t)x(t-\sigma) = 0 \tag{5.6.9}$$

where

$$P(t) = p(t)\frac{g(x(t-\tau))}{x(t-\tau)} \quad \text{and} \quad Q(t) = q(t)\frac{h(x(t-\tau))}{x(t-\tau)}. \tag{5.6.10}$$

Hence $z(t) = x(t) - P(t)x(t-\tau)$, and (5.6.9) becomes

$$z^{(n)}(t) - P(t-\sigma)\frac{Q(t)}{Q(t-\tau)}z^{(n)}(t-\tau) + Q(t)z(t-\sigma) = 0. \tag{5.6.11}$$

For any small $\varepsilon > 0$, $Q(t) \geq q_0(1-\varepsilon)$, and $P(t-\sigma)Q(t)/Q(t-\tau) \geq p_1 M(1-\varepsilon)$ for all large t. Then from (5.6.11) we have

$$z^{(n)}(t) - p_1 M(1-\varepsilon)z^{(n)}(t-\tau) + q_0(1-\varepsilon)z(t-\sigma) \leq 0. \tag{5.6.12}$$

Integrating (5.6.12) n times, we have

$$z(t) \geq p_1 M(1-\varepsilon)z(t-\tau) + \frac{q_0(1-\varepsilon)}{(n-1)!}\int_t^\infty (s-t)^{n-1}z(s-\sigma)\,ds \tag{5.6.13}$$

for all large t. By Lemma 5.1.2 the equation

$$u(t) = p_1 M(1-\varepsilon)u(t-\tau) + \frac{q_0(1-\varepsilon)}{(n-1)!}\int_t^\infty (s-t)^{n-1}z(s-\sigma)\,ds \tag{5.6.14}$$

has an eventually positive solution $u(t)$. Hence the equation

$$\big(u(t) - p_1 M(1-\varepsilon)u(t-\tau)\big)^{(n)} + q_0(1-\varepsilon)u(t-\sigma) = 0$$

has an eventually positive solution. This contradicts (vi) by Lemma 5.6.1. □

We consider the nonlinear neutral differential equation

$$\big(x(t) - p(t)g(x(\tau(t)))\big)^{(n)} + q(t)h(x(\sigma(t))) = 0, \quad t \geq t_0 \tag{5.6.15}$$

where n is an odd integer.

The following is an existence result of nonoscillatory solutions of Eq. (5.6.15).

Theorem 5.6.2. *Assume*

(i) $p,\ q \in C([t_0,\infty), R_+)$;

(ii) $g,\ h \in C(R,R)$, $xg(x) > 0$, $xh(x) > 0$ *as* $x \neq 0$, *and there exists an* $L > 0$ *such that*

$$|g(x) - g(y)| \leq L|x-y|, \quad x,\ y \in [0,1],$$

and h is a nondecreasing function;

(iii) τ, $\sigma \in C([t_0, \infty), R)$, $0 \le t - \tau(t) \le M$, $t \ge t_0$, *where M is a constant, and* $\lim_{t \to \infty} \sigma(t) = \infty$;

(iv) *there exist $\alpha > 0$ and $c \in (0,1)$ such that $Lp(t)e^{\alpha(t-\tau(t))} \le c < 1$ and*

$$Lp(t)e^{\alpha(t-\tau(t))} + \frac{e^{\alpha t}}{(n-1)!} \int_t^\infty (s-t)^{n-1} q(s) h(e^{-\alpha\sigma(s)})\, ds \le 1, \quad t \ge t_0.$$

Then Eq. (5.6.15) has an eventually positive solution $x(t)$ which tends to zero exponentially as $t \to \infty$.

Proof: Denote $T = \min\left\{\inf_{t \ge t_0} \tau(t),\ \inf_{t \ge t_0} \sigma(t)\right\}$. Let BC denote the Banach space of all bounded continuous functions defined on $[T, \infty)$ with the sup norm. Let Ω be the subset of BC defined by

$$\Omega = \{y \in BC \colon 0 \le y(t) \le 1,\ t_0 \le t < \infty\}.$$

Clearly Ω is a bounded, closed and convex subset of BC. Define operators $\mathcal{T}_1$ and $\mathcal{T}_2$ on Ω as follows:

$$(\mathcal{T}_1 y)(t) = \begin{cases} p(t)e^{\alpha t} g(y(\tau(t))e^{-\alpha\tau(t)}), & t \ge t_0 \\ \dfrac{t}{t_0}(\mathcal{T}_1 y)(t_0) + \left(1 - \dfrac{t}{t_0}\right), & T \le t \le t_0, \end{cases}$$

$$(\mathcal{T}_2 y)(t) = \begin{cases} \dfrac{e^{\alpha t}}{(n-1)!} \displaystyle\int_t^\infty (s-t)^{n-1} q(s) h\big(y(\sigma(s))e^{-\alpha\sigma(s)}\big)\, ds, & t \ge t_0 \\ \dfrac{t}{t_0}(\mathcal{T}_2 y)(t_0), \quad T \le t \le t_0. \end{cases}$$

By (iv), for every pair x, $y \in \Omega$ we have $\mathcal{T}_1 x + \mathcal{T}_2 y \in \Omega$. Conditions (ii) and (iv) imply that $\mathcal{T}_1$ is a contraction on Ω. It is easy to see that

$$\left|\frac{d}{dt}(\mathcal{T}_2 y)(t)\right| \le M_1 \quad \text{for } y \in \Omega$$

where M_1 is a positive constant. Thus $\mathcal{T}_2$ is completely continuous. By Krasnoselskii's fixed point theorem there exists a $y \in \Omega$ such that $(\mathcal{T}_1 + \mathcal{T}_2)y = y$.

That is,

$$y(t) = \begin{cases} p(t)e^{\alpha t} g\big(y(\tau(t))e^{-\alpha\tau(t)}\big) \\ \qquad + \dfrac{e^{\alpha t}}{(n-1)!}\displaystyle\int_t^\infty (s-t)^{n-1} q(s) h\big(y(\sigma(s))e^{-\alpha\sigma(s)}\big)\, ds, & t \ge t_0 \\ \dfrac{t}{t_0} y(t_0) + \left(1 - \dfrac{t}{t_0}\right), \quad T \le t \le t_0. \end{cases}$$

It is easy to see that $y(t) > 0$ for $t \ge T$. Set $x(t) = y(t)e^{-\alpha t}$. Then

$$x(t) = p(t)g(x(\tau(t))) + \frac{1}{(n-1)!}\int_t^\infty (s-t)^{n-1} q(s) h(x(\sigma(s)))\, ds, \quad t \ge t_0.$$

This implies that $x(t)$ is a desired solution of (5.6.15). □

Example 5.6.1. Consider

$$\big(x(t) - \tfrac{1}{4}x^3(t-1)\big)' + q(t)x^{\frac{1}{3}}(t) = 0 \tag{5.6.16}$$

where $q(t) = e^{-\frac{2}{3}t} - \frac{3}{4}e^3 e^{-\frac{8}{3}t} > 0$ for all large t. In our notations $p(t) = 1/4$, $g(x) = x^3$, $h(x) = x^{\frac{1}{3}}$, and $L = 3$. Obviously, the hypotheses of Theorem 5.6.2 are satisfied. Therefore, Eq. (5.6.16) has a solution $x(t)$ which tends to zero exponentially as $t \to \infty$. In fact, $x(t) = e^{-t}$ is such a solution of (5.6.16).

We now consider the equation

$$\big(x(t) - g(x(t-\tau))\big)^{(n)} + h(t, x(t-\sigma)) = 0, \quad t \ge t_0 \tag{5.6.17}$$

where n is an odd integer.

Theorem 5.6.3. *Assume that*

(i) $\tau > 0$, $\sigma \ge 0$, $g \in C(R, R)$, $h \in C([t_0, \infty) \times R, R)$;

(ii) *g is nondecreasing, $xg(x) \ge 0$ for $x \in R$, and there exists a $\alpha \in (0,1]$ such that*

$$\limsup_{\substack{t\to\infty \\ |x|\to\infty}} \frac{|g(x)|}{|x|^\alpha} = \begin{cases} M \in [0,\infty) & \text{for } \alpha \in (0,1) \\ c \in (0,1) & \text{for } \alpha = 1; \end{cases}$$

(iii) *h is nondecreasing in* x, $h(t,x)x \geq 0$ *for* $(t,x) \in [t_0,\infty) \times R$, *and for any nonzero constant* β

$$\int_{t_0}^{\infty} s^{n-1} h(s,\beta)\, ds = \infty \cdot \operatorname{sign} \beta;$$

(iv) *there exists a positive integer* N *such that*

$$y^{(n)}(t) + h(t, y(t-\sigma) + g(y(t-\sigma-\tau) \\ + g(y(t-\sigma-2\tau) + \cdots + g(y(t-\sigma-N\tau))\cdots)) \leq 0 \tag{5.6.18}$$

has no eventually positive solutions, and

$$y^{(n)}(t) + h(t, y(t-\sigma) + g(y(t-\sigma-\tau) \\ + g(y(t-\sigma-2\tau) + \cdots + g(y(t-\sigma-N\tau))\cdots)) \geq 0 \tag{5.6.19}$$

has no eventually negative solutions.
Then every solution of Eq. (5.6.17) is oscillatory.

Proof: We prove it for the case that $\alpha = 1$ in condition ii). The case that $\alpha \in (0,1)$ can be similarly proved. Assume the contrary, and let $x(t)$ be an eventually positive solution of Eq. (5.6.17). Set

$$z(t) = x(t) - g(x(t-\tau)). \tag{5.6.20}$$

Then eventually

$$z^{(n)}(t) = -h(t, x(t-\sigma)) \leq 0. \tag{5.6.21}$$

Therefore, $z^{(i)}(t)$ are eventually monotonic, $i = 0, 1, \ldots, n-1$.

Now let us consider the following two possibilities: $z(t) < 0$ and $z(t) > 0$ eventually.

Assume $z(t) < 0$ eventually. Since n is an odd integer, it follows from Eq.(5.6.17) that $z'(t) < 0$ eventually. Then either

$$\lim_{t\to\infty} z(t) = -\infty, \tag{5.6.22}$$

or

$$\lim_{t\to\infty} z(t) = -\ell \in (-\infty, 0). \tag{5.6.23}$$

If (5.6.22) holds, we have from (5.6.20) that $\lim_{t\to\infty} g(x(t-\tau)) = \infty$, and so $x(t) \to \infty$ as $t \to \infty$. This implies that there exists a sequence $\{\xi_k\}$ such that $\lim_{k\to\infty} x(\xi_k) = \infty$ where $x(\xi_k) = \max\{x(t) \colon t_0 \le t \le \xi_k\}$. Hence from condition ii)

$$z(\xi_k) = x(\xi_k) - g(x(\xi_k - \tau)) \ge x(\xi_k)\left[1 - \frac{g(x(\xi_k))}{x(\xi_k)}\right] \longrightarrow \infty \quad \text{as } k \to \infty$$

which contradicts (5.6.22) and so (5.6.23) holds. Thus we have $(-1)^i z^{(i)}(t) > 0$, $i = 1, 2, \ldots, n-1$, eventually. Then eventually

$$z(t) = x(t) - g(x(t-\tau)) < -\frac{\ell}{2}.$$

Hence there exist $t_1 \ge t_0$ and $\beta > 0$ such that $x(t) \ge \beta$ for $t \ge t_1$. Substituting this into Eq. (5.6.17) we obtain

$$z^{(n)}(t) + h(t, \beta) \le 0 \quad \text{for } t \ge t_1 + \sigma. \tag{5.6.24}$$

Multiplying (5.6.24) by t^{n-1}, then integrating it from $t_1 + \sigma$ to t, and letting $t \to \infty$ we obtain from condition iii) that

$$\int_{t_1+\sigma}^{t} s^{n-1} z^{(n)}(s)\, ds \longrightarrow -\infty \quad \text{as } t \to \infty. \tag{5.6.25}$$

On the other hand,

$$\int_{t_1+\sigma}^{t} s^{n-1} z^{(n)}(s)\, ds = F(t) - F(t_1 + \sigma) \tag{5.6.26}$$

where

$$\begin{aligned} F(t) &= t^{n-1} z^{(n-1)}(t) + \sum_{i=0}^{n-2} (-1)^i (i+1)(i+2)\cdots(n-1) t^i z^{(i)}(t) \\ &> (n-1) z(t). \end{aligned} \tag{5.6.27}$$

Combining (5.6.25)–(5.6.27) we obtain that $\lim_{t\to\infty} z(t) = -\infty$, contradicting (5.6.23).

Assume $z(t) > 0$ eventually. Then

$$\begin{aligned} x(t) &= z(t) + g(x(t-\tau)) = z(t) + g(z(t-\tau) + g(x(t-2\tau))) \\ &\geq z(t) + g(z(t-\tau) + g(z(t-2\tau) + \cdots + g(z(t-N\tau))\cdots)). \end{aligned}$$

Substituting this into Eq. (5.6.17) we find that

$$\begin{aligned} &z^{(n)}(t) + h(t, z(t-\sigma) \\ &\qquad + g(z(t-\sigma-\tau)) + g(z(t-\sigma-2\tau)) + \cdots + g(z(t-\sigma-N\tau))\cdots)) \leq 0 \end{aligned}$$

which contradicts condition (iv). □

Theorem 5.6.4. *In addition to condition (i) of Theorem 5.6.3 we assume that*

(i) *g is nondecreasing, $xg(x) \geq 0$ for $x \in R$, and there exists a $d > 0$ such that $g(d) < d$;*

(ii) *h is nondecreasing in x, $h(t,x)x \geq 0$ for $(t,x) \in [t_0, \infty) \times R$, and*

$$\int_{t_0}^{\infty} s^{n-1}|h(s,c)|\, ds < \infty \quad \text{for any } c \neq 0. \tag{5.6.28}$$

Then Eq. (5.6.17) has an eventually positive solution.

Proof: Let $\beta > 0$ be such that $\beta + g(d) < d$. Then by (5.6.28) there exists a $T \geq t_0$ such that

$$\int_{T}^{\infty} s^{n-1} h(s,d)\, ds \leq (n-1)!\,(d - g(d) - \beta). \tag{5.6.29}$$

Now we define a function sequence $\{w_k(t)\}$ as follows: $w_1(t) = \beta$, $t \geq T - m$, and for $k = 1, 2, \ldots$

$$w_{k+1}(t) = \begin{cases} \beta + g(w_k(t-\tau)) \\ \quad + \dfrac{1}{(n-1)!}\displaystyle\int_t^{\infty} (s-t)^{n-1} h(s, w_k(s-\sigma))\, ds, & t \geq T \\ w_{k+1}(T), \quad T - m \leq t \leq T \end{cases} \tag{5.6.30}$$

where $m = \max\{\tau, \sigma\}$. By induction, we can easily show that for $t \geq T - m$,

$$\beta = w_1(t) < w_2(t) \leq \cdots \leq w_k(t) \leq \cdots \leq d.$$

Then $\lim_{k\to\infty} w_k(t) = w(t)$ exists and $\beta \le w(t) \le d$ for $t \ge T-m$. Taking limits on both sides of (5.6.30) and using the Lebesgue's monotone convergence theorem we see that $w(t)$ is a positive solution of Eq. (5.6.17) for $t \ge T$. □

Combining Theorems 5.6.3 and 5.6.4, we obtain the following corollary.

Corollary 5.6.1. *Let $n = 1$. In addition to condition (i) of Theorem 5.6.3, assume that*

(i) *g is nondecreasing, $xg(x) \ge 0$ for $x \in R$, and*

$$\limsup_{\substack{t\to\infty \\ |x|\to\infty}} \frac{|g(x)|}{|x|} = c \in (0,1);$$

(ii) *h is nondecreasing in x, $h(t,x)x > 0$ for $x \ne 0$, and there exists $\beta \in (0,1)$ such that $|h(t,y)|/|y|^\beta$ is nonincreasing in $|y|$.*

Then every solution of (5.6.17) is oscillatory if and only if

$$\int_{t_0}^{\infty} |h(t,c)|\,dt = \infty \quad \text{for any } c \ne 0. \tag{5.6.31}$$

Proof: The necessity follows from Theorem 5.6.4. The sufficiency follows from Theorem 5.6.3. □

Let us consider the equation

$$\left(x(t) - (cx(t-\tau) + px^\alpha(t-\tau))\right)^{(n)} + qx^\beta(t-\sigma) = 0, \quad t \ge t_0. \tag{5.6.32}$$

Here we assume that n is an odd integer; c, p, $\sigma \in [0,\infty)$, $c+p$, q, $\tau \in (0,\infty)$, α and β are quotients of two positive odd integers.

Theorem 5.6.5. *Assume $c \in [0,1)$, $\alpha \in (0,1)$, and $p \in (0,\infty)$. Then every solution of Eq. (5.6.32) is oscillatory.*

Proof: It is obvious that conditions (i), (ii) and (iii) of Theorem 5.6.3 hold. Now we show that the condition (iv) holds also. To this end, we choose a large positive integer N such that $\alpha^N\beta < 1$. If (iv) of Theorem 5.6.3 is not satisfied, then without loss of generality we may assume that the inequality corresponding to (5.6.18) has an eventually positive solution $y(t)$. Hence

$$y^{(n)}(t) + qp^{\frac{\beta(1-\alpha^N)}{1-\alpha}} y^{\alpha^N \beta}(t - \sigma - N\tau) < 0. \tag{5.6.33}$$

We rewrite (5.6.33) in the form

$$y^{(n)}(t) + Q(t)y(t - \sigma - N\tau) < 0 \tag{5.6.34}$$

where

$$Q(t) = qp^{\frac{\beta(1-\alpha^N)}{1-\alpha}} y^{\alpha^N \beta - 1}(t - \sigma - N\tau). \tag{5.6.35}$$

On the other hand, by (5.6.18), we show that $(-1)^i y^{(i)}(t) > 0$ for $i = 0, 1, \ldots, n$ and $\lim_{t\to\infty} y(t) = 0$. It follows that $\lim_{t\to\infty} Q(t) = \infty$. By Theorem 5.3.3 in [4] (5.6.34) has no eventually positive solutions. This contradiction shows that condition (iv) is satisfied. Then all the conditions of Theorem 5.6.3 are satisfied, and so every solution of (5.6.32) is oscillatory. □

Similarly, we have the following result.

Theorem 5.6.6. *Assume $p = 0$, $\beta \in (0, 1)$. Then all solutions of Eq. (5.6.32) are oscillatory.*

5.7. Unstable Type Equations

We consider the existence of positive solutions of higher order neutral delay differential equations of the unstable type

$$\big(x(t) - p(t)x(t - \tau)\big)^{(n)} = q(t)x(t - \sigma), \quad t \geq t_0 \tag{5.7.1}$$

where

$$n \text{ is an even integer}, \ \tau > 0, \ \sigma \geq 0, \ p, \, q \in C([t_0, \infty), R_+). \tag{5.7.2}$$

Theorem 5.7.1. *Eq. (5.7.1) has a positive solution on $[t_0, \infty)$ which tends to infinity as $t \to \infty$.*

Proof: Set

$$p_1(t) = \begin{cases} p(t), & t \geq t_0 \\ p(t_0), & t \leq t_0 \end{cases} \tag{5.7.3}$$

and

$$q_1(t) = \begin{cases} q(t), & t \geq t_0 \\ q(t_0), & t \leq t_0. \end{cases} \tag{5.7.4}$$

Then p_1 and q_1 are nonnegative and continuous on R. Let

$$H(t) = 4\max\left\{q_1(t),\ \frac{(n-1)!}{e^2\tau^n}\overline{p}_1(t)\right\} + 1 \tag{5.7.5}$$

where

$$\overline{p}_1(t) = \max\{p_1(s)\colon s \leq t + 2\tau\}. \tag{5.7.6}$$

Clearly, $\overline{p}_1$ is continuous and nondecreasing on R. Now define

$$y(t) = \exp\left[\int_{t_0-2\tau}^{t} \exp\left(\frac{1}{(n-2)!}\int_{t_0-2\tau}^{s}(s-u)^{n-2}H(u)\,du\right)ds\right]. \tag{5.7.7}$$

Then for $t \geq t_0$

$$\begin{aligned}
p_1(t)\frac{y(t-\tau)}{y(t)} &= p_1(t)\exp\left[-\int_{t-\tau}^{t}\exp\left(\frac{1}{(n-2)!}\int_{t_0-2\tau}^{s}(s-u)^{n-2}H(u)\,du\right)ds\right] \\
&\leq p_1(t)\exp\left[-\tau\exp\left(\frac{1}{(n-2)!}\int_{t_0-2\tau}^{t-\tau}(t-\tau-u)^{n-2}H(u)\,du\right)\right] \\
&\leq p_1(t)\exp\left[-\tau\exp\left(\frac{4(n-1)}{e^2\tau^n}\int_{t-2\tau}^{t-\tau}(t-\tau-u)^{n-2}\overline{p}_1(u)\,du\right)\right] \\
&\leq p_1(t)\exp\left[-\tau\exp\left(\frac{4(n-1)}{e^2\tau^n}\overline{p}_1(t-2\tau)\int_{t-2\tau}^{t-\tau}(t-\tau-u)^{n-2}\,du\right)\right] \\
&\leq p_1(t)\exp\left[-\tau\exp\left(\frac{4}{e^2\tau}p_1(t)\right)\right] \\
&\leq p_1(t)\exp\left(-\frac{4}{e}p_1(t)\right) \leq \frac{1}{4}.
\end{aligned}$$

That is,

$$\frac{1}{4}y(t) \geq p_1(t)y(t-\tau), \quad t \geq t_0. \tag{5.7.8}$$

From (5.7.7) we find that

$$y^{(i)}(t) > 0, \quad i = 0, 1, 2, \ldots, n, \tag{5.7.9}$$

and for $t \geq t_0 - 2\tau$

$$y^{(n)}(t) > H(t)\exp\left[\frac{1}{(n-2)!}\int_{t_0-2\tau}^{t}(t-u)^{n-2}H(u)\,du\right]y(t) \geq H(t)y(t).$$

It follows that for $t \geq t_0 - 2\tau$

$$\frac{q_1(t)y(t-\sigma)}{y^{(n)}(t)} \leq \frac{q_1(t)y(t-\sigma)}{H(t)y(t)} \leq \frac{q_1(t)}{H(t)} \leq \frac{1}{4}.$$

That is,

$$\frac{1}{4}y^{(n)}(t) \geq q_1(t)y(t-\sigma) \quad \text{for } t \geq t_0 - 2\tau. \tag{5.7.10}$$

Integrating (5.7.10) from $t_0 - 2\tau$ to t for $t \geq t_0 - 2\tau$ n times and using (5.7.9), we obtain

$$\frac{1}{4}y(t) \geq \frac{1}{(n-1)!}\int_{t_0-2\tau}^{t}(t-s)^{n-1}q_1(s)y(s-\sigma)\,ds, \quad t \geq t_0 - 2\tau. \tag{5.7.11}$$

Combining (5.7.8) and (5.7.11), we have

$$\frac{1}{2}y(t) \geq p_1(t)y(t-\tau) + \frac{1}{(n-1)!}\int_{t_0-2\tau}^{t}(t-s)^{n-1}q_1(s)y(s-\sigma)\,ds, \quad t \geq t_0. \tag{5.7.12}$$

Let BC denote the Banach space of all bounded and continuous functions defined on R with the sup norm. Set

$$\Omega = \{z \in BC\colon 0 \leq z(t) \leq 1, \ -\infty < t < \infty\}.$$

Then Ω is a bounded, closed and convex subset of BC. Define a mapping $\mathcal{T} : \Omega \to BC$ as

$$(\mathcal{T}z)(t) = \begin{cases} \dfrac{1}{y(t)}\Big[p_1(t)y(t-\tau)z(t-\tau) + \dfrac{\tau}{(n-1)!}(t-t_0+\tau)^{n-1} \\ \quad + \dfrac{1}{(n-1)!}\displaystyle\int_{t_0-2\tau}^{t} (t-s)^{n-1}q_1(s)y(s-\sigma)z(s-\sigma)\,ds\Big], \quad t \geq t_0 \\ (\mathcal{T}z)(t_0), \quad t \leq t_0. \end{cases} \tag{5.7.13}$$

In view of (5.7.5) and (5.7.7) we have for $t \geq t_0$

$$\begin{aligned} y(t) &\geq \exp\left[\int_{t_0-2\tau}^{t} \exp\left(\frac{1}{(n-1)!}(s-t_0+2\tau)^{n-1}\right) ds\right] \\ &\geq e\int_{t_0-2\tau}^{t} \exp\left(\frac{1}{(n-1)!}(s-t_0+2\tau)^{n-1}\right) ds \\ &\geq \frac{e^2}{(n-1)!}\int_{t-\tau}^{t} (s-t_0+2\tau)^{n-1}\,ds \\ &\geq \frac{e^2\tau}{(n-1)!}(t-t_0+\tau)^{n-1}. \end{aligned} \tag{5.7.14}$$

By (5.7.12) – (5.7.14), we see that $0 \leq (\mathcal{T}z)(t) \leq 1$, $t \in R$, which shows that $\mathcal{T}$ maps Ω into itself. Next we will show that $\mathcal{T}$ is a contraction on Ω. In fact, for any $z_1, z_2 \in \Omega$ and $t \geq t_0$

$$\begin{aligned} |(\mathcal{T}z_1)(t) - (\mathcal{T}z_2)(t)| &\leq \frac{1}{y(t)}\Big[p_1(t)y(t-\tau)\,|z_1(t-\tau) - z_2(t-\tau)| \\ &\qquad + \frac{1}{(n-1)!}\int_{t_0-2\tau}^{t} (t-s)^{n-1}p_1(s)y(s-\sigma) \\ &\qquad \times |z_1(s-\sigma) - z_2(s-\sigma)|\,ds\Big] \\ &\leq \frac{1}{2}\|z_1 - z_2\|. \end{aligned}$$

Hence

$$\|\mathcal{T}z_1 - \mathcal{T}z_2\| = \sup_{t\in R} |(\mathcal{T}z_1)(t) - (\mathcal{T}z_2)(t)|$$

$$= \sup_{t \geq t_0} |(\mathcal{T}z_1)(t) - (\mathcal{T}z_2)(t)| \leq \frac{1}{2}\|z_1 - z_2\|$$

which shows that $\mathcal{T}$ is a contraction on Ω. Then by the Banach contraction principle, $\mathcal{T}$ has a fixed point $z \in \Omega$, that is,

$$z(t) = \begin{cases} \dfrac{1}{y(t)}\Bigg[p_1(t)y(t-\tau)z(t-\tau) + \dfrac{\tau}{(n-1)!}(t-t_0+\tau)^{n-1} \\ \qquad + \dfrac{1}{(n-1)!}\displaystyle\int_{t_0-2\tau}^{t}(t-s)^{n-1}q_1(s)y(s-\sigma)z(s-\sigma)\,ds\Bigg], \quad t \geq t_0 \\ z(t_0), \qquad t \leq t_0. \end{cases}$$

Set $x(t) = y(t)z(t)$. Then $x(t)$ is a positive continuous function on R and satisfies

$$x(t) = p_1(t)x(t-\tau) + \frac{1}{(n-1)!}\int_{t_0-2\tau}^{t}(t-s)^{n-1}q_1(s)x(s-\sigma)\,ds + \frac{\tau}{(n-1)!}(t-t_0+\tau)^{n-1}, \quad t \geq t_0.$$

It is easy to see that $x(t)$ is a positive solution of Eq. (5.7.1) and satisfies that $x(t) \to \infty$ as $t \to \infty$. □

Remark 5.7.1. Theorem 5.7.1 is also true if n is an odd integer.

In the following we present some results on existence of a bounded positive solution of Eq. (5.7.1).

Theorem 5.7.2. *Assume there exists an $\alpha \in (0, \infty)$ such that*

$$0 < p(t)e^{\alpha\tau} \leq \overline{p} < 1, \quad t \geq t_0, \tag{5.7.15}$$

and

$$p(t)e^{\alpha\tau} + \frac{1}{(n-1)!}\int_t^{\infty}(s-t)^{n-1}q(s)\exp[\alpha(t-s+\sigma)]\,ds \leq 1, \quad t \geq t_0. \tag{5.7.16}$$

Then Eq. (5.7.1) has a bounded positive solution on $[t_0, \infty)$ which converges to zero exponentially as $t \to \infty$.

Proof: Let $m = \max\{\tau, \sigma\}$ and let BC denote the Banach space of all bounded and continuous functions defined on $[t_0 - m, \infty)$ with the sup norm. Let Ω be the subset of BC defined by

$$\Omega = \{y \in BC\colon 0 \le y(t) \le 1 \text{ for } t \ge t_0 - m\}.$$

Clearly, Ω is a bounded, closed and convex subset of BC. Define two mappings $\mathcal{T}_1$ and $\mathcal{T}_2$ on Ω as follows:

$$(\mathcal{T}_1 y)(t) = \begin{cases} p(t)e^{\alpha\tau}y(t-\tau), & t \ge t_0 \\ \dfrac{t+h}{t_0+h}(\mathcal{T}_1 y)(t_0) + \left(1 - \dfrac{t+h}{t_0+h}\right), & t_0 - m \le t \le t_0 \end{cases} \tag{5.7.17}$$

and

$$(\mathcal{T}_2 y)(t) = \begin{cases} \dfrac{1}{(n-1)!}\displaystyle\int_t^\infty (s-t)^{n-1}p(s)y(s-\sigma)e^{\alpha(t-s+\sigma)}\,ds, & t \ge t_0 \\ (\mathcal{T}_2 y)(t_0), \quad t_0 - m \le t \le t_0 \end{cases} \tag{5.7.18}$$

where h is a constant satisfying $t_0 + h > 0$. In view of (5.7.15) and (5.7.16) for any $y_1, y_2 \in \Omega$, we have $\mathcal{T}_1 y_1 + \mathcal{T}_2 y_2 \in \Omega$. Using (5.7.15) one can verify that $\mathcal{T}_1$ is a contraction on Ω. On the other hand, it is easy to see that $|\frac{d}{dt}(\mathcal{T}_2 y)(t)|$ is uniformly bounded for $y \in \Omega$, and hence $\mathcal{T}_2$ is completely continuous on Ω. Therefore, by Krasnoselskii's fixed point theorem, there exists a $y \in \Omega$ such that $\mathcal{T}_1 y + \mathcal{T}_2 y = y$. That is, for $t \ge t_0$

$$y(t) = p(t)e^{\alpha\tau}y(t-\tau) + \frac{1}{(n-1)!}\int_t^\infty (s-t)^{n-1}q(s)y(s-\sigma)e^{\alpha(t-s+\sigma)}\,ds \tag{5.7.19}$$

and for $t_0 - m \le t \le t_0$

$$y(t) = \frac{t+h}{t_0+h}y(t_0) + \left(1 - \frac{t+h}{t_0+h}\right) > 0. \tag{5.7.20}$$

It follows that $y(t) > 0$ for all $t \ge t_0 - m$. Set

$$x(t) = y(t)e^{-\alpha t}. \tag{5.7.21}$$

Then (5.7.19) becomes

$$x(t) = p(t)x(t-\tau) + \frac{1}{(n-1)!}\int_t^\infty (s-t)^{n-1}q(s)x(s-\sigma)\,ds, \quad t \ge t_0.$$

Consequently, $x(t)$ defined by (5.7.21) is a positive solution of (5.7.1) on $[t_0, \infty)$, and $x(t)$ converges to zero exponentially. $\square$

With Eq. (5.7.1) we associate the linear equation with constant coefficients

$$\big(x(t) - px(t-\tau)\big)^{(n)} = qx(t-\sigma). \tag{5.7.22}$$

We have the following corollary.

Corollary 5.7.1. *Assume that* $0 < p(t) \le p < 1,\ 0 \le q(t) \le q,\ t \ge t_0$, *and the characteristic equation of (5.7.22)*

$$-\lambda^n(1 - pe^{\lambda\tau}) + qe^{\lambda\sigma} = 0$$

has a positive root. Then Eq. (5.7.1) has a bounded positive solution on $[t_0, \infty)$ *which converges to zero exponentially as* $t \to \infty$.

Theorem 5.7.3. *Let* $p(t) \ge 1$. *Assume*

$$\int_{t_0}^{\infty} s^{n-1} q(s)\, ds = \infty. \tag{5.7.23}$$

Then all bounded solutions of Eq. (5.7.1) are oscillatory.

Proof: Let $x(t)$ be a bounded positive solution of Eq. (5.7.1) and set

$$z(t) = x(t) - x(t-\tau). \tag{5.7.24}$$

Then it is not difficult to see that eventually

$$(-1)^i z^{(i)}(t) > 0, \quad i = 0, \ldots, n. \tag{5.7.25}$$

Hence $x(t) > x(t-\tau)$ eventually. Then there exists $M > 0$ such that $x(t) \ge M,\ t \ge t_1$, where t_1 is a sufficiently large number. From (5.7.1)

$$z^{(n)}(t) \ge Mq(t). \tag{5.7.26}$$

We note that (5.7.25) and (5.7.26) lead to

$$\int_{t_0}^{\infty} s^{n-1} q(s)\, ds < \infty. \tag{5.7.27}$$

This contradicts (5.7.23), and hence concludes the theorem. □

Theorem 5.7.4. *Assume $p(t) \equiv 1$. Then Eq. (5.7.1) has a bounded positive solution if and only if*

$$\int_{t_0}^{\infty} s^{n} q(s)\, ds < \infty. \tag{5.7.28}$$

Proof: *Necessity.* Let $x(t)$ be a bounded positive solution of Eq. (5.7.1) and $z(t)$ be defined by (5.7.24). In view of (5.7.25) we have

$$z(t) \geq \frac{1}{(n-1)!} \int_{t}^{\infty} (s-t)^{n-1} q(s) x(s-\sigma)\, ds,$$

or equivalently,

$$x(t) \geq x(t-\tau) + \frac{1}{(n-1)!} \int_{t}^{\infty} (s-t)^{n-1} q(s) x(s-\sigma)\, ds. \tag{5.7.29}$$

As in the proof of Theorem 5.7.3 we have

$$\int_{t_0}^{\infty} s^{n-1} q(s)\, ds < \infty. \tag{5.7.30}$$

Set

$$Q(t) = \frac{1}{(n-2)!} \int_{t}^{\infty} (s-t)^{n-2} q(s)\, ds.$$

Let $t_1 \geq t_0 + m$ be such that (5.7.29) holds for $t \geq t_1 - m$. Since $x(t) \geq x(t-\tau)$ for $t \geq t_1 - m$, there exists an $M > 0$ such that $x(t) \geq M$ for $t \geq t_1 - m$. Thus

$$\begin{aligned} x(t) &\geq x(t-\tau) + \frac{M}{(n-1)!} \int_{t}^{\infty} (s-t)^{n-1} q(s)\, ds \\ &= x(t-\tau) + M \int_{t}^{\infty} Q(s)\, ds, \quad t \geq t_1, \end{aligned}$$

and then

$$x(t_1+n\tau) \geq x(t_1) + M\sum_{i=1}^{n}\int_{t_1+i\tau}^{\infty} Q(s)\,ds.$$

Since $x(t)$ is bounded, it follows that

$$\sum_{i=1}^{\infty}\int_{t_1+i\tau}^{\infty} Q(s)\,ds < \infty. \tag{5.7.31}$$

According to Theorem 3.3.3 we have

$$\int_{t_1}^{\infty} sQ(s)\,ds < \infty \tag{5.7.32}$$

which implies that (5.7.28) holds.

Sufficiency. (5.7.28) implies (5.7.32). By Theorem 3.3.3 we have

$$\sum_{i=0}^{\infty}\int_{t_0+i\tau}^{\infty} Q(s)\,ds < \infty.$$

Choose $t_1 \geq t_0$ so large that

$$\sum_{i=0}^{\infty}\int_{t_1+i\tau}^{\infty} Q(s)\,ds \leq 1.$$

Define a function H as follows:

$$H(t) = \begin{cases} \displaystyle\int_{t}^{\infty} Q(s)\,ds, & t \geq t_1 \\ (t-t_1+\tau)H(t_1)/\tau, & t_1-\tau \leq t \leq t_1 \\ 0, \quad t \leq t_1-\tau. & \end{cases}$$

Then H is a nonnegative and continuous function on R. Set $y(t) = \sum_{i=0}^{\infty} H(t-i\tau)$. Then $y(t)$ is continuous on R and satisfies $0 < y(t) \leq 1$ for $t \geq t_1$, and

$$y(t) = y(t-\tau) + H(t), \quad t \in R.$$

Let Ω be the set of all continuous functions z defined on $[t_0, \infty)$ and satisfying that $0 \le z(t) \le 1$ for every $t \ge t_0$. The set Ω is considered to be endowed with usual pointwise ordering $\le$: $z_1 \le z_2 \iff z_1(t) \le z_2(t)$ for all $t \ge t_0$.

It is easy to see that for any subset A of Ω there exist $\inf A$ and $\sup A$. Define a mapping $\mathcal{T}$ on Ω as follows:

$$(\mathcal{T}z)(t) = \begin{cases} \dfrac{1}{y(t)}\Big[y(t-\tau)z(t-\tau) \\ \qquad + \dfrac{1}{(n-1)!}\displaystyle\int_t^\infty (s-t)^{n-1}q(s)y(s-\sigma)z(s-\sigma)\,ds\Big], & t \ge t_1 \\ \dfrac{t+h}{t_1+h}(\mathcal{T}z)(t_1) + \left(1 - \dfrac{t+h}{t_1+h}\right), & t_0 \le t \le t_1, \end{cases}$$

where h is a constant satisfying $t_1 + h > 0$.

For any $z \in \Omega$ and $t \ge t_1$ we have

$$\begin{aligned} 0 \le (\mathcal{T}z)(t) &\le \frac{1}{y(t)}\left[y(t-\tau) + \frac{1}{(n-1)!}\int_t^\infty (s-t)^{n-1}q(s)\,ds\right] \\ &= \frac{1}{y(t)}[y(t-\tau) + H(t)] = 1, \end{aligned}$$

i.e., $\mathcal{T}\Omega \subset \Omega$. Moreover, $\mathcal{T}$ is a nondecreasing mapping. Therefore, by Knaster's fixed point theorem there exists a $z \in \Omega$ such that $\mathcal{T}z = z$. That is, for $t \ge t_1$

$$z(t) = \frac{1}{y(t)}\left[y(t-\tau)z(t-\tau) + \frac{1}{(n-1)!}\int_t^\infty (s-t)^{n-1}q(s)y(s-\sigma)z(s-\sigma)\,ds\right], \tag{5.7.33}$$

and for $t_0 \le t < t_1$

$$z(t) = \frac{t+h}{t_1+h}z(t_1) + \left(1 - \frac{t+h}{t_1+h}\right) > 0.$$

It follows that $z(t) > 0$ on $[t_0, \infty)$. Set $x(t) = y(t)z(t)$. Then we conclude from (5.7.33) that $x(t)$ is a bounded positive solution of Eq. (5.7.1). □

Theorem 5.7.5. *Assume there exist positive constants M_1 and M_2 such that $1 < p_1 \le p(t) \le p_2$, $t \ge t_0$. Then Eq. (5.7.1) has a bounded positive solution if* *and only if*

$$\int_{t_0}^{\infty} s^{n-1} q(s)\, ds < \infty. \tag{5.7.34}$$

Proof: *Necessity.* Let $x(t)$ be a bounded positive solution of Eq. (5.7.1). Then $x(t) > 0$ for $t \geq t_1 > t_0$. Set

$$z(t) = x(t) - p(t)x(t-\tau). \tag{5.7.35}$$

Then

$$z^{(n)}(t) = q(t)x(t-\sigma) \geq 0, \quad t \geq t_1 + \sigma.$$

According to the boundedness of $x(t)$, we have $(-1)^i z^{(i)}(t) > 0$ for $t \geq t_1 + \sigma$, $i = 0, 1, \ldots, n-1$. Since $x(t) > x(t-\tau)$ for $t \geq t_1 + \sigma$, there exists a constant $M > 0$ such that $x(t) \geq M > 0$ for $t \geq t_1 + \sigma$. From (5.7.1) we have $z^{(n)}(t) \geq Mq(t)$, $t \geq t_1 + \sigma$, which leads to (5.7.34).

Sufficiency. Let $t_1 \geq t_0 + \tau + \sigma$ and $\alpha_0 > \max\{1, \frac{1}{2(M_2-1)}\}$ be such that

$$\frac{1}{(n-1)!} \int_{t+\tau}^{\infty} s^{n-1} q(s)\, ds \leq \frac{(p_1 - 1)(\alpha_0 - 1)}{2p_2\alpha_0} \quad \text{for } t \geq t_1.$$

Consider the Banach space BC of all continuous bounded functions x defined on $[t_0, \infty)$ with the sup norm. Let Ω denote the subset of BC defined by

$$\Omega = \left\{ x \in BC \colon \frac{p_1 - 1}{\alpha_0(p_2 - 1)} \leq x(t) \leq 2p_1,\ t \geq t_0 \right\}.$$

Then Ω is a bounded, closed and convex subset of BC. Define a mapping $\mathcal{T} : \Omega \to BC$ as follows

$$(\mathcal{T}x)(t) = \begin{cases} \dfrac{1}{p(t+\tau)} \Big[p_1 - 1 + x(t+\tau) \\ \qquad - \dfrac{1}{(n-1)!} \displaystyle\int_{t+\tau}^{\infty} (s-t-\tau)^{n-1} q(s) x(s-\sigma)\, ds \Big], & t \geq t_1 \\ (\mathcal{T}x)(t_1), \quad t_0 \leq t \leq t_1. \end{cases}$$

It is not difficult to see that $\mathcal{T}$ maps Ω into Ω and $\mathcal{T}$ is a contraction on Ω. Therefore, $\mathcal{T}$ has a fixed point $x \in \Omega$, that is, $\mathcal{T}x = x$. Clearly, $x(t)$ is a bounded positive solution of Eq. (5.7.1) on $[t_1, \infty)$. $\square$

We now present some bounded oscillation criteria for the even order equation

$$\big(x(t) - p(t)x(t-\tau)\big)^{(n)} = q(t)x(g(t)), \quad t \geq t_0 \tag{5.7.36}$$

where n is an even integer, $\tau > 0$, $p,\ q,\ g \in C([t_0,\infty), R_+)$, g is nondecreasing, $g(t) \leq t$, and $\lim_{t\to\infty} g(t) = \infty$.

Theorem 5.7.6. *Assume that $0 \leq p(t) \leq p < 1$ and*

$$\limsup_{t\to\infty} \int_{g(t)}^{t} (g(t) - g(s))^{n-1} q(s)\, ds > 0. \tag{5.7.37}$$

Then every bounded solution of Eq. (5.7.36) is either oscillatory or tending to zero as $t \to \infty$.

Proof: Let $x(t)$ be a bounded positive solution of (5.7.36), say, $0 < x(t) \leq M$ for $t \geq t_0$. Let $z(t)$ be defined by (5.7.35). Then there exists a $t_1 \geq t_0$ such that $(-1)^k z^{(k)}(t) > 0,\ k = 1, \ldots, n,\ t \geq t_1$, and hence $\lim_{t\to\infty} z(t) = \ell \in (-\infty, \infty)$.

If $\ell < 0$, then $z(t) < 0$ eventually which leads to $\lim_{t\to\infty} x(t) = 0$ and hence $\lim_{t\to\infty} z(t) = 0$. We reach a contradiction.

If $\ell \geq 0$, by Taylor's formula, for $v \geq u \geq t_1$

$$z(u) - z(v) = \sum_{i=1}^{n-2} \frac{(-1)^i}{2!} z^{(i)}(v)(v-u)^i + \frac{(-1)^{n-1}}{(n-1)!} z^{(n-1)}(\xi)(v-u)^{n-1}$$

where $\xi \in (u, v)$. Hence

$$\begin{aligned} z(u) - z(v) &\geq \frac{(-1)^{n-1}}{(n-1)!} z^{(n-1)}(\xi)(v-u)^{n-1} \\ &\geq \frac{(-1)^{n-1}}{(n-1)!} z^{(n-1)}(v)(v-u)^{n-1}, \quad v \geq u \geq t_1, \end{aligned}$$

and then

$$z(g(s)) - z(g(t)) \geq \frac{(-1)^{n-1}}{(n-1)!} z^{(n-1)}(g(t))(g(t) - g(s))^{n-1}, \quad t \geq s \geq t_2, \tag{5.7.38}$$

where t_2 is so large that $g(t) \geq t_1$ for $t \geq t_2$. We rewrite (5.7.38) in the form

$$\int_t^s z'(g(u))g'(u)\, du \geq \frac{(-1)^{n-1}}{(n-1)!} z^{(n-1)}(g(t))(g(t) - g(s))^{n-1}.$$

Multiplying the above inequality by $q(s)$ we have

$$\frac{z^{(n)}(s)}{x(g(s))} \int_t^s z'(g(u))g'(u)\, du \geq \frac{(-1)^{n-1}}{(n-1)!} z^{(n-1)}(g(t))q(s)(g(t) - g(s))^{n-1}.$$

Since $x(g(s)) \geq z(g(s)) \geq z(g(u))$, $u \geq s \geq t_2$, we have

$$z^{(n)}(s) \int_s^t \frac{z'(g(u))g'(u)}{z(g(u))}\, du \geq \frac{(-1)^{n-1}}{(n-1)!} z^{(n-1)}(g(t))(g(t) - g(s))^{n-1}.$$

Integrating it in s from $g(t)$ to t we have

$$\int_{g(t)}^t \frac{z'(g(u))g'(u)}{z(g(u))}\, du + \frac{1}{(n-1)!} \int_{g(t)}^t q(s)(g(t) - g(s))^{n-1}\, ds \leq 0,$$

or

$$\ln z(g(t)) - \ln z(g(g(t))) + \frac{1}{(n-1)!} \int_{g(t)}^t q(s)(g(t) - g(s))^{n-1}\, ds \leq 0. \tag{5.7.39}$$

If $\ell > 0$, (5.7.39) leads to

$$\limsup_{t\to\infty} \int_{g(t)}^t q(s)(g(t) - g(s))^{n-1}\, ds \leq 0$$

contradicting (5.7.37). Therefore, $\lim_{t\to\infty} z(t) = \ell = 0$ which implies that $\lim_{t\to\infty} x(t) = 0$. □

Theorem 5.7.7. *Assume that* $0 \leq p(t) \leq p < 1$, *and*

$$\limsup_{t\to\infty} \int_{g(t)}^t (s - g(t))^{n-1} q(s)\, ds > (n-1)!. \tag{5.7.40}$$

Then every bounded solution of Eq. (5.7.36) is oscillatory.

Proof: Let $x(t)$ be a bounded positive solution of (5.7.36) and $z(t)$ be defined by (5.7.35). Then eventually $(-1)^k z^{(k)}(t) > 0,\ k = 1, \dots, n$, and hence

$$-z(t) + z(g(t)) \geq \int_{g(t)}^{t} \int_{s_1}^{t} \cdots \int_{s_{n-1}}^{t} q(u)x(g(u))\, du\, ds_{n-1} \cdots ds_1$$
$$= \frac{1}{(n-1)!} \int_{g(t)}^{t} (s - g(t))^{n-1} q(s) x(g(s))\, ds. \tag{5.7.41}$$

As before, we have that $z(t) > 0$ eventually. Since $x(t) \geq z(t)$, from (5.7.41) we have

$$0 \geq z(t) + z(g(t)) \left[\frac{1}{(n-1)!} \int_{g(t)}^{t} (s - g(t))^{n-1} q(s)\, ds - 1 \right]$$

which contradicts (5.7.40). □

Theorem 5.7.8. *Assume that $p(t) \geq 1$, and*

$$\limsup_{t \to \infty} \int_{g(t)}^{t} [g(t) - g(s)]^{n-1} q(s)\, ds > 0. \tag{5.7.42}$$

Then every bounded solution of Eq. (5.7.36) is oscillatory.

Proof: Let $x(t)$ be a bounded positive solution and $z(t)$ be defined by (5.7.35). Then $(-1)^i z^{(i)}(t) \geq 0,\ i = 1, \dots, n$, and $\lim_{t \to \infty} z(t) = \ell \in (-\infty, \infty)$. Set $\liminf_{t \to \infty} x(t) = b$. Then $b \geq 0$. If $b = 0$, there exists a sequence $\{t_k\}$ such that $t_k \to \infty$ and $\lim_{k \to \infty} x(t_k) = 0$. Since $\lim_{k \to \infty} z(t_k) \leq 0$, so $\ell \leq 0$. On the other hand,

$$z(t_k + \tau) = x(t_k + \tau) - p(t_k + \tau) x(t_k).$$

Thus $\lim_{k \to \infty} z(t_k + \tau) \geq 0$, and so $\ell \geq 0$. Therefore, $\ell = 0$. Since $z(t) > 0$ we have $x(t) \geq x(t - \tau)$ eventually, which implies that $x(t) \geq b_1 > 0$ eventually. From (5.7.38)

$$\int_{t}^{s} z'(g(u)) g'(u)\, du \geq \frac{(-1)^{n-1}}{(n-1)!} z^{(n-1)}(g(t))(g(t) - g(s))^{n-1}.$$

Multiplying above inequality by $q(s)$ we have

$$\frac{z^{(n)}(s)}{x(g(s))}\int_t^s z'(g(u))g'(u)\,du \geq \frac{(-1)^{n-1}}{(n-1)!}z^{(n-1)}(g(t))q(s)(g(t)-g(s))^{n-1}.$$

Then eventually

$$\frac{z^{(n)}(s)}{b_1}\int_t^s z'(g(u))g'(u)\,du \geq \frac{(-1)^{n-1}}{(n-1)!}z^{(n-1)}(g(t))q(s)(g(t)-g(s))^{n-1}.$$

Integrating it in s from $g(t)$ to t we have

$$\frac{1}{b_1}z^{(n-1)}(g(t))\int_{g(t)}^t z'(g(u))g'(u)\,du \geq \frac{(-1)^{n-1}}{(n-1)!}z^{(n-1)}(g(t))q(s)(g(t)-g(s))^{n-1},$$

or

$$\frac{1}{b_1}[z(g(t))-z(g(g(t)))]+\frac{1}{(n-1)!}\int_{g(t)}^t q(s)(g(t)-g(s))^{n-1}\,ds \leq 0,$$

which implies that

$$\limsup_{t\to\infty}\int_{g(t)}^t q(s)(g(t)-g(s))^{n-1}\,ds = 0.$$

contradicting (5.7.42). □

Example 5.7.1. Consider

$$\big(x(t)-x(t-2\pi)\big)'' = \frac{2\pi}{t-\pi}\,x(t-\pi).$$

It is easy to see that (5.7.42) is satisfied. Therefore, every bounded solution is oscillatory. However, this equation has some unbounded oscillatory solutions. In fact, $x(t) = t\sin t$ is an unbounded oscillatory solution of this equation.

Theorem 5.7.9. *Assume that* $p(t)\equiv p<0,\ \sigma>\tau>0,\ q\in C([t_0,\infty),(0,\infty))$

$$\limsup_{t\to\infty}\frac{q(t)}{q(t-\tau)} = \alpha > 0 \tag{5.7.43}$$

and

$$\limsup_{t\to\infty} \int_{t-(\sigma-\tau)}^{t} (s-(t-(\sigma-\tau)))^{n-1} q(s)\,ds > (n-1)!\,(1-\alpha p). \tag{5.7.44}$$

Then every bounded solution of Eq. (5.7.1) is oscillatory.

Proof: Let $x(t)$ be a bounded positive solution and $z(t)$ be defined by (5.7.35). Then $(-1)^i z^{(i)}(t) > 0,\ i = 0, 1, 2, \ldots, n$. We rewrite Eq. (5.7.1) in the form

$$z^{(n)}(t) = q(t)z(t-\sigma) + p\,\frac{q(t)}{q(t-\tau)}\,z^{(n)}(t-\tau). \tag{5.7.45}$$

From (5.7.43) and (5.7.44) there exists $k > 1$ such that

$$\frac{q(t)}{q(t-\tau)} < k\alpha \quad \text{for all large } t, \tag{5.7.46}$$

and

$$\limsup_{t\to\infty} \int_{t-(\sigma-\tau)}^{t} (s-(t-\sigma-\tau))p(s)\,ds > (n-1)!\,(1-k\alpha p). \tag{5.7.47}$$

From (5.7.45) and (5.7.46) we have that eventually

$$z^{(n)}(t) - k\alpha p z^{(n)}(t-\tau) \ge q(t)z(t-\sigma).$$

Set $w(t) = z(t) - k\alpha p z(t-\tau)$. Then eventually

$$w^{(n)}(t) \ge q(t)z(t-\sigma). \tag{5.7.48}$$

Since $z'(t) < 0$,

$$w(t) \le (1-k\alpha p)z(t-\tau). \tag{5.7.49}$$

Combining (5.7.48) and (5.7.49), we have that eventually

$$w^{(n)}(t) \ge \frac{1}{1-k\alpha p} q(t)w(t-(\sigma-\tau)). \tag{5.7.50}$$

By the boundedness of $w(t)$, we know that eventually $(-1)^i w^{(i)}(t) > 0$, $i = 0, 1, 2, \ldots, n$. Integrating (5.7.50) n times we have

$$\begin{aligned}
-w(t) + w(t - (\sigma - \tau)) \\
&\geq \frac{1}{(n-1)!\,(1 - k\alpha p)} \\
&\qquad \times \int_{t-(\sigma-\tau)}^{t} (s - (t - (\sigma - \tau)))^{n-1} q(s) w(s - (\sigma - \tau))\, ds \\
&\geq \frac{w(t - (\sigma - \tau))}{(n-1)!\,(1 - k\alpha p)} \int_{t-(\sigma-\tau)}^{t} (s - (t - (\sigma - \tau)))^{n-1} q(s)\, ds,
\end{aligned}$$

or

$$0 \geq w(t) + w(t-(\sigma-\tau)) \left[\frac{1}{(n-1)!\,(1 - k\alpha p)} \int_{t-(\sigma-\tau)}^{t} (s-(t-(\sigma-\tau)))^{n-1} q(s)\, ds - 1 \right]$$

which contradicts (5.7.47). □

5.8. Notes

Lemmas 5.1.1–5.1.3 are taken from monograph [110]. Theorem 5.1.1 was first obtained by Gopalamy, Lalli and Zhang [54], the present form is taken from Zhang, Yu and Wang [215]. Theorems 5.1.2 and 5.1.3 are from [54]. Theorem 5.1.4 is adopted from [215]. Theorem 5.1.5 is based on [54]. Theorem 5.1.6 is taken from Yan [174], some further comparison results can be seen from Yu, Wang and Zhang [183], see also Ladas, Qian and Yan [108] for some other results. Theorems 5.2.1, 5.2.2, 5.2.4 and 5.2.5 are taken from Zhang and Yu [211]. Theorem 5.2.3 is adopted from Naito [140]. Theorems 5.2.6–5.2.8 are adopted from Zhang and Gopalamy [204]. Theorem 5.2.9 is from [211]. The material in Section 5.3 is taken from Zhang [192], also see Lalli, Ruan and Zhang [112] for the nonlinear equation case. The materials in Section 5.4 are adopted from Chen [18], also see Jaros and Kusano [90]. Section 5.5 is from Jaros and Kusano [91]. Theorem 5.6.1 is taken from Ladas and Qian [105], Theorems 5.6.2 and 5.6.3 are taken from Zhang and Yu [211]. Section 5.7 is taken from Chen, Yu and Zhang [12], also see Yu and Zhang [185].

6

Oscillation of Systems of Neutral Differential Equations

6.0. Introduction and Preliminaries

Oscillation of systems of n neutral differential equations is an interesting and hard problem. In this chapter we will present some recent contributions.

Here we use the definition for weaker oscillation given in Chapter 1, i.e., a vector solution is said to be oscillatory if at least one of its nontrivial components has arbitrarily large zeros.

For the linear autonomous neutral delay differential system in the form

$$\frac{d^N}{dt^N}\left[X(t) - PX(t-\tau)\right] + \sum_{j=1}^{m} Q_j X(t-\sigma_j) = 0 \tag{6.0.1}$$

where P, Q_j, $j = 1, \ldots, m$, are given $n \times n$ constant matrices, τ, σ_j $j = 1, \ldots, m$, are nonnegative numbers, the above definition is equivalent to a stronger one, i.e., a vector solution of Eq. (6.0.1) is oscillatory if and only if its every component has arbitrarily large zeros.

There is a basic result for the oscillation of system (6.0.1) based on its characteristic equation:

Theorem 6.0.1. *System (6.0.1) is oscillatory if and only if its characteristic equation*

$$det\,[\lambda^N(I - Pe^{-\lambda\tau}) + \sum_{j=1}^{m} Q_j e^{-\lambda\sigma_j}] = 0$$

has no real roots.

Although this is a fundamental result, it is not easy to apply. Therefore, in Section 6.1, we give some explicit conditions for oscillation of (6.0.1) using Lozenskii measures, or logarithmic norms, of matrices. The criteria are sharp in the sense that even for the scalar case they are still the best known up to now.

In Section 6.2 we consider system (6.0.1) for the case that Q_j, $j = 1, \ldots, m$, are variable matrices. Oscillation criteria for (6.0.1) are obtained by comparing (6.0.1) with some systems with constant matrix coefficients. Those results are extended to a kind of nonlinear system and are applied to some Lotka-Volterra models.

The nonautonomous system considered in Section 6.2 is discussed further in Section 6.3. Some comparison results for oscillation of systems with scalar equations are obtained.

Section 6.4 deals with nonlinear systems of the form

$$\frac{d^N}{dt^N}\,[x_i(t) \pm a_i(t)x_i(h_i(t))] = \sum_{j=1}^{n} P_j(t) f_{ij}(x_j(g_{ij}(t))).$$

Some results on the existence of nonoscillatory solutions are given.

For the criteria of oscillation in Section 6.1 we need the following notations and definitions.

For any $n \times n$ real matrix A we denote by $\lambda_i(A)$, $\quad i = 1, \ldots, n$, the eigenvalues of A satisfying

$$\mathrm{Re}\,\lambda_1(A) \geq \mathrm{Re}\,\lambda_2(A) \geq \cdots \geq \mathrm{Re}\,\lambda_n(A).$$

We define $\|A\|_i = \sup\limits_{x \in R^n, x \neq 0} \dfrac{\|Ax\|_i}{\|x\|_i}$, $\quad i = 1, 2, \ldots, \infty$, where $x = (x_1, \ldots, x_n)^T$,

$\|x\|_i = \left(\sum_{j=1}^{n} |x_j|^i\right)^{1/i}$, $i < \infty$, and $\|x\|_\infty = \max_{1\le j\le n}\{|x_j|\}$. For each $i = 1, 2, \ldots, \infty$, the Lozenskii measures $\mu_i(A)$ of A are defined as follows:

$$\mu_i(A) = \lim_{h\to 0+} \frac{\|I + hA\|_i - 1}{h},$$

and $\nu_i(A) = -\mu_i(-A)$, $i = 1, 2, \ldots, \infty$. In general, without specification, we denote by $\mu(A)$ and $\nu(A)$ any pair of $\mu_i(A)$ and $\nu_i(A)$, $i = 1, 2, \ldots, \infty$. It has been shown that $\mu_i(A)$ and $\nu_i(A)$, $i = 1, 2, \ldots, \infty$, exist for any $n \times n$ matrix A and can be explicitly calculated for $i = 1, 2$, and $i = \infty$:

$$\mu_1(A) = \sup_i \Big\{a_{ii} + \sum_{\substack{j=1\\ j\ne i}}^{n} |a_{ij}|\Big\}, \qquad \nu_1(A) = \inf_i \Big\{a_{ii} - \sum_{\substack{j=1\\ j\ne i}}^{n} |a_{ij}|\Big\};$$

$$\mu_2(A) = \lambda_1\left(\frac{1}{2}(A + A^T)\right), \qquad \nu_2(A) = \lambda_n\left(\frac{1}{2}(A + A^T)\right);$$

$$\mu_\infty(A) = \sup_j \Big\{a_{jj} + \sum_{\substack{i=1\\ i\ne j}}^{n} |a_{ij}|\Big\}, \qquad \nu_\infty(A) = \inf_j \Big\{a_{jj} - \sum_{\substack{i=1\\ i\ne j}}^{n} |a_{ij}|\Big\}.$$

For any $n \times n$ matrices A and B and any Lozenskii measures we have

$$\begin{aligned}
&\text{i)} \quad \mu(A+B) \le \mu(A) + \mu(B), \qquad \nu(A+B) \ge \nu(A) + \nu(B);\\
&\text{ii)} \quad \nu(-A) = -\mu(A), \qquad \mu(-A) = -\nu(A);\\
&\text{iii)} \quad \mu(\alpha A) = \alpha\mu(A), \qquad \nu(\alpha A) = \alpha\nu(A), \quad \alpha > 0;\\
&\text{iv)} \quad \mu(A) \ge \operatorname{Re}\lambda_1(A), \qquad \nu(A) \le \operatorname{Re}\lambda_n(A).
\end{aligned} \tag{6.0.2}$$

In Section 6.1, we will obtain some criteria for oscillation by using Lozenskii measures. Since the criteria are given by the general form of Lozenskii measures μ and ν, we will actually have infinitely many different results corresponding to each criterion in the theorems. Moreover, three of them, which are given by μ_i and ν_i, $i = 1, 2, \infty$, can be expressed explicitly. But for the scalar case, where $\mu(A) = \nu(A) = A$, all of them coincide to give the same results.

6.1. Systems with Constant Matrix Coefficients

To simplify the discussion and proofs we first consider a simpler equation

$$\frac{d^N}{dt^N}[X(t) - PX(t-\tau)] + QX(t-\sigma) = 0 \tag{6.1.1}$$

where P, Q are $n \times n$ matrices, τ, $\sigma \geq 0$. The following conditions will be used to determine the oscillation of Eq. (6.1.1):

$$(\mathrm{A}_1) \quad \sum_{k=0}^{\infty}[\nu(P)]^k \,\nu(Q)(k\tau+\sigma)^N \geq \left(\frac{N}{e}\right)^N,$$

$$(\mathrm{A}_2) \quad \sum_k{}^{*}[\mu(P)]^{-(k+1)}\,\nu(Q)[(k+1)\tau-\sigma]^N \geq \left(\frac{N}{e}\right)^N,$$

$$(\mathrm{A}_3) \quad \sum_{k}{}_{*}[\mu(P)]^{-(k+1)}\,\nu(Q)[-(k+1)\tau+\sigma]^N \geq \left(\frac{N}{e}\right)^N$$

where $\sum_k^*$ and $\sum_{*k}$ denote the sums over all the terms for $k \in \mathbb{Z}_+$ such that $(k+1)\tau - \sigma > 0$ and $-(k+1)\tau + \sigma > 0$, respectively.

Remark 6.1.1. The inequality in (A_3) will become strict if the sum has only one term.

As we mentioned in Section 6.0, a solution $X(t)$ of (6.1.1) is oscillatory, if every component of it has arbitrarily large zeros.

Theorem 6.1.1. *Assume N is odd and $\nu(Q) > 0$. Then each of the following is sufficient for (6.1.1) to be oscillatory:*

i) $\mu(P) = \nu(P) = 1$,

ii) $0 < \nu(P) \leq \mu(P) \leq 1$, *and* (A_1) *holds,*

iii) $1 \leq \nu(P) \leq \mu(P)$, *and* (A_2) *holds,*

iv) $\nu(P) < 1 < \mu(P)$, *and* (A_1), (A_2) *hold.*

Theorem 6.1.2. *Assume N is even and $\nu(Q) > 0$. Then each of the following is sufficient for (6.1.1) to be oscillatory:*

i) $0 < \mu(P) \leq 1$, *and* (A_3) *holds,*

ii) $\mu(P) > 1$, *and* (A_2), (A_3) *hold.*

Remark 6.1.2. The condition (A_3) in Theorem 6.1.2 is required in the sense that if the set $\{k \in \mathbb{Z}_+ : -(k+1)\tau + \sigma > 0\}$ is empty and $\nu(P) > 0$, then Eq. (6.1.1) must have a nonoscillatory solution. In fact, the above assumption implies that $r \geq \tau$. If (6.1.1) is oscillatory, then (6.1.2) has no real root. Let

$$F(\lambda) = \lambda^N(I - Pe^{-\lambda\tau}) + Qe^{-\lambda\sigma}.$$

Then

$$\mu(F(0)) = \mu(Q) > 0$$

implies that $\mu(F(\lambda)) > 0$ for all $\lambda \in R$. But

$$\mu(F(\lambda)) \leq \lambda^N (1 - \nu(Pe^{-\lambda\tau})) + \mu(Q)e^{-\lambda\sigma} \to -\infty$$

as $\lambda \to -\infty$) This contradiction shows that all solutions cannot be oscillatory.

The following lemma will be needed in the proofs of the results.

Lemma 6.1.1. *Let A be a $n \times n$ real matrix. If either $\nu(A) > 0$ or $\mu(A) < 0$, then* $\det(A) \neq 0$.

Proof: From (6.0.2), if $\nu(A) > 0$, then $\operatorname{Re}\lambda_n(A) > 0$. Hence $\operatorname{Re}\lambda_i(A) > 0$ for $i = 1, 2, \ldots, n$. Thus

$$\det(A) = \lambda_1(A) \cdots \lambda_n(A) \neq 0.$$

The case that $\mu(A) < 0$ is similar. □

Proof of Theorem 6.1.1: The characteristic equation of (6.1.1) is

$$\det(\lambda^N(I - Pe^{-\lambda\tau}) + Qe^{-\lambda\sigma}) = 0. \tag{6.1.2}$$

i) Assume $\mu(P) = \nu(P) = 1$. Let

$$F(\lambda) = \lambda^N(I - Pe^{-\lambda\tau}) + Qe^{-\lambda\sigma}.$$

Then $\nu(F(0)) = \nu(Q) > 0$, and

$$\nu(F(\lambda)) \geq \nu(\lambda^N(I - Pe^{-\lambda\tau})) + \nu(Q)e^{-\lambda\sigma}.$$

For $\lambda > 0$

$$\begin{aligned} \nu(F(\lambda)) &\geq \lambda^N(1 - \mu(Pe^{-\lambda\tau})) + \nu(Q)e^{-\lambda\sigma} \\ &= \lambda^N(1 - e^{-\lambda\tau}) + \nu(Q)e^{-\lambda\sigma} > 0. \end{aligned} \tag{6.1.3}$$

For $\lambda < 0$

$$\begin{aligned} \nu(F(\lambda)) &\geq |\lambda|^N \nu(-I + Pe^{-\lambda\tau}) + \nu(Q)e^{-\lambda\sigma} \\ &\geq |\lambda|^N(-1 + \nu Pe^{-\lambda\tau}) + \nu(Q)e^{-\lambda\sigma} > 0. \end{aligned}$$

Thus $\nu(F(\lambda)) > 0$ for all $\lambda \in R$. By Lemma 6.1.1 $\det F(\lambda) \neq 0$ for $\lambda \in R$, i.e., (6.1.2) has no real root.

ii) Assume $0 \leq \nu(P) \leq \mu(P) \leq 1$. Clearly $\lambda = 0$ is not a root of (6.1.2). For $\lambda > 0$, by (6.1.3)

$$\nu(F(\lambda)) \geq \lambda^N(1 - \mu(P)e^{-\lambda\tau}) + \nu(Q)e^{-\lambda\sigma} > 0.$$

Hence $\det(F(\lambda)) \neq 0$ for $\lambda > 0$ by Lemma 5.2.1. Let $\lambda = -s$, and denote

$$G(s) = -s^N(I - Pe^{s\tau}) + Qe^{s\sigma}. \tag{6.1.4}$$

Then $\lambda < 0$ is a root of (6.1.2) if and only if $s > 0$ is a root of $\det(G(s)) = 0$. By Lemma 5.2.1, if $\det(G(s)) = 0$ has a root $s > 0$, then $\nu(G(s)) \leq 0$. Since N is odd,

$$0 \geq \nu(G(s)) \geq s^N(-1 + \nu(P)e^{s\tau}) + \nu(Q)e^{s\sigma}$$

which implies that $0 \leq \nu(P)e^{s\tau} < 1$. As a result

$$\begin{aligned} s^N &\geq \nu(Q)e^{s\tau}\,(1-\nu(P)e^{s\sigma})^{-1} \\ &= \sum_{k=0}^{\infty}[\nu(P)]^k\,\nu(Q)e^{s(k\tau+\sigma)} \\ &> \sum_{k=0}^{\infty}[\nu(P)]^k\,\nu(Q)\left[\frac{s(k\tau+\sigma)e}{N}\right]^N. \end{aligned} \tag{6.1.5}$$

The equality cannot hold since $e^{s(k\tau+\sigma)}$ attains its minimal value at a different point s for different k. Thus

$$\sum_{k=0}^{\infty}[\nu(P)]^k\,\nu(Q)(k\tau+\sigma)^N < \left(\frac{N}{e}\right)^N,$$

contradicting (A_1). Therefore, (6.1.2) has no real root.

iii) Assume $1 \leq \nu(P) \leq \mu(P)$. Similar to ii) we see that $\lambda \leq 0$ is not a root of (6.1.2). Assume $\lambda > 0$ is a root of (6.1.2). By Lemma 6.1.1 we have

$$0 \geq \nu(F(\lambda)) \geq \lambda^N(1-\mu(P)\,e^{-\lambda\tau}) + \nu(Q)\,e^{-\lambda\sigma} \tag{6.1.6}$$

which implies that $\mu(P)e^{-\lambda\tau} > 1$. Therefore,

$$\begin{aligned} \lambda^N &\geq \mu(P)\nu(Q)\,e^{\lambda(\tau-\sigma)}(1-[\mu(P)]^{-1}\,e^{\lambda\tau})^{-1} \\ &= \sum_{k=0}^{\infty}[\mu(P)]^{-(k+1)}\,\nu(Q)\,e^{\lambda[(k+1)\tau-\sigma]} \\ &> \sum_k{}^*[\mu(P)]^{-(k+1)}\,\nu(Q)\left[\frac{\lambda((k+1)\tau-\sigma)e}{N}\right]^N, \end{aligned}$$

that is,

$$\sum_k{}^*[\mu(P)]^{-(k+1)}\,\nu(Q)[(k+1)\tau-\sigma]^N < \left(\frac{N}{e}\right)^N,$$

contradicting (A_2).

iv) Clearly, $\lambda = 0$ is not a root of (6.1.2). From the proof of ii) and iii) we see that (A_1) and (A_2) imply that any $\lambda > 0$ and $\lambda < 0$ can not be a root of (6.1.2). □

Proof of Theorem 6.1.2:

i) Assume $\mu(P) \leq 1$. Similar to the proof of Theorem 6.1.1 ii), we see any $\lambda \geq 0$ is not a root of (6.1.2). Assume $\lambda = -s < 0$ is a root of (6.1.2). Then by Lemma 6.1.1 and from (6.1.2)

$$\nu(F(-s)) = \nu\big(s^N(1 - Pe^{s\tau}) + Qe^{s\sigma}\big) \leq 0.$$

Since N is even,

$$0 \geq \nu(F(-s)) \geq s^N\big(1 - \mu(P)e^{s\tau}\big) + \nu(Q)e^{s\sigma}$$

which implies that $\mu(P)e^{s\tau} > 1$. As a result

$$\begin{aligned} s^N &\geq [\mu(P)]^{-1}\nu(Q)\, e^{s(-\tau+\sigma)}(1 - [\mu(P)]^{-1}\, e^{-s\tau})^{-1} \\ &= \sum_{k=0}^{\infty}[\mu(P)]^{-(k+1)}\, \nu(Q)\, e^{s[-(k+1)\tau+\sigma]} \\ &> \sum_{k}{}_{*}\, [\mu(P)]^{-(k+1)}\, \nu(Q) \left[\frac{s[-(k+1)\tau + \sigma]e}{N}\right]^N, \end{aligned}$$

that is,

$$\sum_{k}{}_{*}\, [\mu(P)]^{-(k+1)}\, \nu(Q)[-(k+1)\tau + \sigma]^N < \left(\frac{N}{e}\right)^N.$$

Note that the inequality may become an equality if the sum has only one term, and this contradicts (A_3).

ii) Assume $\mu(P) > 1$. If λ is a real root of (6.1.2), then $\lambda \neq 0$ and $\lambda^N > 0$. By the proof of i), (A_3) implies that $\lambda < 0$ cannot be a root of (6.1.2). By the proof of Theorem 6.1.1 iii), (A_2) implies that $\lambda > 0$ cannot be a root of (6.1.2). □

The above oscillation criteria for Eq. (6.1.1) can be easily extended to the equation

$$\frac{d^N}{dt^N}[X(t) - PX(t-\tau)] + \sum_{i=1}^{m} Q_j X(t-\sigma_j) = 0 \tag{6.1.7}$$

by using the following conditions where $q = \left(\prod_{j=1}^{m} \nu(Q_j)\right)^{1/m}$, $\sigma = \frac{1}{m}\sum_{j=1}^{m}\sigma_j$.

(B$_1$)
$$\sum_{j=1}^{m}\sum_{k=0}^{\infty}[\nu(P)]^k\,\nu(Q_j)(k\tau+\sigma_j)^N \geq \left(\frac{N}{e}\right)^N, \text{ or} \tag{6.1.8}$$

$$mq\sum_{k=0}^{\infty}[\nu(P)]^k\,(k\tau+\sigma)^N \geq \left(\frac{N}{e}\right)^N, \tag{6.1.9}$$

(B$_2$)
$$\sum_{j,k}^{*}[\mu(P)]^{-(k+1)}\,\nu(Q_j)[(k+1)\tau-\sigma_j]^N \geq \left(\frac{N}{e}\right)^N, \text{ or}$$

$$mq\sum_{k}^{*}[\mu(P)]^{-(k+1)}[(k+1)\tau-\sigma]^N \geq \left(\frac{N}{e}\right)^N,$$

(B$_3$)
$$\sum_{*\,j,k}[\mu(P)]^{-(k+1)}\,\nu(Q_j)[-(k+1)\tau+\sigma_j]^N \geq \left(\frac{N}{e}\right)^N, \text{ or}$$

$$mq\sum_{*\,k}[\mu(P)]^{-(k+1)}[-(k+1)\tau+\sigma]^N \geq \left(\frac{N}{e}\right)^N$$

where $\sum_{k}^{*}$, $\sum_{*\,k}$ are defined the same as before; $\sum_{j,k}^{*}$, $\sum_{*\,j,k}$ denote the sums over all the terms for $1 \leq j \leq m$, $k \geq 0$ such that $(k+1)\tau - \sigma_j > 0$ and $-(k+1)\tau + \sigma_j > 0$, respectively. The inequality of (B$_3$) will become strict if the sum has only one term.

Theorem 6.1.3. *Assume N is odd and $\nu(Q_j) \geq 0$ but not all zero, $j = 1, \ldots, m$. Then each one of the following is sufficient for (6.1.7) to be oscillatory:*

i) $\mu(P) = \nu(P) = 1$,

ii) $0 \leq \nu(P) \leq \mu(P) \leq 1$, *and* (B$_1$) *holds,*

iii) $1 \leq \nu(P) \leq \mu(P)$, *and* (B$_2$) *holds,*

iv) $\nu(P) < 1 < \mu(P)$, *and* (B$_1$), (B$_2$) *hold.*

Theorem 6.1.4. *Assume N is even and $\nu(Q_j) \geq 0$ but not all zero, $j = 1, \ldots, m$. Then each one of the following is sufficient for (6.1.7) to be oscillatory:*

i) $0 < \mu(P) \leq 1$, *and* (B_3) *holds,*

ii) $\mu(P) > 1$, *and* (B_2), (B_3) *hold.*

Proof of Theorem 6.1.3 and 6.1.4: Similar to those of Theorem 6.1.1 and 6.1.2. To show the difference we only give an outline of the proof of Theorem 6.1.3 ii) as an example.

Corresponding to (6.1.5) we now have

$$s^N \geq \sum_{j=1}^{m}\sum_{k=0}^{\infty}[\nu(P)]^k\,\nu(Q_j)\,e^{s(k\tau+\sigma_j)} \tag{6.1.10}$$

$$> \sum_{j=1}^{m}\sum_{k=0}^{\infty}[\nu(P)]^k\,\nu(Q_j)\left[\frac{s(k\tau+\sigma_j)e}{N}\right]^N,$$

that is,

$$\sum_{j=1}^{m}\sum_{k=0}^{\infty}[\nu(P)]^k\,\nu(Q_j)[k\tau+\sigma_j]^N < \left(\frac{N}{e}\right)^N,$$

contradicting (6.1.8). From (6.1.10) we also have

$$s^N \geq \left(\sum_{j=1}^{m}\nu(Q_j)\,e^{s\sigma_j}\right)\left(\sum_{k=0}^{\infty}[\nu(P)]^k\,e^{sk\tau}\right)$$

$$\geq m\left(\prod_{j=1}^{m}\nu(Q_j)\,e^{s\sigma_j}\right)^{1/m}\left(\sum_{k=0}^{\infty}[\nu(P)]^k\,e^{sk\tau}\right)$$

$$= mq\sum_{k=0}^{\infty}[\nu(P)]^k\,e^{s(k\tau+\sigma)}$$

$$> mq\sum_{k=0}^{\infty}[\nu(P)]^k\left[\frac{s(k\tau+\sigma)e}{N}\right]^N,$$

that is,

$$mq \sum_{k=0}^{\infty} [\nu(P)]^k (k\tau + \sigma)^N < \left(\frac{N}{e}\right)^N,$$

contradicting (6.1.9). □

The idea in the proofs of the above theorems may also be applied to the case that $\nu(Q_j)$ are not all nonnegative. As an example we give a result for the equations of the form

$$\frac{d^N}{dt^N} [X(t) - PX(t-r)] + \sum_{j=1}^{m} [G_j X(t-\sigma_j) - H_j X(t-\tau_j)] = 0 \qquad (6.1.11)$$

where P, G_j, H_j are $n \times n$ matrices, $\nu(P) > 0$, $\nu(G_j) - \mu(H_j) \geq 0$ and not all zero for $j = 1, \dots, m$, r, σ_j, $\tau_j \geq 0$, $j = 1, \dots, m$.

Theorem 6.1.5. *Assume N is odd. Then each one of the following is sufficient for (6.1.11) to be oscillatory.*

i) $\nu(P) \geq 1$, $\sigma_j < \tau_j < r$, *and (B_2) holds for the case where $\nu(Q_j)$ are replaced by $\nu(G_j) - \mu(H_j)$, $j = 1, \dots, m$. Furthermore,*

$$-\left(\frac{N}{er}\right)^N + a^N \nu(P) + \sum_{j=1}^{m} \left[\nu(G_j) e^{-b_j(r-\sigma_j)} - \mu(H_j) e^{-b_j(r-\tau_j)}\right] > 0 \qquad (6.1.12)$$

where

$$a = \min_{1 \leq j \leq m} \left\{ \frac{1}{\tau_j - \sigma_j} \ln \frac{\nu(G_j)}{\mu(H_j)} \right\},$$

$$b_j = \frac{1}{\tau_j - \sigma_j} \ln \frac{\nu(G_j)(r-\sigma_j)}{\mu(H_j)(r-\tau_j)}, \quad j = 1, \dots, m.$$

ii) $\nu(P) > 0$, $\mu(P) \leq 1$, $\sigma_j > \tau_j > 0$, *and (B_1) holds for the case where $\nu(Q_j)$ are replaced by $\nu(G_j) - \mu(H_j)$, $j = 1, \dots, m$. Furthermore, $\nu(P^{-1}G_j) > 0$, $\mu(P^{-1}H_j) > 0$, $j = 1, \dots, m$ and*

$$-\left(\frac{N}{er}\right)^N + a^{*^N} [\nu(P)]^{-1} + \sum_{j=1}^{m} [\nu(P^{-1}G_j) e^{-b_j^* \sigma_j} - \mu(P^{-1}H_j) e^{-b_j^* \tau_j}] > 0,$$

where

$$a^* = \min_{1 \le j \le m} \left\{ \frac{1}{\sigma_j - \tau_j} \ln \frac{\nu(P^{-1}G_j)}{\mu(P^{-1}H_j)} \right\},$$

$$b_j^* = \frac{1}{\sigma_j - \tau_j} \ln \frac{\nu(P^{-1}G_j)\sigma_j}{\mu(P^{-1}H_j)\tau_j}, \quad j = 1, \ldots, m.$$

Proof: The characteristic equation of Eq. (6.1.11) is

$$\det \left(\lambda^N (I - Pe^{-\lambda r}) + \sum_{j=1}^{m} (G_j e^{-\lambda \sigma_j} - H_j e^{-\lambda \tau_j}) \right) = 0. \qquad (6.1.13)$$

Let

$$F(\lambda) = \lambda^N (I - Pe^{-\lambda r}) + \sum_{j=1}^{m} (G_j e^{-\lambda \sigma_j} - H_j e^{-\lambda \tau_j}).$$

Then $\lambda = 0$ is not a root of (6.1.13) since the assumption before Theorem 6.1.5 implies that

$$\nu(F(0)) = \nu\left(\sum_{j=1}^{m} (G_j - H_j) \right) \ge \sum_{j=1}^{m} (\nu(G_j) - \mu(H_j)) > 0.$$

Hence $\det F(0) \ne 0$.

i) Assume $\nu(P) \ge 1$ and $\sigma_j < \tau_j < r$. If $\lambda > 0$ is a root of (6.1.13), then by Lemma 6.1.1

$$0 \ge \nu(F(\lambda)) \ge \lambda^N (1 - \mu(P) e^{-\lambda r}) + \sum_{j=1}^{m} \left(\nu(G_j) e^{-\lambda \sigma_j} - \mu(H_j) e^{-\lambda \tau_j} \right)$$

$$\ge \lambda^N (1 - \mu(P) e^{-\lambda r}) + \sum_{j=1}^{m} \left(\nu(G_j) - \mu(H_j) \right) e^{-\lambda \tau_j}.$$

This is a similar inequality to (6.1.6) for Eq. (6.1.1). By a similar discussion we can get a contradiction to (B_2) where $\nu(Q_j)$ are replaced by $\nu(G_j) - \mu(H_j)$,

$j = 1, \ldots, m$. If $\lambda < 0$ is a root of (6.1.13), let $\lambda = -s$, and denote

$$\phi(s) = -s^N(Ie^{-sr} - P) + \sum_{j=1}^{m}(G_j e^{-s(r-\sigma_j)} - H_j e^{-s(r-\tau_j)}), \tag{6.1.14}$$

$$\alpha(s) = -s^N(Ie^{-sr} - P),$$

$$\beta_j(s) = G_j e^{-s(r-\sigma_j)} - H_j e^{-s(r-\tau_j)}, \quad j = 1, \ldots, m.$$

Then $\det(\phi(s)) = 0$, and since N is odd,

$$\nu(\alpha(s)) \geq s^N(-e^{-sr} + \nu(P)) > 0,$$

$$\nu(\beta_j(s)) \geq \nu(G_j)e^{-s(r-\sigma_j)} - \mu(H_j)e^{-s(r-\tau_j)} \triangleq \ell_j(s), \quad j = 1, \ldots, m.$$

$\nu(\beta_j(s)) \geq 0$ if and only if

$$s \leq \frac{1}{\tau_j - \sigma_j} \ln \frac{\nu(G_j)}{\mu(H_j)} \triangleq a_j, \qquad j = 1, \ldots, m.$$

Set $a = \min_{1 \leq j \leq m} \{a_j\}$. We have $\nu(\phi(s)) > 0$ for $0 < s \leq a$. Consider the case that $s > a$. Then

$$\nu(\alpha(s)) \geq -\left(\frac{N}{er}\right)^N + a^N \nu(P),$$

and since

$$\frac{d}{ds}\ell_j(s) = -\nu(G_j)(r - \sigma_j)e^{-s(r-\sigma_j)} + \mu(H_j)(r - \tau_j)e^{-s(r-\tau_j)},$$

we see that the minimum of ℓ_j are attained at

$$s_j = \frac{1}{\tau_j - \sigma_j} \ln \frac{\nu(G_j)(r - \sigma_j)}{\mu(H_j)(r - \tau_j)} = b_j,$$

and thus

$$\nu(\beta_j(s)) \geq \ell_j(s) \geq \ell_j(s_j) = \nu(G_j)e^{-b_j(r-\sigma_j)} - \mu(H_j)e^{-b_j(r-\tau_j)},$$

$$j = 1, \ldots, m.$$

Therefore by (6.1.14) and (6.1.12)

$$\begin{aligned}\nu(\phi(s)) &\geq \nu(\alpha(s)) + \sum_{j=1}^{m} \nu(\beta_j(s)) \\ &\geq -\left(\frac{N}{er}\right)^N + a^N \nu(P) + \sum_{j=1}^{m} \left[\nu(G_j)e^{-b_j(r-\sigma_j)} - \mu(H_j)e^{-b_j(r-\tau_j)}\right] \\ &> 0\end{aligned}$$

contradicting that $\det(\phi(s)) = 0$.

ii) By (6.0.2) $\nu(P) > 0$ implies that P^{-1} exists. (6.1.13) is equivalent to

$$\det\left(-\lambda^N(I - P^{-1}e^{\lambda r}) + \sum_{j=1}^{m} \left[P^{-1}G_j e^{\lambda(r-\sigma_j)} - P^{-1}H_j e^{\lambda(r-\tau_j)}\right]\right) = 0.$$

With $\lambda = -s$, we have

$$\det\left(s^N(I - P^{-1}e^{-sr}) + \sum_{j=1}^{m} \left[P^{-1}G_j e^{-s(r-\sigma_j)} - P^{-1}H_j e^{-s(r-\tau_j)}\right]\right) = 0. \tag{6.1.15}$$

Then we have a duality between (6.1.13) and (6.1.15) as follows:

$$(P, G_j, H_j, \sigma_j, \tau_j) \longleftrightarrow (P^{-1}, P^{-1}G_j, P^{-1}H_j, r - \sigma_j, r - \tau_j).$$

Using this duality and part i) we obtain the desired result. □

Remark 6.1.3. A special case for Eq. (6.1.7) is that P and Q_j, $j = 1, \ldots, m$, are symmetric matrices. Since for any symmetric matrix A, $\mu_2(A) = \lambda_1(A)$, $\nu_2(A) = \lambda_n(A)$ and $\mu(A) \geq \lambda_1(A)$, $\nu(A) \leq \lambda_n(A)$ for any Lozenskii measures, then if we use μ_2 and ν_2 in the previous theorems, they will give the best results among those using all the Lozenski measures.

For another special case, we obtain a criterion for oscillation of the following delay differential system of first order

$$X(t) + Q_0 X(t) + \sum_{j=1}^{m} Q_j X(t - \sigma_j) = 0 \tag{6.1.16}$$

where $Q_j, j = 0, \dots, m$, are $n \times n$ matrices, and $\sigma_j \geq 0, j = 1, \dots, m$. The main result of this section is given in the following.

Theorem 6.1.6. *Assume $\nu(Q_j) \geq 0,\ j = 0, 1, \dots, m$, and $\sum_{j=1}^{m} \nu(Q_j) > 0$. Then either one of the following guarantees that (6.1.16) is oscillatory:*

i) $\sum_{j=1}^{m} \nu(Q_j) e^{\nu(Q_0)\sigma_j} \sigma_j e \geq 1,$ (6.1.17)

ii) $\left(\prod_{j=1}^{m} \nu(Q_j) \right)^{1/m} \exp(\frac{\nu(Q_0)}{m} \sigma^*) \sigma^* e \geq 1$, *where* $\sigma^* = \sum_{j=1}^{m} \sigma_j$. (6.1.18)

The above inequalities become strict if $m = 1$.

Proof: The characteristic equation of (6.1.16) is

$$\det \left(\lambda I + Q_0 + \sum_{j=1}^{m} Q_j e^{-\lambda \sigma_j} \right) = 0. \tag{6.1.19}$$

Let $F(\lambda) = \lambda I + Q_0 + \sum_{j=1}^{m} Q_j e^{-\lambda \sigma_j}$. Then $\det \big(F(\lambda)\big) \neq 0$ for $\lambda > 0$. If (6.1.19) has a real root $\lambda_0 < 0$, then let $\lambda_0 = -s_0$, and let

$$G(s) = -sI + Q_0 + \sum_{j=1}^{m} Q_j e^{s\tau_j}.$$

Then $\det \big(G(s_0)\big) = 0$ implies that $\nu\big(G(s_0)\big) \leq 0$. Hence

$$0 \geq \nu\big(G(s_0)\big) \geq -s_0 + \nu(Q_0) + \sum_{j=1}^{m} \nu(Q_j) e^{s_0 \sigma_j} \tag{6.1.20}$$

$$s_0 - \nu(Q_0) \geq \sum_{j=1}^{m} \nu(Q_j) e^{\nu(Q_0)\sigma_j} e^{(s_0 - \nu(Q_0))\sigma_j}.$$

Obviously, $s_0 - \nu(Q_0) > 0$ and then

$$1 \geq \sum_{j=1}^{m} \nu(Q_j)e^{\nu(Q_0)\sigma_j} e^{(s_0 - \nu(Q_0))\sigma_j} / (s_0 - \nu(Q_0))$$

$$> \sum_{j=1}^{m} \nu(Q_j)e^{\nu(Q_0)\sigma_j} \sigma_j e,$$

contradicting (6.1.17). Note that this last inequality may become an equality if $m = 1$. Therefore, (6.1.19) has no real root, and (6.1.16) is oscillatory.

The case corresponding to (6.1.18) can be similarly proved. □

6.2. Systems with Variable Matrix Coefficients

Now we turn to the oscillation of the linear nonautonomous system

$$\frac{d^N}{dt^N}[X(t) - PX(t-\tau)] + \sum_{j=1}^{m} Q_j(t)X(t-\sigma_j) = 0 \tag{6.2.1}$$

where $P \in R^{n\times n}, Q_j \in C(R_+, R^{n\times n}), j = 1,\ldots,m;\ \tau, \sigma_j, j = 1,\ldots,m$, are nonnegative constants. Furthermore, assume $Q_j(t) = \big((q_j)_{\ell k}(t)\big)$, where $(q_j)_{\ell k}(t),\ \ell, k = 1,\ldots,n,\ j = 1,\ldots,m$, are continuous, bounded and do not change signs on R_+.

As we mentioned in Section 6.0, a solution $X(t)$ of (6.2.1) is oscillatory if at least one of its nontrivial components has arbitrarily large zeros.

It is known (see [79]) that all solutions of Eq. (6.2.1) have the exponential order and hence the Laplace transforms of the solutions exist. Denote

$$(\bar{q}_j)_{\ell k} = \inf_{t\in R_+} (q_j)_{\ell k}(t),\ (\hat{q}_j)_{\ell k} = \sup_{t\in R_+} (q_j)_{\ell k}(t),\ \ell, k = 1,\ldots,n,\ j = 1,\ldots,m.$$

For any $(q_j^*)_{\ell k} \in \big((\bar{q}_j)_{\ell k}, (\hat{q}_j)_{\ell k}\big),\ \ell, k = 1,\ldots,n,\ j = 1,\ldots,m$, define $Q_j^* = \big((q_j^*)_{\ell k}\big)$.

We will determine oscillation properties of (6.2.1) by comparing (6.2.1) with the following equation

$$\frac{d^N}{dt^N}[X(t) - PX(t-\tau)] + \sum_{j=1}^{m} Q_j^* X(t-\sigma_j) = 0. \tag{6.2.2}$$

This is done in the following theorem.

Theorem 6.2.1. *If for every $(q_j^*)_{\ell k} \in \big((\bar{q}_j)_{\ell k}, (\hat{q}_j)_{\ell k}\big)$, $\ell, k = 1, \ldots, n$, $j = 1, \ldots, m$, Eq. (6.2.2), is oscillatory, then Eq. (6.2.1) is oscillatory.*

Proof: We only prove it for the case that $N = 1$, since the general proof is similar. Assume the contrary. Then there exists an eventually nontrivial solution $X(t)$ of (6.2.1) and $t_0 \geq 0$ such that $x_i(t)$ has eventually constant signs for $t \geq t_0 - r$, $i = 1, \ldots, n$, where $r = \max\{\tau, \sigma_1, \ldots, \sigma_m\}$. Without loss of generality we may assume $x_1(t) > 0$ for $t \geq t_0 - r$.

The Laplace transforms of $X(t)$ exists for $s \geq s_0$, where $s_0 \in (-\infty, \infty)$. Let

$$u(s) = \mathcal{L}[X(t)] \triangleq \int_{t_0}^{\infty} e^{-st} X(t)dt, \quad s \geq s_0,$$

and $u(s) = \big(u_1(s), \ldots, u_n(s)\big)^T$. Then multiplying (6.2.1) by e^{-st}, integrating from t_0 to ∞, and using the Mean Value Theorem we find that there exist $(q_j^*)_{\ell k} \in \big((\bar{q}_j)_{\ell k}, (\hat{q}_j)_{\ell k}\big)$, $\ell, k = 1, \ldots, n$, $j = 1, \ldots, n$, such that $Q_j^* = \big((q_j^*)_{\ell k}\big)$ and

$$F(s)u(s) = \varphi(s), \quad s \geq s_0.$$

where

$$\begin{aligned} F(s) =& sI - sPe^{-s\tau} + \sum_{j=1}^{m} Q_j^* e^{-s\sigma_j}, \\ \varphi =& X(t_0) - PX(t_0 - \tau) + sPe^{-s\tau} \int_{t_0 - r}^{t_0} e^{-st} X(t)dt \\ & - \sum_{j=1}^{m} Q_j^* e^{-s\sigma_j} \int_{t_0 - \tau_j}^{t_0} e^{-st} X(t)dt. \end{aligned}$$

Obviously, $F(s)$ and $\varphi(s)$ are continuous on R. Since (6.2.2) is oscillatory for the above Q_j^*, $j = 1, \ldots, m$, we have

$$\det\big(F(s)\big) \neq 0 \quad \text{for} \quad s \in R.$$

Hence $F^{-1}(s)$ exists and

$$u(s) = F^{-1}(s)\varphi(s), \quad s \geq s_0 \tag{6.2.3}$$

where $F^{-1}(s)\varphi(s)$ is continuous on R. We claim that $s_0 = -\infty$. Otherwise, from the definition of $u(s)$ we have $\lim_{s \to s_0^+} |u(s)| = \infty$, contradicting (6.2.3). Thus, $s_0 = -\infty$ implies that (6.2.3) holds for all $s \in R$.

It is easy to see that there exist $a_1 > 0$ and $a_2 > 0$ such that

$$|\varphi(s)| \leq a_1 e^{a_2|s|}, \quad s \in R, \tag{6.2.4}$$

and since $\det \big(F(s)\big) > 0$,

$$|F^{-1}(s)| \leq \frac{b}{\det F(s)} \, |F(s)|^{n-1} \tag{6.2.5}$$

where b is a constant depending only on n. It is also seen that there exist $C_1 > 0$ and $C_2 > 0$ such that

$$|F(s)| \leq C_1 e^{C_2|s|}, \quad s \in R.$$

Moreover, since $\det \big(F(s)\big)$ is a polynomial of s and $e^{-s\tau_j}$, $j = 1, \ldots, m$, and $\det \big(F(s)\big) > 0$ for $s \in R$, we know that $\lim_{s \to -\infty} \det \big(F(s)\big) = \infty$, and hence there exists an $s_1 < 0$ such that $\det \big(F(s)\big) \geq 1$ for $s \leq s_1$. Thus from (6.2.5) we have

$$|F^{-1}(s)| \leq bC_1^{n-1} e^{(n-1)C_2|s|}, \quad s \leq s_1. \tag{6.2.6}$$

Noting that $u_1(s) = L[x_1](s) > 0$, from (6.2.3), (6.2.4) and (6.2.6) we get that there exist $d_1 > 0$ and $d_2 > 0$ such that

$$0 \leq u_1(s) \leq |u(s)| \leq d_1 e^{d_2|s|}, \quad s \leq s_1.$$

Therefore,

$$0 \leq \limsup_{s \to -\infty} e^{-d_2|s|} u_1(s) \leq d_1. \tag{6.2.7}$$

But for $t_1 > \max\{t_0, d_2\}$ we obtain that

$$\begin{aligned}
e^{-d_2|s|} u_1(s) &= e^{-d_2|s|} \int_{t_0}^{\infty} e^{-st} x_1(t) dt \\
&\geq e^{-d_2|s|} \int_{t_1}^{\infty} e^{-st} x_1(t) dt \\
&\geq e^{-(d_2 - t_1)|s|} \int_{t_1}^{\infty} x_1(t) dt \longrightarrow \infty
\end{aligned}$$

as $s \to -\infty$ since $x_1(t) > 0$, contradicting (6.2.7). The proof is complete. □

By combining Theorems 6.2.1 and 6.1.6, we can discuss the following equation

$$X'(t) + Q_0(t)X(t) + \sum_{j=1}^{m} Q_j(t)X(t-\sigma_j). \tag{6.2.8}$$

Assume $Q_j(t) = ((q_j)_{\ell r}(t))$, $j = 0, \ldots, m$, where $(q_j)_{\ell k}(t)$, $\ell, k = 1, \ldots, n$, are continuous, bounded, and do not change signs on R_+, $\sigma_j > 0$, $j = 1, \ldots, m$. Denote

$$(\bar{q}_j)_{\ell k} = \inf_{t \in R_+} ((q_j)_{\ell k}(t)), \quad (\hat{q}_j)_{\ell k} = \sup_{t \in R_+} ((q_j)_{\ell k}(t)).$$

For any $(q_j^*)_{\ell k} \in ((\bar{q}_j)_{\ell k}, (\hat{q}_j)_{\ell k})$, define $Q_j^* = ((q_j^*)_{\ell k})$.

Corollary 6.2.1. *Assume for any* $(q_j^*)_{\ell k} \in ((\bar{q}_j)_{\ell k}, (\hat{q}_j)_{\ell k})$, $\ell, k = 1, \ldots, n$, $j = 0, \ldots, m$, *we have*

i) $\nu(Q_j^*) \geq 0$, $j = 0, \ldots, m$, *and* $\sum_{j=1}^{m} \nu(Q_j^*) > 0$,

and one of the following conditions holds:

ii) $\sum_{j=1}^{m} \nu(Q_j^*) e^{\nu(Q_0^*)\sigma_j} \sigma_j e \geq 1$, *or*

ii)′ $\left(\prod_{j=1}^{m} \nu(Q_j^*)\right)^{1/m} \exp\left(\frac{\nu(Q_0^*)}{m}\sigma^*\right)\sigma^* e \geq 1$, *where* $\sigma^* = \sum_{j=1}^{m} \sigma_j$,

the above inequalities become strict if $m = 1$.

Then Eq. (6.2.8) is oscillatory.

For the scalar case we have the following result.

Consider the scalar equation

$$x'(t) + q_0(t)x(t) + \sum_{j=1}^{m} q_j(t)x(t-\sigma_j) = 0 \tag{6.2.9}$$

where $q_j(t)$, $j = 0, \ldots, m$, are continuous, bounded and do not change signs, $0 \leq \bar{q}_j \leq q_j(t) \leq \hat{q}_j$, $j = 0, \ldots, m$, and $\sigma_j > 0$, $j = 1, \ldots, m$. Then our results reduce to the following

Corollary 6.2.2. *Assume that the equation*

$$x'(t) + \bar{q}_0 x(t) + \sum_{j=1}^{m} \bar{q}_j x(t - \sigma_j) = 0 \tag{6.2.10}$$

is oscillatory. Thus Eq. (6.2.9) is also oscillatory. In particular, assume

$$\sum_{j=1}^{m} \bar{q}_j e^{\bar{q}_0 \sigma_j} \sigma_j e \geq 1 \quad \text{or}$$

$$\Big(\prod_{j=1}^{m} \bar{q}_j\Big)^{1/m} e^{\frac{q_0}{m}\sigma^*} \sigma^* e \geq 1, \quad \text{where} \quad \sigma^* = \sum_{j=1}^{m} \sigma_j,$$

the inequalities become strict if $m = 1$. *Then Eq. (6.2.9) is oscillatory.*

Proof: The characteristic equation of (6.2.10) is

$$F(\lambda) \equiv \lambda + \bar{q}_0 + \sum_{j=1}^{m} \bar{q}_j e^{-\lambda \sigma_j} = 0.$$

Eq. (6.2.10) is oscillatory implies that $F(\lambda) > 0$ for all $\lambda \in R$. Then for any $q_j^* \in [\bar{q}_j, \hat{q}_j]$, $j = 0, \ldots, m$, we see that

$$\lambda + q_0^* + \sum_{j=1}^{m} q_j^* e^{-\lambda \sigma_j} \geq \lambda + \bar{q}_0 + \sum_{j=1}^{m} \bar{q}_j e^{-\lambda \sigma_j} > 0.$$

This means that the equation

$$x'(t) + q_0^* x(t) + \sum_{j=1}^{m} q_j^* x(t - \sigma_j) = 0$$

is oscillatory. By Theorem 6.2.1, Eq. (6.2.9) is oscillatory. The rest of the proof is immediate from the first part and Corollary 6.2.1. □

Finally we show how our results may be extended to certain nonlinear systems of the form

$$X'(t) + \sum_{j=1}^{m} Q_j(t, X(t)) X(t - \sigma_j) = 0 \tag{6.2.11}$$

where $\tau_j > 0$, $j = 1, \ldots, m$. Let $Q_j(t, X) = \big((q_j)_{\ell k}(t, X)\big)$, $j = 0, \ldots, m$, be $n \times n$ matrices, where the $(q_j)_{\ell k}\big(t, X(t)\big)$ are continuous, uniformly bounded, and do not change signs on R_+ for all solutions $X(t)$.

Denote

$$(\bar{q}_j)_{\ell k} = \inf_{t \in R_+} \Big(\big((q_j)_{\ell k}(t, X(t)\big)\Big), \quad (\hat{q}_j)_{\ell k} = \sup_{t \in R_+} \Big(\big((q_j)_{\ell k}(t, X(t)\big)\Big)$$

for $\ell, k = 1, \ldots, n$, $j = 1, \ldots, m$, where the infima and suprema are taken over all solutions of the system (6.2.11) and for any $(q_j^*)_{\ell k} \in \big((\bar{q}_j)_{\ell k}, (\hat{q}_j)_{\ell k}\big)$ define $Q_j^* = \big((q_j^*)_{\ell k}\big)$, $j = 0, \ldots, m$.

Theorem 6.2.2. *i) If the conditions of Theorem 6.2.1 or Corollary 6.2.1 are satisfied, then system (6.2.11) is oscillatory.*

The proofs are essentially the same as Theorem 6.2.1, Corollary 6.2.1, and so we omit them here.

We remark that if we can give a priori upper and lower bounds for all solutions, then with some general assumptions we can easily show the uniform boundedness of the functions $(q_j)_{\ell k}\big(t, X(t)\big)$ in (6.2.11) which are required by Theorem 6.2.2.

As an application of our results, we derive oscillation properties for Lotka-Volterra models in the system (predator-prey and competition) cases.

We consider the system of delay logistic equations

$$\dot{N}_i(t) = N_i(t)[a_i - \sum_{j=1}^{m} b_{ij} N_j(t - \sigma)], \quad i = 1, \ldots, m \tag{6.2.12}$$

with $N_i(t) = \varphi_i(t)$, $t \in [-\sigma, 0]$, where

$$\sigma \in (0, \infty), \quad a_i, b_{ij} \in R \quad \text{for} \quad i, j = 1, \ldots, m, \quad \varphi_i \in C([-\sigma, 0], R_+) \quad \text{and}$$
$$\varphi_i(0) > 0, \quad i = 1, \ldots, m.$$

We assume that (6.2.12) has an equilibrium $N^* = (N_1^*, \ldots, N_m^*)^T$ with positive components. Set $N_i(t) = N_i^* e^{x_i(t)}$ for $t \geq 0$ and $i = 1, \ldots, m$, then the $x_i(t)$

satisfy

$$\dot{x}_i(t) + \sum_{j=1}^{m} p_{ij}\left(e^{x_j(t-\tau)} - 1\right) = 0, \quad i = 1, \ldots, m, \tag{6.2.13}$$

where $p_{ij} = b_{ij}N_j^*$, $i, j = 1, \ldots, m$. As shown in (see [76], Section 5.4) we see that every solution of (6.2.13) satisfies

$$\lim_{t\to\infty} x(t) = 0, \quad i = 1, \ldots, m, \tag{6.2.14}$$

if the matrix $P = (p_{ij})$ satisfies

$$\nu_\infty(P) = \min_{1\le j\le m} \{N_j^*(b_{jj} - \sum_{i=1, i\neq j}^{m} |b_{ij}|)\} > 0.$$

Now we are ready to derive criteria for oscillation.

Theorem 6.2.3. *Assume $\nu_\infty(P) > 0$, and for some $i = 1, 2, \ldots,$ or ∞,*

$$\nu_i(P)\sigma e > 1. \tag{6.2.15}$$

Then (6.2.13) is oscillatory, and hence every solution of (6.2.12) is oscillatory about N^*.

Proof: Rewrite (6.2.13) in the form

$$\dot{x}_j(t) + \sum_{j=1}^{m} p_{ij}^*(t)x_j(t-\sigma) = 0, \quad i = 1, \ldots, m \tag{6.2.16}$$

where

$$p_{ij}^*(t) = p_{ij}\,\frac{e^{x_j(t-\sigma)} - 1}{x_j(t-\sigma)} \quad \text{for} \quad i, j = 1, \ldots, m,$$

and by (6.2.14), $\lim_{t\to\infty} p_{ij}^*(t) = p_{ij}$, $i, j = 1, \ldots, m$. Define the matrix $P^*(t) = \left(p_{ij}^*(t)\right)$. Then $\lim_{t\to\infty} P^*(t) = P$.

By (6.2.14) and the continuity of the Lozinskii measures we see that there is a $T_0 \ge 0$ such that $\nu_i\left(P^*(t)\right)\sigma e > 1$, $t \ge T_0$. Then by Theorem 6.2.1, we obtain the desired result. □

6.3. Comparison with Scalar Equations

Here we investigate the systems of first order neutral delay differential equations of the form

$$\frac{d}{dt}(X(t) - PX(t-\tau)) + \sum_{j=1}^{m} Q_j(t)X(t-\sigma_j(t)) = 0. \tag{6.3.1}$$

We assume that

(i) P is an $n \times n$ diagonal matrix with diagonal elements $p_1, p_2, \ldots, p_n$ such that $p_i \geq 0$ for $i = 1, \ldots, n$ and $\tau > 0$;
(ii) for each $j = 1, \ldots, m$, $\quad Q_j(t) = \big((q_j)_{\ell k}(t)\big)$ satisfy

$$(q_j)_{\ell k}(t) \in C([t_0, \infty), R_+), \quad \ell, k = 1, \ldots, n;$$

(iii) $\sigma_j \in C([t_0, \infty), R_+)$ and $\lim_{t\to\infty}\big(t - \sigma_j(t)\big) = \infty; \quad j = 1, \ldots, m.$

Theorem 6.3.1. *Assume that* $0 \leq p_i \leq 1, \quad i = 1, \ldots, n,$ *and* $q_j(t) \geq 0,$ $j = 1, \ldots, m,$ *where*

$$q_j(t) = \min_{1\leq \ell \leq n}\Big\{(q_j)_{\ell k}(t) - \sum_{\substack{k=1\\k\neq \ell}}^{n} |(q_j)_{\ell k}(t)|\Big\}, \quad j = 1, \ldots, m. \tag{6.3.2}$$

Further assume that there exists a nonnegative N *such that every solution of the scalar equation*

$$u'(t) + \sum_{j=1}^{m} q_j(t) \sum_{h=0}^{N} p_*^h u\big(t - h\tau - \sigma_j(t)\big) = 0 \tag{6.3.3}$$

is oscillatory. Then every solution of system (6.3.1) is oscillatory, where $p_* = \min_{1\leq i\leq n} p_i$.

Proof: Assume the contrary. Then (6.3.1) has a nonoscillatory solution $X(t) = \big(x_1(t), \ldots, x_n(t)\big)^T$, i.e., there exists $t_1 \geq t_0$ such that $X(t) \neq 0$, and $x_i(t) \neq 0$ if it is nontrivial for $t \geq t_1, \quad i = 1, \ldots, n$. Set $\delta_i =$ sign $x_i(t)$, i.e.,

$\delta_i = 1$, if $x_i(t) \geq 0$ and $\delta_i = -1$ if $x_i(t) < 0$. Then

$$Y(t) = \big(y_1(t), \dots, y_n(t)\big)^T = \big(\delta_1 x_1(t), \dots, \delta_n x_n(t)\big)^T$$

is a solution of the system

$$\big(Y(t) - PY(t-\tau)\big)' + \sum_{j=1}^{m} \overline{Q}_j(t) Y\big(t - \sigma_j(t)\big) = 0 \tag{6.3.4}$$

where $\overline{Q}_j(t) = \big((\overline{q}_j)_{\ell k}\big)$, and $(\overline{q}_j)_{\ell k} = \frac{\delta_k}{\delta_\ell}\,(q_j)_{\ell k}, \quad \ell,\ k = 1, \dots, n, \quad j = 1, \dots, m.$ Set

$$z_\ell(t) = y_\ell(t) - p_\ell y_\ell(t-\tau),$$
$$v(t) = \sum_{\ell=1}^{n} z_i(t), \quad w(t) = \sum_{\ell=1}^{n} y_i(t).$$

From (6.4.4) we have

$$z'_\ell(t) + \sum_{j=1}^{m} \sum_{k=1}^{n} (\overline{q}_j)_{\ell k}(t) y_k\big(t - \sigma_j(t)\big) = 0, \quad \ell = 1, \dots, n,$$

and hence

$$v'(t) + \sum_{j=1}^{m} \sum_{k=1}^{n} \left(\sum_{\ell=1}^{n} (\overline{q}_j)_{\ell k}(t) \right) y_k\big(t - \sigma_j(t)\big) = 0.$$

By the definition of $q_j(t)$ we find

$$v'(t) + \sum_{j=1}^{m} q_j(t) w\big(t - \sigma_j(t)\big) \leq 0. \tag{6.3.5}$$

Since $q_j(t) \geq 0, \quad w\big(t - \sigma_j(t)\big) \geq 0$, (6.4.5) implies that $v'(t) < 0$ and thus $v(t)$ is decreasing. Set $\lim_{t \to \infty} v(t) = \ell$. Then $\ell \in [-\infty, \infty)$. We claim that $\ell \geq 0$. Otherwise, there exist $i \in \{1, \dots, n\}$ and $\alpha > 0$ such that for $t \geq T \geq t_0$

$$z_i = y_i(t) - p_i y_i(t-\tau) \leq -\alpha. \tag{6.3.6}$$

If $p_i \in (0,1)$, then (6.3.6) leads to $\lim_{t\to\infty} y_i(t) = 0$ as before, consequently, $\lim_{t\to\infty} z_i(t) = 0$, a contradiction. If $p_i = 1$, then (6.3.6) implies that

$$y_i(T + n\tau) \le y(T) - n\alpha \to \infty, \quad \text{as} \quad n \to \infty,$$

a contradiction again.

The above discussion shows that $v(t) > 0$ for all large t. Since $v(t) \le w(t) - p_* w(t-\tau)$, we obtain $w(t) \ge \sum_{h=0}^{N} p_*^h v(t - h\tau)$. Substitute it into (6.4.5) we have

$$v'(t) + \sum_{j=1}^{m} q_j(t) \sum_{h=0}^{N} p_*^h v\big(t - h\tau - \sigma_j(t)\big) \le 0.$$

By Corollary 2.6.5 Eq. (6.3.3) has a positive solution, contradicting the assumption. □

Remark 6.3.1. For the case that $n = 1$, Theorem 6.3.1 leads to the fact that every solution of the neutral delay equation

$$\big(x(t) - px(t-\tau)\big)' + \sum_{j=1}^{m} q_j(t) x\big(t - \sigma_j(t)\big) = 0 \tag{6.3.7}$$

is oscillatory if every solution of the delay equation

$$y'(t) + \sum_{j=1}^{m} q_j(t) \sum_{h=0}^{N} p^h y\big(t - h\tau - \sigma_j(t)\big) = 0 \tag{6.3.8}$$

is oscillatory, where $p \in [0,1]$, $q_j(t) \ge 0$, $\sigma_j(t) \ge 0$, and $\lim_{t\to\infty}\big(t - \sigma_j(t)\big) = 0$, $j = 1, \ldots, m$.

Theorem 6.3.2. *Suppose $p_i > 1$ for $i = 1, \ldots, n$, $q_j(t) \ge 0$, where $q_j(t)$ are defined by (6.3.2), $j = 1, \ldots, m$. Assume there exists a $j_0 \in \{1, \ldots, m\}$ and a nonnegative integer N such that $\int_T^\infty q_{j_0}(t)dt = \infty$, and every solution of the equation*

$$u'(t) - \sum_{j=1}^{m} q_j(t+\tau) \sum_{h=0}^{N} (p^*)^{-(h+1)} u\big(t + (h+1)\tau - \sigma_j(t+\tau)\big) = 0 \tag{6.3.9}$$

is oscillatory. Then every solution of system (6.3.1) is oscillatory, where $p^* = \max_{1 \le i \le n} p_i$.

Proof: Assume $X(t) = \big(x_1(t), \ldots, x_n(t)\big)^T$ is a nonoscillatory solution of (6.3.1). As in the proof of Theorem 6.3.1.

$$y(t) = \big(y_1(t), \ldots, y_n(t)\big)^T = \big(\delta_1 x_1(t), \ldots, \delta_n x_n(t)\big)^T$$

is a nonoscillatory solution of the system (6.3.4). Thus

$$\big(y_\ell(t) - p_\ell y_\ell(t-\tau)\big)' + \sum_{j=1}^{m} \sum_{k=1}^{n} (\overline{q}_j)_{\ell k}(t) y_k\big(t - \sigma_j(t)\big) = 0, \quad \ell = 1, \ldots, n. \tag{6.3.10}$$

Rewrite (6.3.10) in the form

$$\big(y_\ell(t) - \frac{1}{p_\ell} y_\ell(t+\tau)\big)' = \frac{1}{p_\ell} \sum_{j=1}^{m} \sum_{k=1}^{n} (\overline{q}_j)_{\ell k}(t+\tau) y_k\big(t+\tau - \sigma_j(t+\tau)\big), \quad \ell = 1, \ldots, n. \tag{6.3.11}$$

Set

$$v(t) = \sum_{\ell=1}^{n} \big(y_\ell(t) - \frac{1}{p_\ell} y_\ell(t+\tau)\big), \quad w(t) = \sum_{\ell=1}^{n} y_\ell(t).$$

Summing up both sides of (6.3.11) for $\ell = 1, \ldots, n$ we obtain

$$v'(t) \ge \frac{1}{p^*} \sum_{j=1}^{m} q_j(t+\tau) w\big(t+\tau - \sigma_j(t+\tau)\big) \ge 0. \tag{6.3.12}$$

Hence $\lim_{t\to\infty} v(t) = \ell \in (-\infty, \infty]$. We claim that $\ell > 0$. Otherwise, we get $w(t) \le \frac{1}{p_*} w(t+\tau)$ for all large t, where $p_* = \min_{1 \le t \le n} p_i$, which implies that $\lim_{t\to\infty} w(t) = \infty$. Integrating (6.3.12) we see

$$v(t) - v(T) \ge \frac{1}{p^*} \sum_{j=1}^{m} \int_T^t q_j(s+\tau) w\big(s+\tau - \sigma_j(s+\tau)\big) ds \to \infty, \quad t \to \infty,$$

a contradiction. The above discussion shows that eventually

$$0 < v(t) \le w(t) - \frac{1}{p^*} w(t+\tau).$$

and hence

$$w(t) \ge \sum_{h=0}^{N} (p^*)^{-h} v(t + h\tau).$$

Substituting it into (6.3.12) we have

$$v'(t) \ge \sum_{j=1}^{m} q_j(t+\tau) \sum_{h=0}^{N} (p^*)^{-(h+1)} v\big(t + (h+1)\tau - \sigma_j(t+\tau)\big).$$

This implies that (6.3.9) has an eventually positive solution, contradicting the assumption. □

6.4. Existence of Nonoscillatory Solutions

In this section, we consider the existence of nonoscillatory solutions of nonlinear systems of neutral differential equations of the form

$$\frac{d^N}{dt^N}\left[x_1(t) + (-1)^r a_i(t) x_i\big(h_i(t)\big)\right] = \sum_{j=1}^{n} p_{ij}(t) f_{ij}\big(x_j(g_{ij}(t))\big),$$
$$i = 1, \ldots, n, \quad n \ge 2, \quad N \ge 2, \quad r \in \{0, 1\}. \tag{6.4.1$_r$}$$

subject to the hypotheses

$$\begin{aligned} a_i &: [t_0, \infty) \to [0, \beta_i], \quad t_0 \ge 0, \quad 0 < \beta_i < 1, \\ h_i, p_{ij}, g_{ij} &: [t_0, \infty) \to R \quad \text{and} \quad f_{ij} : R \to R, \quad 1 \le i, \quad j \le n \end{aligned} \tag{6.4.2}$$

are continuous functions;

$$h_i(t) < t \quad \text{for} \quad t \ge t_0, \quad \lim_{t\to\infty} h_i(t) = \infty, \quad \lim_{t\to\infty} g_{ij}(t) = \infty,$$
$$1 \le i, \; j \le n; \tag{6.4.3}$$

$f_{ij}(u)u > 0$ for $u \neq 0$ and f_{ij} are nondecreasing functions, $1 \leq i,\ j \leq n$; (6.4.4)

$$\lim_{t\to\infty} a_i(t)\left(\frac{h_i(t)}{t}\right)^k = \overline{a}_{ik} \in [0, \beta_i] \tag{6.4.5}$$

for $1 \leq i \leq n$ and every $k \in \{1, \dots, N-1\}$.

Let $t_1 \geq t_0$. Denote

$$t_2 = \min\left\{\inf_{t\geq t_1} h_i(t), \quad \inf_{t\geq t_1} g_{ij}(t), \quad 1 \leq i,\ j \leq n\right\}.$$

A function $X = (x_1, \dots, x_n)^T$ is a solution of the system (6.4.1_r), if there exists a $t_1 \geq t_0$ such that X is continuous on $[t_2, \infty)$, $x_i(t) + (-1)^r a_i(t) x_i\big(h_i(t)\big)$, $1 \leq i \leq n$, are N-times continuously differentiable on $[t_1, \infty)$, and X satisfies (6.4.1_r) on $[t_1, \infty)$.

As mentioned before, a solution $X = (x_1, \dots, x_N)^T$ of (6.4.1_r) is nonoscillatory if there exists an $a \geq t_0$ such that its every nontrivial component is different from zero for all $t \geq a$.

Our aim in this section is to give conditions for the system (6.4.1_r) to possess nonoscillatory solutions $X = (x_1, \dots, x_n)^T$ with the asymptotic behavior

$$\lim_{t\to\infty} \frac{x_i(t)}{t^{k_i}} = c_i \neq 0, \quad \text{sgn } c_i = \text{ sgn } c_1$$

or

$$\lim_{t\to\infty} \frac{x_i(t)}{t^{k_i}} = 0, \quad \lim_{t\to\infty} \frac{x_i(t)}{t^{k_i-1}} = \infty$$

for some $k_i \in \{1, \dots, N-1\}$, $1 \leq i \leq n$.

Denote

$$\gamma(t) = \max\left\{\sup\left\{s \geq t : h_i(s) \leq t,\ g_{ij}(s) \leq t \quad \text{for} \quad 1 \leq i,\ j \leq n\right\}\right\},$$

$$H_i(0,t) \equiv t, \quad H_i(k,t) = H_i\big(k-1, h_i(t)\big), \quad 1 \leq i \leq n, \quad k = 1, 2, \dots \tag{6.4.6}$$

$$A_i(0,t) \equiv 1, \quad A_i(k,t) = \prod_{j=0}^{k-1} a_i\big(H_i(j,t)\big), \quad 1 \leq i \leq n, \quad k = 1, 2, \dots \tag{6.4.7}$$

$$(p_{ij})^+_{k_i}(t) = \max\{(-1)^{N-k_i} p_{ij}(t), 0\} \quad \text{and}$$
$$(p_{ij})^-_{k_i}(t) = \max\{-(-1)^{N-k_i} p_{ij}(t), 0\}, \quad t \geq t_0, \tag{6.4.8}$$
$$1 \leq i, j \leq n, \quad k_i \in \{1, \ldots, N-1\}.$$

Note that

$$\begin{aligned} (-1)^{N-k_i} p_{ij}(t) &= (p_{ij})^+_{k_i}(t) - (p_{ij})^-_{k_i}(t), \\ |p_{ij}(t)| &= (p_{ij})^+_{k_i}(t) + (p_{ij})^-_{k_i}(t), \quad 1 \leq i,\ j \leq n. \end{aligned} \tag{6.4.9}$$

For $T \geq t_0$, $\ell(t) = (\ell_1(t), \ldots, \ell_n(t))^T$, $\lambda = (\lambda_1, \ldots, \lambda_n)^T$ we denote

$$\begin{aligned} G_i(T, k, \ell, \lambda) &= \int_T^{\gamma(T)} t^{N-k-1} \sum_{j=1}^n \ell_j(t) f_{ij}\big((\lambda_j \gamma(T))^k\big) dt \\ &+ \int_{\gamma(T)}^{\infty} t^{N-k-1} \sum_{j=1}^n \ell_j(t) f_{ij}\big((\lambda_j (g_{ij}(t)))^k\big) dt, \quad 1 \leq i \leq n. \end{aligned}$$

Theorem 6.4.1. *Let the assumptions (6.4.2)-(6.4.5) hold and let* $k_i \in \{1, \ldots, N-1\}$, $1 \leq i \leq n$. *Denote* $|p_i|(t) = (|p_{i\ell}(t)|, \ldots, |p_{in}(t)|)^T$. *If for some* $b = \left(\frac{b_1}{k_1!}, \ldots, \frac{b_n}{k_n!}\right)^T$ *with* $b_i > 0$,

$$\lim_{T \to \infty} G_i(T, k_i, |p_i|, b) = 0, \tag{6.4.10}$$

then there exists $(\bar{c}_1, \ldots, \bar{c}_n)$ *with* $\bar{c}_i > 0$, $1 \leq i \leq n$, *such that the system* $(6.4.1_r)$ *has a positive solution* $X = (x_1, \ldots, x_n)^T$ *satisfying*

$$\lim_{t \to \infty} \frac{x_i(t)}{t^{k_i}} = \bar{c}_i > 0, \quad 1 \leq i \leq n. \tag{6.4.11}$$

Proof: Choose $c_i > 0$, $1 \leq i \leq n$, be constants satisfying $c_i \in (\frac{1+\beta_i}{2}\, b_i, b_i)$. Let $k_i \in \{1, \ldots, N-1\}$, $1 \leq i \leq n$. We put $b_i = c_i + d_i$. Then

$$0 < d_i < c_i \,\frac{1 - \beta_i}{1 + \beta_i}\,. \tag{6.4.12}$$

Let $T \geq t_0$ be such that

$$T_0 = \min\left\{\inf_{t\geq T} h_i(t),\ \inf_{t\geq T} g_{ij}(t);\quad \ell \leq i,\ j \leq n\right\} \geq t_0 \tag{6.4.13}$$

and

$$G_i(T, k_i, |p_i|, b) < d_i, \quad 1 \leq i \leq n. \tag{6.4.14}$$

We denote by $C[T_0, \infty)$ the locally convex space of all continuous vector functions $X = (x_i, \ldots, x_n)^T$ defined on $[T_0, \infty)$ which are constant on $[T_0, T]$, with the topology of uniform convergence on any compact subinterval of $[T_0, \infty)$. Thus $C[T_0, \infty)$ is a Frechet space.

(I) Let $r = 0$. We consider the closed convex subset S_0 of $C[T_0, \infty)$ defined by

$$\begin{aligned} S_0 = \{Y = (y_1, \ldots, y_n)^T \in C[T_0, \infty);\ y_i(t) = c_i\,\frac{T^{k_i}}{k_i!} \quad \text{for} \\ t \in [T_0, T],\quad \frac{1}{k_i!}\,(c_i - d_i)t^{k_i} \leq y_i(t) \leq \frac{1}{k_i!}\,b_i t^{k_i} \quad \text{for} \quad t \geq T, \quad 1 \leq i \leq n\}. \end{aligned} \tag{6.4.15}$$

For each $Y \in S_0$ we define functions x_i, $1 \leq i \leq n$, by

$$x_i(t) = \begin{cases} \frac{y_i(T)}{1+a_i(T)}, & t \in [T_0, T], \\ \sum\limits_{k=0}^{n_i(t)-1} (-1)^k A_i(k,t) y_i\big(H_i(k,t)\big) & \\ \quad + (-1)^{n_i(t)} A_i\big(n_i(t), t\big)\, \frac{y_i(T)}{1+a_i(T)}, & t \geq T, \end{cases} \tag{6.4.16}$$

where $n_i(t)$, $1 \leq i \leq n$, are the least positive integers such that $T_0 < H_i\big(n_i(t), t\big) \leq T$.

We easily verify that $x_i(t) \in C[T_0, \infty)$, $1 \leq i \leq n$, and they satisfy the functional equations

$$x_i(t) + a_i(t) x_i\big(h_i(t)\big) = y_i(t), \quad t \geq T, \quad 1 \leq i \leq n. \tag{6.4.17}$$

Let $n_i(t) = 2m_i + 1$ or $n_i(t) = 2m_i + 2$, $m = 0, 1, \ldots$, $1 \leq i \leq n$. Then (6.4.16) together with $Y \in S_0$, (6.4.2) and (6.4.3) implies

$$x_i(t) \geq \frac{1}{k_i!}\left((c_i - d_i)t^{k_i} - a_i(t) b_i\big(h_i(t)\big)^{k_i}\right.$$

$$+ A_i(2,t)[(c_i - d_i)\big(H_i(2,t)\big)^{k_i} - a_i\big(H_i(2,t)\big)b_i\big(H_i(3,t)\big)^{k_i}] + \dots$$

$$+ A_i(2m_i,t)\big[(c_i - d_i)\big(H_i(2m_i,t)\big)^{k_i}$$

$$- a_i\big(H_i(2m_i,t)\big)b_i\big(h_i(2m_i+1,t)\big)^{k_i}]\Big)$$

$$\geq \frac{1}{k_1!}\,[(c_i - d_i) - \beta_i b_i][t^{k_i} + A_i(2,t)\big(H_i(2,t)\big)^{k_i} + \dots$$

$$+ A_i(2m_i,t)\big(H_i(2m_i,t)\big)^{k_i}]$$

$$\geq \frac{1}{k_i!}\,[c_i(1-\beta_i) - d_i(1+\beta_i)]t^{k_i} > 0, \quad t \geq T, \quad 1 \leq i \leq n.$$

Taking into account $Y \in S_0$ and the last inequality we obtain from (6.4.17)

$$0 < \frac{1}{k_i!}\,[c_i(1-\beta_i) - d_i(1+\beta_i)]t^{k_i} \leq x_i(t) \leq y_i(t) \leq \frac{1}{k_i!}\,b_i t^{k_i}. \tag{6.4.18}$$

We define an operator $\mathcal{T} = (\mathcal{T}_1, \dots, \mathcal{T}_n)^T : \; S_0 \to C[T_0, \infty)$ by

$$\mathcal{T}_i Y(t) = \begin{cases} \dfrac{c_i T^{k_i}}{k_i!}, & t \in [T_0, T], \\[2ex] \dfrac{c_i t^{k_i}}{k_i!} + (-1)^{N-k_i} \displaystyle\int_T^t \frac{(t-s)^{k_i-1}}{(k_i-1)!} \int_s^\infty \frac{(u-s)^{N-k_i-1}}{(N-k_i-1)!} \\[2ex] \quad \times \displaystyle\sum_{j=1}^n p_{ij}(u) f_{ij}\big(x_j(g_{ij}(u))\big)du\, ds, & t \geq T, 1 \leq i \leq n. \end{cases} \tag{6.4.19}$$

We shall show that the operator $\mathcal{T}$ is continuous and maps S_0 into a compact subset of S_0.

(i) We prove that $\mathcal{T}(S_0) \subset S_0$. From (6.4.19) in view of (6.4.9), (6.4.4), (6.4.15), (6.4.12) and (6.4.14) we conclude that

$$\mathcal{T}_i Y(t) \leq \frac{c_i t^{k_i}}{k_i!} + \int_T^t \frac{(t-s)^{k_i-1}}{(k_i-1)!(N-k_i-1)!}\, G_i(T, k_i, |p_i|, b)ds$$

$$\leq \frac{c_i t^{k_i}}{k_i!} + d_i \int_T^t \frac{(t-s)^{k_i-1}}{(k_i-1)!} ds$$

$$\leq \frac{1}{k_i!}\, b_i t^{k_i}, \quad t \geq T, \quad 1 \leq i \leq n, \tag{6.4.20}$$

$$\mathcal{T}_i(Y) \geq \frac{c_i t^{k_i}}{k_i!} - \int_T^t \frac{(t-s)^{k_i-1}}{(k_i-1)!(N-k_i-1)!} G_i(T, k_i, |p_i|, b) ds$$

$$\geq \frac{c_i t^{k_i}}{k_i!} - d_i \int_T^t \frac{(t-s)^{k_i-1}}{(k_i-1)!} ds$$

$$\geq \frac{(c_i - d_i)}{k_i!} t^{k_i}, \quad t \geq T, \quad 1 \leq i \leq n.$$

(ii) We prove that the operator $\mathcal{T}$ is continuous. Let $Y_m = (y_{1m}, \ldots, y_{nm})^T \in S_0$, $m = 1, 2, \ldots$, and $y_{im} \to y_i$ for $m \to \infty$, $1 \leq i \leq n$ in the space $C[T_0, \infty)$. Denote

$$x_{im}(t) = \sum_{k=0}^{n_i(t)-1} (-1)^k A_i(k,t) y_{im}\big(H_i(k,t)\big) + (-1)^{n_i(t)} A_i\big(n_i(t), t\big)$$

$$\times \frac{y_{im}(T)}{1 + a_i(T)}, \quad t \geq T, \quad 1 \leq i \leq n, \quad m = 1, 2, \ldots .$$

Using (6.4.19) we obtain

$$|\mathcal{T}_i Y_m(t) - \mathcal{T}_i Y(t)| \leq \int_T^t \frac{(t-s)^{k_i-1}}{(k_i-1)!} \int_T^\infty \frac{(u-s)^{N-k_i-1}}{(N-k_i-1)!}$$

$$\times \sum_{j=1}^n |p_{ij}(u)| \, |f_{ij}\big(x_{jm}(g_{ij}(u))\big) - f_{ij}\big(x_j(g_{ij}(u))\big)| du \, ds$$

$$\leq \frac{(t-T)^{k_i}}{k_i!} \int_T^\infty G_i^m(u) du, \quad 1 \leq i \leq n \tag{6.4.21}$$

where

$$G_i^m(t) = t^{N-k_i-1} \sum_{j=1}^n |p_{ij}(u)| \, |f_{ij}\big(x_{jm}(g_{ij}(t))\big) - f_{ij}\big(x_j(g_{ij}(t))\big)|,$$

$$t \geq T, \quad 1 \leq i \leq n.$$

We easily see that

$$\lim_{m\to\infty} G_i^m(t) = 0 \quad \text{for} \quad t \geq T \quad \text{and} \quad G_i^m(t) \leq M_i(t), \quad \text{where}$$

$$M_i(t) = \begin{cases} 2t^{N-k_i-1} \sum_{j=1}^{n} |p_{ij}(u)| f_{ij}\left(\frac{b_j}{k_j!}(g_{ij}(t))^{k_j}\right), & t \geq \gamma(T), \\ 2t^{N-k_i-1} \sum_{j=1}^{n} |p_{ij}(u)| f_{ij}\left(\frac{b_j}{k_j!}(\gamma(T))^{k_j}\right), & t \in [T, \gamma(T)],\ 1 \leq i \leq n. \end{cases}$$

By virtue of (6.4.10) and the Lebesgue dominated convergence theorem we conclude that $\mathcal{T}_i Y_m(t) \to \mathcal{T}_i Y(t)$ in $C[T_0, \infty)$ for $m \to \infty$, $1 \leq i \leq n$. This implies the continuity of $\mathcal{T}$.

(iii) $\mathcal{T}(S_0)$ is relatively compact. This follows from the Arzela-Ascoli theorem and the observation that for $(y_1, \ldots, y_n)^T \in S_0$, $\left((\mathcal{T}_1 Y(t))', \ldots, (\mathcal{T}_n Y(t))'\right)^T$ is given by

$$|(\mathcal{T}_i Y(t))'| \leq \frac{c_i t^{k_i-1}}{(k_i-1)!} + \frac{t^{k_i-1}}{(k_i-1)!(N-k_i-1)!} G_i(T, k_i, |p_i|, b)$$

$$\leq \frac{b_i t^{k_i-1}}{(k_i-1)!}, \quad t \geq T, \quad K \in \{1, \ldots, N-1\},\ 1 \leq i \leq n.$$

Then by the Schauder-Tychonov fixed point theorem there exists a $\overline{Y} = (\overline{y}_1, \ldots, \overline{y}_n)^T \in S_0$ such that $(\mathcal{T}_1 \overline{Y}, \ldots, \mathcal{T}_n \overline{Y})^T = (\overline{y}_1, \ldots, \overline{y}_n)^T$. The components of $\overline{X} = (\overline{x}_1, \ldots, \overline{x}_n)^T$ satisfy the system

$$\overline{y}_i(t) = \frac{c_i t^{k_i}}{k_i!} + (-1)^{N-k_i} \int_T^t \frac{(t-s)^{k_i-1}}{(k_1-1)!} \int_s^\infty \frac{(u-s)^{N-k_i-1}}{(N-k_i-1)!} \tag{6.4.22}$$

$$\times \sum_{j=1}^{n} p_{ij}(u) f_{ij}(\overline{x}_j(g_{ij}(u)))\, du\, ds, \quad t \geq T,\ 1 \leq i \leq n.$$

where $\overline{y}_i(t) = \overline{x}_i(t) + a_i(t)\overline{x}_i(h_i(t))$, $1 \leq i \leq n$, and $(\overline{x}_i, \ldots, \overline{x}_n)^T$ is a solution of (6.4.10) on $[T_0, \infty)$.

Differentiating (6.4.22) k_i-times we obtain

$$\overline{y}_i^{(k_i)}(t) = c_i + (-1)^{N-k_i} \int_t^\infty \frac{(u-t)^{N-k_i-1}}{(N-k_i-1)!} \sum_{j=1}^{n} P_{ij}(u) f_{ij}(\overline{x}_j(g_{ij}(u)))\, du,$$

$$t \geq T, \quad 1 \leq i \leq n,$$

which implies that $\lim_{t\to\infty} \overline{y}_i^{(k_i)}(t) = c_i > 0, \quad 1 \le i \le n$. The last relation is equivalent to

$$\lim_{t\to\infty} \frac{\overline{y}_i(t)}{t^{k_i}} = c_i \ (>0), \quad 1 \le i \le n. \tag{6.4.23}$$

Then (6.4.23) together with (6.4.5) implies (6.4.11), where $\overline{c}_i = c_i/(1+\overline{a}_{ik_i}), \quad 1 \le i \le n$.

(II) Let $r = 1$. We consider the closed convex subset S_i of $C[T_0,\infty)$ defined by

$$S_1 = \{Z = (z_1,\ldots,z_n)^T \in C[T_0,\infty): \quad z_i(t) = \frac{c_i T^{k_i}}{k_i!} \text{ for } \quad t \in [T_0,T],$$
$$\frac{1}{k_i!}\,(c_i - d_i)t^{k_i} \le z_i(t) \le \frac{1}{k_i!}\,b_i t^{k_i} \quad \text{for} \quad t \ge T, \quad 1 \le i \le n\}. \tag{6.4.24}$$

For each $Z \in S_1$ we define

$$x_i(t) = \begin{cases} \frac{z_i(T)}{1-a_i(T)}\,, & t \in [T_0,T], \\ \sum_{k=0}^{n_i(t)-1} A_i(k,t) z_i\big(H_i(k,t)\big) + A_i\big(n_i(t),t\big)\,\frac{z_i(T)}{1-a_i(T)}\,, & t \ge T, \end{cases} \tag{6.4.25}$$

where $n_i(t), \quad 1 \le i \le n$, are the same as in the case (I).

We easily verify that $x_i(t) \in C[T_0,\infty), \quad 1 \le i \le n$ and they satisfy the functional equations

$$x_i(t) - a_i(t)x_i\big(h_i(t)\big) = z_i(t), \quad t \ge T, \quad 1 \le i \le n. \tag{6.4.26}$$

From (6.4.25), (6.4.26) (6.4.2) and (6.4.3) we obtain

$$\begin{aligned} \frac{1}{k_i!}\,(c_i - d_i)t^{k_i} &\le x_i(t) \\ &\le \frac{1}{k_i!}\,b_i[t^{k_i} + \beta_i\big(h_i(t)\big)^{k_i} \\ &\qquad + \cdots + \beta_i^{n_i(t)}\big(H_i(n_i(t),t)\big)^{k_i}] \\ &\le \frac{b_i t^{k_i}}{k_i!(1-\beta_i)}\,. \end{aligned} \tag{6.4.27}$$

We define an operator $\overline{T} = (\overline{T}_1, \ldots, \overline{T}_n)^T : S_1 \to C[T_0, \infty)$ by (6.4.19) in which we replace $y_i(t)$ by $z_i(t)$, $1 \le i \le n$.

Proceeding similarly as in the case (I) we prove that the operator T is continuous and maps S_1 into a compact subset of S_1. Then by the Schauder-Tychonov theorem there exists a fixed point $\overline{Z} = (\overline{z}_1, \ldots, \overline{z}_n)^T \in S_1$ such that the components of $(\overline{x}_1, \ldots, \overline{x}_n)^T$ are solutions of the system $(6.4.1_1)$ on $[T_0, \infty)$ with the property

$$\lim_{t\to\infty} \frac{\overline{z}_i(t)}{t^{k_i}} = c_i > 0, \quad 1 \le i \le n \tag{6.4.28}$$

where $\overline{z}_i(t) = \overline{x}_i(t) - a_i(t)\overline{x}_i\big(h_i(t)\big)$, $1 \le i \le n$. Then (6.4.28) together with (6.4.27) and (6.4.5) implies

$$\lim_{t\to\infty} \frac{\overline{x}(t)}{t^{k_i}} = \frac{c_i}{1 - \overline{a}_{ik_i}} = \overline{c}_i > 0, \quad 1 \le i \le n.$$

The proof of Theorem 6.4.1 is complete. □

Theorem 6.4.2. *Let the assumptions (6.4.2) – (6.4.4) hold and* $k_i \in \{1, \ldots, N-1\}$,
$1 \le i \le n$. *Denote* $p_i^{\pm}(t) = \big(p_n^{\pm}(t), \ldots, p_{in}^{\pm}(t)\big)^T$, $1 \le i \le n$. *Assume*

(i) *for some* $a = \big(\frac{a_1}{(k_i-1)!}, \ldots, \frac{a_n}{(k_i-1)!}\big)^T$ *with* $a_i > 0$, $1 \le i \le n$

$$\lim_{T\to\infty} G_i(T, k_i, p_i^+, a) = 0, \tag{6.4.29}$$

(ii) *for some* $b = \big(\frac{b_1}{(k_i-1)!}, \ldots, \frac{b_n}{(k_i-1)!}\big)^T$ *with* $b_i > 0$, $1 \le i \le n$

$$\lim_{T\to\infty} G_i(T, k_i - 1, p_i^-, b) = 0, \quad \text{and} \tag{6.4.30}$$

(iii) *for some* $h \in \{1, \ldots, n\}$, $C = \big(0, \ldots, 0, \frac{C_k}{(k_i-1)!}, 0, \ldots, 0\big)^T$ *with* $c_h > 0$

$$G_i(T, k_i - 1, p_i^+, c) = \infty, \quad 1 \le i \le n. \tag{6.4.31}$$

Then there exists a positive solution of the system $(6.4.1_r)$ *with the property*

$$\lim_{t\to\infty} \frac{x_i(t)}{t^{k_i}} = 0, \quad \lim_{t\to\infty} \frac{x_i(t)}{t^{k_i-1}} = \infty. \tag{6.4.32}$$

Proof: Let $a_i, b_i,\quad 1 \le i \le n$, be positive constants. Then we choose δ_i such that $0 < 2\delta_i = \min\{a_i, b_i\},\quad 1 \le i \le n$. We put $2\delta_i/(1-\beta_i) = \overline{\delta}_i$, and $\overline{\delta} = \left(\frac{\delta_1}{(k_i-1)!}, \ldots, \frac{\delta_n}{(k_i-1)!}\right)^T$.

Let $T \ge \max\{\gamma(t_0), 1\}$ be such that (6.4.13) holds and

$$G_i(T, k_i, p_i^+, \overline{\delta}) < \delta_i(N - k_i - 1)!, \tag{6.4.33}$$

$$G_i(T, k_i - 1, p_i^-, \overline{\delta}) < \frac{\delta_i}{2(N - k_i)!}, \qquad 1 \le i \le n. \tag{6.4.34}$$

Let $C[T_0, \infty)$ be the space defined in the proof of Theorem 1. We consider the closed convex subset S of $C[T_0, \infty)$ defined by

$$S = \Bigg\{ Z = (z_1, \ldots, z_n)^T \in C[T_0, \infty) : \; z_i(t) = \delta_i \, \frac{T^{k_i - 1}}{(k_i - 1)!} \text{ for } t \in [T_0, T],$$

$$\frac{\delta_i t^{k_i - 1}}{2(k_i - 1)!} \le z_i(t) \le \frac{2\delta_i t^k}{(k_i - 1)!} \text{ for } t \ge T, \;\; 1 \le i \le n \Bigg\}. \tag{6.4.35}$$

With every $(z_1, \ldots, z_n)^T \in S$ we associate the functions $(x_1, \ldots, x_n)^T$ defined by the formula (6.4.25). From (6.4.25) in view of (6.4.26), (6.4.35), (6.4.2) and (6.4.3) we obtain

$$\begin{aligned} z_i(t) &\le x_i(t) \\ &\le \frac{2\delta_i}{(k_i - 1)!} \Big[t^{k_i} + \beta_i \big(h_i(t)\big)^{k_i} + \ldots \\ &\quad + \beta_i^{n_i(t) - 1} \big(H_i(n_i(t) - 1, t)\big)^{k_i} + \beta_i^{n_i(t)} \big(H_i(n_i(t)), t\big)^{k_i} \Big] \\ &\le \frac{\overline{\delta}_i t^{k_i}}{(k_i - 1)!}, \quad t \ge T, \quad 1 \le i \le n. \end{aligned} \tag{6.4.36}$$

Define an operator $\mathcal{T} = (\mathcal{T}_1, \ldots, \mathcal{T}_n)^T : S \to C[T_0, \infty)$ by

$$\mathcal{T}_i Z(t) = \begin{cases} \dfrac{\delta_i T^{k_i-1}}{(k_i-1)!}, & t \in [T_0, T], \\ \dfrac{\delta_i t^{k_i-1}}{(k_i-1)!} + (-1)^{n-k_i} \displaystyle\int_T^t \dfrac{(t-s)^{k_i-1}}{(k_i-1)!} \int_T^\infty \dfrac{(u-s)^{N-k_i-1}}{(N-k_i-1)!} & \\ \quad \times \displaystyle\sum_{j=1}^n p_{ij}(u) f_{ij}(x_j(g_{ij}(u)))\, du\, ds, & t \geq T, \quad 1 \leq i \leq n. \end{cases} \tag{6.4.37}$$

Clearly, S is a closed convex subset of $C[T_0, \infty)$. We show that $\mathcal{T}(S) \subset S$. Let $Z = (z_1, \ldots, z_n)^T \in S$. From (6.4.37) in view of (6.4.9) and (6.4.36) we get for $t \geq T, \quad 1 \leq i \leq n:$

$$\begin{aligned} \mathcal{T}_i Z(t) &\leq \frac{\delta_i t^{k_i-1}}{(k_i-1)!} + \int_T^t \frac{(t-s)^{k_i-1}}{(k_i-1)!(N-k_i-1)!} G_i(T, k_i, p_i^+, \overline{\delta}) \\ &\leq \frac{\delta_i t^{k_i-1}}{(k_i-1)!} + \frac{\delta_i (t-T)^{k_i}}{k_i!} \\ &\leq \frac{2\delta_i t^{k_i}}{(k_i-1)!}, \quad t \geq T, \quad 1 \leq i \leq n. \end{aligned} \tag{6.4.38}$$

Using (6.4.4), (6.4.35) and (6.4.34) we easily derive that

$$\begin{aligned} &\int_T^t \int_s^\infty \frac{(u-s)^{N-k_i-1}}{(N-k_i-1)!} \sum_{j=1}^n (p_{ij})_{k_i}^-(u) f_{ij}(x_j(g_{ij}(u)))\, du\, ds \\ &\qquad\leq \frac{1}{(N-k_i)!} G_i(T, k_i - 1, p_i^-, \overline{\delta}) \\ &\qquad\leq \frac{\delta_i}{2}, \quad 1 \leq i \leq n. \end{aligned} \tag{6.4.39}$$

From (6.3.37) with regard to (6.4.9) and (6.4.39) we obtain for $k_i \geq 2$, $1 \leq i \leq n$

$$\begin{aligned} \mathcal{T}_i Z(t) \geq \frac{\delta_i t^{k_i-1}}{(k_i-1)!} &- \int_T^t \frac{(t-s)^{k_i-1}}{(k_i-1)!} \\ &\times \int_s^\infty \frac{(u-s)^{N-k_i-1}}{(N-k_i-1)!} \sum_{j=1}^n (p_{ij})_{k_i}^-(u) f_{ij}(x_j(g_{ij}(u)))\, du\, ds \end{aligned}$$

$$\geq \frac{\delta_i t^{k_i-1}}{(k_i-1)!} - \int_T^t \frac{(t-\sigma)^{k_i-2}}{(k_i-2)!}$$

$$\times \int_T^\sigma \int_s^\infty \frac{(u-s)^{N-k_i-1}}{(N-k_i-1)!} \sum_{j=1}^n (p_{ij})_{k_i}^-(u) f_{ij}\big(x_j(g_{ij}(u))\big)\, du\, ds\, d\sigma$$

$$\geq \frac{\delta_i t^{k_i-1}}{(k_i-1)!} - \frac{\delta_i}{2} \int_T^t \frac{(t-\sigma)^{k_i-2}}{(k_i-2)!}\, d\sigma$$

$$\geq \frac{\delta_i t^{k_i-1}}{2(k_i-1)!}, \qquad t \geq T, \quad 1 \leq i \leq n.$$

If $k_i = 1$, then from (6.4.37) in virtue of (6.4.39) we have

$$\mathcal{T}_i Z(t) \geq \delta_i - \int_T^t \int_s^\infty \frac{(u-s)^{N-2}}{(N-2)!} \sum_{j=1}^n (p_{ij})_1^-(u)$$

$$\times f_{ij}\big(x_j(g_{ij}(u))\big)\, du\, ds$$

$$\geq \frac{1}{2}\, \delta_i, \qquad t \geq T, \quad 1 \leq i \leq n.$$

We have proved that $\mathcal{T}(S) \subset S$.

Proceeding similarly as in the proof of Theorem 6.4.1 we obtain that the operator $\mathcal{T}$ is continuous and $\mathcal{T}(S)$ has a compact closure. Therefore, by the Schauder-Tychonov fixed point theorem there exists a $\overline{Z} = (\overline{z}_1, \ldots, \overline{z}_n)^T \in S$ such that $(\mathcal{T}_1\overline{Z}, \ldots, \mathcal{T}_n\overline{Z})^T = (\overline{z}_1, \ldots, \overline{z}_n)^T$ and the components $(\overline{x}_1, \ldots, \overline{x}_n)^T$ satisfy that for $t \geq T$ the system

$$\overline{z}_i(t) = \frac{\delta_i t^{k_i-1}}{(k_i-1)!} + (-1)^{N-k_i} \int_T^t \frac{(t-s)^{k_i-1}}{(k_i-1)!} \int_s^\infty \frac{(u-s)^{N-k_i-1}}{(N-k_i-1)!}$$

$$\times \sum_{j=1}^n p_{ij}(u) f_{ij}\big(\overline{x}_j(g_{ij}(u))\big)\, du\, ds; \qquad 1 \leq i \leq n, \tag{6.4.40}$$

where $\overline{z}_i(t) = \overline{x}_i(t) - a_i(t)\overline{x}_i\big(h_i(t)\big)$.

Differentiating (6.4.40) $(k_i - 1)$-times and k_i-times, we get

$$\overline{z}_i^{(k_i-1)}(t) = \delta_i + (-1)^{N-k_i} \int_T^t \int_s^\infty \frac{(u-s)^{N-k_i-1}}{(N-k_i-1)!}$$

$$\times \sum_{j=1}^{n} p_{ij}(u) f_{ij}\big(\overline{x}_j(g_{ij}(u))\big)\, du\, ds, \quad t \geq T, \quad 1 \leq i \leq n, \tag{6.4.41}$$

$$\overline{z}_i^{(k_i)}(t) = (-1)^{N-k_i} \int_t^{\infty} \frac{(u-t)^{N-k_i-1}}{(N-k_i-1)!}$$
$$\times \sum_{j=1}^{n} p_{ij}(u) f_{ij}\big(\overline{x}_j(g_{ij}(u))\big) du\, ds, \quad t \geq T, \quad 1 \leq i \leq n, \tag{6.4.42}$$

respectively.

Then (6.4.42) implies that

$$\lim_{t\to\infty} \overline{z}_i^{(k_i)}(t) = 0. \tag{6.4.43}$$

From (6.4.41) on the basis of (6.4.9), (6.4.39), (6.4.3), (6.4.36) and (6.4.35) we conclude

$$\overline{z}_i^{(k_i-1)}(t) \geq \delta_i + \int_T^t \int_s^{\infty} \frac{(u-s)^{N-k_i-1}}{(N-k_i-1)!} \big[(p_{ih})^+_{k_i}(u) f_{ih}\big(\overline{x}_h(h_{ih}(u))\big)\big]$$
$$- \sum_{j=1}^{n} p_{ij}(u)^-_{k_i} f_{ij}\big(\overline{x}_j(g_{ij}(u))\big)\, du\, ds$$
$$\geq \frac{\delta_i}{2} + \int_T^t \frac{(u-T)^{N-k_i}}{(N-k_i)!} (p_{ih})^+_{k_i}(u) f_{ih}\big(\overline{x}_h(h_{ih}(u))\big) du$$
$$\geq \frac{\delta_i}{2} + \int_T^t \frac{(u-T)^{N-k_i}}{(N-k_i)!} (p_{ih})^+_{k_i}(u) f_{ih}\Big(\frac{\delta_h\big(g_{ih}(u)\big)^{k_h-1}}{2(k_h-1)!}\Big)\, du.$$

The last inequality together with (6.4.31) implies that

$$\lim_{t\to\infty} \overline{z}_i^{(k_i-1)}(t) = \infty. \tag{6.4.44}$$

By L'Hospital's rule, (6.4.43) and (6.4.44)

$$\lim_{t\to\infty} \frac{\overline{z}_i(t)}{t^{k_i}} = 0, \quad \lim_{t\to\infty} \frac{\overline{z}(t)}{t^{k_i-1}} = \infty. \tag{6.4.45}$$

Then from (6.4.45) in view of (6.4.2), (6.4.3) and (6.4.36) we get

$$\lim_{t\to\infty} \frac{\overline{x}_i(t)}{t^{k_i}} = 0, \quad \lim_{t\to\infty} \frac{\overline{x}_i(t)}{t^{k_i-1}} = \infty,$$

respectively.

The proof of Theorem 6.4.2 is complete. □

6.5. Notes

Theorem 6.0.1 is given by Orino and Györi [5]. The content in Section 6.1 is from Erbe and Kong [36]. Section 6.2 is based on the work by Kong and Freedman [93]. Section 6.3 is extracted from Wang [164]. Section 6.4 is a modification of the results by Marusiak [131].

7

Boundary Value Problems for Second Order Functional Differential Equations

7.0. Introduction

In this chapter we shall discuss certain boundary value problems (BVPs) associated with second order functional differential equations. Although the study of the existence and location of zeros of the solutions of ordinary differential equations is fundamental in the study of BVPs for such equations, the relation between solutions to certain BVPs for FDEs and the oscillatory behavior of such solutions is less clear. In spite of this, we shall present a number of approaches to the study of such problems, which, in some sense, parallels the corresponding techniques for BVPs associated with ODEs. Although we do not claim in any sense to be complete, it is hoped that these techniques will give an idea of what may be obtained. In Sections 7.1 and 7.2 we establish existence via Lipschitz and Nagumo-type methods, respectively. In Sections 7.3 and 7.4 we employ topological methods and Section 7.5 deals with existence and uniqueness for a certain type of singular problem.

7.1. Lipschitz Type Conditions

In this section, we shall consider the existence and uniqueness of solutions of the BVP

$$(\rho(t)\,x'(t))' = f(t, x_t, x'(t)) \tag{7.1.1}$$

$$x_0 = \varphi, \qquad x'(T) = \eta \tag{7.1.2}$$

where

$$x_t(u) = x(t+u) \qquad \text{for } u \in J = [-r, 0],$$

the function f is defined on the set

$$[0,T] \times D \times R^n, \qquad D \subset C(J),\ T > 0,$$

and where $C(J)$ is the space of all continuous bounded functions on J with values in R^n. We assume $C(J)$ is endowed with the norm $\|\cdot\|_J$ defined by the formula

$$\|\varphi\|_J = \sup_{u \in J} |\varphi(u)|.$$

Moreover, (φ, η) belongs to $D \times R^n$ and ρ is a positive continuous function defined on $[0,T]$.

The following lemma will play an important role in the proof of our results.

Lemma 7.1.1. *Let $g \in C([0,T], R^n)$, ξ, $\eta \in R^n$. Then the problem*

$$\begin{cases} (\rho(t)\,y'(t))' = g(t) \\ y(0) = \xi, \quad y'(T) = \eta \end{cases} \tag{7.1.3}$$

has a unique solution

$$y(t) = \xi + \rho(T)\,\eta\,\Phi(t) - \int_0^t \Phi(u) g(u)\,du - \Phi(t)\int_t^T g(u)\,du \tag{7.1.4}$$

where $\Phi(T) = \int_0^t \frac{ds}{\rho(s)}$.

Proof: By differentiating (7.1.4) we can easily verify that (7.1.4) is a solution of (7.1.3).

From (7.1.4) we have

$$y'(t) = \frac{\rho(T)\eta}{\rho(t)} - \frac{1}{\rho(t)} \int_t^T g(u)\,du \tag{7.1.5}$$

$$|y(t)| \leq |\xi| + |\eta|\rho(T)\Phi(T) + 2\Phi(T)T \max_{0\leq s\leq T} |g(s)| \tag{7.1.6}$$

and

$$|y'(t)| \leq \frac{\rho(T)|\eta|}{\nu} + \frac{T}{\nu} \max_{0\leq t\leq T} |g(t)| \tag{7.1.7}$$

where $\nu = \min_{0\leq t\leq T} \rho(t)$. □

Theorem 7.1.1. *Let $f \in C([0,T] \times C(J) \times R^n, R^n)$ and satisfy the Lipschitz condition*

$$|f(t,\varphi,\rho) - f(t,\overline{\varphi},\overline{\rho})| \leq \alpha_1 \|\varphi - \overline{\varphi}\|_J + \alpha_2 |\rho - \overline{\rho}| \tag{7.1.8}$$

where $|\cdot|$ is the norm on R^n and

$$\left(2\Phi(T)\alpha_1 + \frac{\alpha_2}{\nu}\right) T < 1. \tag{7.1.9}$$

Then, for every $(\varphi,\eta) \in C(J)\times R^n$ there exists a unique solution of the boundary value problem (7.1.1) and (7.1.2).

Proof: Let $(\varphi,\eta) \in C(J)\times R^n$ and consider the set S of all continuous functions $x : J \cup [0,T] \to R^n$ which are continuously differentiable on $[0,T]$ and are such that $x_0 = \varphi$ for any $x \in S$. Let x, $y \in S$ and define the distance in S by

$$\delta(x,y) = \max\{\max_{0\leq t\leq T} |x(t) - y(t)|,\ 2\Phi(T)\,\nu \max_{0\leq t\leq T} |x'(t) - y'(t)|\}. \tag{7.1.10}$$

With respect to this distance function the set S is a complete metric space. Let $x \in S$ and consider

$$\begin{cases} (\rho(t)\,y'(t))' = f(t,x_t,x'(t)) = g(t) \\ y(0) = \xi = \varphi(0), \quad y'(T) = \eta. \end{cases} \tag{7.1.11}$$

By Lemma 7.1.1 Eq. (7.1.11) has a unique solution $g(t)$ defined on $[0,T]$.

Define

$$\widetilde{x}(t) = \begin{cases} \varphi(t), & t \in J \\ y(t), & t \in (0,T]. \end{cases} \tag{7.1.12}$$

Then $\widetilde{x} \in S$. Therefore we define an operator P as follows.

$$\widetilde{x} = P(x) \qquad \text{for } x \in S. \tag{7.1.13}$$

Then $P(S) \subset S$. From (7.1.12) we have

$$\begin{aligned} (\rho(t)\,\widetilde{x}'(t))' &= f(t, x_t, x'(t)), \quad t \in [0,T] \\ \widetilde{x}_0 &= \varphi, \quad \widetilde{x}'(T) = \eta. \end{aligned} \tag{7.1.14}$$

We shall show that P is a contraction on S. In fact, let x_1, $x_2 \in S$ defined for $t \in J \cup [0,T]$ and

$$z(t) = \widetilde{x}_1(t) - \widetilde{x}_2(t) \qquad \text{for } t \in J \cup [0,T].$$

Then

$$(\rho(t)\,z'(t))' = f(t, x_{1t}, x_1'(t)) - f(t, x_{2t}, x_2'(t)), \quad t \in [0,T]$$

$$z_0 \equiv 0, \quad z'(T) = 0.$$

In view of Lemma 7.1.1, (7.1.6) and (7.1.7) we have

$$|z(t)| \leq 2\Phi(T)T \max_{0 \leq t \leq T} |f(t, x_{1t}, x_1'(t)) - f(t, x_{2t}, x_2'(t))|$$

and

$$|z'(t)| \leq \frac{T}{\nu} \max_{0 \leq t \leq T} |f(t, x_{1t}, x_1'(t)) - f(t, x_{2t}, x_2'(t))|.$$

By definition, we have

$$\begin{aligned} \delta(\widetilde{x}_1, \widetilde{x}_2) &= \max\{\max_{0 \leq t \leq T} |z(t)|,\ 2\Phi(T)\,\nu \max_{0 \leq t \leq T} |z'(t)|\} \\ &\leq \max\{2\Phi(T)T \max_{0 \leq t \leq T} |f(t, x_{1t}, x_1'(t)) - f(t, x_{2t}, x_2'(t))|, \end{aligned}$$

$$2\Phi(T)T \max_{0\le t\le T} |f(t,x_{1t},x_1'(t)) - f(t,x_{2t},x_2'(t))|\},$$

$$\le 2\Phi(T)T\big[\alpha_1 \max_{0\le t\le T} \|x_{1t} - x_{2t}\|_J + \alpha_2 \max_{0\le t\le T} |x_1'(t) - x_2'(t)|\big]$$

$$= 2\Phi(T)T\big[\alpha_1 \max_{0\le t\le T} |x_1(t) - x_2(t)| + \alpha_2 \max_{0\le t\le T} |x_1'(t) - x_2'(t)|\big]$$

$$\le 2\Phi(T)T\left(\alpha_1 + \frac{\alpha_2}{2\Phi(T)\nu}\right) \max\{ \max_{0\le t\le T} |x_1(t) - x_2(t)|,$$

$$2\Phi(T)\,\nu \max_{0\le t\le T} |x_1'(t) - x_2'(t)|\}$$

$$= 2\Phi(T)T\left(\alpha_1 + \frac{\alpha_2}{2\Phi(T)\nu}\right) \delta(x_1,x_2). \tag{7.1.15}$$

The conditions (7.1.9) and (7.1.15) imply that P is a contraction on S. That is, there is a unique element $x \in S$ such that $P(x) = x$. This x is a solution of the boundary value problem (7.1.1) and (7.1.2). The proof is complete. □

One would be interested in finding the largest possible interval in which the preceding result is valid. In the following we consider this problem in one dimension. We modify the definition of the norm by introducing a weight function. That is, we define the distance in S by

$$\delta(x,y) = \max\left\{ \max_{0\le t\le T} \frac{|x(t)-y(t)|}{w(t)}, \max_{0\le t\le T} \frac{|x'(t)-y'(t)|}{v(t)} \right\}, \tag{7.1.16}$$

where w and v are positive continuous functions on $[0,T]$.

Theorem 7.1.2. *If*

$$(\rho(t)\,u'(t))' + \alpha_2 u'(t) + \alpha_1 u(t) = 0 \tag{7.1.17}$$

where α_1, α_2 are as in (7.1.8), has a solution with $u(0) = 0$, $u'(t) > 0$ on $[0,T]$, then (7.1.1) and (7.1.2) has a unique solution. This result is best possible.

Proof: By continuous dependence of solutions on parameters, there exists α sufficiently close to but less than 1 such that

$$(\rho(t)w'(t))' + \frac{1}{\alpha}(\alpha_2 w'(t) + \alpha_2 w(t)) = 0 \tag{7.1.18}$$

has a solution with $w(0) > 0$, $w'(t) > 0$ on $[0, T]$. Integrating (7.1.18) from t to T with $t < T$ we have

$$\rho(T)w'(T) + \frac{1}{\alpha}\int_t^T (\alpha_2 w'(s) + \alpha_1 w(s))\, ds = \rho(t)w'(t)$$

or

$$\frac{1}{\alpha}\int_t^T (\alpha_2 w'(s) + \alpha_1 w(s))\, ds \le \rho(t)w'(t). \tag{7.1.19}$$

Now we consider the map defined by (7.1.13) and calculate $\delta(\widetilde{x}_1, \widetilde{x}_2)$ according to the definition (7.1.16) and using the formula (7.1.4). It is easy to see that

$$\begin{aligned}
&\frac{|\widetilde{x}_1(t) - \widetilde{x}_2(t)|}{w(t)} \\
&\quad \le \frac{1}{w(t)}\int_0^t \Phi(u)\,|f(u, x_{1u}, x_1'(u)) - f(u, x_{2u}, x_2'(u))|\, du \\
&\quad + \frac{\Phi(t)}{w(t)}\int_t^T |f(u, x_{1u}, x_1'(u)) - f(u, x_{2u}, x_2'(u))|\, du \\
&\quad \le \frac{1}{w(t)}\int_0^t \Phi(u)\left[w(u)\,\frac{\alpha_1\|x_{1u} - x_{2u}\|_J}{w(u)} + v(u)\,\frac{\alpha_2|x_1'(u) - x_2'(u)|}{v(u)}\right] du \\
&\quad + \frac{\Phi(t)}{w(t)}\int_t^T \left(\alpha_1 w(u)\frac{\|x_{1u} - x_{2u}\|_J}{w(u)} + \alpha_2 v(u)\frac{|x_1'(u) - x_2'(u)|}{v(u)}\right) du \\
&\quad \le \frac{\delta(x_1, x_2)}{w(t)}\left[\int_0^t \Phi(u)\{\alpha_1 w(u) + \alpha_2 v(u)\}\, du\right. \\
&\quad \left. + \Phi(t)\int_t^T (\alpha_1 w(u) + \alpha_2 v(u))\, du\right].
\end{aligned} \tag{7.1.20}$$

Similarly, from (7.1.5) we obtain

$$\frac{|\widetilde{x}_1'(t) - \widetilde{x}_2'(t)|}{v(t)} \le \frac{\delta(x_1, x_2)}{v(t)\rho(t)}\int_t^T (\alpha_1 w(u) + \alpha_2 v(u))\, du. \tag{7.1.21}$$

We require that

$$\int_0^t \Phi(u)[\alpha_1 w(u) + \alpha_2 v(u)]\, du + \Phi(t)\int_t^T (\alpha_1 w(u) + \alpha_2 v(u))\, du \le \alpha w(t) \tag{7.1.22}$$

and

$$\frac{1}{v(t)\rho(t)}\int_t^T (\alpha_1 w(u) + \alpha_2 v(u))\,du \le \alpha < 1. \tag{7.1.23}$$

If we choose $v(t) = w'(t)$, then (7.1.23) follows from (7.1.19). Integrating (7.1.19) from 0 to $t < T$ we can obtain (7.1.22). With this choice of α, w, v (7.1.22) and (7.1.23) hold. This implies that

$$\delta(\widetilde{x}_1, \widetilde{x}_2) \le \alpha\,\delta(x_1, x_2), \qquad \alpha < 1.$$

That is, P defined by (7.1.13) is a contraction. Therefore (7.1.1)–(7.1.2) has a unique solution.

If T is so large that (7.1.17) has a solution $u(t)$ such that $u(0) = 0$, $u'(t) > 0$ for $t \in [0, T)$ and $u'(T) = 0$, then $u(t)$ is a nontrivial solution of (7.1.17) with $u(0) = 0$, $u'(T) = 0$. On the other hand $u(t) \equiv 0$ is a solution of this boundary value problem which violates the uniqueness of the solutions. This shows that the result is best possible.

Next we consider the general linear boundary value problem

$$\begin{cases} (\rho(t)x'(t))' = f(t, x_t, x'(t)) \\ x_0 = \varphi, \quad ax(T) + bx'(T) = A \in R^n \end{cases} \tag{7.1.24}$$

where a, b, A are given. We assume that

$$\rho(T)\Phi(T) + \frac{b}{a} \ne 0. \tag{7.1.25}$$

We first prove the following.

Lemma 7.1.2. *Let*

$$g \in C([0, T], R^n), \qquad \xi,\ A \in R^n,\ a,\ b \in R.$$

Then the problem

$$\begin{cases} (\rho(t)y'(t))' = g(t) \\ y(0) = \xi, \quad ay(T) + by'(T) = A \end{cases} \tag{7.1.26}$$

has a unique solution

$$\begin{aligned} y(t) = \xi + \rho(T) \int_0^t \frac{ds}{\rho(s)} \Bigg\{ \left(\frac{A}{a} - \xi \right) \left(\rho(T)\Phi(T) + \frac{b}{a} \right)^{-1} \\ + \left(\int_0^T \frac{1}{\rho(s)} \int_s^T g(u)\, du\, ds \right) \left(\rho(T)\Phi(T) + \frac{b}{a} \right)^{-1} \Bigg\} \\ - \int_0^t \frac{1}{\rho(s)} \int_s^T g(u)\, du\, ds. \end{aligned} \tag{7.1.27}$$

In fact, by differentiating we can prove that (7.1.27) satisfies (7.1.26). The uniqueness is obvious.

For the sake of convenience we put

$$\left| \left(\rho(T)\Phi(T) + \frac{b}{a} \right)^{-1} \right| = c.$$

From (7.1.27) it is not difficult to see that

$$\begin{aligned} |y(t)| \leq |\xi|(1 + c\rho(T)\Phi(T)) + \rho(T)\Phi(T) \left| \frac{A}{a} \right| \\ + (1 + c\rho(T)\Phi(T)) + \Phi(T) \max_{0 \leq t \leq T} |g(u)| \end{aligned} \tag{7.1.28}$$

and

$$\begin{aligned} |y'(t)| &\leq \left(\frac{c}{|a|} |A| + c|\xi| + cT\Phi(T) \max_{0 \leq u \leq T} |g(u)| \right) \frac{\rho(T)}{\nu} + \frac{T}{\nu} \max_{0 \leq u \leq T} |g(u)| \\ &\leq \frac{\rho(T)}{\nu} \left(\frac{c}{|a|} |A| + c|\xi| \right) + \frac{T}{\nu} (c\Phi(T)\rho(T) + 1) \max_{0 \leq u \leq T} |g(u)|. \end{aligned} \tag{7.1.29}$$

Similar to Theorem 7.1.1 we can prove the following result.

Theorem 7.1.3. *Let $f \in C([0,T] \times C(J) \times R^n,\ R^n)$ and assume (7.1.8) holds. Further suppose that*

$$(1 + c\rho(T)\Phi(T)) + \left(\alpha_1 \Phi(T) + \frac{\alpha_2}{\nu} \right) < 1.$$

Then, for every $(\varphi, A) \in C(J) \times R^n$, there is a unique solution of (7.1.24).

Remark. Theorem 7.1.2 is also true for problem (7.1.24).

7.2. Nagumo Type Condition

We consider the following boundary value problem

$$\begin{cases} x''(t) + f(t, x_t, x'(t)) = 0 \\ x(t) = h(t), \quad -r \le t \le 0, \ h(0) = 0, \ x(T) = 0 \end{cases} \tag{7.2.1}$$

and

$$|x(t)| \le \varphi(t)$$

where $f \in C([0,T] \times C(J) \times R^n,\ R^n)$ and φ is a given specified function.

The following lemmas will be used to prove the existence result for (7.2.1).

Lemma 7.2.1. *Let $x\colon [0,T] \to R^n$ be an absolutely continuous function with an absolutely continuous derivative. Assume that*

$$|\langle x'(t), x''(t)\rangle| \le H(|x'(t)|)|x'(t)| \tag{7.2.2}$$

for almost all $t \in [0,T]$ where $\langle \cdot, \cdot \rangle$ denotes the scalar product, $H : R^+ \to R^+ \backslash \{0\}$ is continuous and satisfies

$$\int_0^\infty \frac{s^2\, ds}{H(s)} = \infty. \tag{7.2.3}$$

Then

$$|x'(t)| \le g\left(\int_0^T |x'(t)|^2\, dt\right), \qquad t \in [0,T] \tag{7.2.4}$$

where g is defined by

$$\int_{\sqrt{\frac{z}{T}}}^{g(z)} \frac{s^2\, ds}{H(s)} = z. \tag{7.2.5}$$

Proof: Let $E = \{t \in [0,T] \colon |x'(t)| \ne 0\}$. Then, for *a.e.* $t \in E$

$$|x'(t)| \ge \frac{|\langle x'(t), x''(t)\rangle|}{H(|x'(t)|)}$$

and hence, for $v,\ u \in [0,T]$

$$\begin{aligned}
\int_0^T |x'(t)|^2\,dt = \int_E |x'(t)|^2\,dt &\geq \int_E \frac{|x'(t)|\,|\langle x'(t), x''(t)\rangle|}{H(|x'(t)|)}\,dt \\
&= \int_0^T \frac{|x'(t)|\,|\langle x'(t), x''(t)\rangle|}{H(|x'(t)|)}\,dt \\
&\geq \left| \int_u^v \frac{|x'(t)|\,|\langle x'(t), x''(t)\rangle|}{H(|x'(t)|)}\,dt \right| \\
&= \left| \int_{|x'(u)|}^{|x'(v)|} \frac{s^2\,ds}{H(s)} \right|.
\end{aligned}$$

By the Mean Value Theorem there exists $u \in [0,T]$ such that

$$T|x'(u)|^2 = \int_0^T |x'(t)|^2\,dt$$

and the continuity of x' implies the existence of $v \in [0,T]$ such that

$$|x'(v)| = \max_{t\in[0,T]} |x'(t)|.$$

Set

$$w = \int_0^T |x'(t)|^2\,dt.$$

Then (7.2.5) gives

$$w = \int_{\sqrt{w/T}}^{g(w)} \frac{s^2\,ds}{H(s)} \geq \left| \int_{\sqrt{w/T}}^{|x'(v)|} \frac{s^2\,ds}{H(s)} \right|.$$

Hence

$$|x'(v)| \leq g(w) \tag{7.2.6}$$

which implies that (7.2.4) holds.

The Green's function associated with the boundary value problem

$$x''(t) = 0, \qquad x(0) = 0 = x(T)$$

is

$$G(t,s) = \begin{cases} \left(1-\frac{t}{T}\right)s, & 0 \le s \le t \le T \\ t\left(1-\frac{s}{T}\right), & 0 \le t \le s \le T. \end{cases} \tag{7.2.7}$$

Then the boundary value problem (7.2.1) is equivalent to the integral equation

$$x(t) = \int_0^T G(t,s) f(s, x_s, x'(s))\, ds. \tag{7.2.8}$$

Let B be the Banach space of all continuously differentiable functions $x\colon [0,T] \to R^n$ with the norm

$$\|x\|^* = \max\{\sup_{t\in[0,T]} |x(t)|, \sup_{t\in[0,T]} |x'(t)|\}.$$

The main existence result is the following.

Theorem 7.2.1. *Assume that there exist $\varphi \in C^2([0,T], R^+ - \{0\})$ and $F \in C([0,T] \times R \times R, R)$ satisfying the following condition: for any (t,u,v) with $t \in [0,T]$, $\|u\| = \varphi(T)$, $\langle u(0), v\rangle = \|u\|\varphi'(t)$, one has*

(i) $\langle u(0), f(t,u,v)\rangle \le \varphi(t)F(t,\varphi(t),\varphi'(t)) + |v|^2 - \varphi'(t)^2$

(ii) $\varphi''(t) + F(t,\varphi(t),\varphi'(t)) \le 0$.

Assume moreover, that there exist numbers $a \in [0,1)$ and $b \ge 0$ such that, for any (t,u,v) with $t \in [0,T]$, $\|u\| \le \varphi(t)$ and $v \in R^n$

(iii) $\langle u(0), f(t,u,v)\rangle \le a|v|^2 + b$

(iv) $|\langle v, f(t,u,v)\rangle| \le H(|v|)|v|$

where $H : R^+ \to R^+ - \{0\}$ is an increasing continuous function which satisfies the condition

$$\int_0^\infty \frac{s^2\, ds}{H(s)} = \infty. \tag{7.2.9}$$

Then the boundary value problem (7.2.1) has at least one solution x such that $|x(t)| \le \varphi(t)$ for all $t \in [0,T]$.

Proof: Define a mapping $S : B \to B$ by

$$(Sy)(t) = \int_0^T G(t,s) f(s, y_s, y'(s))\, ds. \tag{7.2.10}$$

By (iii) and (iv), f maps bounded subsets of $[0,T] \times C(J) \times R^n$ into bounded subsets of R^n, so the map S defined by (7.2.10) is compact.

Now define $A : B \to B$ and the set Ω by

$$(Ax)(t) = -\int_0^T \alpha^2 G(t,s) x(s)\, ds, \qquad \alpha \neq 0 \tag{7.2.11}$$

and

$$\Omega_\eta = \{x \in B,\ \forall t \in [0,T],\ |x(t)| < \varphi(t),\ |x'(t)| < \eta\} \tag{7.2.12}$$

for a certain $\eta > 0$.

Clearly, A is completely continuous, Ω_η is open and bounded in B.

We will show that $I - A$ is one-to-one and that if η and α are chosen large enough, the equation

$$x = \lambda S x + (1-\lambda) A x, \qquad \lambda \in (0,1) \tag{7.2.13}$$

has no solution which belongs to $\partial\Omega_\eta$. Then S has a fixed point $x \in \overline{\Omega}$ by Theorem 1.4.8.

In order to prove the linear operator $I - A$ is one-to-one, we must show that $x = Ax$ implies $x = 0$. In fact, if x satisfies the equation $x - Ax = 0$, then it is a solution of the boundary value problem

$$\begin{aligned} x''(t) &= \alpha^2\, x(t) \\ x(0) &= 0 = x(T). \end{aligned} \tag{7.2.14}$$

This problem has the unique solution $x = 0$. Therefore the operator $I - A$ is one-to-one.

Finally, we prove that if $x \in \partial\Omega_\eta$ is a solution of (7.2.13), then there exists no $\xi \in (0,T)$ such that either $|x(t)|^2 - \varphi^2(t)$ reaches the maximum value 0 at $t = \xi$ or $|x'(\xi)| = \eta$.

Assume that $|x(t)|^2 - \varphi^2(t)$ reaches the maximum value 0 at $\xi \in (0,T)$. Then

$$|x(\xi)| = \varphi(\xi), \qquad \langle x(\xi), x'(\xi)\rangle - \varphi(\xi)\varphi'(\xi) = 0$$

$$\langle x_\xi(0), x'(\xi)\rangle - \varphi(\xi)\varphi'(\xi) = 0, \qquad \langle u(0), x'(\xi)\rangle - \varphi(\xi)\varphi'(\xi) = 0 \quad (7.2.15)$$

$$J \equiv \langle u(0), x'(\xi)\rangle + |x'(\xi)|^2 - \varphi''(\xi)\varphi(\xi) - \varphi'^2(\xi) \le 0 \qquad (7.2.16)$$

Since x is a solution of (7.2.13), That is,

$$x''(t) + \lambda f(t, x_t, x'(t)) = (1-\lambda)\alpha^2 x(t). \qquad (7.2.17)$$

For $\lambda \in (0,1)$, by our assumptions and (7.2.15) we have

$$\begin{aligned} J &\equiv \langle u(0), -\lambda f(t,u,v)\rangle + |v|^2 - \varphi''\varphi - \varphi'^2 + (1-\lambda)\alpha^2\langle u(0), u(0)\rangle \\ &= -\langle u(0), f(t,u,v)\rangle + (1-\lambda)\alpha^2\varphi^2 + |v|^2 - \varphi''\varphi - \varphi'^2 \\ &\ge \lambda[-\langle u(0), f(t,u,v)\rangle + \varphi F + |v|^2 - \varphi'^2] \\ &\qquad + (1-\lambda)\left[|v|^2 + \varphi\{\alpha^2\varphi - \varphi''\} - \frac{\langle u(0), v\rangle^2}{\|u\|^2}\right] \\ &= J_1 + (1-\lambda)J_2. \end{aligned}$$

But $J_1 \ge 0$ by (i) and $J_2 \ge 0$ by the Cauchy-Schwarz inequality, if α^2 is large enough. Then $J > 0$ which contradicts (7.2.16).

By Theorem 1.4.8, S defined by (7.2.10) has a fixed point y. Then

$$x(t) = \begin{cases} h(t), & -r \le t \le 0 \\ y(t), & 0 \le t \le T \end{cases} \qquad (7.2.18)$$

is a solution of the boundary value problem (7.2.1).

Now we will show that for any solution $x(t)$ of (7.2.13) $|x(t)| \le \varphi(t)$, $|x'(t)|$ has an upper bound which does not depend on $\lambda \in [0,1]$.

Let $x(t)$ be a solution of (7.2.13), multiply both sides of (7.2.17) by $x(t)$, and integrate by parts over $[0,T]$. We have

$$\langle x(t), x''(t)\rangle + \lambda\langle u(0), f(t,u,v)\rangle = (1-\lambda)\alpha^2\langle u(0), u(0)\rangle$$

$$-\int_0^T |x'(t)|^2\,dt + \lambda\int_0^T \langle u(0), v(t,u,v)\rangle\,dt = (1-\lambda)\alpha^2\int_0^T |x(t)|^2\,dt$$

or

$$-\int_0^T |x'(t)|^2\,dt + \lambda bT + \alpha\lambda \int_0^T |x'(t)|^2\,dt \geq 0.$$

Hence

$$\int_0^T |x'(t)|^2\,dt \leq K$$

and by Lemma 7.2.1

$$|x'(t)| \leq g(K), \qquad t \in [0,T].$$

Choose $\eta > g(K)$. The fixed point y of S belongs to $\overline{\Omega}$. Thus $|x(t)| \leq \varphi(t)$. The proof is complete. □

Example 7.2.1. Consider the linear equation

$$x''(t) + q(t)L(x_t) + m(t)x'(t) = 0 \tag{7.2.19}$$

where $q(t)$, $m(t)$ are bounded functions with bounds Q and M respectively and $L(\varphi)$ is a linear bounded operator in $C(J)$. We shall show that the assumptions of Theorem 7.2.1 are satisfied for (7.2.19). In fact, let $\|L\|$ be the norm of L. Set $A = Q\|L\|$ and define F by

$$F(t,x,y) = \left(A + \frac{M^2}{4}\right)\|x\| + M|y|. \tag{7.2.20}$$

Thus we have

$$\langle u(0), f(t,u,v)\rangle = q(t)\,\langle L(u), u(0)\rangle + m(t)u(0)v \leq A\|u\|^2 + Mvu(0) \tag{7.2.21}$$

$$\begin{aligned} 0 &\leq \left[\frac{M}{2}\|u\| - \frac{|\langle u(0), v\rangle|}{\|u\|} + |v|\right]^2 \\ &\leq \frac{M^2}{4}\|u\|^2 + \frac{|\langle u(0), v\rangle|^2}{\|u\|^2} + |v|^2 - M|\langle u(0), v\rangle| + M\|u\|\,|v| \\ &\quad - 2\frac{|\langle u(0), v\rangle|}{\|u\|}|v| \end{aligned}$$

$$\leq \frac{M^2}{4}\|u\|^2 + \frac{|\langle u(0), v\rangle|}{\|u\|}\left\{\frac{|\langle u(0), v\rangle|}{\|u\|} - 2|v|\right\}$$
$$+ |v|^2 - M|\langle u(0), v\rangle| + M\|u\|\,|v|$$
$$\leq \frac{M^2}{4}\|u\|^2 - \frac{|\langle u(0), v\rangle|^2}{\|u\|^2} + |v|^2 - M|\langle u(0), v\rangle| + M\|u\|\,|v|.$$

Thus

$$\langle u(0), f(t,u,v)\rangle \leq A\|u\|^2 + \frac{M^2}{4}\|u\|^2 - \frac{|\langle u(0), v\rangle|^2}{\|u\|^2} + |v|^2 + M\|u\|\,|v|$$
$$= \|u\|\left\{\left(A + \frac{M^2}{4}\right)\|u\| + M|v|\right\} + |v|^2 - \frac{|\langle u(0), v\rangle|^2}{\|u\|^2}$$
$$= \varphi F(t, \varphi, \varphi') + |v|^2 - \varphi'^2. \tag{7.2.22}$$

That is, condition (i) is satisfied for $\varphi = \|u\|^2$; condition (ii) is also satisfied, as may be easily verified.

For (iii) we notice that

$$\langle u(0), f(t,u,v)\rangle = q(t)\langle L(u), u(0)\rangle + m(t)u(0)v$$
$$\leq A\|u\|^2 + Mv|u(0)| \leq A^* + B^*|v| \tag{7.2.23}$$

where $A^* = AC$, $B^* = MC$, C is the bound of φ. If $|v| < 1$, then (iii) is obvious from (7.2.23). If $|v| > 1$, then (iii) follows from the inequality

$$B^* \geq B^*|v| - B_1|v|^2, \qquad \text{for each } B_1 \geq 0.$$

Indeed, we have

$$A^* + B^*|v| = A + B_1|v|^2 + B^*|v| - B_1|v|^2 \leq A^* + B^* + B_1|v|^2.$$

Finally, it is easy to see that (iv) is satisfied.

Hence (7.2.19) with

$$x(t) = h(t), \qquad -r \leq t \leq 0,\ h(0) = 0,\ x(T) = 0$$

and

$$x(t) \le \varphi(t)$$

has a solution.

7.3. Leray-Schauder Alternative

We shall now consider the boundary value problem

$$x''(t) = f(t, x_t, x'(t)), \qquad t \in [0,T] \tag{7.3.1}$$

$$\alpha_0 x_0 - \alpha_1 x'(0) = \varphi \in C([-r,0], R^n) \tag{7.3.2}$$

$$\beta_0 x(T) + \beta_1 x'(T) = A \in R^n$$

where $f \in C([0,T] \times C(J) \times R^n, R^n)$, $C(J) = C([-r,0])$, α_0, α_1, β_0, β_1 are nonnegative real constants such that

$$\ell = \alpha_0\, \beta_0 T + \alpha_0 \beta_1 + \alpha_1 \beta_0 \neq 0. \tag{7.3.3}$$

We assume that $\alpha_0 > 0$. In the case where $\alpha_0 = 0$, (7.3.2) is replaced by

$$x_0 = x(0)$$

$$-\alpha_1 x'(0) = \varphi \tag{7.3.4}$$

$$\beta_0\, x(T) + \beta_1\, x'(T) = A$$

We omit the discussion for this case here.

The key tool is the Leray-Schauder alternative. This method reduces the problem of existence of solutions of a BVP to the establishment of suitable *a priori* bounds for solutions of these problems.

The Green's function with respect to the BVP

$$x''(t) = 0, \qquad t \in [0,T]$$

$$\alpha_0 x(0) - \alpha_1 x'(0) = 0 \tag{7.3.5}$$

$$\beta_0 x(T) + \beta_1 x'(T) = 0$$

is

$$G(t,s) = \frac{1}{\ell}\begin{cases} (\beta_0 t - \beta_0 T - \beta_1)(\alpha_0 s + \alpha_1), & 0 \le s \le t \le T \\ (\alpha_0 t + \alpha_1)(\beta_0 s - \beta_0 T - \beta_1), & 0 \le t \le s \le T. \end{cases} \tag{7.3.6}$$

The following is an existence result for BVP (7.3.1) and (7.3.2).

Theorem 7.3.1. *Assume that there exists a constant K such that*

$$\|x\|_{[-r,T]} \le K \quad \text{and} \quad \|x'\|_{[0,T]} \le K$$

for every solution x of the BVP for the equation

$$x''(t) = \lambda f(t, x_t, x'(t)), \qquad t \in [0,T] \tag{7.3.1$_\lambda$}$$

with (7.3.2) and $\lambda \in (0,1)$. Then the BVP (7.3.1) and (7.3.2) has at least one solution.

Proof: First we assume that $\varphi(0) = 0$. We let X denote the space $C^1[0,T]$ with the norm $\|x\|_1 = \max\{\|x\|_{[0,T]}, \|x'\|_{[0,T]}\}$. Then, the set $C = \{x \in X : \alpha_0 x(0) - \alpha_1 x'(0) = 0\}$ is a subspace of X.

Define a mapping $F : C \to X$ as follows:

$$Fx(t) = \int_0^T G(t,s)\, f(s, x_s, x'(s))\, ds + \frac{1}{\ell}(\alpha_1 + \alpha_0 t)A, \qquad t \in [0,T] \tag{7.3.7}$$

where

$$x_s(\theta) = \begin{cases} x(s+\theta), & \theta \ge -s \\ \frac{1}{\alpha_0}[\alpha_1 x'(0) + \varphi(s+\theta)], & \theta < -s. \end{cases} \tag{7.3.8}$$

Clearly, $F(C) \subset C$.

We shall show that F is completely continuous.

Indeed, let B be a bounded subset of C. Then there exists $b \ge 0$ such that $\|x\|_1 \le b$, $x \in B$. Moreover, for any t_1, t_2 in $[0,T]$ and $x \in B$ we have

$$|x(t_1) - x(t_2)| = \left[\sum_{i=1}^{n} (x_i(t_1) - x_i(t_2))^2\right]^{\frac{1}{2}} \le b\sqrt{n}\, |t_1 - t_2|.$$

Therefore, B is an equicontinuous set.

Now we consider the subset $\widehat{B}$ of the space $C(J)$ defined by

$$\widehat{B} = \{x_t\colon x \in B,\ t \in [0,T]\}$$

where x_t is defined by (7.3.8). We shall show that there exists a compact subset D of $C(J)$ such that

$$\widehat{B} \subseteq D.$$

In order to show that, it suffices to prove that the set $\widehat{B}$ is uniformly bounded and equicontinuous. In fact, for any $x \in B$ and $t \in [0,T]$

$$\sup_{\theta\in[-r,0]} |x_t(\theta)| = \sup_{\theta\in[-r,0]} |x(t+\theta)| \le \widehat{b},$$

where

$$\widehat{b} = \max\left\{b,\ \frac{|\alpha_1|}{|\alpha_0|}\,(b + \|\varphi\|_{[-r,0]})\right\}$$

and hence $\widehat{B}$ is uniformly bounded.

Moreover, for any $x \in B$, $t \in [0,T]$ and θ_1, θ_2 in $[-r,0]$ we have

$$|x_t(\theta_1) - x_t(\theta_2)| = \begin{cases} |x(t+\theta_1) - x(t+\theta_2)|, & \text{if } t+\theta_1 \ge 0,\ t+\theta_2 \ge 0 \\ |\frac{1}{\alpha_0}[\varphi(t+\theta_1) + \alpha_1 x'(0)] & \\ \qquad -x(t+\theta_2)|, & \text{if } t+\theta_1 < 0,\ t+\theta_2 \ge 0 \\ |\frac{1}{\alpha_0}[\varphi(t+\theta_1) - \varphi(t+\theta_2)]|, & \text{if } t+\theta_1 < 0,\ t+\theta_2 < 0 \\ |x(t+\theta_1) - \frac{1}{\alpha_0}[\varphi(t+\theta_2) & \\ \qquad +\alpha_1 x'(0)]|, & \text{if } t+\theta_1 > 0,\ t+\theta_2 \le 0. \end{cases}$$

For any $\varepsilon > 0$, if $t+\theta_1 \ge 0$, $t+\theta_2 \ge 0$, by the equicontinuity of B, there exists $\delta = \delta(\varepsilon)$ such that for every $x \in B$, $t \in [0,T]$, θ_1, $\theta_2 \in [-r,0]$ we have

$$|x_t(\theta_1) - x_t(\theta_2)| = |x(t+\theta_1) - x(t+\theta_2)| \le \varepsilon$$

provided $|\theta_1 - \theta_2| < \delta(\varepsilon)$.

If $t+\theta_1<0$, $t+\theta_2\geq 0$, then

$$\left|\frac{1}{\alpha_0}[\varphi(t+\theta_1)+\alpha_1 x'(0)]-x(t+\theta_2)\right| \leq \left|\frac{1}{\alpha_0}[\varphi(t+\theta_1)-\varphi(0)]\right|+|x(t+\theta_2)-x(0)|.$$

But, by the uniform continuity of the function φ on $[-r,0]$, the equicontinuity of the set B and for the given $\varepsilon>0$, there exist $\delta'(\varepsilon)>0$ and $\delta''(\varepsilon)>0$ so that

$$|\varphi(s_1)-\varphi(s_2)|<\frac{\varepsilon\,\alpha_0}{2}, \quad \text{provided } |s_1-s_2|<\delta'(\varepsilon)$$

and

$$|x(s_1)-x(s_2)|<\frac{\varepsilon}{2}, \quad \text{provided } |s_1-s_2|<\delta''(\varepsilon).$$

Hence, if $|\theta_1-\theta_2|<\min\{\delta'(\varepsilon),\,\delta''(\varepsilon)\}=\widehat{\delta}$, then

$$|t+\theta_1-t-\theta_2|<\widehat{\delta} \implies |t+\theta_1-0|<\delta'(\varepsilon), \qquad |t+\theta_2-0|<\delta''(\varepsilon)$$

and so

$$\left|\frac{1}{\alpha_0}\,[\varphi(t+\theta_1)-\varphi(0)]\right|+|x(t+\theta_2)-x(0)|<\frac{\varepsilon}{2}+\frac{\varepsilon}{2}=\varepsilon.$$

The rest of the cases are similar. Therefore, for every $\varepsilon>0$ there exists $\delta=\delta(\varepsilon)$ so that

$$|x_t(\theta_1)-x_t(\theta_2)|<\varepsilon, \quad \text{provided } |\theta_1-\theta_2|<\delta(\varepsilon)$$

for every $x\in B$ and $t\in[0,T]$, which proves the equicontinuity of the set $\widehat{B}$. So, there exists a compact subset D of $C(J)$ such that

$$\widehat{B}\subseteq D.$$

Now we consider a bounded sequence $\{x_\nu\}$ in C. Then, as is known, for every $t\in[0,T]$ the sequence $\{x_{t\nu}\}$ is bounded in $C(J)$ and, moreover, there exists a compact subset D in $C(J)$ such that $x_{t\nu}\in D$ for every ν and $t\in[0,T]$. Thus, if d is the bound of $\{x_\nu\}$, it is obvious that the set $V=[0,T]\times D\times\overline{B}(0,d)$ is compact in $[0,T]\times C(J)\times R^n$, where $\overline{B}(0,d)$ is the closed ball with center 0 and radius d in R^n. We set

$$\theta=\max\{|f(t,u,v)|\colon (t,u,v)\in V\}$$

and

$$\int_0^T |G(t,s)|\,ds \le K_1, \qquad \int_0^T |G_t(t,s)|\,ds \le K_2$$

and set

$$\widehat{K} = \max\{\theta K_1, \theta K_2\}.$$

Then we have

$$\|Fx_\nu\|_{[0,T]} \le \widehat{K}, \qquad \|(Fx_\nu)'\|_{[0,T]} \le \widehat{K}.$$

Hence, for any t_1, t_2 in $[0,T]$ and arbitrary ν we have

$$|Fx_\nu(t_1) - Fx_\nu(t_2)| = \left|\int_{t_1}^{t_2} (Fx_\nu)'(s)\,ds\right| \le \widehat{K}\,|t_2 - t_1|$$

$$|(Fx_\nu)'(t_1) - (Fx_\nu)'(t_2)| = \left|\int_{t_1}^{t_2} (Fx_\nu)''(s)\,ds\right| \le \theta|t_1 - t_2|,$$

i.e., the sequences $\{Fx_\nu\}$, $\{(Fx_\nu)'\}$ are equicontinuous. That is, F is completely continuous.

Finally, we observe that by assumptions the set $\mathcal{E}(F) = \{x \in C\colon x = \lambda Fx$ for some
$\lambda \in (0,1)\}$ is bounded. Hence the assumptions of Theorem 1.4.10 are satisfied and hence the operator F has a fixed point $x \in C$. Then it is clear that the function

$$z(t) = \begin{cases} x(t), & t \in [0,T] \\ \frac{1}{\alpha_0}\,[\varphi(t) + \alpha_1 x'(0)], & t \in [-r,0] \end{cases} \tag{7.3.9}$$

is a solution of the BVP (7.3.1) and (7.3.2).

If $\varphi(0) \ne 0$, by the transformation

$$y = x - \frac{\varphi(0)}{\alpha_0}$$

the BVP (7.3.1) and (7.3.2) is reduced to the BVP

$$y''(t) = f(t, y_t + \tfrac{\varphi(0)}{\alpha_0}, y'(t)) \equiv \widehat{f}(t, y_t, y'(t)), \qquad t \in [0,T]$$

$$\alpha_0 y_0 - \alpha_1 y'(0) = \varphi - \varphi(0) \equiv \widehat{\varphi}$$

$$\beta_0 y(T) + \beta_1 y'(T) = A + \frac{\beta_0}{\alpha_0}\varphi(0), \tag{7.3.10}$$

where, obviously, $\widehat{\varphi}(0) = 0$. The proof is complete. □

Theorem 7.3.1 reduces the BVP (7.3.1) and (7.3.2) to the problem of finding $\max_{[-r,T]} |x(t)|$ and $\max_{[0,T]} |x'(t)|$, where x is the solution of $(7.3.1)_\lambda$ and (7.3.2). In the following we will show some results for these bounds.

Theorem 7.3.2. *Assume that*
(H_1) *There exists a constant $M > 0$ such that for every $(t,u,v) \in [0,T] \times C(J) \times R^n$ with $|u(0)| > M$ and $\langle u(0), v\rangle = 0$ implies*

$$\langle u(0), f(t,u,v)\rangle > 0. \tag{7.3.11}$$

Then every solution x of the BVP $(7.3.1)_\lambda$ and (7.3.2) satisfies

$$\max_{t\in[-r,T]} |x(t)| \le M_0 \equiv \frac{2}{\alpha_0}\|\varphi\|_{[-r,0]} + \widehat{M_0} \tag{7.3.12}$$

where

$$\widehat{M_0} = \begin{cases} \max\{M, \frac{1}{\alpha_0}|\varphi(0)|, \frac{1}{\beta_0}|A|\}, & \text{if } \alpha_0,\ \beta_0 \ne 0 \\ \max\{M, \frac{1}{\alpha_0}|\varphi(0)|\}, & \text{if } \beta_0 = 0. \end{cases} \tag{7.3.13}$$

Proof: Assume that $\lambda \in (0,1]$. Let x be a solution of $(7.3.1)_\lambda$ and (7.3.2) and

$$r(t) = \frac{1}{2}\langle x(t), x(t)\rangle = \frac{1}{2}|x(t)|^2, \qquad t \in [0,T].$$

Suppose that the function γ takes its maximum value at a point $\xi \in (0,T)$. Then we have

$$r'(\xi) = \langle x(\xi), x'(\xi)\rangle = 0$$

and

$$0 \ge r''(\xi) = \lambda\langle x(\xi), f(\xi, x_\xi, x'(\xi))\rangle + |x'(\xi)|^2$$

or

$$\langle x_\xi(0), x'(\xi)\rangle = 0 \tag{7.3.14}$$

and

$$0 \geq r''(\xi) = \lambda \langle x_\xi(0), f(\xi, x_\xi, x'(\xi))\rangle + |x'(\xi)|^2. \tag{7.3.15}$$

Since $\lambda > 0$, conditions (7.3.14), (7.3.15) and (H_1) imply $|x(\xi)| \leq M$. That is, if the function $|x(t)|$ takes its maximum value at a point $\xi \in (0,T)$, then $|x(\xi)| \leq M$.

Now we suppose that $|x(t)|$ on $[0,T]$ takes its maximum value at the point 0. Then from the first boundary condition we have

$$\begin{aligned} 0 \geq \alpha_1 r'(0) &= \langle x(0), \alpha_1 x'(0)\rangle \\ &= \langle x(0), \alpha_0 x(0) - \varphi(0)\rangle \\ &= \alpha_0 |x(0)|^2 - \langle x(0), \varphi(0)\rangle \\ &\geq \alpha_0 |x(0)|^2 - |x(0)|\,|\varphi(0)| \\ &= |x(0)|\{\alpha_0 |x(0)| - |\varphi(0)|\}. \end{aligned}$$

Hence

$$\alpha_0 |x(0)| - |\varphi(0)| \leq 0.$$

Since $\alpha_0 > 0$,

$$|x(0)| \leq \frac{1}{\alpha_0}|\varphi(0)|.$$

Assume that $|x(t)|$ on $[0,T]$ takes its maximum value at the point T. Then, by using the same arguments as above we have

$$|x(T)| \leq M, \quad \text{if } \beta_0 = 0$$

$$|x(T)| \leq \frac{1}{\beta_0}\,|A|, \quad \text{if } \beta_0 > 0.$$

Therefore,

$$|x(T)| \leq \max\left\{M, \frac{1}{\beta_0}|A|\right\}$$

and thus the theorem is proved, if $0 < \lambda \leq 1$. The case $\lambda = 0$ in $(7.3.1)_\lambda$ is simple and gives the same bounds as in the case $\lambda \neq 0$.

Summarizing the above, if $|x(t)|$ takes its maximum value at a point $\xi \in [0, T]$, then

$$\max_{t\in[0,T]} |x(t)| \leq \widehat{M_0} = \begin{cases} \max\{M, \frac{1}{\alpha_0}|\varphi(0)|, \frac{1}{\beta_0}|A|\}, & \text{if } \alpha_0, \beta_0 \neq 0 \\ \max\{M, \frac{1}{\alpha_0}|\varphi(0)|\}, & \text{if } \beta_0 = 0. \end{cases}$$

Next we shall prove that x_0 is bounded on $C(J)$. Let x be a solution of the BVP $(7.3.1)_\lambda$ and (7.3.2). From the first boundary condition we have, if $\alpha_1 > 0$

$$\begin{aligned} |x'(0)| &\leq \frac{1}{\alpha_1}(|\varphi(0)| + \alpha_0\widehat{M_0}) \\ &\leq \frac{1}{\alpha_1}(\|\varphi\|_{[-r,0]} + \alpha_0\widehat{M_0}) \end{aligned}$$

and consequently, from the boundary condition

$$\begin{aligned} \|x\|_{[-r,0]} &\leq \frac{1}{\alpha_0}\left[\|\varphi\|_{[-r,0]} + \alpha_1|x'(0)|\right] \\ &\leq \frac{2}{\alpha_0}\|\varphi\|_{[-r,0]} + \widehat{M_0}. \end{aligned}$$

Set

$$M_0 = \frac{2}{\alpha_0}\|\varphi\|_{[-r,0]} + \widehat{M_0},$$

then

$$\|x\|_{[-r,T]} \leq M_0.$$

The proof is complete. □

Next we will establish a priori bounds for $\max_{t\in[0,T]}|x'(t)|$ if an a priori bound for $\max_{[-r,T]}|x(t)|$ is given.

Theorem 7.3.3. *Assume that there exists a constant M_0 such that*

$$\max_{t\in[-r,T]} |x(t)| \leq M_0 \tag{7.3.16}$$

for every solution x of the BVP (7.3.1)$_\lambda$ and (7.3.2), $\lambda \in [0,1]$. Also, we suppose that the continuous function $f : [0,T] \times C(J) \times R^n \to R^n$ satisfies the following conditions

(H$_2$) $$\langle u(0), f(t,u,v)\rangle \leq K_1|v|^2 + K_2$$

(H$_3$) $$|\langle v, f(t,u,v)\rangle| \leq (K_1'|v|^2 + K_2')\,|v|$$

for any $(t,u,v) \in [0,T] \times C(J) \times R^n$ with $\|u\|_{[-r,0]} \leq M_0$, where K_1, K_2, K_1', K_2' are positive constants such that

$$K_1 < 1 \quad \text{and} \quad K_1' < \frac{1}{8M_0}(1-K_1)^2.$$

Then, there exists a constant M_1 independent of λ, such that

$$\max_{t\in[0,T]} |x'(t)| \leq M_1$$

for every solution x of BVP (7.3.1)$_\lambda$ and (7.3.2).

Proof: For a solution $x \in C^2[0,T]$ satisfying the hypotheses, let $M = \|x'\|$ and let $t_0 \in [0,T]$ such that $|x'(t_0)| = M$. We will show that M can be bounded independently of x.

If $\sigma \in C^2[0,T]$, we obtain, by a Taylor expansion,

$$\sigma(t_0+\mu) - \sigma(t_0) = \mu\sigma'(t_0) + \int_{t_0}^{t_0+\mu} \sigma''(s)(t_0+\mu-s)\,ds \tag{7.3.17}$$

provided that $t_0 + \mu \in [0,T]$. Set $\sigma(t) = \int_0^t |x'(s)|^2 ds$, then

$$\int_{t_0}^{t_0+\mu} |x'(s)|^2\,ds = \mu|x'(t_0)|^2 + 2\int_{t_0}^{t_0+\mu} \langle x'(s), x''(s)\rangle\,(t_0+\mu-s)\,ds. \tag{7.3.18}$$

On the other hand, integrating by parts we have

$$\int_{t_0}^{t_0+\mu} |x'(s)|^2\,ds = \langle x(t_0+\mu), x'(t_0+\mu)\rangle - \langle x(t_0), x'(t_0)\rangle - \int_{t_0}^{t_0+\mu} \langle x(s), x''(s)\rangle\,ds.$$

By (7.3.16), (H_2) and (H_3), we have

$$\int_{t_0}^{t_0+\mu} |x'(s)|\, ds \le 2M_0M + K_1M^2|\mu| + K_2|\mu|. \tag{7.3.19}$$

Combining (7.3.18) and (7.3.19) and using (H_3) we obtain

$$|\mu|M^2 \le 2M_0M + |\mu|K_1M^2 + |\mu|K_2 + |\mu|^2K_1'M^3 + |\mu|^2K_2'M. \tag{7.3.20}$$

Let us assume that $8M_0 \le M(1-K_1)T$; if this is not the case, the desired bound for $\|x'\|$ is already obtained. Let us choose $|\mu|$ such that $|\mu|M = 4M_0/(1-K_1)$. With the above restriction on M, we then have $|\mu| \le \frac{T}{2}$ and the sign of μ can be chosen so that $t_0 + \mu \in [0,T]$, which guarantees the validity of the Taylor expansion used above. With that choice of $|\mu|$ the relation (7.3.20) becomes

$$2M_0M \le |\mu|K_2 + |\mu|^2K_1'M^3 + |\mu|^2K_2'M$$

since $|\mu| \le \frac{T}{2}$ and hence

$$2M_0M \le \frac{T}{2}K_2 + \frac{M_0^2}{(1-K_1)^2}K_1'M + 2TM_0\frac{K_2'}{1-K_1}.$$

Since $K_1' < \frac{1}{8M_0}(1-K_1)^2$, we have

$$M \le \frac{T(1-K_1)}{4M_0} \cdot \frac{K_2(1-K_1) + 4M_0K_2'}{(1-K_1)^2 - 8M_0K_1'}.$$

This proves that there exists a constant M_1 independent of λ such that

$$\max_{t\in[0,T]} |x'(t)| \le M_1$$

for every solution x of BVP $(7.3.1)_\lambda$ and (7.3.2), $\lambda \in [0,1]$. □

Theorem 7.3.4. *Assume that there exists a constant M_0 such that*

$$\max_{t\in[-r,T]} |x(t)| \le M_0 \tag{7.3.21}$$

for every solution x of the BVP $(7.3.1)_\lambda$ and (7.3.2), $\lambda \in [0,1]$. Further assume that

(H$_4$) *there exist $q \in C([0,T],(0,\infty))$ and $\psi([0,\infty),(0,\infty))$ such that*

$$|f(t,u,v)| \le q(t)\,\psi(|v|), \qquad (t,u,v) \in [0,T] \times B_{M_0} \times R^n \tag{7.3.22}$$

where ψ is nondecreasing, B_{M_0} is the closed ball in $C(J)$ with center 0 and radius M_0.

If $\alpha_1 > 0$, then there exists a constant M_1, independent of λ, such that

$$\max_{t\in[0,T]} |x'(t)| \le M_1$$

for every solution x of BVP (7.3.1)$_\lambda$ and (7.3.2).

Proof: Let x be a solution of the BVP (7.3.1)$_\lambda$ and (7.3.2), $\lambda \in [0,1]$. Then, since $\alpha_1 > 0$, the first boundary condition gives

$$|x'(0)| \le \frac{1}{\alpha_1}(|\varphi(0)| + \alpha_0\, M_0) \equiv c.$$

From the Cauchy-Schwarz inequality, if $|x'(t)| \ne 0$, we have

$$|x'(t)|' = \frac{|\langle x'(t), x''(t)\rangle|}{|x'(t)|} \le |x''(t)|.$$

Clearly this inequality is also true if $|x'(t)| = 0$.

From (H$_4$)

$$|x'(t)|' \le |x''(t)| \le q(t)\,\Omega(|x'(t)|) \qquad \text{for every } t \in [0,T].$$

Thus

$$|x'(t)| \le k + \int_0^t q(s)\,\Omega(|x'(s)|)\,ds, \qquad t \in [0,T]$$

where $k = |x'(0)|$. By Bihari's inequality we have

$$|x'(t)| \le G^{-1}\left[G(k) + \int_0^T q(s)\,ds\right] = M_1, \qquad t \in [0,T]$$

where

$$G(s) = \int_\varepsilon^s \frac{dt}{\Omega(t)}, \qquad \varepsilon > 0,\ s \ge 0,$$

and G^{-1} is the inverse mapping of G, which is supposed such that

$$\left(G(k) + \int_0^T q(s)\,ds\right) \in \operatorname{Dom} G^{-1}.$$

Hence

$$\max_{t\in[0,T]} |x'(t)| \le M_1.$$

The proof is complete. □

Remark 7.3.1. (H_4) can be replaced by a Bernstein-Nagumo type condition:
$(H_4)'$ There is an increasing function $\psi = [0,\infty) \to (0,\infty)$ such that $\frac{1}{\psi}$ is integrable on $[0,\infty)$

$$|f(t,u,v)| \le \psi(|v|) \tag{7.3.23}$$

for $t \in [0,T]$, $\|u\|_{[-r,0]} \le M_0$ and

$$\int_c^\infty \frac{ds}{\psi(s)} > T \tag{7.3.24}$$

where

$$c = \frac{1}{\alpha_1}\,(|\varphi(0)| + \alpha_0\, M_0).$$

Combining the above results we have the following propositions.

Theorem 7.3.5. *Assume that* (H_1)–(H_3) *hold. Then BVP (7.3.1) and (7.3.2) has at least one solution.*

Theorem 7.3.6. *Assume that* (H_1) *and* (H_4) *hold. Then BVP (7.3.1) and (7.3.2) with* $\alpha_1 > 0$ *has at least one solution.*

Example 7.3.1. Consider the equation

$$x''(t) = g(t)x(t)F(x_t) + h(t)x'(t)|x'(t)|, \qquad t \in [0,1] \tag{7.3.25}$$

where $g : [0,1] \to R$ is continuous and positive, $F : C(J) \to R$ is continuous and positive and maps bounded sets into bounded sets and $h : [0,1] \to R$ is continuous with $\max_{t\in[0,1]} |h(t)| < 1$.

Consider the following boundary conditions

$$\begin{aligned} x_0 - x'(0) &= \varphi \\ x(1) + x'(1) &= A. \end{aligned} \tag{7.3.26}$$

In the notations of (7.3.1) and (7.3.2)

$$f(t,u,v) = g(t)u(0)F(u) + h(t)v|v|$$

$$T = 1, \qquad \alpha_0 = \alpha_1 = \beta_0 = \beta_1 = 1.$$

It is not difficult to prove that (H_1)–(H_3) are satisfied. Therefore, BVP (7.3.25) and (7.3.26) has a solution by Theorem 7.3.5.

7.4. Topological Transversality Method

In this section we consider the system

$$x''(t) = f(t, x(\sigma_1(t)), \ldots, x(\sigma_k(t)), x'(\sigma_{k+1}(t)), \ldots, x'(\sigma_{k+m}(t))), \quad t \in I \tag{7.4.1}$$

where $I = [a,b]$ $(a < b)$, $f \in C[I \times (R^n)^{m+k} \to R^n]$, σ_i, $i = 1, \ldots, k+m$, are real valued functions defined on I, such that the set $\{t \in I\colon \sigma_i(t) = a \text{ or } \sigma_i(t) = b,\ i = k+1, \ldots, k+m\}$ is finite.

We suppose that

$$-\infty < r_a = \min_{1\le i\le k+m} \min_{t\in I} \sigma_i(t) < a$$

and

$$b < r_b = \max_{1\le i\le k+m} \max_{t\in I} \sigma_i(t) < +\infty$$

and we set $E(a) = [r_a, a]$, $E(b) = [b, r_b]$ and $J = [r_a, r_b]$.

Consider the following boundary conditions:

$$\begin{aligned} -\alpha_0 x(t) + \alpha_1 x'(t) &= q_1(t), \quad t \in E(a) \\ \beta_0 x(t) + \beta_1 x'(t) &= q_2(t), \quad t \in E(b) \end{aligned} \tag{7.4.2}$$

where q_1, q_2 are R^n valued functions defined and differentiable on $E(a)$, $E(b)$ respectively, also α_0, α_1, β_0, β_1 are nonnegative real constants such that

$$\ell = \alpha_0\beta_0(b-a) + \alpha_0\beta_1 + \alpha_1\beta_0 \neq 0. \tag{7.4.3}$$

By a solution of the boundary value problem (7.4.1) and (7.4.2) we mean a function $x \in C(J, R^n) \cap C^1(I, R^n) \cap C^1(E(a) \cup E(b), R^n)$ which is piecewise twice differentiable on I, satisfies the equation (7.4.1) for $t \in I$ and the boundary conditions (7.4.2) for $t \in E(a) \cup E(b)$.

Let B be the space

$$B = C(J, R^n) \cap C^1(I, R^n) \cap C^1(E(a) \cup E(b), R^n)$$

with norm

$$\|x\| = \max\{\max_{t\in J}|x(t)|, \max_{t\in E(a)\cup E(b)}|x'(t)|, \max_{t\in I}|x'(t)|\}, \qquad x \in B.$$

The following is an existence result for the boundary value problem (7.4.1) and (7.4.2).

Theorem 7.4.1. *Let $f \in C[I \times (R^n)^{k+m}, R^n]$ and assume that*
(H$_1$) *There exists a constant $M > 0$ such that, if for every $v \in R^n$ with $|v| > m$ and*
$(t, u_1, \ldots, u_{k+m}) \in I \times (R^n)^{k+m}$ *the following conditions are fulfilled:*

(i) *if $\sigma_j(t) = t$ for a $j \in \{k+1, \ldots, k+m\}$, then $\langle v, u_j\rangle = 0$,*

(ii) *if $\sigma_i(t) = \sigma_j(t) \in I$ for a $i \in \{1, \ldots, k\}$ and a $j \in \{k+1, \ldots, k+m\}$, then $\langle u_i, u_j\rangle = 0$, and*

$$\langle v, f(t, \widehat{u}_1, \ldots, \widehat{u}_k, u_{k+1}, \ldots, u_{k+m})\rangle > 0$$

where

$$\widehat{u}_i = \begin{cases} v, & \text{if } \sigma_i(t) = t \\ u_i, & \text{if } \sigma_i(t) \neq t, \qquad i = 1, \ldots, k. \end{cases}$$

(H_2) *There exists an index $j \in \{k+1, \ldots, k+m\}$ such that $\sigma_j(t) \leq t$, $t \in I$ and*

$$|f(t, u_1, \ldots, u_{k+m})| \leq g(t)\, \Omega(|u_j|)$$

for every $(t, u_1, \ldots, u_{k+m}) \in I \times (R^n)^{k+m}$, with $(|u_1|, \ldots, |u_k|) \in [0, M_0]^k$ and where the functions g and Ω are nonnegative continuous on $[0, \infty)$, Ω is nondecreasing.

(H_3) $\quad \alpha_0 = 0 \implies q_1(a) = 0, \quad \beta_0 = 0 \implies q_2(b) = 0 \quad$ *and* $\quad \alpha_1 > 0.$

Then the boundary value problem (7.4.1) and (7.4.2) has at least one solution.

To prove this result we will use a fixed point theorem in Chapter 1. We divide the proof into the following lemmas.

The following basic results can be found in any elementary differential equations text book.

Lemma 7.4.1. *The Green's function for the homogeneous boundary value problem*

$$x''(t) = 0, \qquad t \in I$$

$$-\alpha_0 x(a) + \alpha_1 x'(a) = 0$$

$$\beta_0 x(b) + \beta_1 x'(b) = 0$$

exists and is given by the formula

$$G(t,s) = \frac{1}{\ell} \begin{cases} (\beta_0 t - \beta_0 b - \beta_1)(\alpha_0 s - \alpha_0 a + \alpha_1), & s \leq t \\ (\beta_0 s - \beta_0 b - \beta_1)(\alpha_0 t - \alpha_0 a + \alpha_1), & t \leq s, \end{cases}$$

where ℓ is defined by (7.4.3). Also the following inequalities hold:

$$\int_a^b |G(t,s)|\,ds \le K_1, \qquad \int_a^b |G_t(t,s)|\,ds \le K_2,$$

where K_1, K_2 are constants depending on α_0, α_1, β_0, β_1, a and b.

We may easily establish the following fact.

Lemma 7.4.2. *The boundary value problem*

$$\begin{aligned} &x''(t) = 0 \\ &-\alpha_0 x(t) + \alpha_1 x'(t) = q_1(t), \qquad t \in E(a) \\ &\beta_0 x(t) + \beta_1 x'(t) = q_2(t), \qquad t \in E(b) \end{aligned}$$

has a solution $w : J \to R^n$ defined by

$$w(t) = \begin{cases} \left(w(a) + \frac{1}{\alpha_1}\int_a^t q_1(s)\exp(-\frac{\alpha_0}{\alpha_1}(s-a))\,ds\right)\exp(\frac{\alpha_0}{\alpha_1}(t-a)), & t < a,\ \alpha_1 \ne 0; \\ -\frac{1}{\alpha_0} q_1(t), \qquad t < a,\ \alpha_1 = 0; \\ \frac{1}{\ell}\left(\beta_0(b-t)q_1(a) + \beta_1 q_1(a) - \alpha_1 q_2(b) - \alpha_0(t-a)q_2(b)\right), & t \in I; \\ \left(w(b) + \frac{1}{\beta_1}\int_b^t q_2(s)\exp(\frac{\beta_0}{\beta_1}(s-b))\,ds\right)\exp(-\frac{\beta_0}{\beta_1}(t-b)), & t > b,\ \beta_1 \ne 0; \\ \frac{1}{\beta_0} q_2(t), \qquad t > b,\ \beta_1 = 0. \end{cases}$$

We consider, for $\lambda \in [0,1]$,

$$x''(t) = \lambda f(t, x(\sigma_1(t)), \ldots, x(\sigma_k(t)), x'(\sigma_{k+1}(t)), \ldots, x'(\sigma_{k+m}(t))), \quad t \in I. \tag{7.4.1$_\lambda$}$$

In the following lemma we establish an a priori bound for $\max_{t\in I}|x(t)|$, where x is a solution of the boundary value problem (7.4.1)$_\lambda$ and (7.4.2), $\lambda \in [0,1]$.

Lemma 7.4.3. *Assume that* (H$_1$) *and* (H$_3$) *hold. Then every solution x of the* *boundary value problem* $(7.4.1)_\lambda$ *and (7.4.2),* $\lambda \in [0,1]$, *satisfies*

$$\max_{t\in I}|x(t)| \le M_0 := \begin{cases} \max\left\{M, \frac{1}{\alpha_0}|q_1(a)|, \frac{1}{\beta_0}|q_2(b)|\right\}, & \textit{if } \alpha_0,\, \beta_0 \neq 0 \\ \max\left\{M, \frac{1}{\beta_0}|q_2(b)|\right\}, & \textit{if } \alpha_0 = 0 \\ \max\left\{M, \frac{1}{\alpha_0}|q_1(a)|\right\}, & \textit{if } \beta_0 = 0. \end{cases} \tag{7.4.4}$$

Proof: For any $\lambda \in [0,1]$, let x be a solution of $(7.4.1)_\lambda$ and (7.4.2) and

$$\varphi(t) = \frac{1}{2}\langle x(t), x(t)\rangle = \frac{1}{2}|x(t)|^2, \qquad t \in I$$

where $\langle \cdot,\cdot\rangle$ denotes the Euclidean inner product.

If φ takes its maximum value at a point $\xi \in (a,b)$, then we have

$$\varphi'(\xi) = \langle x(\xi), x'(\xi)\rangle = 0$$

and

$$\varphi''(\xi) = |x'(\xi)|^2 + \lambda\langle x(\xi), f(\xi, x(\sigma_1(\xi)),\dots, \\ x(\sigma_k(\xi)), x'(\sigma_{k+1}(\xi)),\dots, x'(\sigma_{k+m}(\xi)))\rangle \le 0. \tag{7.4.5}$$

Consider the case that $\lambda > 0$ first. Then the above relations and (H_1) imply that $|x(\xi)| \le M$.

If x takes its maximum value at the point a, then from the boundary condition,

$$-\alpha_0 x(a) + \alpha_1 x'(a) = q_1(a)$$

and, if $\alpha_0 = 0$, we obtain $\alpha_1 x'(a) = q_1(a)$. By (H_3), $q_1(a) = 0$ and, because $\alpha_1 > 0$, we have $x'(a) = 0$. Thus $\varphi'(a) = 0$. If $|x(a)| > M$, by (7.4.6) and (H_1), we obtain $\varphi''(a) > 0$. Hence, there exists a positive ε such that $\varphi'(t) > 0$ for $t \in (a, a+\varepsilon)$, which means that the function φ is increasing near a which contradicts the fact that $\varphi(a)$ is maximum. Therefore $|\varphi(a)| \le M$.

If $\alpha_0 > 0$, $\alpha_1 > 0$, since $\varphi(a)$ is the maximum value of φ, we have

$$0 \ge \alpha_1\varphi'(a) = \langle x(a), \alpha_1 x'(a)\rangle$$

$$= \langle x(a), q_1(a) + \alpha_0 x(a) \rangle$$
$$= \langle x(a), q_1(a) \rangle + \alpha_0 \, |x(a)|^2$$
$$\geq -|x(a)| \, |q_1(a)| + \alpha_0 \, |x(a)|^2$$
$$= -|x(a)| \, [|q_1(a)| - \alpha_0 \, |x(a)|].$$

Hence, $|q_1(a)| - \alpha_0 |x(a)| \geq 0$ and consequently,

$$|x(a)| \leq \frac{1}{\alpha_0} \, |q_1(a)|.$$

This inequality is also true for $\alpha_1 = 0$.

If $|x(a)|$ is the maximum value of the function $|x(t)|$, then

$$|x(a)| \leq \max \left\{ M, \frac{1}{\alpha_0} \, |q_1(a)| \right\}.$$

Similarly, if $|x(b)|$ is the maximum value of $|x(t)|$ on I, then

$$|x(b)| \leq \max \left\{ M, \frac{1}{\beta_0} \, |q_2(b)| \right\}.$$

Thus, for $\lambda \in (0, 1]$, the lemma is proved.

Now we consider (7.4.1) for $\lambda = 0$. Then $x(t) = c_1 t + c_2$ for some c_1, c_2 in R^n and $t \in I$. If $|x(t)|$ takes its maximum at $\xi \in (a, b)$, then (7.4.6) yields $|x'(\xi)| = 0$ and hence $x(t) = c_2$. The boundary conditions give $-\alpha_0 c_2 = q_1(a)$ and $\beta_0 c_2 = q_2(b)$. Hence either

$$c_2 = -\frac{1}{\alpha_0} \, q_1(a), \qquad c_2 = \frac{1}{\beta_0} \, q_2(b)$$

and $|x(t)| \leq M_0$, $t \in (a, b)$ in this case.

Next assume that $|x(a)|$ is a maximum value of $|x(t)|$ on $[a, b]$. If $\alpha_0 = 0$, then $\alpha_1 x'(a) = q_1(a)$ and by (H_3), $x'(a) = 0$. Thus $x(t) = c_2$, $t \in I$. The boundary condition at $t = b$ gives $\beta_0 c_2 = q_2(b)$ and thus

$$|x(t)| = |c_2| = \frac{1}{\beta_0} \, |q_2(b)|, \qquad t \in I.$$

Finally, we consider the case $\alpha_0 > 0$. But then the previous arguments apply again and yield

$$|x(a)| \leq \frac{1}{\alpha_0} |q_1(a)|.$$

Consequently, we find that when the maximum occurs at $t = a$, then

$$|x(a)| \leq \max\left\{ \frac{1}{\alpha_0} |q_1(a)|, \frac{1}{\beta_1} |q_2(b)| \right\}.$$

The same bounds hold also if $|x(t)|$ takes its maximum at $t = b$ on I. The lemma is proved. □

The following lemma concerns a priori bounds for $\max_{t \in I} |x'(t)|$.

Lemma 7.4.4. *We assume that there exists a constant M_0 such that*

$$\max_{t \in I} |x(t)| \leq M_0$$

for every solution of the boundary value problem $(7.4.1)_\lambda$ *and (7.4.2),* $\lambda \in [0, 1]$ *and assume* (H_2) *holds.*

Then, if $\alpha_1 > 0$, there exists a constant M_1 independent of λ, such that

$$\max_{t \in I} |x'(t)| \leq M_1$$

for every solution x of the boundary value problem $(7.4.1)_\lambda$ *and (7.4.2).*

Proof: Let x be a solution of the boundary value problem $(7.4.1)_\lambda$ and (7.4.2), $\lambda \in [0, 1]$. Since $\alpha_1 > 0$, the first of the boundary conditions implies that

$$\alpha_1 |x'(a)| \leq |\alpha_0| \, |x(a)| + |q_1(a)|$$

or

$$|x'(a)| \leq \frac{1}{\alpha_1} (|q_1(a)| + \alpha_0 M_0).$$

Let

$$C = \begin{cases} \max\left(\frac{1}{\alpha_1}(|q_1(a)| + \alpha_0 M_0),\ \frac{1}{\beta_1}(|q_2(b)| + \beta_0 M_0),\ \sup\limits_{t\in E(a)} |w'(t)|\right), & \beta_1 > 0, \\ \max\left(\frac{1}{\alpha_1}(|q_1(a)| + \alpha_0 M_0),\ \sup\limits_{t\in E(a)} |w'(t)|\right), & \beta_1 = 0 \end{cases} \tag{7.4.6}$$

Without loss of generality we assume that M_0 is large enough so that $\sup\limits_{t\in E(a)} |w(t)| \leq M_0$. From (7.4.2) and ($H_2$), for every $t \in I$ we have

$$|x''(t)| \leq g(t)\,\Omega(|x'(\sigma_j(t))|).$$

From the Cauchy-Schwarz inequality, if x' is a nonzero function defined on I, we have

$$|x'(t)|' = \frac{|\langle x'(t), x''(t)\rangle|}{|x'(t)|} \leq |x''(t)|,$$

and thus

$$|x'(t)|' \leq g(t)\,\Omega(x'(\sigma_j(t))|), \quad \text{for every } t \in I, \text{ with } x'(t) \neq 0.$$

It is obvious that the last inequality remains valid even when $x'(t) = 0$ for $t \in I$. Therefore, we have

$$|x'(t)|' \leq g(t)\,\Omega(|x'(\sigma_j(t))|), \quad \text{for every } t \in I,$$

and so

$$\begin{aligned} |x'(t)| &\leq |x'(a)| + \int_a^t g(s)\,\Omega(|x'(\sigma_j(s))|)\,ds \\ &\leq C + \int_a^t g(s)\,\Omega(|x'(\sigma_j(s))|\,ds. \end{aligned}$$

Furthermore, we define

$$\phi(r) = \sup_{\theta\in E(a)\cup[a,r]} |x'(\theta)| \qquad \text{for } r \in [a,b]$$

which is continuous, since $\alpha_1 > 0$ implies $w'_-(a) = w'_+(a)$. So, for $\tau \in [a,t]$, $t \in [a,b]$ and $\sigma_j(t) \leq t$, $t \in I$, we get

$$|x'(\tau)| \leq C + \int_a^t g(s)\,\Omega(\phi(s))\,ds$$

and hence

$$\sup_{\tau\in[a,t]} |x'(\tau)| \leq C + \int_a^t g(s)\,\Omega(\phi(s))\,ds.$$

Since $x(t) = w(t)$, $t \in E(a)$ and from the choice of C, we obtain

$$\phi(t) \leq C + \int_a^t g(s)\,\Omega(\phi(s))\,ds, \qquad t \in I.$$

By Bihari's inequality we have

$$\phi(t) \leq G^{-1}\left(G(C) + \int_a^b g(s)\,ds\right) := M_1$$

where G is supposed such that

$$\left(G(C) + \int_a^b g(s)\,ds\right) \in \operatorname{Dom} G^{-1}.$$

Hence $\sup_{t\in I} |x'(t)| \leq M_1$. The proof is complete. □

Proof of Theorem 7.4.1: By Lemma 7.4.3 and Lemma 7.4.4, there exists a constant d such that $\|x\| \leq d$ for every solution x of the boundary value problem $(7.4.1)_\lambda$ and (7.4.2), $\lambda \in [0,1]$. Consider the subspace X of B given by

$$X = \{u \in B\colon u \text{ is piecewise twice differentiable on } I\}$$

and define

$$T_\lambda x = \lambda\,Lx + w \tag{7.4.7}$$

where

$$Lx(t) = \begin{cases} \int_a^b G(t,s) f(s, x(\sigma_1(s)), \ldots, x(\sigma_k(s)), x'(\sigma_{k+1}(s)), \\ \qquad \ldots, x'(\sigma_{k+m}(s))), \qquad t \in I, \\ e^{(\alpha_0/\alpha_1)(t-a)}\, Lx(a), \qquad t < a,\ \alpha_1 \neq 0 \\ 0, \qquad t < a,\ \alpha_1 = 0 \\ e^{-(\beta_0/\beta_1)(t-b)}\, Lx(b), \qquad t > b,\ \beta_1 \neq 0 \\ 0, \qquad t > b,\ \beta_1 = 0. \end{cases} \tag{7.4.8}$$

It is clear that (7.4.8) defines a mapping $T_\lambda \colon X \to X$ for every $\lambda \in [0,1]$. Define

$$U = \{u \in X : \|u\| < d + \|w\| + 1\}$$

and

$$H \colon [0,1] \times \overline{U} \longrightarrow X$$

where

$$H(\lambda, u) = T_\lambda\, u = \lambda\, Lu + w.$$

We shall show that the mapping H is a compact homotopy. To this end it is enough to prove that the mapping $L : X \to X$, defined by (7.4.9), is compact.

Indeed, let $\{h_\nu\}$ be a bounded sequence in X, i.e. $\|h_\nu\| \leq C$, for all ν. Then we can prove that

$$\|Lh_\nu\| \leq \widehat{C} := \max\{FK, FKC_1\}$$

where

$$F = \max_{\substack{t \in I \\ |u_i| \leq C}} |f(t, u_1, \ldots, u_{k+m})|, \qquad i = 1, \ldots, k+m$$

$K = \max\{K_1, K_2\}$, K_1, K_2 are the constants of Lemma 7.4.1,

and

$$C_1 = \max\left\{ \max_{t\in E(a)} \left|e^{\frac{\alpha_0}{\alpha_1}(t-a)}\right|, \max_{t\in E(a)} \left|\frac{\alpha_0}{\alpha_1} e^{\frac{\alpha_0}{\alpha_1}(t-a)}\right|, \right.$$

$$\left. \max_{t\in E(b)} \left|e^{-\frac{\beta_0}{\beta_1}(t-b)}\right|, \max_{t\in E(b)} \left|-\frac{\beta_0}{\beta_1} e^{-\frac{\beta_0}{\beta_1}(t-b)}\right| \right\}. \tag{7.4.9}$$

In (7.4.9) C_1 can be appropriately adjusted when either α_1 or β_1 is zero.

For all t_1, $t_2 \in J$ and arbitrary ν we have

$$|Lh_\nu(t_1) - Lh_\nu(t_2)| = \left|\int_{t_1}^{t_2} (Lh_\nu)'(s)\,ds\right| \le \widehat{C}|t_1 - t_2|,$$

$$|(Lh_\nu)'(t_1) - (Lh_\nu)'(t_2)| = \left|\int_{t_1}^{t_2} (Lh_\nu)''(s)\,ds\right| \le \widehat{h}|t_1 - t_2|,$$

where

$$\widehat{h} = \max\left\{\widehat{C}, \left(\frac{\alpha_0}{\alpha_1}\right)^2 e^{(\alpha_0/\alpha_1)(r_a - a)}, \left(\frac{\beta_0}{\beta_1}\right)^2 \widehat{C}\right\}.$$

Hence the sequences $\{Lh_\nu\}$ and $\{(Lh_\nu)'\}$ are equicontinuous.

Moreover, it is obvious that $H(0, \partial\overline{U}) = \{w\}$. Thus, since $w \in X$, it is clear that $H(0, \partial\overline{U}) \subset X$. On the other hand, for every $x \in \partial\overline{U}$, i.e., for every $x \in X$ with $\|x\| = d + \|w\| + 1$, we cannot have $x = H(\lambda, x)$ for some $\lambda \in [0, 1]$. Indeed, in that case x must be a solution of the boundary value problem $(7.4.1)_\lambda$ and (7.4.2) and hence $\|x\| \le d$. Therefore,

$$x \ne H(\lambda, x) \qquad \text{for all } x \in \partial\overline{U},\ \lambda \in [0, 1].$$

The assumptions of the fixed point theorem are then satisfied and the function $H(1, \cdot)$ has a fixed point in X. This means that there exists at least one $x \in X$ such that $x = H(1, x)$, or $x = T_1 x = Lx + w$, which implies that x is a solution of the boundary value problem of (7.4.1) and (7.4.2). □

Example 7.4.1. Consider the boundary value problem

$$x''(t) = x^3(t) + \frac{x'(t)(x'(t-1))^2}{1 + t + |x'(t)|} \sin(x'(t+1)) - 1, \quad t \in [0, 1]$$

$$-\alpha_0 x(t) + \alpha_1 x'(t) = q_1(t), \quad t \in [-1, 0],$$

$$\beta_0 x(t) + \beta_1 x'(t) = q_2(t), \quad t \in [1, 2].$$

It is easy to check that (H_1) holds with $M = 1$. Also, if we set

$$f(t, u_1, u_2, u_3, u_4) = u_1^3 + \frac{u_2 u_3^2}{1 + t + |u_2|} \sin u_4 - 1,$$

clearly, for every $(t, u_1, u_2, u_3, u_4) \in [0, 1] \times R^4$ with $|u_1| \in [0, 1]$, we have

$$|f(t, u_1, u_2, u_3, u_4)| \leq 2 + u_3^2,$$

and consequently, (H_2) holds with $\Omega(s) = 2 + s^2$.

In order to have a solution for the above boundary value problem it is enough to check that

$$G(C) + 1 \in \operatorname{Dom} G^{-1},$$

where C is defined by (7.4.6).

For example, if $\alpha_0 = 1$, $\alpha_1 = 1$, $q_1(t) = 0$ and $\beta_1 = 0$, the above problem has at least one solution since $0 < C = 1 < \sqrt{2}\,\tan(\sqrt{2}\,(\frac{\pi}{2} - 1))$.

7.5. Boundary Value Problems for Singular Equations

In this section we consider the second order functional differential equation

$$x'' + f(t, x(\tau(t))) = 0, \qquad 0 \leq t \leq 1 \tag{7.5.1}$$

under the assumption

(H1) $f(t, x) : (0, 1) \times (0, \infty) \to (0, \infty)$ is continuous and decreasing in x for each fixed t and integrable on $[0, 1]$ in t for each fixed x. And

$$\lim_{x \to 0+} f(t, x) = \infty \quad \text{uniformly on compact subsets of } (0, 1)$$

and

$$\lim_{x \to \infty} f(t, x) = 0 \quad \text{uniformly on compact subsets of } (0, 1).$$

(H2) $\tau(t)$ is continuous on $[0,1]$ satisfying

$$\inf_{t\in[0,1]} \tau(t) < 1 \qquad \text{and} \qquad \sup_{t\in[0,1]} \tau(t) > 0.$$

From (H_2) we see that the set $E = \{t \in [0,1] : \tau(t) \in (0,1)\}$ satisfies $\operatorname{mes} E > 0$. According to the continuity of $\tau(t)$ there exist closed intervals $I \subset E$ and $J \subset (0,1)$ with $\operatorname{mes} I > 0$ such that $t \in I$ means $\tau(t) \in J$. Denote $a = \min\{0, \inf_{t\in[0,1]} \tau(t)\}$, $b = \max\{1, \sup_{t\in[0,1]} \tau(t)\}$.

The boundary conditions considered here are

$$\begin{aligned} \alpha x(t) - \beta x'(t) &= \mu(t), \quad t \in [a,0] \\ \gamma x(t) + \delta x'(t) &= \nu(t), \quad t \in [1,b] \end{aligned} \tag{7.5.2}$$

which satisfy

(H3) α, β, γ, δ are nonnegative constants, with

$$\rho = \gamma\beta + \alpha\gamma + \alpha\delta > 0;$$

$\mu(t)$ and $\nu(t)$ are continuous functions defined on $[a,0]$ and $[1,b]$, respectively with $\mu(0) = \nu(1) = 0$, and satisfying $\mu(t) \geq 0$ for $\beta = 0$ and $\int_t^0 e^{-\frac{\alpha}{\beta}s} \mu(s)\, ds \geq 0$ for $\beta > 0$ and $\nu(t) \geq 0$ for $\delta = 0$ and $\int_1^t e^{\frac{\gamma}{\delta}s} \nu(s)\, ds \geq 0$ for $\delta > 0$.

Note that boundary condition (7.5.2) gives that

$$\begin{cases} \alpha x(0) - \beta x'(0) = 0 \\ \gamma x(1) + \delta x'(1) = 0. \end{cases} \tag{7.5.3}$$

If $\tau(t) \equiv t$, this coincides with the usual linear BVP for an ODE. By solving the linear equations in (7.5.2) we see that (7.5.2) is equivalent to

$$x(t) = \begin{cases} e^{\frac{\alpha}{\beta}t}\left(\dfrac{1}{\beta}\displaystyle\int_t^0 e^{-\frac{\alpha}{\beta}s} \mu(s)\, ds + x(0)\right), & \beta > 0 \\[2ex] \dfrac{1}{\alpha}\, \mu(t), & \beta = 0 \end{cases} \qquad t \in [a,0] \tag{7.5.4}$$

and

$$x(t) = \begin{cases} e^{-\frac{\gamma}{\delta}t}\left(\frac{1}{\delta}\int_1^t e^{\frac{\gamma}{\delta}s}\,\nu(s)\,ds + e^{\frac{\gamma}{\delta}}x(1)\right), & \delta > 0 \\ \frac{1}{\gamma}\,\nu(t), & \delta = 0 \end{cases} \qquad t \in [1, b]. \quad (7.5.5)$$

(7.5.4) and (7.5.5) imply that any function $x(t)$ satisfying (7.5.2) with $x(0) \geq 0$ and $x(1) \geq 0$ will be nonnegative on $[a, 0] \cup [1, b]$ (if $x(0) > 0$, $x(1) > 0$, then $x(t) > 0$ on $[a, 0] \cup [1, b]$).

By a positive solution of the problem (7.5.1) and (7.5.2) we mean a function in $C[a, b] \cap C^2(0, 1)$ which is nonnegative in (a, b) and positive in $(0, 1)$ and which satisfies the equation (7.5.1) and the boundary condition (7.5.2).

Remark 7.5.1. As we will see later for the case that $\beta > 0$ (or $\delta > 0$) the solutions are actually in $C[a, b] \cap C^1[a, 1] \cap C^2(0, 1)$ (or $C[a, b] \cap C^1[0, b] \cap C^2(0, 1)$).

Although the expressions for boundary functions are different for different values of β, δ, the conclusion concerning existence and uniqueness of a positive solution are exactly the same. In the sequel, we will state the results for the general case and give the proofs only for the case that $\beta > 0$ and $\delta > 0$, since the other cases can be done in a similar way. A fixed point theorem on cones will be used which we outline below.

Lemma 7.5.1. *Let X be a Banach space, K a normal cone in X, D a subset of K such that if x, y are elements of D, $x \leq y$, then $\langle x, y \rangle$ is contained in D, and let $T : D \to K$ be a continuous decreasing mapping which is compact on any closed order interval contained in D. Suppose that there exists $x_0 \in D$ such that T^2x_0 is defined (where $T^2x_0 = T(Tx_0)$) and furthermore Tx_0, T^2x_0 are (order) comparable to x_0. Then T has a fixed point in D provided either*

(I) *$Tx_0 \leq x_0$ and $T^2x_0 \leq x_0$ or $Tx_0 \geq x_0$ and $T^2x_0 \geq x_0$, or*

(II) *The complete sequence of iterates $\{T^n x_0\}_{n=0}^{\infty}$ is defined and there exists $y_0 \in D$ such that $Ty_0 \in D$ and $y_0 \leq T^n x_0$ for all n.*

We shall now give an existence theorem for boundary value problems (7.5.1) and (7.5.2) under the hypotheses (H1)–(H3) and $\beta > 0$, $\delta > 0$.

Let $g_1 : [a,b] \to [0,\infty)$ be defined by

$$g_1(t) = \begin{cases} t-a & \text{if } a \le t \le \frac{1}{2} \\ b-t & \text{if } \frac{1}{2} \le t \le b, \end{cases}$$

and for $\theta > 0$, $g_\theta(t)$ is defined by $g_\theta = \theta g_1$. We will assume further that

(H4) $0 < \int_0^1 f(t, g_\theta(\tau(t)))dt < \infty$ for all $\theta > 0$.

With the assumption (H3) the boundary value problem

$$\begin{aligned} x'' &= 0 \\ \alpha x(0) - \beta x'(0) &= 0 \\ \gamma x(1) + \delta x'(1) &= 0 \end{aligned} \tag{7.5.6}$$

has a Green's function $G : [0,1] \times [0,1] \to [0,\infty)$ given by

$$G(t,s) = \begin{cases} \dfrac{1}{\rho}(\gamma + \delta - \gamma t)(\beta + \alpha s), & 0 \le s \le t \\ \dfrac{1}{\rho}(\beta + \alpha t)(\gamma + \delta - \gamma s), & t \le s \le 1. \end{cases}$$

It is clear that $G(t,s) > 0$ for $(t,s) \in (0,1) \times (0,1)$, and $G(t,s)$ satisfies condition (7.5.3). We seek to transform (7.5.1) and (7.5.2) into an integral equation via the use of the Green's function and then find a positive solution by using Lemma 7.5.1.

Denote by X the Banach space of real-valued continuous functions defined on $[a,b]$ with supremum norm $\|\cdot\|$, K be the cone in X of nonnegative functions $x(s)$ which are differentiable on $[a,0] \cup [1,b]$ and satisfy the differential equations

$$\alpha x(s) - \beta x'(s) = k\mu(s), \qquad s \in [a,0]$$

and

$$\gamma x(s) + \delta x'(s) = h\nu(s), \qquad s \in [1,b]$$

for some k, $h \in R$. Obviously, K is a normed cone. Define a subset of K by

$$D = \{\varphi \in K : \varphi(t) \ge g_\theta(t) \text{ for some } \theta > 0,\ t \in [a,b]\}.$$

Then we can define an operator $T : D \to K$ by

$$T\varphi(t) = \begin{cases} e^{\frac{\alpha}{\beta}t}\left(\frac{1}{\beta}\int_t^0 e^{-\frac{\alpha}{\beta}s}\mu(s)\,ds + \varphi(0)\right), & t \in [a,0] \\ \int_0^1 G(t,s)f(s,\varphi(\tau(s)))\,ds, & t \in (0,1) \\ e^{-\frac{\gamma}{\delta}t}\left(\frac{1}{\delta}\int_1^t e^{\frac{\gamma}{\delta}s}\nu(s)\,ds + e^{\frac{\gamma}{\delta}}\varphi(1)\right), & t \in [1,b]. \end{cases} \tag{7.5.7}$$

Noting that $G(t,s)$ is the Green's function for the boundary value problem (7.5.6) and that the functions in (7.5.4) and (7.5.5) satisfy the condition (7.5.3) we see that any fixed point φ of the operator T is in $C^1[a,b] \cap C^2(0,1)$, and hence satisfies

$$\varphi(0) = \int_0^1 G(0,s)f(s,\varphi(\tau(s)))\,ds$$

and

$$\varphi(1) = \int_0^1 G(1,s)f(s,\varphi(\tau(s)))\,ds.$$

Thus for the fixed point φ of the operator T

$$T\varphi(t) = \begin{cases} e^{\frac{\alpha}{\beta}t}\left(\frac{1}{\beta}\int_t^0 e^{-\frac{\alpha}{\beta}s}\mu(s)\,ds + \int_0^1 G(0,s)f(s,\varphi(\tau(s)))\,ds\right), & t \in [a,0] \\ \int_0^1 G(t,s)f(s,\varphi(\tau(s)))\,ds, & t \in (0,1) \\ e^{-\frac{\gamma}{\delta}t}\left(\frac{1}{\delta}\int_1^t e^{\frac{\gamma}{\delta}s}\nu(s)\,ds + e^{\frac{\gamma}{\delta}}\int_0^1 G(1,s)f(s,\varphi(\tau(s)))\,ds\right), & t \in [1,b]. \end{cases} \tag{7.5.8}$$

To show the existence of the fixed point of the operator T we would like in the following to use the definition of the operator T given by (7.5.8) rather than (7.5.7). Observe that if $\varphi \in D$, then $T\varphi \in K$, and that

$$(T\varphi)'' = -f(t,\varphi(\tau(t))) < 0, \qquad t \in (0,1),$$

then $T\varphi$ is concave down. Considering that for $\varphi \in D$

$$T\varphi(t) \geq e^{\frac{\alpha}{\beta}a}\int_0^1 G(0,s)f(s,\varphi(\tau(s)))\,ds := m \geq 0, \quad t \in [a,0]$$

and

$$T\varphi(t) \geq e^{-\frac{\gamma}{\delta}(b-1)} \int_0^1 G(1,s)f(s,\varphi(\tau(s)))\,ds := n \geq 0, \quad t \in [1,b] \tag{7.5.9}$$

where $m > 0$ if $a < 0$ and $n > 0$ if $b > 1$, we see that $T\varphi \in D$, so that if we pick any function in D the complete sequence of iterates is defined. Furthermore, if φ is a positive solution of (7.5.1) and (7.5.2) of class $C^1[a,b] \cap C^2(0,1)$, then it must belong to D. With this we conclude that $\varphi \in D$ is a solution of (7.5.1) and (7.5.2) if and only if $T\varphi = \varphi$.

Next we give an a priori bound for the set of the solutions of (7.5.1) and (7.5.2) (assuming the set is not empty).

Theorem 7.5.1. *Assume (H1)–(H4) hold. Then there exists an $R > 0$ such that $\|\varphi\| \leq R$ for every positive solution φ of (7.5.1) and (7.5.2).*

Proof: Assume the contrary. Then there exists a sequence of solutions $\{\varphi_n\}_{n=1}^{\infty}$ such that $\varphi_n(t) > 0$ for $t \in (a,b)$, $\|\varphi_n\| \leq \|\varphi_{n+1}\|$ and $\lim_{n\to\infty} \|\varphi_n\| = \infty$. From (7.5.2) this means that there exist t_n $(n = 1, 2, \ldots) \in [0,1]$ such that $\varphi_n(t_n) \to \infty$. We can show then that for the closed intervals $I \subset E$ given in (H2) we have $\varphi_n(\tau(t)) \to \infty$ as $n \to \infty$ for $t \in I$ uniformly. Suppose $\tau(t) \in J = [c,d] \subset (0,1)$ for $t \in I$. Then for each n, the fact that the graph of φ_n is concave down on $(0,1)$ implies that if $c \in (0,t_n)$, then

$$\varphi_n(c)/\varphi_n(t_n) \geq c/t_n \geq c,$$

and if $c \in (t_n, 1)$, then

$$\varphi_n(c)/\varphi_n(t_n) \geq (1-c)/(1-t_n) \geq 1-c.$$

Hence $\varphi_n(c) \to \infty$ as $n \to \infty$. Similarly, we can show that $\varphi_n(d) \to \infty$ as $n \to \infty$. Noting that

$$\varphi_n(\tau(t)) \geq \min\{\varphi_n(c),\ \varphi_n(d)\} \qquad \text{for } \tau(t) \in [c,d],$$

we know that $\varphi_n(\tau(t)) \to \infty$ as $n \to \infty$ for $t \in I$ uniformly. By (H1) there exists a n_0 such that if $n \geq n_0$ then

$$f(t, \varphi_n(\tau(t))) \leq \frac{1}{M|I|}$$

where $M = \max\{G(t,s) : (t,s) \in [0,1] \times [0,1]\}$, $|I|$ is the length of I.

Then we show that $\varphi_n(0) \geq \ell_1 > 0$ for all n. Otherwise, without loss of generality, we may assume $\varphi_n(0) > 0$ and $\varphi_n(0) \to 0$ as $n \to \infty$. (7.5.2) gives that $\varphi_n'(0) \to 0$ as $n \to \infty$, i.e., there exists a $L > 0$ such that $\varphi_n'(0) \leq L$ for all n. Since $\varphi_n''(t) < 0$ for $t \in (0,1)$, we have $\varphi_n'(t) \leq L$ for all n and $t \in (0,1)$. This together with $\varphi_n(0) \to 0$ as $n \to \infty$ gives that $\varphi_n(t)$ are uniformly bounded on $[0,1]$, which contradicts that $\|\varphi_n\| \to \infty$ as $n \to \infty$. Similarly, we conclude that $\varphi_n(1) \geq \ell_2 > 0$ for all n. In view of the fact that $\varphi_n(t)$ is concave down on $(0,1)$ we see that there exists a $\theta > 0$, which is independent of n, such that

$$\varphi_n(t) \geq g_\theta(t), \qquad t \in (0,1) \setminus I.$$

Therefore for $t \in [0,1]$ and $n \geq n_0$,

$$\begin{aligned}
\varphi_n(t) = T\varphi_n(t) &= \int_0^1 G(t,s) f(s, \varphi_n(\tau(s)))\, ds \\
&= \left(\int_I + \int_{[0,1]\setminus I} \right) G(t,s) f(s, \varphi_n(\tau(s)))\, ds \\
&\leq \int_I M \cdot \frac{1}{M|I|}\, dt + \int_0^1 G(t,s) f(s, g_\theta(s))\, ds \\
&\leq 1 + M \int_0^1 f(s, g_\theta(s))\, ds < \infty,
\end{aligned}$$

contradicting $\varphi_n(t_n) \to \infty$. This completes the proof. □

Theorem 7.5.2. *Assume (H1)–(H4) hold. Then the BVP (7.5.1), (7.5.2) has at least one positive solution.*

Proof: For each n, let

$$\psi_n(t) = \begin{cases} \dfrac{a-t}{a} \displaystyle\int_0^1 G(0,s) f(s,n)\, ds, & t \in [a,0] \\ \displaystyle\int_0^1 G(t,s) f(s,n)\, ds, & t \in (0,1) \\ \dfrac{b-t}{b-1} \displaystyle\int_0^1 G(1,s) f(s,n)\, ds, & t \in [1,b]. \end{cases} \tag{7.5.10}$$

Then $\psi_n(t)$ is continuous on $[a,b]$, and from (H1), it follows that $\psi_{n+1} \le \psi_n$, $\psi_n(t) > 0$ for $t \in (a,b)$, and $\lim_{n\to\infty} \psi_n(t) = 0$ uniformly on $[a,b]$. Define $f_n : (0,1) \times [0,\infty) \to (0,\infty)$ by

$$f_n(t,s) = f\big(t, \max(s, \psi_n(\tau(t)))\big).$$

and observe that f_n is continuous and by (H1), for $(t,s) \in (0,1) \times (0,\infty)$,

$$f_n(t,s) \le f(t,s) \qquad \text{and} \qquad f_n(t,s) \le f(t, \psi_n(\tau(t))).$$

Now define operators $T_n : K \to K$ by

$$T_n\varphi(t) = \begin{cases} e^{\frac{\alpha}{\beta}t}\Big(\frac{1}{\beta}\int_t^0 e^{-\frac{\alpha}{\beta}s}\mu(s)\,ds \\ \qquad + \int_0^1 G(t,s) f_n(s,\varphi(\tau(s)))\Big)ds, & t \in [a,0] \\ \int_0^1 G(t,s) f_n(s,\varphi(\tau(s)))\,ds, & t \in (0,1) \\ e^{-\frac{\gamma}{\delta}t}\Big(\frac{1}{\delta}\int_1^t e^{\frac{\gamma}{\delta}s}\nu(s)\,ds \\ \qquad + e^{\frac{\gamma}{\delta}} \int_0^1 G(1,s) f_n(s,\varphi(\tau(s)))\,ds\Big), & t \in [1,b]. \end{cases} \tag{7.5.11}$$

It is easy to see that T_n is a compact and decreasing mapping on K and that $T_n(0) \ge 0$, $T_n^2(0) \ge 0$. By Lemma 7.5.1, T_n must have a fixed point φ_n in K. Similar to the proof of Theorem 7.5.1 we can show that there exists an $R > 0$ such that $\|\varphi_n\| < R$ for all n, we omit the detail here. We also claim that there exists a $k > 0$ such that $\sup_{t\in[0,1]} \{\varphi_n(t)\} \ge k$ for all n. If this were not true, then by going to a subsequence if necessary, we may assume that $\lim_{n\to\infty} \varphi_n(t) = 0$ uniformly on $[0,1]$. Let

$$m = \inf\{G(t,s) \,:\, (t,s) \in I \times I\} > 0$$

where I is defined in (H2). According to (H1) there exists a $\delta > 0$ such that $t \in I$ implies that $f(t,\delta) \ge \frac{1}{m|I|}$. By assumption there exists a n_0 such that $n \ge n_0$ implies that $0 < \varphi_n(\tau(t)) < \delta$ for $t \in I$ since $\tau(t) \in J \subset (0,1)$. The definition of ψ_n shows that there exists a $n_1 \geq n_0$ such that $\psi_n(\tau(t)) < \delta$ for $t \in I$ and $n \geq n_1$. Then for $t \in I$ and $n \geq n_1$,

$$\begin{aligned}
\varphi_n(t) &= \int_0^1 G(t,s) f_n(s, \varphi_n(\tau(s)))\, ds \\
&\geq \int_I G(t,s) f_n(s, \varphi_n(\tau(s)))\, ds \\
&\geq m \int_I f(s, \max(\varphi_n(\tau(s)), \psi_n(\tau(s))))\, ds \\
&\geq m \int_I f(s, \delta)\, ds \geq 1
\end{aligned}$$

contradicting that $\lim_{n\to\infty} \varphi_n(t) = 0$ on $[0,1]$.

From the property of $\varphi_n(t)$, it is easy to see that there is a $\theta > 0$, independent of n such that $\varphi_n(t) \geq g_\theta(t),\ t \in [a,b]$. Let

$$p(t) = \begin{cases} e^{\frac{\alpha}{\beta}t}\left(\dfrac{1}{\beta}\displaystyle\int_t^0 e^{-\frac{\alpha}{\beta}s}\mu(s)\, ds + g_\theta(0)\right), & t \in [a,0] \\ g_\theta(t), & t \in (0,1) \\ e^{-\frac{\gamma}{\delta}t}\left(\dfrac{1}{\delta}\displaystyle\int_1^t e^{\frac{\gamma}{\delta}s}\nu(s)\, ds + e^{\frac{\gamma}{\delta}} g_\theta(1)\right), & t \in [1,b], \end{cases}$$

and

$$g(t) = \begin{cases} e^{\frac{\alpha}{\beta}t}\left(\dfrac{1}{\beta}\displaystyle\int_t^0 e^{-\frac{\alpha}{\beta}}\mu(s)\, ds + R\right), & t \in [a,0] \\ R, & t \in (0,1) \\ e^{-\frac{\gamma}{\delta}t}\left(\dfrac{1}{\delta}\displaystyle\int_1^t e^{\frac{\gamma}{\delta}s}\nu(s)\, ds + e^{\frac{\gamma}{\delta}} R\right), & t \in [1,b]. \end{cases}$$

Then $p, q \in K$, and $\varphi_n \in \langle p, q\rangle$. It is easy to see that T is compact on this interval. Then by going to a subsequence if necessary, we may assume that $\lim_{n\to\infty} T\varphi_n$ exists, and we denote it by φ^*.

The last step is to show that $\lim_{n\to\infty}(T\varphi_n - \varphi_n) = 0$, since, if this is true, then we have that $\varphi^* \in \langle p, q\rangle$ and that

$$T\varphi^* = T\left(\lim_{n\to\infty} T\varphi_n\right) = T\left(\lim_{n\to\infty} \varphi_n\right) = \lim_{n\to\infty} T\varphi_n = \varphi^*.$$

To see that $\lim_{n\to\infty}(T\varphi_n - \varphi_n) = 0$, choose $\theta > 0$ such that for all n, $\varphi_n(t) \geq g_\theta(t)$, $t \in [a, b]$. Let $\varepsilon > 0$ be given and choose δ such that $0 < \delta < 1$ and

$$2M\left[\int_0^\delta f(s, g_\theta(\tau(s)))\, ds + \int_{1-\delta}^1 f(s, g_\theta(\tau(s)))\, ds\right] < \varepsilon.$$

Then there exists a n_0 such that if $n \geq n_0$, then

$$\psi_n(\tau(t)) \leq g_\theta(\tau(t)), \qquad t \in [\delta, 1-\delta].$$

Noting that

$$f_n\big(t, \varphi_n(\tau(t))\big) = f\big(t, \varphi_n(\tau(t))\big) \qquad \text{for } t \in (\delta, 1-\delta)$$

we see that for $t \in [0, 1]$,

$$\begin{aligned}
T\varphi_n(t) - \varphi_n(t) &= T\varphi_n(t) - T_n\varphi_n(t) \\
&= \int_0^\delta G(t,s)\left[f\big(s, \varphi_n(\tau(s))\big) - f_n\big(s, \varphi_n(\tau(s))\big)\right] ds \\
&\quad + \int_{1-\delta}^1 G(t,s)\left[f\big(s, \varphi_n(\tau(s))\big) - f_n\big(s, \varphi_n(\tau(s))\big)\right] ds
\end{aligned}$$

Thus for $t \in [0, 1]$ and $n \geq n_0$,

$$\begin{aligned}
|T\varphi_n(t) - \varphi_n(t)| &\leq 2M\left[\int_0^\delta f\big(s, \varphi_n(\tau(s))\big)\, ds + \int_{1-\delta}^1 f\big(s, \varphi_n(\tau(s))\big)\, ds\right] \\
&\leq 2M\left[\int_0^\delta f\big(s, g_\theta(\tau(s))\big)\, ds + \int_{1-\delta}^1 f\big(s, g_\theta(\tau(s))\big)\, ds\right] < \varepsilon.
\end{aligned}$$

Observe that for $t \in [a, 0)$

$$\begin{aligned}
T\varphi_n(t) - \varphi_n(t) &= T\varphi_n(t) - T_n\varphi_n(t) \\
&= \int_0^1 G(0,s)\left[f\big(s, \varphi(\tau(s))\big) - f_n\big(s, \varphi(\tau(s))\big)\right] ds \\
&= T\varphi_n(0) - \varphi_n(0).
\end{aligned}$$

Hence we have for $t \in [a, 0)$,

$$|T\varphi_n(t) - \varphi_n(t)| < \varepsilon. \tag{7.5.12}$$

Similarly we can show that (7.5.12) holds for $t \in (1, b]$. This means that $\|T\varphi_n - \varphi_n\| < \varepsilon$ for $n \geq n_0$, and this completes the proof of the theorem. □

We next obtain some conditions for uniqueness of solutions of the boundary value problem (7.5.1) and (7.5.2) in several different cases.

Theorem 7.5.3. *Assume $f : (0, 1] \times (0, \infty) \to (0, \infty)$ is continuous and non-increasing in x for each t. Let $\tau(t) : [0, 1] \to [a, b]$ be continuous. Then the boundary value problem (7.5.1) and (7.5.2) has at most one positive solution in $C^1[a, b] \cap C^2(0, 1)$ if either one of the following conditions is satisfied:*

(i) *$\tau(t) \leq t$ for $t \in [0, 1]$ and for each $t_0 \in [0, 1]$ such that $\tau(t_0) = t_0$, the initial value problem*

$$\begin{aligned} &x'' + f(t, y(\tau(t))) = 0 \\ &x(t) = \eta(t), \quad t \in N_1 \\ &x'(t_0) = x_1 \end{aligned} \tag{7.5.13}$$

has at most one solution in a right neighborhood N_2 of t_0, where N_1 is a left neighborhood of t_0, $\eta(t)$ is defined and continuous on N, $x_1 \in R$;

(ii) *$\tau(t) \geq t$ for $t \in [0, 1]$ and for each $t_0 \in [0, 1]$ such that $\tau(t_0) = t_0$, the initial value problem (7.5.13) has at most one solution in a left neighborhood N_2 of t_0 where N_1 in (7.5.13) is a right neighborhood of t_0;*

(iii) *$\tau(t) \equiv t$ for $t \in [0, 1]$;*

(iv) *$\tau(t) < t$ for $t \in [0, 1]$;*

(v) *$\tau(t) > t$ for $t \in [0, 1]$.*

Proof: (i) Let $\varphi_1(t)$, $\varphi_2(t)$ be two different solutions of (7.5.1) and (7.5.2). We consider the following cases:

1. $\varphi_1(t) > \varphi_2(t)$ for $t \in [0, 1]$. Then $\sigma(t) = \varphi_1(t) - \varphi_2(t) > 0$ for $t \in [a, b]$. From (7.5.2) we have that $\sigma'(0) > 0$ and $\sigma'(1) < 0$. But (7.5.1) gives that on $(0, 1)$

$$\sigma''(t) = -f(t, \varphi_1(\tau(t))) + f(t, \varphi_2(\tau(t))) \geq 0,$$

i.e., $\sigma'(t)$ is nondecreasing. This is impossible.

2. $\varphi_1(t) > \varphi_2(t)$ for $t \in [0, t_0)$ for some $t_0 \in (0, 1]$ and $\varphi_1(t_0) = \varphi_2(t_0)$. Then $\varphi_1(t) > \varphi_2(t)$ for $t \in [a, t_0)$. Denote

$$w(t) = \varphi_1(t)\varphi_2'(t) - \varphi_1'(t)\varphi_2(t).$$

Then $w(0) = 0$, $w(t_0) \geq 0$, and on $(0, t_0)$

$$w'(t) = -\varphi_1(t)f(t, \varphi_2(\tau(t))) + \varphi_2(t)f(t, \varphi_1(\tau(t))) \leq 0.$$

This implies that $w(t) \equiv 0$ on $[0, t_0]$. We claim that $\varphi_1(t) \equiv \varphi_2(t)$ on $[0, t_0]$. In fact, for $t \in (0, t_0]$,

$$\varphi_2'/\varphi_2 = \varphi_1'/\varphi_1$$

which follows that $\varphi_2(t)/\varphi_1(t) = \varphi_2(t_0)/\varphi_1(t_0) = 1$, or $\varphi_2(t) = \varphi_1(t)$, contradicting that $\varphi_1(t) > \varphi_2(t)$ on $(0, t_0)$.

3. $\varphi_1(0) = \varphi_2(0)$, $t_0 = \inf\{t \in [0, 1] : \varphi_1(t) \neq \varphi_2(t)\}$, and there exists a sequence $\{t_n\}$ such that $t_n > t_0$, $t_n \to t_0$ and $\varphi_1(t_n) = \varphi_2(t_n)$. In this case $\varphi_1(t_0) = \varphi_2(t_0)$ and $\varphi_1'(t_0) = \varphi_2'(t_0)$. This implies that $\tau(t_0) = t_0$. For otherwise, $\tau(t_0) < t_0$. Then there is a right neighborhood N of t_0 such that $\tau(t) < t_0$ for $t \in N$. In N (1.1) becomes

$$\begin{gathered} x'' + f(t, \varphi(\tau(t))) = 0 \\ x(t_0) = \varphi(t_0), \quad x'(t_0) = \varphi'(t_0) \end{gathered} \tag{7.5.14}$$

where $\varphi := \varphi_1 = \varphi_2$, and φ_1 and φ_2 are solutions of (7.5.14) in N. This means $\varphi_1 \equiv \varphi_2$ in N, contradicting the definition of t_0.

For the case that $\tau(t_0) = t_0$, (7.5.1) and (7.5.2) gives the boundary value problem

$$\begin{aligned} & x'' + f(t, x(\tau(t))) = 0 \\ & x(t) = \varphi(t), \quad t \in [a, t_0) \\ & x'(t_0) = \varphi'(t_0) \end{aligned} \tag{7.5.15}$$

where $\varphi := \varphi_1 = \varphi_2$, and φ_1 and φ_2 are solutions of (7.5.15) on $[a, b]$. According to the uniqueness assumption on the initial value problem (7.5.13), we conclude that $\varphi_1(t) \equiv \varphi_2(t)$ in a right neighborhood N_2 of t_0, contradicting the definition of t_0.

4. $\varphi_1(0) = \varphi_2(0)$, $t_0 = \inf\{t \in [0,1] : \varphi_1(t) \neq \varphi_2(t)\}$, and there exists a $t_1 \in (t_0, 1]$ such that $\varphi_1(t) > \varphi_2(t)$ for $t \in (t_0, t_1)$. Similar to the discussion in parts 1 and 2 we can show this case is still impossible so we omit it here.

This completes the proof of uniqueness since if there were two different positive solutions of (7.5.1) and (7.5.2), then one of the above four cases would occur and this is impossible.

(ii) Similar reasoning as in (i) gives the proof.

(iii) For the case $\tau(t) \equiv t$, $t \in [0,1]$, the only different part of the proof from (i) is part 3. In this case, since t_0 is an accumulation point of the set of the points where φ_1, φ_2 coincide, there exist t_1, $t_2 \in (t_0, 1)$ such that $t_1 < t_2$, $\varphi_1(t_1) = \varphi_2(t_1)$, $\varphi_1(t_2) = \varphi_2(t_2)$, and $\varphi_1(t) \neq \varphi_2(t)$ for $t \in (t_1, t_2)$. Without loss of generality we assume $\varphi_1(t) > \varphi_2(t)$ for $t \in (t_1, t_2)$. Let $\sigma(t) = \varphi_1(t) - \varphi_2(t)$. Then $\sigma'(t_1) \geq 0$, $\sigma'(t_2) \leq 0$, and for $t \in (t_1, t_2)$,

$$\sigma''(t) = -f(t, \varphi_1(t)) + f(t, \varphi_2(t)) \geq 0.$$

We conclude from this that $\sigma'(t) \equiv 0$ for $t \in (t_1, t_2)$, and then $\sigma(t) \equiv 0$ on $[t_1, t_2]$, contradicting the assumption.

(iv) and (v) are immediate from (i) and (ii) since there does not exist $t_0 \in [0,1]$ such that $\tau(t_0) = t_0$.

7.6. Notes

The results in Section 7.2 are based on Ntouyas [142]; see also Fabry [46]. The treatment in Sections 7.3 and 7.4 is based on Tsamatos and Ntouyas [161]; see also Lee and O'Regan [120–122]. BVPs for singular second order ODEs are also discussed in O'Regan [146], Baxley [8] and Taliaferro [158]. The treatment in Section 7.5 is based on Erbe and Kong [36]; see also Gatica et al [50] for the ODE case.

References

1. R.P. Agarwal, *Existence uniqueness and iterative methods for right focal point boundary value problems for differential equations with deviating arguments*, Ann. Polon. Math. **LII** (1991), 211-230.

2. R.P. Agarwal, *Boundary value problems for higher order differential equations*, World Scientific, Singapore, 1986.

3. R.P. Agarwal and Y.M. Chow, *Finite difference methods for boundary value problems of differential equations with deviating arguments*, Comp. Math. Appl. **12A** (1986), no. 11, 1143-1153.

4. W.E. Aiello, *The existence of nonoscillatory solutions to a generalized nonautonomous delay logistic equation*, J. Math. Anal. Appl. **149** (1990), 114-123.

5. O. Arino and I. Gyori, *Necessary and sufficient conditions for oscillation of a neutral differential system with several delays*, J. Differential Equations **81** (1989), 98-105.

6. D.D. Bainov and D.P. Mishev, *Oscillation Theory for Neutral Differential Equations with Delay*, Adam Hilger, 1991.

7. D.D. Bainov, A.D. Myskis and A.I. Zahariev, *Necessary and sufficient conditions for oscillation of the solutions of linear functional differential equations of neutral type with distributed delay*, J. Math. Anal. Appl. **148** (1990), no. 1, 263-273.

8. John V. Baxley, *Some singular nonlinear boundary value problems*, SIAM J. Math. Anal. **22** (1991), 463-479.
9. S.R. Bernfeld and V. Lakshmikantham, *An Introduction to Nonlinear Boundary Value Problems*, Academic Press, New York, 1974.
10. J.E. Bowcock and Y.H. Yu, *Sharp conditions for oscillations caused by retarded and advanced perturbations*, Bull. Austral. Math. Soc. **37** (1988), 429-435.
11. M. Cecchi and M. Marini, *Asymptotic decay of solutions of a nonlinear second order differential equation with deviating argments*, J. Math. Anal. Appl. **138** (1989), no. 2, 371-383.
12. Ming-Po Chen, J.S. Yu and B.G. Zhang, *The existence of positive solutions for even order neutral differential equations*, preprint.
13. Ming-Po Chen and B.G. Zhang, *Oscillation criteria for a class of perturbed Schrödinger equations*, Hiroshima Math. J., to appear.
14. Ming-Po Chen and B.G. Zhang, *The existence of the bounded positive solutions of delay difference equations*, Pan American Math. J. **3** (1993), no. 1, 79–94.
15. S. Chen and Q. Huang, *On a conjecture of Hunt and Yorke*, J. Math. Anal. Appl. **159** (1991), 469-484.
16. Shaozhu Chen and Qingguang Huang, *Necessary and sufficient conditions for the oscillations of solutions to systems of neutral functional differential equations*, Funkcial. Ekvac. **33** (1990), no. 3, 427-440.
17. Yongshao Chen, *Existence of oscillatory solutions of second order functional differential equations*, Ann. Diff. Eqs. **6** (1990), no. 4, 389-394.
18. Yongshao Chen, *Classification of nonoscillatory solutions of neutral differential equations*, Funkcial. Ekvac., to appear.
19. Yongshao Chen, *Oscillation and asymptotic behavior of solutions of first order linear functional differential equations with oscillatory coefficients*, Acta Mathematicae Applicatae SINICA **12** (1989), no. 1, 96-104.
20. Yuan Ji Cheng, *Oscillation in linear differential equations with continuous distribution of time lag*, Math. Nachr. **153** (1991), 217-229.
21. Q. Chuanxi and G. Ladas, *Oscillations of higher order neutral differential equations with variable coefficients*, Math. Nachr. **150** (1991), 15-24.
22. Q. Chuanxi and G. Ladas, *Linearized oscillations for equations with positive and negative coefficients*, Hiroshima Math. J. **20** (1990), no. 2, 331-340.
23. Q. Chuanxi and G. Ladas, *Oscillations of first order neutral equations with variable coefficients*, Mh. Math. **109** (1990), 103-111.
24. Q. Chuanxi and G. Ladas, *Oscillatory behavior of difference equations with positive and negative coefficients*, Le Matematiche. V. **XLIV** (1988).

25. Q. Chuanxi, G. Ladas, B.G. Zhang and T. Zhao, *Sufficient conditions for oscillation and existence of positive solutions*, Applicable Analysis **35** (1990), no. 4, 187-194.

26. Kenneth L. Cooke and Joseph Wiener, *A survey of differential equations with piecewise continuous arguments*, Delay Differential Equations and Dynamical Systems (Claremont CA 1990) 1-15, Lecture Notes in Math 1475, Springer, Berlin (1991).

27. Q. Cui and G.J. Lawson, *Study on models of single population: an expansion of the logistic and exponential equations*, J. Theor. Biol. **98** (1982), 645-659.

28. K. DeNevers and K. Schmitt, *An application of shooting method to boundary value problems for second order delay equations*, J. Math. Anal. Appl. **36** (1971), 588-597.

29. M. Paul Devasahayam, *Existence of monotone solutions for functional differential equations*, J. Math. Anal. Appl. **118** (1986), 487-495.

30. Y.I. Domshlak and A.I. Aliev, *On oscillatory properties of the first order differential equations with one or two retarded arguments*, Hiroshima Math. J. **18** (1988), no. 1, 31-46.

31. Michael S. Du and Man Kam Kwong, *Sturm comparison theorems for second order delay equations*, J. Math. Anal. Appl. **152** (1990), no. 2, 305-323.

32. J. Dugundji and A. Granas, *Fixed point theory, V.I.*, Monographi Matematyczne, PNW Warszawa, 1982.

33. A. Elbert and I.P. Stavroulakis, *Oscillations of first order differential equations with deviating arguments*, Tech. Rep. N-172, Ioannina, Greece (1990).

34. A.R.M. El-Namoury, *On the solution of a boundary value problem with retarded arguments*, J. Natur. Sci. Math. **27** (1987), no. 2, 51-56.

35. L.H. Erbe, *Oscillatory criteria for second order nonlinear delay equations*, Canad. Math. Bull. **16** (1973), no. 1, 49-56.

36. L.H. Erbe and Q. Kong, *Boundary value problems for singular second order functional differential equations*, Journal of Computational and Applied Mathematics, to appear.

37. L.H. Erbe and Q. Kong, *Explicit conditions for oscillation of solutions to neutral differential systems*, Hiroshima Math. J. **24** (1994).

38. L.H. Erbe and Q. Kong, *Oscillation and nonoscillation properties of neutral differential equations*, Canad. J. Math. **46** (1994), 284–297.

39. L.H. Erbe and Q. Kong, *Oscillation results for second order neutral differential equations*, Funkcial Ekvac. **35** (1992), 545-555.

40. L.H. Erbe and Q. Kong, *Some necessary and sufficient conditions for oscillation of neutral differential equations*, preprint.

41. L.H. Erbe, W. Krawcewicz and J. Wu, *Leray Schauder degree for semilinear Fredholm maps and periodic boundary value problems of neutral equations*, Nonlinear Analysis **15** (1990), 747-764.

42. L.H. Erbe and B.G. Zhang, *Oscillation of discrete analogues of delay equations*, Differential and Integral Equations **2** (1989), no. 3, 300-309.

43. L.H. Erbe and B.G. Zhang, *Oscillation of second order neutral differential equations*, Bull. Austral. Math. Soc. **39** (1989), no. 1, 71-80.

44. L.H. Erbe and B.G. Zhang, *Oscillation for first order linear differential equations with deviating arguments*, Differential and Integral Equations **1** (1988), no. 3, 305-314.

45. L.H. Erbe and B.G. Zhang, *Oscillation of second order linear difference equations*, Chinese Mathematical Journal **16** (1988), no. 4, 239-252.

46. C. Fabry, *Nagumo conditions for systems of second order differential equations*, J. Math. Anal. Appl. **107** (1985), 132-143.

47. Mohd Faheem and M. Rama Mohana Rao, *A boundary value problem for nonautonomous nonlinear functional differential equations of delay type with L_2-initial functions*, Funkcial. Ekvac. **30** (1987), no. 2-3, 237-249.

48. K. Farrell, E.A. Grove and G. Ladas, *Neutral delay differential equations with positive and negative coefficients*, Applicable Analysis **27** (1988), 181-197.

49. J.M. Ferreira, *Oscillatory behavior in linear retarded functional differential equations*, J. Math. Anal. Appl. **128** (1987), no. 2, 332-346.

50. J.A. Gatica, Vladimir Oliker and Paul Waltman, *Singular nonlinear boundary value problems for second order ordinary differential equations*, J. Differential Equations **1** (1989), 62-78.

51. K. Gopalsamy, *Stability and Oscillation in Delay Differential Equations of Population Dynamics*, Kluwer Academic, Netherlands, 1992.

52. K. Gopalsamy, Xue-zhong He and Lizhi Wen, *On a periodic neutral logistic equation*, Glasgow Math. J. **33** (1991), 281-286.

53. K. Gopalsamy and G. Ladas, *On the oscillation and asymptotic behavior of* $N'(t) = N(t)[a + bN(t-2) - cN^2(t-2)]$, Quart. Appl. Math. **XLVIII** (1990), 433-440.

54. K. Gopalsamy, B.S. Lalli and B.G. Zhang, *Oscillation in odd order neutral differential equations*, Czech. Math. J. **42** (1992), no. 117, 313-323.

55. K. Gopalsamy, B.S. Lalli and B.G. Zhang, *A note on the nonoscillation of* $\frac{d}{dt}x(t) - p(t)x(t-\tau) + Q(t)x(t-0) = 0$, Applicable Analysis, to appear.

56. K. Gopalsamy and B.G. Zhang, *Comparison theorem and oscillation of the neutral equations*, preprint.

57. K. Gopalsamy and B.G. Zhang, *Oscillation and nonoscillation in first order neutral differential equations*, J. Math. Anal. Appl. **151** (1990), no. 1, 42-57.

58. K. Gopalsamy and B.G. Zhang, *On delay differential equations with impulses*, J. Math. Anal. Appl. **139** (1989), no. 1, 110-121.

59. K. Gopalsamy and B.G. Zhang, *On a neutral delay logistic equation*, Dynamics and Stability of Systems **2** (1988), no. 4, 183-195.

60. S.R. Grace and B.S. Lalli, *Oscillations on nonlinear second order neutral delay differential equations*, Rad. Math. **3** (1987), 77-84.

61. J. Graef, P.W. Spikes and M.K. Grammatikopoulos, *On the behavior of solutions of a first order nonlinear neutral delay differential equations*, Applicable Analysis **40** (1991), 111-121.

62. J. Graef, P.W. Spikes and B.G. Zhang, *A sufficient condition of convergence to zero for solutions of second order functional differential equations*, Applicable Analysis **23** (1986), 11-21.

63. M.K. Grammatikopoulos, G. Ladas and A. Meimaridou, *Oscillation and asymptotic behavior of second order neutral differential equations*, Proceedings of the Int'l Conference on Theory and Application of Differential Equations, Pan American University, Edinburg, Texas, U.S.A., May 20-23, 1985.

64. M.K. Grammatikopoulos, Y.G. Sficas and I.P. Stavroulakis, *Necessary and sufficient conditions for oscillations of neutral equations with several coefficients*, J. Differential Equations **76** (1988), 294-311.

65. A. Granas, R. Guenther and J. Lee, *Nonlinear boundary value problems for ordinary differential equations*, Dissertationes Mathematicae, Warszawa, 1985.

66. A. Granas, R. Guenther and J. Lee, *Topological transversality II. Applications to Neumann problem for $y'' = f(t, y, y')$*, Pacific J. Math. **104** (1983), 95-109.

67. L.J. Grimm and K. Schmitt, *Boundary value problems for delay differential equations*, Bull. Amer. Math. Soc. **74** (1986), 997-1000.

68. E.A. Grove, M.R.S. Kuenovic and G. Ladas, *Sufficient conditions for oscillation and nonoscillation of neutral equations*, J. Differential Equations **68** (1987), 373-382.

69. E.A. Grove, G. Ladas and J. Schinas, *Sufficient conditions for the oscillation of delay and neutral delay equations*, Canad. Math. Bull. **31** (1988), no. 4, 459-466.

70. R.B. Guenther and J.W. Lee, *Topological transversality and differential equations*, Contemporary Mathematics **72** (1988), 121-130.

71. Chaitan P. Gupta, *Existence and uniqueness theorems for boundary value problems involving reflection of the arguments*, Nonlinear Analysis **11** (1987), no. 9, 1075-1083.

72. I. Gyori, *Existence and growth of oscillatory solutions of first order unstable type delay differential equations*, Nonlinear Analysis **13** (1989), no. 7, 739-751.

73. I. Gyori, *Oscillations of retarded differential equations of the neutral and the mixed type*, J. Math. Anal. Appl. **141** (1989), 1-20.

74. I. Gyori, *Oscillation conditions in scalar linear delay differential equations*, Bull. Austral Math. Soc. **34** (1986), 1-9.

75. I. Gyori, *On the oscillatory behavior of solutions of certain nonlinear and linear delay differential equations*, Nonlinear Analysis **8** (1984), no. 5, 429-439.

76. I. Gyori and G. Ladas, *Oscillation Theory of Delay Differential Equations with Applications*, Clarendon Press, Oxford, 1991.

77. I. Gyori and I.P. Stavroulakis, *Positive solutions of functional differential equations*, Bollettino U.M.I. **7** (1989), no. 3-B, 185-198.

78. A. Halanay, *Differential Equations: Stability, Oscillations and Time Lags*, Academic Press, New York, 1966.

79. J.K. Hale, *Theory of Differential Equations*, Springer Verlag, New York, 1977.

80. Alexander Hascak, *Disconjugacy and multipoint boundary value problems for linear differential equations with delay*, Czechoslovak Math. J. **39** (1989), no. 114, No. 1, 70-77.

81. S. Heikkila, J. Mooney and S. Seikkala, *Existence, uniqueness and comparison results for nonlinear boundary value problems involving deviating arguments*, J. Differential Equations **41** (1981), 320-333.

82. S. Heikkila, *On well-posedness of a boundary value problem involving deviating arguments*, Funkcial. Ekvac. **28** (1985), 222-232.

83. Q. Huang and S. Chen, *Oscillation of neutral differential equations with periodic coefficients*, Proc. Amer. Math. Soc. **110** (1990), 997-1001.

84. Wengang Huang, *Necessary and sufficient conditions for oscillation of first order functional differential equations*, J. Math. Anal. Appl. **148** (1990), 360-370.

85. B.R. Hunt and J.A. Yorke, *When all solutions of* $x' = \sum_{j=1}^{n} q_j(t)x(x - T_j(t))$ *oscillate*, J. Differential Equations **53** (1984), 139-145.

86. L.K. Jackson, *Subfunctions and second order differential inequalities*, Advances in Math. **2** (1968), 307-363.

87. Jaroslav Jaros, *An oscillation test for a class of linear neutral differential equations*, J. Math. Anal. Appl. **159** (1991), 406-417.

88. Jaroslav Jaros, *On characterization of oscillation in first order linear neutral differential equations*, Funkcial. Ekvac. **341** (1991), no. 2, 331-342.

89. J. Jaros and T. Kusano, *Oscillation properties of first order nonlinear functional differential equations of neutral type*, Differential and Integral Equations **4** (1991), no. 2, 425-436.

90. J. Jaros and T. Kusano, *Asymptotic behavior of nonoscillating solutions of nonlinear differential equations of neutral type*, Funcialaj. Eckvac. **32** (1989), 251-262.

91. J. Jaros and T. Kusano, *Existence of oscillatory solutions for functional differential equations of neutral type*, Acta Math. Univ. Comenianae **V.L.X.2** (1991), 185-194.

92. C. Jian, *On the oscillation of linear differential equations with deviating arguments*, Math. in Practice and Theory **1** (1991), 32-40.

93. J. Kalinowski, *Two point boundary value problems for some systems of ordinary differential equations of second order with deviating arguments*, Ann. Polon. Math. **30** (1974), 71-76.

94. Y. Kitamura and T. Kusano, *Oscillation and asymptotic behavior of solutions of first order functional differential equations of neutral type*, Funkcial. Ekvac. **33** (1990), 325-343.

95. V.B. Kolmanovskii and V.R. Nosov, *Stability of Functional Differential Equations*, Academic Press, 1986.

96. Q. Kong and H.I. Freedman, *Oscillations in delay differential systems*, Differential and Integral Equations **6** (1993), 1325–1336.

97. R.G. Koplatadze and T.A. Chanturia, *On oscillatory and monotonic solutions of first order differential equations with retarded arguments*, Differenial'nye Uravneniya **8** (1982), 1462-1465.

98. R.G. Koplatadze, *Monotone and oscillatory solutions of nth order differential equations with retarded argument*, Math. Bohem. **116** (1991), no. 3, 296-308.

99. Y. Kuang, B.G. Zhang and T. Zhao, *Qualitative analysis of a nonautonomous nonlinear delay differential equation*, Tohoku Mathematical J. **43** (1991), no. 4, 509-528.

100. M.R. Kulenovic and M.K. Grammatikopoulos, *Some comparison and oscillation results for first order differential equations and inequalities with a deviating argument*, J. Math. Anal. Appl. **131** (1988), 67-84.

101. Man Kam Kwong, *Oscillation of first order delay equations*, J. Math. Anal. Appl. **156** (1991), 274-286.

102. M.K. Kwong and T. Patula, *Comparison theorem for first order linear delay equations*, J. Differential Equations **70** (1987), 275-292.

103. S.M. Labovskii, *Positive solutions of a two point boundary value problem for a linear singular functional differential equation*, Differentsial'nye Uravneniya **24** (1988), no. 10, 1695-1704,1836.

104. G. Ladas, Ch.G. Philos and Y.G. Sticas, *Oscillation in neutral equations with periodic coefficients*, Proc. Amer. Math. Soc. **113** (1991), 123-134.

105. G. Ladas and C. Qian, *Linearized oscillations for odd order neutral differential equations*, J. Differential Equations **88** (1990), 238-247.

106. G. Ladas and C. Qian, *Oscillation in differential equations with positive and negative coefficients*, Canad. Math. Bull. **33** (1990), no. 4, 442-451.

107. G. Ladas, C. Qian and J. Yan, *Oscillations of higher order neutral differential equations*, Portugaliae Mathematica **48** (1991), 291-307.

108. G. Ladas, Chuanix Qian and Jurang Yan, *A comparison result for the oscillation of delay differential equation*, Proc. Amer. Math. Soc. **114** (1992), no. 4, 939-947.

109. G. Ladas and S.W. Schults, *On oscillations of neutral equations with mixed arguments*, Hiroshima Math. J. **19** (1989), no. 2, 409-429.

110. G.S. Ladde, V. Lakshmikantham and B.G. Zhang, *Oscillation Theory of Differential Equations with Deviating Arguments*, Marcel Dekker, New York, 1987.

111. B.S. Lalli, *Oscillations of nonlinear second order neutral delay differential equations*, Bull. of the Institute Math. **18** (1990), 233-238.

112. B.S. Lalli, S. Ruan and B.G. Zhang, *Oscillation theorems for nth order neutral functional differential equations*, Ann. Differential Equations **8** (1992), 401–413.

113. B.S. Lalli and B.G. Zhang, *Attractivity and oscillation for neutral equations*, Proceedings of International Conference on Differential Equations **1**, 682–686, (Barcelona, Aug. 1991, Spain)..

114. B.S. Lalli and B.G. Zhang, *Boundary value problems for second order functional differential equations*, Ann. Differential Equations **8** (1992), 261-268.

115. B.S. Lalli and B.G. Zhang, *Oscillation and comparison theorems for certain neutral difference equations*, Austr. Math. Soc. J., Ser. B. **34** (1992), 245-256.

116. B.S. Lalli and B.G. Zhang, *Oscillation and nonoscillation of some neutral differential equations of odd order*, Int'l. J. of Math. and Math. Sci. **15** (1992), 509-515.

117. B.S. Lalli and B.G. Zhang, *Oscillation of first order neutral differential equations*, Applicable Analysis **39** (1990), 265-274.

118. B.S. Lalli, B.G. Zhang and Li Juan Zhao, *On oscillations and existence of positive solutions of neutral difference equations*, J. Math. Anal. Appl. **158** (1991), no. 1, 213-233.

119. B.S. Lalli, B.G. Zhang and T. Zhao, *Oscillations of certain second order nonlinear differential equations*, Periodica Mathematica Hungarica **25** (1992), no. 2, 167–177.

120. John W. Lee and Donal O'Regan, *Existence results for differential delay equation I*, J. Differential Equations, to appear.

121. John W. Lee and Donal O'Regan, *Existence results for differential delay equation II*, Nonlinear Analysis **17** (1991), 683-702.

122. J.W. Lee and D. O'Regan, *Topological transversality. Applications to initial value problems*, Ann. Polon. Math. **48** (1988), 247-252.

123. S. Leela and M.N. Oguztoreli, *Periodic boundary value problem for differential equations with delay and monotone iterative method*, J. Math. Anal. Appl. **122** (1987), no. 2, 301-307.

124. Bing Tuan Li, *An estimate for distance between delay adjacent zeros of solution of first order differential equations*, Acta Mathematicae Applicatae SINICA **13** (1990), no. 4, 467-472.

125. Guanghua Li, *Oscillatory behavior of solutions to a generalized, nonautonomous, delay logistic equations*, Ann. Diff. Eqs. **7** (1991), no. 4, 432-438.

126. Guanghua Li, *Existence criteria for nonoscillatory solutions to varying deviating neutral equations*, J. Huaihua Teachers' College **9** (1990), no. 1, 45-54.

127. Shenlin Li and Lizhi Wen, *Functional Differential Equations*, Hunan Scientific Press (in Chinese), 1987.

128. Yong Li, *A theorem on the existence of solutions to differential difference equations*, Acta. Aci. Natur. Univ. Jilin (1988), no. 2, 25-29.

129. Lu Wudu, *The existence and asymptotic behavior of nonoscillatory solutions to the second order nonlinear neutral equations*, preprint.

130. Lu Wudu, *Oscillation of high order neutral differential equations with oscillatory coefficient*, Acta Mathematicae Applicatae SINICA **7** (1991), no. 2, 135-142.

131. Lu Wudu, *Oscillation of solutions to second order differential equations of advanced type*, Chinese Ann. Math. **12A (Supplement)** (1991), 133-138.

132. Lu Wudu, *Existence of nonoscillatory solutions of first order nonlinear neutral equations*, J. Austral. Math. Soc., Ser. B **32** (1990), 180-192.

133. Lu Wudu, *Necessary and sufficient condition for oscillations of neutral differential equations*, Ann. Diff. Eqs. **6** (1990), 417-430.

134. Pavol Marusiak, *On unbounded nonoscillatory solutions of systems of neutral differential equations*, Czech. Math. J. **42** (1992), 117-128.

135. R.M. Mathsen, Xu Yuantong and Wang Qiru, *Asymptotics and oscillation for first order neutral functional differential equations*, Rocky Mountain Mathematics Journal, to appear.

136. J. Mawhin, *Topological degree methods in nonlinear boundary value problems*, CBMS Regional Conference in Mathematics **40** (1979).

137. D.P. Mishev and D.D. Bainov, *Oscillation properties of the solutions of a class of hyperbolic equations of neutral type*, Funkcial. Ekvac. **29** (1986), 213-218.

138. A.D. Myskis, *Linear Differential Equations with Retarded Argument*, Moscow, Nanka, 1972.

139. V.A. Nadareishvili, *On oscillation and nonoscillation of solutions of first order linear differential equations with deviating arguments*, Differentisial'nye Uravneniya **25** (1989), 611-616.

140. Yuki Naito, *Asymptotic behavior of decaying nonoscillatory solutions of neutral differential equations*, Funkcial. Ekvac. **35** (1992), 95-110.

141. M. Nashed and J.S.W. Wong, *Some variations of a fixed point theorem of Krasnoselskii and applications to nonlinear integral equations*, J. Math. Mech. **18** (1969), 666-677.

142. S. Ntouyas, *On a boundary value problem for functional differential equations*, Acta Math. Hung. **48** (1986), no. 1–2, 87-93.

143. S. Ntouyas and P. Tsamatos, *Existence of solution of boundary value problems for functional differential equations*, Internat. J. Math. and Math. Sci. **14** (1991), 509–516.

144. S. Ntouyas and P. Tsamatos, *On well-posedness of boundary value problems involving deviating arguments*, Funkcial. Ekvac. **35** (1992), 137–147.

145. A.G. O'Farrell and D. O'Regan, *Existence results for some initial and boundary value problem*, Proc. Amer. Math. Soc. **110** (1990), 661-673.

146. Donal O'Regan, *Singular second order boundary value problem*, Nonlinear Analysis, TMA **12** (1990), 1097-1109.

147. Donal O'Regan, *Some new results for second order boundary value problem*, J. Math. Anal. Appl. **2** (1990), 548-570.

148. P.K. Palamides, *A topological principle and its application on a nonlinear two point boundary value problem for functional differential equations*, Arch. Math. (Basel) **49** (1987), no. 1, 44-65.

149. Ch.G. Philos, *On oscillations of some difference equations*, Funkcial. Ekvac. **34** (1991), 157-172.

150. Ch.G. Philos, *Oscillations of first order linear retarded differential equations*, J. Math. Anal. Appl. **157** (1991), no. 1, 17-33.

151. X.Z. Qian, *A necessary and sufficient condition of the oscillation of a class of neutral differential equations*, Acta Math. Appl. SINICA **12** (1989), 418-429.

152. J. Ruan, *Types and criteria of nonoscillatory solutions for second order linear neutral differential difference equations*, Chinese Ann. Math., Ser. A **8** (1987), 114-124.

153. J. Ruan, *Behavior of solutions to first order differential inequalities with deviating arguments and its applications*, Kexue Tonghao **29** (1984), 1125-1227.

154. S. Ruan, *Oscillations for first order neutral differential equations with variable coefficients*, Bull. Austral. Math. Soc. **43** (1991), 147-152.

155. S. Ruan, *Oscillations of nth order functional differential equations*, Computers Math. Applic. **21** (1990), no. 2–3, 95-102.

156. K. Schmitt, *Comparison theorems for second order delay differential equations*, Rocky Mountain J. of Math. **I** (1971), no. 3, 459-467.

157. Y.G. Sficas and S.K. Ntouyas, *A boundary value problem for n^{th} order functional differential equations*, Nonlinear Analysis **5** (1981), 325-335.

158. S. Taliaferro, *A nonlinear singular boundary value problem*, Nonlinear Analysis, TMA **3** (1979), 897-904.

159. A. Tarski, *A lattice theoretical fixed point theorem and its applications*, Pacific J. Math. **5** (1955), 285-309.

160. Mu Zhong Tang, *Oscillatory property of solutions for second order functional differential equation to be changed into an "integrally small" coefficient*, Chinese Sci. Bull. **36** (1991), no. 2, 89-94.

161. P. Tsamatos and S. Ntouyas, *Existence of solutions of boundary value problems for differential equations with deviating arguments, via the topological transversality method*, Proc. Royal Soc. Edinburgh **118A** (1991), 79-89.

162. Kh. Turaev and T. Turdiev, *Formulation of a boundary value problem for first order functional differential equations of neutral type*, Izv. Akad. Nauk UZSSR Ser. Fiz-Mat. Nauk (1988), no. 4, 54-57.

163. Huai Zhong Wang and Li Yong, *Differential equations with delay and advance*, Dongbei Shuxue **4** (1988), no. 3, 371-378.

164. Kai Xun Wang, *Oscillation of a class of nonlinear neutral differential equations of higher order*, Chinese Ann. Math. **12A** (1991), no. 1, 104-113.

165. Lian Wen Wang, *Classification of nonoscillation solutions to neutral functional differential equations of higher order*, Chinese Ann. Math. **12A (Supplement)** (1991), 60-65.

166. Lian Wen Wang, *Oscillation of first order nonlinear neutral differential equations*, Acta Mathematicae Applicatae SINICA **14** (1991), no. 3, 348-359.

167. Lian Wen Wang, *The comparison theorems of oscillations for systems of one order neutral differential equations*, Ann. Diff. Eqs. **6** (1990), no. 3, 319-330.

168. Z.C. Wang, *A necessary and sufficient conditions for the oscillation of higher order neutral equations*, Tohoku. Math. J. **41** (1989), 575-588.

169. Z.C. Wang, *A necessary and sufficient condition for the oscillation of neutral equation* $\frac{d}{dx}(x(t) + px(t-2)) + qx(t-v) - hx(t-s) = 0$, Kexue Tongbao **33** (1988), 1452-1454.

170. Junjei Wei, *A necessary and sufficient condition for the oscillation of the first order differential equation with deviating arguments*, Acta. Math. Sinica **32** (1989), 632-638.

171. Lizhi Wen, *Asymptotic behavior and oscillation of the second order functional differential equations*, Scientia SINICA, No. 1 (1986).

172. J. Wiener and A. Aftabizadeh, *Boundary value problems for differential equations with reflection of the arguments*, Int'l. J. Math. and Math. Sci. **8** (1985), 151-163.

173. J. Yan, *Asymptotic behavior and oscillation of* n^{th} *order nonlinear delay differential equations*, Acta Math. Sinica **33** (1990), no. 4, 537-543.

174. J. Yan, *Comparison theorems for the oscillation of higher order neutral equations and applications*, Scientia Sinica, Ser. A **12** (1990), 1256-1266.

175. J.S. Yu, *Linearized oscillation of the second order delay differential equations*, preprint.

176. J.S. Yu, *On the neutral delay differential equations with positive and negative coefficients*, Acta. Math. Sinica **34** (1991).

177. J.S. Yu, *The existence of positive solutions to the neutral delay differential equations*, Proceeding of Differential Equations, Academic Press (in Chinese) (1991), 263-269.

178. J.S. Yu, *On the oscillation of second order neutral differential equations*, Kexue Tongbao **34** (1989), 1754-1755.

179. J.S. Yu and Z.C. Wang, *Some further results on oscillation of neutral differential equations*, Bull. Austral. Math. Soc. **46** (1992), 149-157.

180. J.S. Yu, Z.C. Wang and Q. Chuanxi, *Oscillation and nonoscillation of neutral differential equations*, Bull. Austral. Math. Soc., to appear.

181. J.S. Yu, Z. Wang and C. Qian, *Oscillation and nonoscillation of neutral delay differential equations*, Bull. Austra. Math. Soc. **45** (1991), no. 2, 195-200.

182. J.S. Yu, Z.C. Wang and X.Z. Qian, *Oscillation and nonoscillation of first order neutral differential equations*, preprint.

183. J.S. Yu, Z.C. Wang and B.G. Zhang, *Further comparison theorems for higher order neutral differential equations*, submitted.

184. J.S. Yu, Z.C. Wang, B.G. Zhang and X.Z. Qian, *Oscillations of differential equations with deviating arguments*, Pan American Math. J. **2** (1992), no. 2, 59-78.

185. J.S. Yu and B.G. Zhang, *The existence of positive solution for second order neutral differential equations with unstable type*, Sys. Sci. and Math., to appear.

186. J.S. Yu, B.G. Zhang and Z.C. Wang, *Oscillation of delay difference equations*, Applicable Analysis, to appear.

187. Yanghong Yu, *Oscillation of first order neutral delay differential equations*, Kexue Tongbao **34** (1989), 158.

188. Yuanhong Yu and Baotong Cui, *Oscillation of solutions of hyperbolic equations of neutral type*, Acta. Math. Appl. SINICA **15** (1992), no. 1, 105-111.

189. B.G. Zhang, *Oscillation and asymptotic behavior of second order difference equations*, J. Math. Anal. Appl. **173** (1993), 58–68.

190. B.G. Zhang, *On the second order nonlinear neutral differential equations*, preprint.

191. B.G. Zhang, *Forced oscillation of first order differential equations with deviating arguments*, Acta. Math. Appl. SINICA **15** (1992), no. 1, 105-111.

192. B.G. Zhang, *Riccati techniques for oscillation of neutral equations*, Chinese Ann. Math. A. V.**12**, (Supplement), (1991), 92-97.

193. B.G. Zhang, *On the oscillation of solutions of reduced wave equations*, Mathematica Applicata **3** (1990), no. 1, 94-95.

194. B.G. Zhang, *Oscillation of direct control systems with delay*, Kexue Tongbao **35** (1990), no. 13, 1037.

195. B.G. Zhang, *Oscillation of first order neutral differential equations*, J. Math. Anal. Appl. **139** (1989), no. 2, 311-318.

196. B.G. Zhang, *Oscillation of second order neutral differential equations*, Kexue Tongbao **34** (1989), no. 8, 563-566.

197. B.G. Zhang, *Forced oscillation of neutral equations*, Kexue Tongbao **34** (1989), no. 24, 1845-1848.

198. B.G. Zhang, *On oscillation of a kind of integro differential equation with delay*, Atti Acc. Lincei Rend. **32** (1988), no. 3, 437-444.

199. B.G. Zhang, *A survey of the oscillation of solutions to first order differential equations with deviating arguments*, Ann. Diff. Eqs. **2** (1986), no. 1, 65-86.

200. B.G. Zhang, *Oscillation for a kind of second order functional differential equation with deviating arguments*, Kexue Tongbao **23** (1985), 1770-1774.

201. B.G. Zhang, *On oscillation of differential inequalities and equations with deviating arguments*, Ann. Diff. Eqs. **1** (1985), no. 2, 209-218.

202. B.G. Zhang, *Oscillation of the solution of the first order advanced type differential equations*, Science Exploration **3** (1982), 79-82.

203. B.G. Zhang and K. Gopalsamy, *Oscillation and nonoscillation of delay differential equations with positive and negative coefficients*, preprint.

204. B.G. Zhang and K. Gopalsamy, *Oscillation and nonoscillation of neutral delay differential equations*, Proceedings of the First World Congress on Nonlinear Analytits., Florida, Aug. 19-26, 1992.

205. B.G. Zhang and K. Gopalsamy, *Oscillation and nonoscillation in higher order neutral equations*, Math. Phys. **25** (1991), no. 2, 152-165.

206. B.G. Zhang and K. Gopalsamy, *Global attractivity and oscillations in a periodic delay logistic equation*, J. Math. Anal. Appl. **150** (1990), 274-283.

207. B.Z. Zhang and K. Gopalsamy, *Global attractivity in the delay logistic equation with variable parameters*, Math. Proc. Camb. Phil. Soc. **107** (1990), 579-590.

208. B.G. Zhang and K. Gopalsamy, *Oscillation and nonoscillation in a nonautonomous delay logistic equations*, Quart. Appl. Math. **XLVI** (1988), no. 2, 267-273.

209. B.G. Zhang and B.S. Lalli, *On the existence of positive decaying solutions of neutral equation*, J. Sys. Sci. & Math. Scis. **13** (1993), no. 2, 167–170.

210. B.G. Zhang and N. Parhi, *Oscillation of first order differential equations with piecewise constant arguments*, J. Math. Anal. Appl. **139** (1989), no. 1, 23-35.

211. B.G. Zhang and J.S. Yu, *Oscillation and nonoscillation of odd order neutral differential equations*, preprint.

212. B.G. Zhang and J.S. Yu, *Oscillation and nonoscillation for neutral differential equations*, J. Math. Anal. Appl. **172** (1993), no. 1, 11–23.

213. B.G. Zhang and J.S. Yu, *On the existence of asymptotically decaying positive solutions of second order neutral differential equations*, J. Math. Anal. Appl. **166** (1992), 1-11.

214. B.G. Zhang and J.S. Yu, *The existence of positive solutions of neutral differential equations*, Scientia Sinica (1992), no. 8, 785–790.

215. B.G. Zhang, J.S. Yu and Z.C. Wang, *Oscillations of higher order neutral differential equations*, Proceeding of the Second G.J. Butler Memorial Conference on Differential Equations and Population Biology, June 17-20, 1992, Univ. of Alberta, Canada.

216. Feng Qin Zhang, Ju Rang Yan and Qian Chuanxi, *Limit boundary value problems of first order equations with piecewise constants arguments*, Rad. Mat. **6** (1990), no. 2, 347-355.

217. Detang Zhou, *On some problems on oscillation of functional differential equations of first order*, J. Shandong University **25** (1990), no. 4, 434-442.

218. Detang Zhou, *Oscillations of solutions to neutral functional differential equations*, Acta Mathematicae Applicatae SINICA **13** (1990), no. 3, 358-367.

219. Detang Zhou, *Negative answer to a problem of Gyori*, J. Shandong University **24** (1989), no. 4, 117-121.

Index